AF615054

Philosophy of Mathematics and Deductive Structure in Euclid's *Elements*

Philosophy of Mathematics and Deductive Structure in Euclid's *Elements*

Ian Mueller

Dover Publications, Inc.
Mineola, New York

Copyright

Bibliographical Note

This Dover edition, first published in 2006, is an unabridged republication of the work originally published by the MIT Press, Cambridge, Massachusetts, in 1981.

Library of Congress Cataloging-in-Publication Data

Mueller, Ian.
Philosophy of mathematics and deductive structure in Euclid's Elements / Ian Mueller.
p. cm.
Originally published: Cambridge, Mass. : MIT Press, c1981.
Includes bibliographical references and index.
ISBN 0-486-45300-6 (pbk.)
1. Euclid. Elements. 2. Mathematics–Philosophy. 3. Logic, Symbolic and mathematical. I. Title.

QA31.E9M83 2006
510.1–dc22

2006047444

Manufactured in the United States of America
Dover Publications, Inc., 31 East 2nd Street, Mineola, N.Y. 11501

Contents

Acknowledgments

This book has taken a long time to write, perhaps too long a time, given the extensive institutional support I have received in writing it. Serious research for it was begun under a grant from the American Council of Learned Societies supplemented by the University of Chicago and continued for a second consecutive year because of the generosity of the same university. The book was completed during a third sabbatical which was spent in the ideal working conditions provided by the Center for Hellenic Studies in Washington, D.C. I am in no position to explain, let alone to justify, the generosity of these institutions. I can only record my gratitude to them for the opportunity to pursue my research and writing without interruption. A number of individuals encouraged me in my endeavors, of whom I would like to mention particularly the late Glenn Morrow, Anne Burnett, Benson Mates, and Gregory Vlastos. I would also like to thank Karl J. Weintraub, Dean of the Division of the Humanities at the University of Chicago, who consistently supported my attempts to find time for research and provided money to pay for the drafting of the illustrations in the book. William Tait and Wilbur Knorr read the manuscript through in its final stages, and provided the kind of constructive criticism from which an author can only benefit. Finally, I want to thank my family. My daughters, Maria and Monica, did not know why I was obsessed with my work, but they accepted my obsession and the uprooting caused by the sabbatical leaves. My wife Janel did understand it, and encouraged me to continue working in every way she could. For such support no amount of thanks is sufficient. Nor is the finished book a satisfactory indication of what this support has meant to me. Nevertheless, I dedicate this book to my family, Janel, Maria, and Monica, with love and gratitude.

Introduction

The reader of English who wishes to know something about Euclid's *Elements* is normally referred to the monumental three-volume translation and commentary by T. L. Heath. Although time has not stood still in Euclid studies since the last revision of this work over half a century ago, Heath's *Elements* remains a basic reference work. However, it is a rather cumbersome tool for someone neither already familiar with the *Elements* nor willing to expend a great deal of labor to become familiar with them. The principal aims of this book are to give a survey of the contents of the *Elements* for such persons, and to provide an understanding of the classical Greek conception of mathematics and its foundations and of the similarities and differences between that conception and our own. For this purpose it seemed best to concentrate attention on the *Elements* themselves and, in particular, to look at propositions in the work in terms of their use in the work. I have, accordingly, been relatively sparing in my use of other ancient materials, introducing them only to support interpretations advanced or because they have been invoked by others in support of contrary interpretations. I have emphasized philosophical, foundational, and logical questions, rather than certain kinds of historical and mathematical questions. To be sure, this division cannot be maintained as a sharp dichotomy, and I certainly hope to have provided a historically plausible representation of the mathematical content of the *Elements*. But in general, despite a heavy use of symbolism and frequent comparisons with later mathematical work, I have not tried to describe the content in the mathematically most elegant way; nor have I discussed the so-called prehistory of the *Elements*, except in cases where doing so seemed relevant to the interpretation of the *Elements* themselves. I have, however, indicated in several bibliographical notes what seem to me the most significant or useful discussions of more strictly historical or mathematical questions.

A fundamental organizing principle of the *Elements* is mathematical subject matter. The following list gives a reasonably precise conception of Euclid's arrangement of the *Elements* in these terms:

Books I and II: plane rectilineal geometry
Book III: the circle
Book IV: regular polygons
Book V: the theory of proportion

Book VI: plane geometry with proportions
Books VII–IX: arithmetic
Book X: irrational lines
Book XI: elementary solid geometry
Book XII: the method of exhaustion
Book XIII: regular polyhedra

By and large I have attempted to observe Euclid's subject divisions, but I have not followed him closely in his arrangement of subjects, in order to make clearer points of comparison and contrast with modern analogues of these subjects and to bring out important deductive relationships; for example, to show the complete or virtually complete independence of a book from a predecessor. However, in keeping with the policy of examining propositions in terms of their use, it has sometimes seemed advisable to treat the applications of a subject in connection with the subject or to postpone the treatment of materials until their application is studied. For example, I discuss the few applications of arithmetic at the end of my treatment of arithmetic in chapter 2, and I deal with book X in the middle of the chapter on book XIII, the only place where book X is applied. There are many other smaller-scale rearrangings, the purpose of which has always been to bring out logical and conceptual features of the *Elements*. Although there can be no question of explaining all of these details of organization in an introduction and although the table of contents gives an overview of the order in which topics are taken up, it is perhaps worthwhile to describe briefly what is done in the various chapters of the book.

The focus of chapter 1 is Euclid's development of the rudiments of plane geometry in the first 45 propostions of book I. On the whole the mathematics of book I is quite simple, and I have usually taken it for granted that a reader can reconstruct the essentials of the proof of a proposition from an indication of the propositions used to prove it. My major concern is with the deductive structure of book I, which I argue is organized around the proof of 1,45, and with the starting points of the book, its explicit definitions, postulates, and common notions and its implicit presuppositions. I have attempted to characterize the axiomatic method used in book I and to compare it with its modern analogue, Hilbert's famous presentation of geometry. I discuss this presentation in section 1.1, where I introduce the notation and concepts of modern logic which I use throughout the book. In chapter 1 I try to establish that the differences between Hilbert's and Euclid's geometry stem from a fundamental contrast between the dominant role of structure in modern mathematics and its virtual absence in ancient mathematics. This contrast is a basic presupposition of the remainder of the book, although further argumentation in support of it is also given.

In section 1.3 I raise in a preliminary way the important question of the relevance of algebraic ideas to the interpretation of the *Elements*. This is a question to which I return throughout the book because it arises in connection with different parts of the *Elements*. I introduce the question in connection with the first seven propositions of book II, because these are the easiest propositions in terms of which to explain the issues involved. However, since I believe that these propositions have to be understood in terms of their use, I postpone interpreting them until their uses have been explained. In general I argue that although algebraic ideas are useful for simplifying complex geometric materials in the *Elements* for the modern reader, the use of these ideas is historically unjustified and philosophically misleading. I also attempt to show that a strictly geometric reading of the *Elements* is sufficiently plausible to render the importation of algebraic ideas unnecessary.

In chapter 2 I move immediately to books VII–IX of the *Elements* because their subject, arithmetic, is developed from scratch by Euclid and plays a fundamental role in modern foundational studies. As in chapter 1, my major concerns are deductive structure and foundations, but unfortunately the deductive structure of VII–IX is much less linear than that of I and their foundations are almost entirely tacit rather than explicit. I have discussed these topics in a relatively formal way which should be perspicuous to anyone with reasonable facility in mathematics and logic; I have in this case tried to include all details, so that no substantive mathematical knowledge is presupposed. Foundations are treated in the first part of section 2.1, where the basic differences between ancient and modern arithmetic are described. The remainder of the section is devoted to characterizing the principal parts of book VII and the reasons for Euclid's arrangement of it. In section 2.2 I deal with books VIII and IX. I first present the content of propositions 1–10 of book VIII in a more perspicuous way than Euclid does, in order to make clear the core of Euclid's own elaborate proofs. I then discuss the deductive structure of VIII,11–IX,20 and the mathematical significance of these propositions, before turning to the curiously elementary set of propositions which Euclid inserts at the end of book IX just before the famous last proposition on perfect numbers. In section 2.3 I treat two apparent applications of algebraic laws in the arithmetic books and then the applications of arithmetic in the subsequent books, all of which are in fact in book X.

The theory of proportion of book V has often been compared to Dedekind's theory of the real numbers. In section 3.1 I explain the point of the comparison and the limits of its viability, and then consider the foundations of Euclid's theory. The treatment of foundations is greatly simplified because of the similarity between them and the foundations of Euclid's arithmetic. Here again I adopt a relatively formal mode of presentation, as I do in a brief account of the content and deductive structure of book V with which section 3.1 concludes. In section 3.2 I consider a series of problems in the interpretation of book V. I argue first that there is no reason to connect Euclid's theory of proportionality with calculation, an argument which also weighs against algebraic readings of the *Elements*. Secondly I argue that magnitudes are geometric objects only and do not include numbers; and thirdly, that Euclid did not attempt to formulate what is normally called the Archimedean condition. I conclude with a brief, somewhat technical, discussion of the relative logical strength of this condition and some other related ones such as density and continuity.

In chapters 4 and 5 I return to the subject of plane geometry as it is presented in books III, IV, and VI. It is reasonably clear that Euclid postpones the treatment of similarity for rectilineal figures and of proportionality for as long as he can, namely until the end of his development of plane geometry. The mathematically more elegant procedure would be to begin with the theory of proportion and similarity and to treat congruence as a special case of similarity. I argue that Euclid is quite consciously not adopting this procedure, at least to the extent of recasting proofs based on proportionality to avoid the concept. In chapter 4, which includes a rather lengthy discussion of the apparently algebraic propositions VI,28 and 29 and their relation to Babylonian mathematics, I give a first example of such a recasting, Euclid's proof of the Pythagorean theorem (I,47) which appears to be reworked from the proof of a more general theorem concerning similar figures (VI,31). In the same chapter I discuss Euclid's treatment of proportions in plane geometry in book VI and argue that it is not at all what one would expect if Euclid were in some way concerned with the calculation of areas. The argument is developed further in connection with solid geometry in section 6.2.

In chapter 5 I conclude my account of Euclid's plane geometry with a discussion of books III and IV. In the latter Euclid treats problems of inscription and superscription involving the circle and rectilineal figures. Most of the book is elementary and I discuss it very briefly in section 5.2. However, the inscription of a regular pentagon in a circle is perhaps the most complex argument in all of Euclid's plane geometry. I argue that the complexity arises from the avoidance of the theory of proportion and that allegedly algebraic ideas in Euclid's proof arise from a geometric analysis aimed at avoiding the theory. The same motive is invoked to explain other parts of book III, notably the curious treatment of equality for circles and the use of similar segments. The first part of chapter 5 is devoted to Euclid's account of the geometry of the circle. Here I have focused mainly on deductive structure and pointed out some of the logical peculiarities in Euclid's argumentation.

In Euclid's treatment of solid geometry book XI corresponds roughly to book I, but it lacks an explicit axiomatic foundation. In section 6.1 I discuss Euclid's approach to the foundations of solid geometry and in section 6.2 his treatment of volumes. I argue that Euclid is principally concerned to establish conditions for the equality of two solids of the same kind and the volume relationship between similar solids expressed in terms of a nonquantitative relation between the sides, and not to establish anything like formulas for computing volumes. It is also shown that Euclid consistently fails to follow a line of argument making maximal use of his own proportion-theoretic apparatus, relying instead on elaborate geometric constructions. In section 6.3 I discuss the use of the method of exhaustion in book XII, a method which is closely related to the integral calculus. Here again the exposition takes on a more formal character. For although Euclid approaches each application of the method individually, they follow a common formal pattern, and without comprehension of the pattern it is easy to get lost in the details of the complex argumentation. For this reason I first characterize the method in a general way and then, after explaining the significant differences between it and the integral calculus, I show how Euclid's proofs are applications of it.

Book XIII, Euclid's treatment of the regular solids and their relation to the sphere, is the analogue of book IV. Its geometric material is complex but elementary, and I describe it fairly quickly in sections 7.1 and 7.3. Section 7.2 is a discussion of the complications which arise from Euclid's attempt to characterize the relationship between the edge of a regular icosahedron and the diameter of a circumscribing sphere, a characterization which leads back to book X, "la croix des mathématiciens." I argue that the content of book X is purely classificatory and that the schema of classification arises entirely from the treatment of the icosahedron. The discussion of books X and XIII leads back again to the question of algebra and also book II. Section 7.3 includes a summary account of book II.

I hope that this descriptive outline provides a sufficient indication of the development of the book to orient the reader. However, although the argumentation is cumulative, a major purpose of the book is to provide analyses and discussions of individual propositions and concepts. To facilitate access to these analyses and discussions and also to make reading the book easier, I have provided appendices in which are listed the special symbols and additional propositions I have introduced as well as all the assumptions and propositions of the *Elements*, together with indications of where in the present work they are discussed.

Bibliographical Note

I have used the standard edition of Euclid put out by J. L. Heiberg, now being published in a different form under the editorship of E. S. Stamatis. Normally in discussions of textual questions I simply cite 'Heiberg', with the understanding that I am referring to Heiberg's version of the passage being treated or to a footnote on the passage by him. When more explicit information is needed I cite 'Euclid, *Opera*' and give references by volume and page to the old edition. (The pagination of the old edition is reproduced in the margins of the new, which also maintains the volume divisions of the old.) The scholia, which make up most of volume V of the *Opera*, are cited as '*Scholia*' followed by either page and line numbers or the number assigned to a scholium by Heiberg. In editing the *Elements* Heiberg followed, wherever he thought he could, a single manuscript called P by him, which he took to embody a version of the text predating an edition of it by Theon (fourth century A.D.). In many cases a decision not to follow P was influenced by Heiberg's conception of the logical structure of the *Elements*. In order to eliminate this influence I have chosen to follow the main text (i.e., the text independent of additions in the margin) of P, except in cases of obvious scribal error, and to indicate problems which arise from doing so.

All works referred to are listed in the bibliography. In the notes I refer to works by author's name or, where more than one work by an author is listed in the bibliography, by name and short title. There are some exceptions to this policy. One is Heath's three-volume translation of the *Elements* with commentary, which I cite as 'Heath' followed by volume and page number. I have used Heath's translation with occasional emendations, and in cases where his account of an issue seemed to be sufficient or to represent a commonly held view, I have simply referred the reader to his discussion. However, I have not always indicated where my interpretations diverge from his. Other modern translations with useful notes which I have consulted with profit are those of C. Thaer (German) and A. Frajese and L. Maccioni (Italian). Unfortunately, I did not have satisfactory access to the earlier Italian translation of F. Enriques; nor did I consult pre-Heiberg editions and translations, except on isolated points. I have benefited greatly from E. J. Dijksterhuis's two-volume book on the *Elements*, which is also cited by author's name, volume, and page. One other book which is cited by author's name only is B. L. van der Waerden's *Science Awakening*, undoubtedly the most stimulating recent book on ancient mathematics and probably now the standard account of the prehistory of the *Elements*.

Philosophy of Mathematics and Deductive Structure in Euclid's *Elements*

1 Plane Rectilineal Geometry

1.1 Hilbert's Geometry and Its Interpretation

Of all the differences between Greek and modern mathematics, the most fundamental concerns the role of geometry in each. One might say that the history of nineteenth-century mathematics is the history of the replacement of geometry by algebra and analysis. There is no geometric truth which does not have a nongeometric representation, a representation which is usually much more compact and useful. Indeed, many mathematicians might prefer to say that traditional or descriptive geometry is simply an interpretation of certain parts of modern algebra. For such people geometry is of no real "mathematical" interest.[1] The marginal position of geometry in modern mathematics is a complete contrast to its central position in the *Elements* and other classical Greek mathematical texts. One could almost say that Greek mathematics is nothing but a variety of forms of geometry. The extent to which this assertion is true is one interpretative crux to which this book is addressed. However, the most elementary part of Euclid's geometry will be my first concern here. And although it would be possible and enlightening to contrast this with algebraic treatments of corresponding subjects, it is more useful to consider modern treatments of elementary Euclidean geometry which do not invoke algebra in an essential way. The outstanding and most influential work in this relatively narrow field is undoubtedly Hilbert's *Grundlagen der Geometrie*, first published in 1899. I shall simply quote from and paraphrase the beginning of this work.

> 1. The elements of geometry and the five groups of axioms. Explanation. We consider three distinct systems of objects: we call the objects in the first system *points* and designate them by $A, B, C, \ldots$; we call the objects of the second system *straight lines* and designate them by $a, b, c, \ldots$; we call the objects of the third system *planes* and designate them by $\alpha, \beta, \gamma, \ldots$.
>
> We consider these points, straight lines, and planes to be in certain relations to one another and designate these relations by words like 'lie', 'between', 'congruent', 'parallel', 'continous'; the exact and, for mathematical purposes, complete description of these relations is accomplished by the *axioms of geometry*.

Hilbert goes on to describe the five groups of axioms, each of which "expresses certain associated fundamental facts of our intuition." He then gives the axioms of the first group, the axioms connecting points, lines, and planes together. I give here the first three of these axioms and the axioms of group II, the axioms of order, in English and then in logical notation.[2]

I,1 For any two points A, B, there is always a straight line a associated with both of the two points A, B.

$\forall A \forall B [A \neq B \rightarrow \exists a [\mathscr{L}(A, a) \,\&\, \mathscr{L}(B, a)]]$.

($\mathscr{L}(A, a)$ should be read as 'A lies on a'.)

I,2 For any two points A, B, there is not more than one straight line associated with both of the two points A, B.

$$\forall A \forall B [A \neq B \rightarrow \forall a \forall b [\mathscr{L}(A, a) \,\&\, \mathscr{L}(B, a) \,\&\, \mathscr{L}(A, b) \,\&\, \mathscr{L}(B, b) \rightarrow a = b]].$$

I,3 On a straight line there are always at least two points. There are at least three points which do not lie on one straight line.

$$\forall a \exists A \exists B [A \neq B \,\&\, \mathscr{L}(A, a) \,\&\, \mathscr{L}(B, a)].$$
$$\exists A \exists B \exists C [A \neq B \,\&\, B \neq C \,\&\, A \neq C \,\&\, \forall a \neg [\mathscr{L}(A, a) \,\&\, \mathscr{L}(B, a) \,\&\, \mathscr{L}(C, a)]].$$

II,1 If a point A is between a point B and a point C, then A, B, C are three distinct points on a straight line and A is between C and B.

$$\forall A \forall B \forall C [\mathscr{B}(A, B, C) \rightarrow A \neq B \,\&\, B \neq C \,\&\, A \neq C \,\&\, \exists a [\mathscr{L}(A, a) \,\&\, \mathscr{L}(B, a) \,\&\, \mathscr{L}(C, a)] \,\&\, \mathscr{B}(A, C, B)].$$

($\mathscr{B}(A, B, C)$ is read as 'A is between B and C'.)

II,2 Given two points A and C there is always at least one point B on the straight line AC such that C is between A and B.

The straight line AC is defined to be the straight line the existence and uniqueness of which follow for given distinct points A and C from axioms I,1 and I,2. In the logical formulation of the axiom the phrase 'the straight line AC' is represented by $\overline{AC}$:

$$\forall A \forall C \exists B [A \neq C \rightarrow \mathscr{L}(B, \overline{AC}) \,\&\, \mathscr{B}(C, A, B)].$$

An essential feature of a defined term is that its use can be avoided in favor of the terms in its definition. Axiom II,2 could be stated

Given two points A and C there is always at least one point B and a straight line a such that A, B, C lie on a and C is between A and B.

In fact because of II,1 it would be sufficient to write

II,2′ Given two points A and C there is always at least one point B such that C is between B and A.

$\forall A \forall C [A \neq C \rightarrow \exists B \mathscr{B}(C, B, A)]$.

The third axiom of the group is

II,3 For any three points on a line, not more than one of them is between the other two.

$$\forall A \forall B \forall C \forall a [A \neq B \,\&\, B \neq C \,\&\, A \neq C \,\&\, \mathscr{L}(A, a) \,\&\, \mathscr{L}(B, a) \,\&\, \mathscr{L}(C, a) \,\&\, \mathscr{B}(A, B, C) \rightarrow \neg \mathscr{B}(B, A, C) \,\&\, \neg \mathscr{B}(C, A, B)].$$

Because of II,1 I will be able to use the simpler formulation

$$\forall A \forall B \forall C [\mathscr{B}(A, B, C) \rightarrow \neg \mathscr{B}(B, A, C) \,\&\, \neg \mathscr{B}(C, A, B)].$$

After stating this axiom Hilbert gives the following explanation:

We consider two points A and B on a straight line a. We call the system of both points A and B a segment and designate this by AB or BA. The points between A and B are called points of the segment AB or points lying within the segment AB. . . .

He then gives the next axiom.

II,4 Let A, B, C be three points not lying on a straight line and a a straight line . . .[3] which meets none of the points A, B, C: if the straight line a goes through a point of the segment AB then it certainly also goes either through a point of the segment AC or through a point of the segment BC.

There are difficulties involved in rendering this axiom in logical notation. Hilbert apparently thinks of the notion of a system as a logical notion like our notion of a pair or couple. It would be possible to follow him here, but it seems simpler to avoid the notion of segment altogether. The following symbolization of II,4 accomplishes this purpose:

$$\forall A \forall B \forall C \forall a [A \neq B \,\&\, B \neq C \,\&\, A \neq C \,\&\, \neg \mathscr{B}(A, B, C) \,\&\, \neg \mathscr{B}(B, A, C) \,\&\, \neg \mathscr{B}(C, A, B) \,\&\, \neg \mathscr{L}(A, a) \,\&\, \neg \mathscr{L}(B, a) \,\&\, \neg \mathscr{L}(C, a) \,\&\, \exists D [\mathscr{B}(D, A, B) \,\&\, \mathscr{L}(D, a)] \rightarrow \exists E [\mathscr{L}(E, a) \,\&\, [\mathscr{B}(E, A, C) \vee \mathscr{B}(E, B, C)]]].$$

Here is the first proof in the *Grundlagen*:

Theorem 3. Given two points A and C there is always at least one point D on the straight line AC which lies between A and C.

Proof: According to axiom I,3 there is a point E outside the straight line AC [fig. 1.1] and according to axiom II,2 there is a point F on AE such that E is a point of the segment AF. According to the same axiom and according to axiom II,3 there is a point G on FC which does not lie on the segment FC. According to axiom II,4 the straight line EG must then intersect the segment AC in a point D.

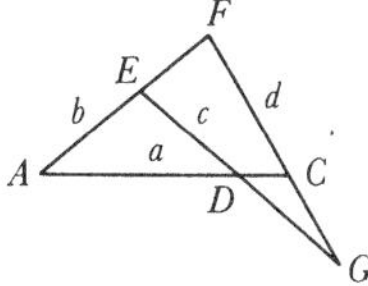

Figure 1.1

It would be possible to represent this proof written in English prose as a finite sequence of logical formulas each of which is either an axiom or a syntactic transformation of previous formulas in the sequence in accordance with fixed rules. If the rules were standard ones, such a representation would require more than 100 such formulas and would be virtually unintelligible unless read in the light of Hilbert's proof. However, the possibility of such a representation has an effect on the philosophical interpretation of Hilbert's geometry, to which I now turn.

Hilbert's *Grundlagen* is open to several such interpretations, all compatible with his prose explanations. One is based on his characterization of the axioms as expressions of fundamental facts of our intuition. Here intuition might be construed psychologically, so that facts of our intuition would be, or rest upon, features of the human mind. On the other hand, intuition might be interpreted as insight into reality, so that facts of our intuition would be facts in a more straightforward sense. Hilbert himself seems to have held this view, as did most of his contemporaries.[4] In his well-known description of outstanding mathematical problems he described geometrical figures as "signs representing the memory images of spatial intuition."[5] The obvious question is how to connect this conception of geometry with the axiomatic method of the *Grundlagen*. On this question Hilbert wrote,

> The application of geometrical signs in rigorous proof presupposes an exact knowledge and complete mastery of the axioms which underlie those figures; and therefore, in order that these geometrical figures may be incorporated into the general treasury of mathematical signs, a rigorous axiomatic investigation of their intuitive content is necessary.[6]

In other words, Hilbert saw rigorous axiomatization as a necessary feature of mathematics. In this opinion he was undoubtedly influenced by earlier work on the foundations of the calculus, work which resulted in a thorough axiomatization of the subject and the elimination of any need to rely upon intuition in proofs. However, there is a very important difference between the calculus and geometry with respect to the role of intuition. In the calculus reliance on intuition led into blind alleys in connection with curves (functions) for which no intuitive picture exists. Rigorous axiomatization was required for a satisfactory treatment of these curves. On the other hand, in elementary geometry reliance on intuition led into no blind alleys. Hilbert's contemporary, Felix Klein, justified the need for axiomatization on the grounds that intuition alone might lead to a false conclusion:

The significance of these axioms of betweenness [axioms of group II] must not be underestimated. They are just as important as any of the other axioms, if we wish to develop geometry as a really logical science, which, *after* the axioms are selected, no longer needs to have recourse to intuition and to figures for the deduction of its conclusions. Such recourse is, however, stimulating, and will of course always remain a necessary aid in research. Euclid, who did not have these axioms, always had to consider different cases with the aid of figures. Since he placed so little importance on correct geometric drawing, there is real danger that a pupil of Euclid may, because of a falsely drawn figure, come to a *false* conclusion. It is in this way that the numerous so-called *geometric sophisms* arise.[7]

Klein went on to give an example of a sophism proving that all triangles are isosceles. Perhaps a "pupil of Euclid" might stumble on such a proof; but probably he, and certainly an interested mathematician, would have no trouble in figuring out the fallacy on the basis of intuition and figures alone. And in the history of Euclidean geometry no such fallacious arguments are to be found. There are indeed many instances of tacit assumptions being made, but these assumptions were always true. In Euclidean geometry, conceived as the description of intuitively grasped truth, precautions to avoid falsehood are really unnecessary. Indeed, although Hilbert's axiomatization decreases the chances of an invocation of a tacit assumption, it increases the chances of clerical mistakes because of the complexity of the material. Such mistakes become almost inevitable.

The common nineteenth-century conception of geometry as descriptive of an intuitive content provides very little justification for Hilbert's axiomatization. It is not surprising then that the enormous mathematical influence of the *Grundlagen* gave impetus to new philosophical interpretations of geometry. One of these was stated clearly by Poincaré in his review of the first edition of the *Grundlagen*. After quoting briefly from its beginning, he said,

Here are the reflections which these assertions inspire us to make: the expressions 'lie on', 'pass through', etc., are not intended to evoke images; they are simply synonyms of the word 'determine'. The words 'point', 'straight line', and 'plane' should not produce any sensible representation in the mind. They could with indifference designate objects of any nature whatever, provided that one can establish a correspondence among these objects to that there corresponds to each system of two of the objects called points one and only one object called a line [and so on].

Thus Hilbert has, so to speak, tried to put the axioms in such a form that they could be applied by someone who did not understand their meaning because he had never seen a point, a straight line, or a plane. Reasoning should, according to him, be capable of being carried out according to purely mechanical rules, and for doing geometry it suffices to apply these rules to the axioms slavishly without knowing what they mean. In this way one could build up all of geometry, I will not say without understanding anything at all since one must grasp the logical sequence of the propositions, but at least without perceiving anything. One could give the axioms to a logic machine, for example the *logical piano* of Stanley Jevons, and one would see all of geometry emerge from it.

It is the same concern which has inspired certain Italian scholars, like Peano and Padoa, who tried to develop a *pasigraphy*, that is to say a kind of universal algebra in which all reasoning is replaced by symbols or formulas.[8]

At the time Poincaré wrote his review Hilbert would not have accepted this extreme formulation of what I shall call the formalist conception of geometry. But, as Poincaré himself pointed out, the formulation is admirably suited to Hilbert's description of his goal in the foreword to the first edition of the *Grundlagen*:

The following investigation is a new attempt to choose for geometry a *simple* and *complete* set of *independent* axioms and to deduce from these the most important geometrical theorems in such a manner as to bring out as clearly as possible the significance of the different groups of axioms and the scope of the conclusions to be derived from the individual axioms.[9]

Moreover, in pursuing his goal Hilbert was led to consider arithmetic interpretations of his axioms and also systems of axioms having no intuitive geometric meaning. There eventually developed around Hilbert a school of formalist mathematicians (or metamathematicians) who concerned themselves primarily with the study of axiom systems formulated in the logical notation used above, a notation which is the descendant of the pasigraphy of Peano and Padoa.

The logical formulation of the axioms brings out clearly the point of view expressed by Poincaré. To understand the logical formulation of I,1, for example, one has to understand the logical signs $\forall$, $\exists$, $\neq$, &, and $\rightarrow$, the notion of a relation holding between two objects, the use of variables, and logical syntax. There is nothing more to understand, since all the logical formulation by itself says is that, for any two things of the kind indicated by uppercase letters, there is a thing of the kind indicated by lowercase letters such that some relation called $\mathscr{L}$ connects each of the first two things to the third. But in a sense one need not even understand this much. For, as Poincaré suggests, a machine could be constructed, in principle at least, which presented with the axioms of geometry in logical notation, would in time grind out any particular theorem.

There is no more reason to attribute understanding of a "logical sequence" to this machine than there is to attribute understanding of messages to a teletype machine. The teletype machine and the logic machine are constructed to respond to specific input signals in specific ways. A person trained to apply purely mechanical rules to axioms is not in his performance of this task significantly different from a machine.

It is sometimes thought that formalism deprives mathematics of the meaningfulness and content which it apparently has. But in fact no philosopher of mathematics of the twentieth century seems to have maintained that mathematics is simply the application of rules of inference to logical formulas. Hilbert looked on formalization as a means of solving certain mathematical questions, notably the question of consistency, but he regarded mathematics itself as the study of ideal objects created by the intellect to simplify treatment of the empirically and intuitively given.[10] A more extreme kind of formalism has been advocated by one of Hilbert's students, Haskell Curry. He defines mathematics as "the science of formal systems."[11] For him mathematics is not meaningless, but the content of mathematics is provided by formal axiom systems. For example, the question whether all pairs of Euclidean straight line segments are Archimedean is for Curry the question whether a formula expressing what Hilbert called the Archimedean axiom[12] is obtainable by slavish application of logical rules to the axioms of geometry. One way to answer this question might be to apply the rules slavishly or to construct a machine to do it.[13] However, the motivating idea behind formalism is that such questions should be answerable by direct consideration of the axioms and logical rules themselves. The science of formal systems would then be like the attempt to determine whether a certain chess position could automatically produce victory by combinatorial reasoning rather than by moving the pieces in a variety of ways. The science of formal systems is a branch of mathematics, but it is not a replacement for mathematics, as Curry's definition suggests it might be. One important reason would seem to be that even though most well-formulated mathematical questions can be translated into questions about formal systems, they seem to require for their solution reasoning typical of the branches from which they were translated. I shall give one example to illustrate the point. Hilbert's first problem was the evaluation, in terms of Cantor's theory of transfinite numbers, of the size of the set of all real numbers. Cantor himself had made a conjecture on this question. In the late thirties Gödel, probably the foremost metamathematician of the twentieth century, provided a partial solution to the problem by showing that the negation of the logical formulation of Cantor's conjecture was not derivable by the rules of logic from a standard axiomatization of set theory. Although Gödel's result can be described in formalistic terms, a perusal of its proof shows its set-theoretic nature. What Gödel showed is that Cantor's conjecture is true if the real numbers satisfy a condition called constructibility.[14]

A similar point can be made about the formalist interpretation of the *Grundlagen*. As Poincaré realized, Hilbert made clear the possibility of mechanizing elementary geometry. However, the possibility of mechanization is quite different from the actual replacement of ordinary mathematical reasoning with mechanical theorem-proving. Hilbert's proof of theorem 3 and even its logical formulation are only representations of ordinary mathematical reasoning, not substitutes for it. In making this point I do not intend to deny the mathematical and philosophical significance of the possibility of mechanizing mathematics but only to deny the necessity of accepting formalism as the correct interpretation of the *Grundlagen*.

At the time of the *Grundlagen* the two interpretations I have been discussing appeared to be the only alternatives. It was felt that the axioms of geometry must either be descriptive of some reality or meaningless formulas. This dichotomy can be seen at work in the passage from Poincaré's review of the *Grundlagen* which was quoted above. There Poincaré moves directly from the observation that the axioms require no specific interpretation to the conclusion that the axioms can be said to have no meaning at all. The same dichotomy is found in a standard criticism of the doctrine of implicit definition, the doctrine that the axioms themselves define the nonlogical terms occurring in them. The criticism involves pointing out that the axioms have many possible realizations in which the nonlogical terms get quite different interpretations; hence, it is concluded, the axioms by themselves leave the meaning of the nonlogical terms quite indefinite.[15]

The existence of these alternative realizations provides the basis for what seems to me to be the correct interpretation of Hilbert's presentation of geometry and of many other parts of mathematics. An example may help in making the interpretation clear. The axioms I,1–3 and II,1–4 are true of the Euclidean plane when the objects designated by uppercase letters are construed as points, those designated by lowercase letters as straight lines, and $\mathscr{L}$ and $\mathscr{B}$ are taken to mean 'lies on' and 'between', respectively. They are also true under other geometric interpretations, e.g., when the points are taken to be points on the surface of a half sphere not including the great circle determining the half sphere, and straight lines are construed as arcs of great circles on this surface, with $\mathscr{L}$ and $\mathscr{B}$

still interpreted as 'lies on' and 'between'. Moreover, there are nongeometric interpretations making the axioms true. One involves taking a point as an ordered pair $\langle x, y\rangle$ of real numbers and a straight line as an ordered triple of numbers $\langle x, y, z\rangle$, not both x and y being 0. (In this interpretation the triple $\langle x, y, z\rangle$ is identified with the triple $\langle u, v, w\rangle$ if there is a real number r such that $x = ru, y = rv$, and $z = rw$.) $\langle x, y\rangle$ is said to lie on $\langle u, v, w\rangle$ if $ux + vy = w$; $\langle u, v\rangle$ is said to be between $\langle w, x\rangle$ and $\langle y, z\rangle$ if there are reals r,s,t, such that $ru + sv = rw + sx = ry + sz = t$ and either $w < u < y$ or $w > u > y$ or $x < v < z$ or $x > v > z$. Under this interpretation the axioms become truths of elementary algebra or real number theory.[16]

As we have seen, the existence of such alternative interpretations has led some people to the conclusion that the axioms of the *Grundlagen* have no determinate content. And, of course, the meaning of the nonlogical terms is not determinate in the sense in which the meaning of 'between' for points in the Euclidean plane is. But the terms are not contentless either. In the case of $\mathscr{B}$, for example, Hilbert's axioms tell us that it designates a relation which holds only between three distinct objects and holds among A, B, C if it holds among A, C, B, and so on. A common word for the kind of information about $\mathscr{B}$ given by the axioms is 'structural'. I shall say that the content of Hilbert's axioms is structural and that Hilbertian geometry and many other parts of modern mathematics are the study of structure.

A more precise account of this conception of structure depends upon a more precise account of the notion of an interpretation. For the axioms I have been discussing, an interpretation is an ordered quadruple $\langle P, S, L, B\rangle$, with P and S disjoint sets of objects, L a two-place relation connecting members of P with members of S, B a three-place relation connecting members of P. The axioms are true under such an interpretation if they are true when the uppercase variables of the axioms are taken to refer to, or range over, P, the lowercase variables to refer to S, $\mathscr{L}$ to designate L, and $\mathscr{B}$ to designate B.[17] I have already indicated the existence of different interpretations $\langle P, S, L, B\rangle$ under which the axioms are true, and quite clearly there are interpretations under which they are not true. Interpretations of the latter kind will assign to $\mathscr{L}$ or $\mathscr{B}$ features which are incompatible with the restrictions placed on them by the axioms. Obviously the axioms exclude such interpretations and, therefore, give some content to the letters $\mathscr{L}$ and $\mathscr{B}$. If two interpretations make exactly the same assertions involving $\mathscr{L}$ and $\mathscr{B}$ true, any intuitive differences between them are irrelevant to the axiom system of which they are an interpretation. For both interpretations have the same structure. There are interpretations which agree in making axioms I,1–3, II,1–4 true but disagree on the truth value of the parallel postulate:

$$\forall a \forall A[\neg \mathscr{L}(A, a) \rightarrow \forall b \forall c[\mathscr{L}(A, b) \ \& \ \mathscr{L}(A, c) \rightarrow \exists B[\mathscr{L}(B, a) \ \& \ [\mathscr{L}(B, b) \vee \mathscr{L}(B, c)]] \vee b = c]].$$

Hence, even though I,1–3 and II,1–4 exclude certain interpretations, they do not determine a unique structure. Although most axiom systems of interest in modern mathematics fail to determine a unique structure,[18] they can be said to determine the common structure of all interpretations under which they are true.

The structural interpretation of mathematics is relatively recent. It throws new light on the significance of logic, which can be considered to be the theory of structure-preserving inference. For the rules of logic permit the derivation from an axiom system of exactly those assertions which are true under all the interpretations under which the axioms are true. In other words, logical derivations simply bring to light features of the structure characterized by the axioms. On the other hand, there is little reason to attempt to connect logic and structure with mathematical psychology. Most mathematical reasoning may well involve intuition or some kind of imagining, but such intuition is as irrelevant to the content of mathematics as it is to the correctness of the reasoning.

The irrelevance of intuition for the study of structure necessitates the axiomatic method or something like it. One specifies the structure under consideration by specifying the conditions which it fulfills, i.e., by giving the axioms which determine it. In some cases, the axioms are the only characterization of the structure. For example, in algebra a group is defined to be any system of objects satisfying certain axioms. In other cases the specification of axioms is an attempt to characterize precisely a roughly grasped structure. Peano's axiomatization of arithmetic and Hilbert's *Grundlagen* are examples. However, the important point is that in either case axiomatization is required for specifying the structure under consideration. One cannot be said to know what a structure is until its relevant features have been characterized explicitly.

I hope that enough has been said to make the structural interpretation of modern geometry reasonably clear. For in this book I will be contrasting the Greek use of the axiomatic method with the modern one and arguing that Greek mathematics should not be interpreted in terms of structure. In order to begin bringing out these points I turn now to the first book of Euclid's *Elements*.

1.2 Book I of the *Elements*[19]

Euclid does not favor his readers with any informal explanations in Hilbert's manner. He begins directly with a list of assertions divided into three groups and labelled 'definitions' (*horoi*), 'postulates' (*aitēmata*), and 'common notions' (*koinai ennoiai*). He then proceeds directly to the statement and proof of 48 propositions (*protaseis*), which divide into two kinds: those which describe a task (1–3, 9–12, 22, 23, 31, 42, 44–46), and those which make an assertion (the rest). At some time, perhaps after Euclid, the Greeks developed a terminology to mark this distinction, calling the tasks problems (*problēmata*) and the assertions theorems (*theōrēmata*).[20] For each of these propositions Euclid gives a proof which has a very stylized form and can easily be divided into parts. Proposition 1 illustrates this form. I quote it in full, indicating its parts with the Greek terms for referring to them.

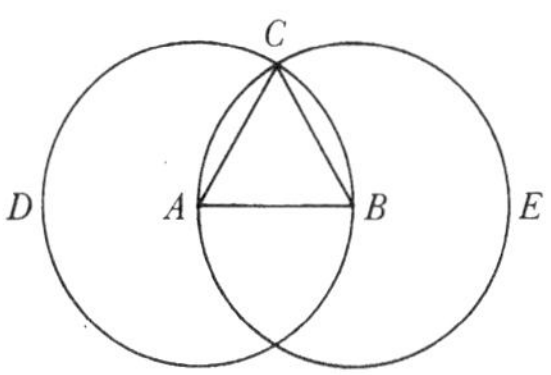

Figure 1.2

Protasis On a given finite straight line to construct an equilateral triangle.

Ekthesis Let *AB* [fig. 1.2] be the given finite straight line.

Diorismos Thus it is required to construct an equilateral triangle on the straight line *AB*.

Kataskeuē With center *A* and distance *AB* let the circle *BCD* have been described; again with center *B* and distance *BA* let the circle *ACE* have been described; and from the point *C* in which the circles cut one another to the points *A*, *B* let the straight lines *CA*, *CB* have been joined.

Apodeixis Now since the point *A* is the center of the circle *CDB*, *AC* is equal to *AB*. Again, since the point *B* is the center of the circle *CAE*, *BC* is equal to *BA*. But *CA* was also proved equal to *AB*; therefore each of the straight lines *CA*, *CB* is equal to *AB*. And things which are equal to the same thing are equal to one another; therefore *CA* is also equal to *CB*. Therefore the three straight lines *CA*, *AB*, *BC* are equal to one another.

Sumperasma Therefore the triangle *ABC* is equilateral; and it has been constructed on the given finite straight line. Which was required to be done (Q.E.F.).

In the case of a theorem the only differences are (1) the *diorismos* is an assertion beginning "I say that," (2) the *sumperasma* repeats the *protasis*, and (3) the *sumperasma* ends with "Which was required to be proved" (Q.E.D.).

The Greek word *apodeixis* means 'proof', but from a modern point of view everything following the *protasis*, except the *diorismos*, is proof. For clarity I shall use the word 'proof' in the modern sense, and *apodeixis* in the technical Greek sense. I shall use the word 'proposition' ambiguously, sometimes as a translation of the word *protasis*, sometimes to refer to the whole sequence from *protasis* to *sumperasma*. It is interesting to compare Euclid's proposition 1 with Hilbert's theorem 3.

Euclid formulates his *protasis* in a perfectly general way, whereas Hilbert's use of variables in stating theorem 3 might appear to be less general. In fact this appearance is misleading, since the apparently free A and C in theorem 3 are tacitly universally quantified. Euclid sometimes uses free variables in this way in *lēmmata.*[21] His refusal to use them in propositions sometimes produces sentences so complicated that it is difficult to believe that anyone could understand them without reading the *ekthesis* and *diorismos*.

Hilbert omits anything corresponding to the *ekthesis*, the setting-out of the straight line AB. This omission is merely a matter of compression. In the natural logical representation of Hilbert's proof, the first step would be the "setting-out" of two points A and C. ("Let A and C be two distinct points.") Hilbert simply takes this step for granted. The *ekthesis* is never taken for granted in Euclid's proofs, and normally the *diorismos* isn't either. This statement of what has to be proved or done on the basis of the *ekthesis* is not, of course, a logically necessary part of a proof. The validity of an argument does not depend upon stating in advance what is to be proved. The *diorismos* should be seen as an expository device designed to make the proof easier to follow.[22]

In the logical representation of Hilbert's proof the last steps would involve what is called conditionalization and universal generalization. The representation would begin with the setting-out of the two points A and C, i.e., with the assumption

$$A \neq C.$$

A series of inferences would lead to the conclusion, based on this assumption, that

$$\exists \mathrm{D}\mathscr{B}(D, A, C).$$

By conditionalization one would infer the conditional conclusion

$$A \neq C \rightarrow \exists D\mathscr{B}(D, A, C),$$

based on no assumptions except axioms, and then by two steps of universal generalization the expression of theorem 3,

$$\forall A \forall C(A \neq C \rightarrow \exists D\mathscr{B}(D, A, C)).$$

In Euclid's proofs of theorems the transition from the end of the *apodeixis* to the *sumperasma* looks very much like a compressed representation of this kind of inference in which one moves directly from the conclusion based on the *ekthesis* to the fully universalized proposition which has been proved. The step of conditionalization is so natural that it is difficult to know whether or not to count it as an inference of which the Greeks were conscious. The evidence from their logical writings suggests that they did not grasp very clearly the difference between an inference and a conditional proposition.[23] However, the question of conditionalization is undoubtedly of less philosophical importance than the question of universal generalization. Although in logic the word *sumperasma* means 'conclusion', it is probably a mistake to think of the *sumperasma* as being inferred from some other proposition by generalization. The word *sumperasma* can also mean 'completion' or 'finish', and in the case of a problem like proposition 1 it is clear that the *sumperasma* merely sums up what has taken place in the proof. There is no good reason to think of the *sumperasma* of a theorem any differently. It merely completes the proof by summarizing what has been established.[24]

Thus in a Euclidean proposition what is proved is stated three times, first generally in the *protasis*, then in terms of a particular example in the *ekthesis-diorismos*, and then in summary in the *sumperasma*. The explanation for this logical redundancy would seem to be connected with the difficulty of grasping the idea of generalization. So far as I know, the basic method of proof in every historical form of mathematics in which proof has played an explicit role has involved the setting-out of an apparently particular case and arguing on the basis of it. I do not know of any way to demonstrate that this form of proof is essential to mathematics, but there is no reason to think that any Greek mathematician could envisage a genuine alternative. In this sense at least, the *ekthesis* represents a necessary part of a Greek proof. It is natural to ask about the legitimacy of such a proof. How can one move from an argument based upon a particular example to a general conclusion, from an argument about the straight line AB to a conclusion about any straight line? I do not believe that the Greeks ever answered this question satisfactorily, but I suspect that the threefold repetition of what is to be proved reflects a sense of the complexity of the question. The *protasis* is formulated without letters to make the generality of what is being proved apparent. The *ekthesis* starts the proof, but, before the proof is continued, the *diorismos* insists that it is only necessary to establish something particular to establish the *protasis*. When the particular thing has been established, the *sumperasma* repeats what was insisted upon in the *diorismos*. Of course, insisting that the particular argument is sufficient to establish the general *protasis* is not a justification, but it does amount to laying down a rule of mathematical proof: to prove a particular case is to count as proving a general proposition.

Here it is helpful to contrast the move to the *sumperasma* in a Euclidean proof with a logical step of universal generalization. The statement of the logical rule involved includes a precise specification of the conditions under which generalization is permitted. Although these conditions are stated formalistically, they have a justification which makes clear that universal generalization is permitted only when no special assumptions have been made about the particulars in terms of which the proof was carried on. In the case of Hilbert's theorem 3, the permissibility of such generalization is made clear by the fact that the conditional to which generalization is applied depends upon no assumptions in which the letters *A* and *C* occur as "free" variables. Since no such assumptions are made, there is a sense in which *A* and *C* are not particular points at all. There is certainly no geometric way to pick them out from the system of points. Thus logic and the structural interpretation of mathematics make it possible to give a clear and reasonable account of ordinary mathematical reasoning. However, there is no reason to suppose the Greeks to have had anything like modern logic to represent actual mathematical argument, and the Euclidean style makes it look as though a proof is thought of as being carried out with respect to a particular object, but in a way assumed to be generalizable.[25] In the absence of something like the rules of logic there is no uniform procedure for checking the correctness of this assumption in individual cases. Rather one must rely on general mathematical intelligence.

This difference between Euclid's and Hilbert's geometry is reflected in the difference between Euclid's first three postulates and their analogues in the *Grundlagen*. Consider Euclid's first postulate and Hilbert's first axiom. Hilbert asserts the existence of a straight line for any two points, as part of the characterization of the system of points and straight lines he is treating. Euclid demands the possibility of drawing the straight-line segment connecting the two points when the points are given. This difference is essential. For Hilbert geometric axioms characterize an existent system of points, straight lines, etc. At no time in the *Grundlagen* is an object brought into existence, constructed. Rather its existence is inferred from the axioms. In general Euclid produces, or imagines produced, the objects he needs for a proof, the two circles, the point *C*, and the straight lines *AC* and *BC* in the proof of proposition 1. It seems fair to say then that in the geometry of the *Elements* there is no underlying system of points, straight lines, etc. which Euclid attempts to characterize. Rather, geometric objects are treated as isolated entities about which one reasons by bringing other entities into existence and into relation with the original objects and one another. The emphasis on construction in Greek geometry is connected with the absence of absolute existence assertions like the second part of Hilbert's axiom I,3, which asserts the existence of three noncollinear points. In the geometry of the *Elements* the existence of one object is always inferred from the existence of another by means of a construction.

Hilbert's method of handling existence is common throughout modern mathematics. Indeed, the method is simply a consequence of the use of ordinary logic and the axiomatic method. There have been critics of this method who have argued that in mathematics existence must be established by a construction.[26] Some of these critics have even identified mathematical existence with constructibility. It is important to realize that, on the whole, modern discussions of existence and constructibility have not concerned elementary geometry but rather branches of mathematics in which infinite sets or sequences play a crucial role. Nevertheless, it is possible to see in Euclid's first three postulates something like an identification of existence and constructibility.[27] However, it is difficult to construe this identification as a matter of conscious choice in the absence of an available alternative to it. And it seems that a reasonable alternative requires the conception of geometric objects as constituting a system. For if one presupposes the existence of a system of objects it makes sense to ask what properties an object or objects might have without worrying about producing an instance of the property. For example, it makes sense in connection with the *Grundlagen* to ask whether there exists a straight line segment with endpoint A and equal to a given segment; but if, as in the *Elements*, the given segment is not assumed to have any relation to other objects, one can only ask whether such an equal can be produced. Euclid's restriction of proofs of existence to constructions would seem then to be a natural outgrowth of his conception of geometric objects, and not the result of the conscious adoption of a constructivist philosophy of mathematics. This conception may well have its roots in the practical or empirical origins of geometry. For in practical applications geometric objects are given in isolation and not as part of a spatial system.

The situation is quite different with respect to Euclid's restriction of the means of construction to those contained in the first three postulates. I shall designate these means as straightedge and compass, although Euclid makes no mention of mechanical instruments, and the first three propositions of book I can be said to show that the straightedge can not be marked and the compass not lifted off of a plane surface without collapsing. (The propositions also show that these restrictions are not really limitations.) The demonstration that a whole series of constructions can be carried out with straightedge and compass is clearly one of the purposes of the early books of the *Elements*. Why Euclid chose precisely these means does not seem to be directly ascertainable.[28] The reason is certainly not that only these means were considered legitimate. For mathematicians both before and after Euclid make use of other constructions without indicating any serious doubts about their legitimacy. What can be shown is how the important purposes of Euclid's geometry and the conception of geometric objects already described lead rather naturally to the first three postulates.

To show this one cannot examine book I in the order of presentation, because the deceptive smoothness of this order provides little insight into the underlying structure of the book. Much more insight is obtained by examining the central proposition or propositions of book I and showing how the book builds to its or their proof. In fact, almost the entire content of book I can be explained by reference to the construction of a parallelogram in a given angle and equal (in area) to a given rectilineal figure in proposition 45. This proposition makes it possible to represent any rectilineal area as a rectangle. Euclid could have proved a stronger result, namely that any rectilineal area can be represented as a rectangle with a given base. (Compare I,44.) From our point of view this result would be more interesting, since the areas of rectangles on equal bases are proportional to the lengths of their sides. For the Greeks, however, the important representation of an area seems to be as a square, and I,45 is sufficient for Euclid to be able to show in II,14 how to construct a square equal to any given rectilineal figure. This proposition represents the true culmination of the geometry of the area of rectilineal figures. Euclid postpones it to book II because its proof involves methods which he introduces there and which he wishes, presumably for purposes of exposition, to separate from the methods of book I.

What I propose to do then is to describe briefly how the analysis of the conditions of solution of the problem of I,45 leads back to the more elementary constructions of book I and to the theorems needed to justify those constructions. In most cases, of course, the analysis of one proposition leads back to several preconditions and becomes extremely complicated unless one assumes that some of the conditions are independently known to be satisfied or satisfiable. This complexity makes it natural to suppose that analysis is only applicable at the later stages of deductive theorizing when there is a body of established results to which new problems and assertions can be reduced. However, it is important to realize that one can know a result in mathematics on the basis of considerations which in other contexts will not count as proof; and one might take for granted a construction, e.g, the bisection of an angle or straight line, without any notion of justifying it by reduction to more elementary constructions. It seems to be generally accepted that most or all of the propositions of book I were known in this sense long before 300 B.C. The point of view adopted here may be expressed by saying that this knowledge and the desire to prove I,45 by themselves suffice to account for much of book I.

Obviously, even if this point of view is correct, it leaves much out of consideration. No attempt is made to explain why there is any desire to prove at all. Nor is any account given of the form or forms which the knowledge of the mathematical content of book I took in pre-Euclidean times. More importantly perhaps, no explanation is offered for Euclid's interest in proving I,45 or II,14 rather than, e.g., justifying some formula or procedure for computing the area of an arbitrary rectilineal figure. Finally, no attempt is made to account for the particular base which Euclid chooses for his reduction; it seems reasonably clear that he could have accomplished the same task more elegantly using the concepts of similarity and proportionality. The person interested in these questions will find hypothetical answers to them in most works on the history of Greek mathemematics. For my purposes it is sufficient to record the limitation of my concern to the logical and foundational aspects of the book I which has come down to us.

The basic idea of the proof of I,45 is to divide the given rectilineal area into triangles $t_1, t_2, t_3, \ldots$ and to construct a series of parallelograms $p_1, p_2, p_3, \ldots,$ as in fig. 1.3, with p_i equal in area to t_i and angle $G_iH_iH_{i+1}$ equal to the given angle. Quite clearly p_{i+1} must be constructed not only in the given angle but also on the given straight line $G_{i+1}H_{i+1}$. This task (I,44) is reduced to the construction in a given angle of a parallelogram equal to a given triangle (I,42) by means of

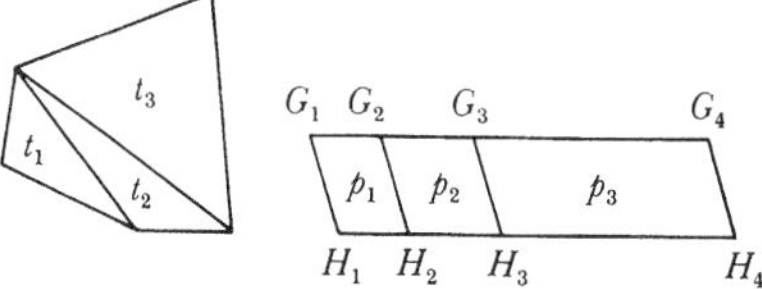

Figure 1.3

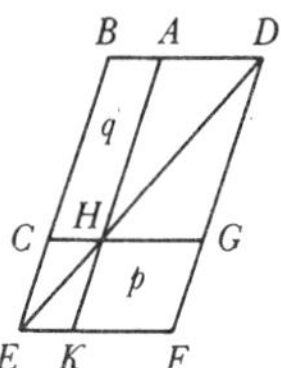

Figure 1.4

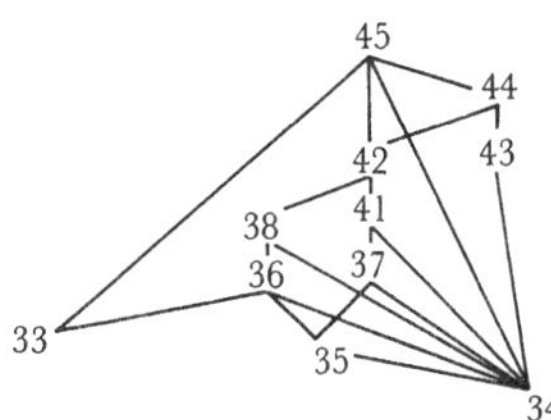

Figure 1.5

the fundamental I,43, which depends heavily on the theory of parallels and says, in terms of fig. 1.4, that p and q are equal in area. Hence, to construct the parallelogram p on GH one need only construct the parallelogram $ABCH$ in the angle AHC equal to the given angle and, with C on GH extended, draw DG parallel to AH and meeting BA extended at D. If DH extended meets BC at E, the figure can be completed by drawing EK through E parallel to CH, and extending it and DG until they meet at F. I,43 depends only on the equality of the two triangles produced by the diameter of a parallelogram (I,34), i.e., of EDB to DEF, of HDA to DHG, and of EHC to HEK; two subtractions of equals from equals leave the equal parallelograms q and p. I,42 reduces rather easily to I,41, a result which we would normally express in terms of formulas for the areas of triangles and parallelograms; these formulas were known, at least for right-angled figures, many centuries before Euclid.[29] Rather than give an analysis of Euclid's proof, of which several variations are possible, I shall simply indicate (fig. 1.5) the chain of logical dependencies among propositions 33–45. The only propositions between 33 and 45 not included in this diagram are 39 and 40. 40 is considered by Heiberg to be an interpolation.[30] Whether or not he is right, 39 and 40 are partial converses of 37 and 38 and, therefore, in a natural position in the sense that Euclid often does prove a converse of a theorem immediately after the theorem for no apparent reason except that it is the converse. In this sense it would of course be equally natural for 39 to follow 37 immediately and 40 38. To say that the position of a proposition is natural is not to deny that some other position might not be equally so. Nor is it generally possible to explain why Euclid proves some converses and not others. 40 is not used in the *Elements* and 39 is not used until VI,2. But there are other converses of book I propositions used in book VI which are not proved. For example, the converse of I,15 is used in the proof of VI,14.

The order of the propositions referred to in fig. 1.5 is virtually forced by the deductive relations. In cases where there is some slack, Euclid's order is easy enough to understand. 43 immediately precedes 44 as a kind of lemma. 38 is proved immediately after 37, 36 immediately before because of considerations of symmetry. Finally, 33 precedes 34 because 33 is the last preliminary to the discussion of parallelograms. It is not, of course, possible to say that the author of book I could not have hit upon some other proof of I,45 using different propositions; one can only say that Euclid's proof proceeds naturally enough.

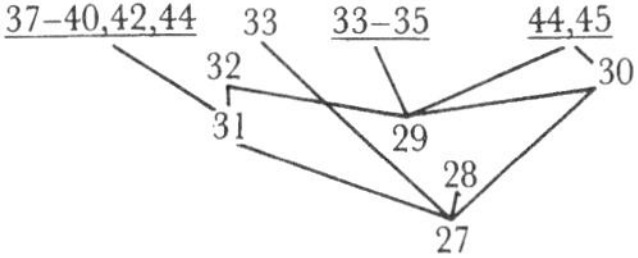

Figure 1.6

Obviously the dependencies described so far do not exhaust the deductive build-up culminating in I,45; there is a downward chain from 33 and 34, and propositions subsequent to them make use of earlier propositions. 27–32 contain Euclid's treatment of the elementary properties of parallel straight lines. The connections among these propositions and their connections with 33–45 is represented in fig. 1.6.[31] Of the propositions 27–32, 28 and 32 are not used in book I but are used subsequently: 32 first in II,9; 28 in IV,7. 28 is a natural completion of 27; the converses of both are combined into one proposition in 29. If there is any reason why 29 follows rather than precedes 27 and 28, it is perhaps because 29 is the first invocation of the parallel postulate, postulate 5. In fact, every proposition in book I after 29 except for 31 depends upon this postulate. 31, the drawing through a given point of a parallel to a straight line, is a preliminary construction for 32, the position of which is sufficiently explained by the importance of this well-known theorem in the *Elements* and in Greek geometry generally; Euclid proves it as soon as he has the materials to do so. The only serious problem in the arrangement of 27–32 is the location of 30, the seemingly innocuous assertion that parallelism is a transitive relation. Heath (vol. I, p. 316), taking a cue from Proclus (376.14–25),[32] explains the location on the grounds that it entails the uniqueness of the parallel constructed in I,31. But Euclid does not in general show any concern with the question whether a constructed object is unique; and 30 is such an oblique expression of uniqueness that one would expect Euclid to be more explicit in drawing this consequence of 30 if he were concerned to establish it.[33] Euclid uses 30 in a fairly explicit way in I,45. Perhaps it is sufficient to explain the location of 30 by saying that Euclid realized he needed it and proved it as soon as he could.

27 marks the beginning of what Heath calls the second section of book I. There are, of course, many applications of propositions from the first section in the rest of book I, but only two serve to explain the position of the proposition employed. One is the use of 26 (first part) in 34, which I will discuss shortly, the other the uses of 23 in 31 and 42, only the second of which seems unavoidable.[34] 23 precedes 27 because it does not involve parallels. It must precede 24 and 25 since it is used in the proof of 24, which in turn is used in 25. The problem of 23 is to copy the angle *DCE* on *AB* at *A* (fig. 1.7). To solve it Euclid forms the triangle *DCE* and copies it on the straight line *AB*. Since Euclid is attempting to establish the possibility of copying an angle, copying a triangle must in this case mean copying its

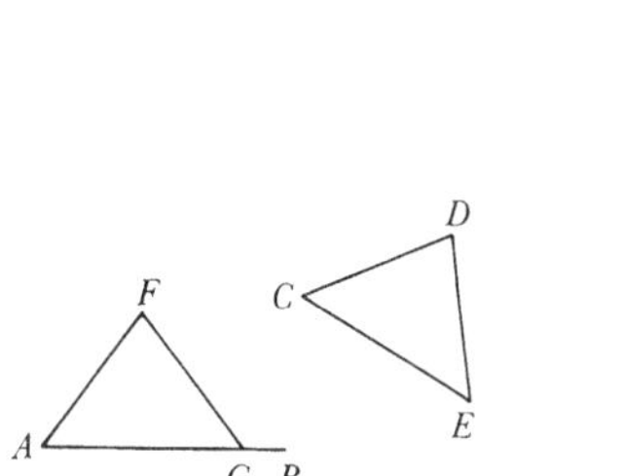

Figure 1.7

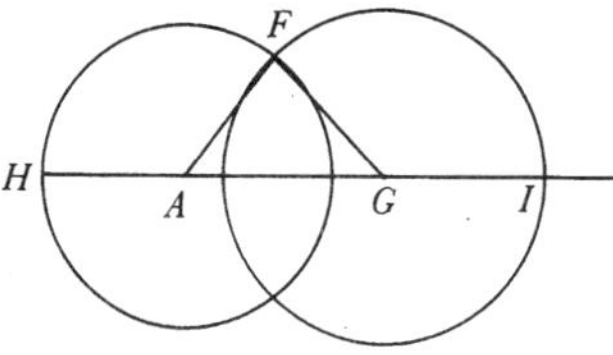

Figure 1.8

sides. Hence Euclid needs I,8 to infer the equality of the appropriate angles from the equality of the corresponding sides of the two triangles. The procedure for constructing the triangle *AGF* with sides equal to *CE*, *ED*, *CD* is given in proposition 22. Euclid imagines an infinitely long straight line (fig. 1.8) on which he marks off *HA*, *AG*, *GI* equal, respectively, to *CD*, *CE*, *ED*. He then draws circles with centers *A* and *G*, radii *AH* and *GI*, and argues that their intersection point *F* determines with *AG* the desired triangle. But here the question of the existence of *F* arises. It is intuitively clear that there would be no such *F* if *GI* (*DE*) were not shorter than *HG* (*CE* + *CD*), or *AH* (*CD*) were not shorter than *AI* (*CE* + *ED*), or *AG* (*CE*) were not shorter than *HA* + *GI* (*CD* + *DE*). Hence, for 22 Euclid needs I,20 as a precondition (*diorismos* in the second sense of this word).

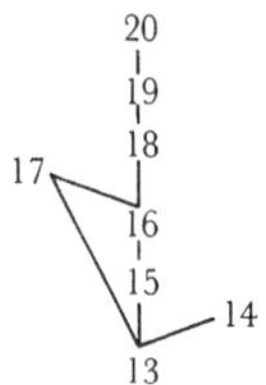

Figure 1.9

I shall not here describe the sequence of propositions leading to I,20, but simply indicate the deductive relationships diagramatically (fig. 1.9). The only logical play in this arrangement is provided by 14 and 17. 14 is not used until I,45, but its position is explained by its being the converse of 13. 17 is not used until III,16, but it is related to I,16 in exactly the way that the second part of I,32 is related to the first. In this respect 17 is very like 28, which is not used until IV,7 and is related to 27 in the same way that the last two parts of 29 are related to the first part. It is not unreasonable to suppose that both 17 and 28 were inserted into the main deductive development of book I when the subsequent need for them was realized.

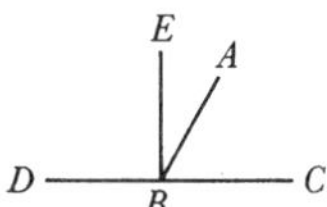

Figure 1.10

Proposition 13 says in effect that if a straight line *AB* set up on the straight line *DC* is not perpendicular to *DC*, then it makes with *DC* angles equal to two right angles (fig. 1.10). In the *kataskeuē* of this proposition Euclid erects a perpendicular *EB* to *DC* at *B*, which by definition 10 makes two right angles with *DC*. The construction of the perpendicular is given by I,11. In establishing I,11 Euclid uses the bisection of an angle (I,9). I,10, the bisection of a straight line, is used in I,16 and 42; but I,12, the dropping of a perpendicular to a straight line, is not invoked until II,12. The sequence of constructions I,9–12 has no real logical or geometric necessity. The four problems could have been solved in any order, since they all depend upon the fact that for a straight line *s* from the vertex of an isosceles triangle to its base the following three conditions are equivalent: *s* bisects the vertex angle; *s* bisects the base; *s* is perpendicular to the base. Presumably the close relation of all four accounts for Euclid's proving them together, but there seems to be no way to account for the particular sequence he adopts.

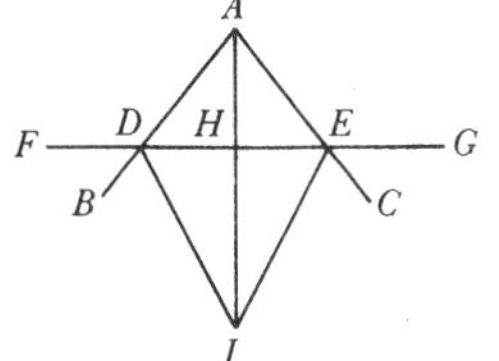

Figure 1.11

For future reference I give an alternative proof of 9–12 showing their interrelationship and their independence of I,8, a proposition which Euclid invokes in the proofs of 9, 11, and 12. In 9 one starts from an angle *BAC*, in 10 from a straight line *DE*, in 11 from a straight line *FG* and a point *H* on it, and in 12 from a straight line *FG* and a point *A* not on it (fig. 1.11). For 9, 10, and 12 it is easy to construct an isosceles (and, for 10, equilateral) triangle with vertex *A* and base *DE*, and to construct the equilateral triangle *DIE* on the other side of *DE* from *A*. By I,5 angle *ADE* equals *AED* and *IDE* equals *IED*. Hence, angle *ADI* equals *AEI*. By I,4 angle *DAH* equals *EAH* (so *DAE* has been bisected); by I,4 again, *DH* equals *HE* (so *DE* has been bisected), and angle *AHD* equals *AHE* (so *AH* is perpendicular to *DE*). For 11 one makes *DH* equal to *HE*, constructs the equilateral triangle *ADE*, and argues by I,4 that angle *AHD* equals *AHE*.

I,4 is one of three "congruence" theorems for triangles in book I, the others being 8 and 26.[35] Only in 4 does Euclid state the equality of the two triangles involved. Heath (vol. I, p. 262), following Proclus (269.27–270.4), attributes this difference to the fact that in 8 and 26 the equality of the triangles is an easy inference from I,4. However, the proof of I,34 shows that Euclid is conscious of the difference and does not take the inference for granted.[36] In 34 Euclid wishes to prove that if *AB*, *CD* and *AD*, *BC* are two pairs of parallel straight lines, then they are equal, and so are the angles *DAB*, *BCD*, and so are the triangles *ABD*, *DCB* (fig. 1.12). Euclid infers the equality of angle *ABD* to *CDB* and of *CBD* to *ADB* (and hence of *ABC* to *CDA*) using I,29. Since the side *BD* is common to the two triangles, Euclid is able to infer the equality of the angles *DAB*, *BCD* and of the pairs of parallel lines from I,26. If Euclid took I,26 to include a tacit assertion of the equality of the triangles, he would be finished. In fact, he argues for their equality in terms of I,4.

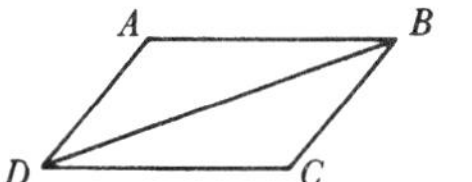

Figure 1.12

In general, Euclid's treatment of the three congruence theorems is quite perplexing. In I,4 Euclid is given the equality of *AB*, *DE* and of *AC*, *DF* and of angles *BAC*, *EDF* (fig. 1.13). He proceeds by what is usually called the method of superposition (*epharmodzein*), a method depending upon intuitive ideas about the coincidability of equal straight lines and angles and upon the possibility of motion without deformation. Euclid places the triangle *ABC* on *DEF*, making *AB* coincide with *DE*. He then argues that *AC* and *DF* will coincide. From the coincidence of *B*, *E* and *C*, *F* Euclid concludes that *BC* and *EF* will also coincide. This complete coincidence guarantees the equality of these two sides, of the remaining pairs of angles, and of the two triangles.

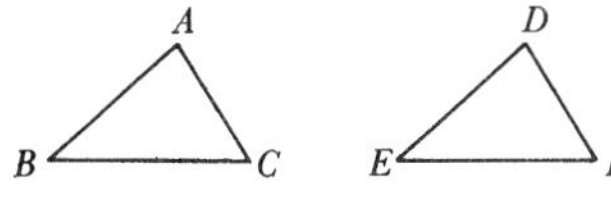

Figure 1.13

In I,8 Euclid again proceeds by superposition. Given the equality of the three pairs of sides, he places *BC* on *EF*. To establish the coincidence of *A* and *D* he invokes I,7, which rules out the possibility of having two pairs of equal straight lines *EA*, *ED* and *FA*, *FD* on the same side of *EF* unless *A* and *D* coincide. The proof of I,7 invokes I,5, which depends upon I,4. Since I,6, which is not used until II,4, is the converse of I,5, it seems reasonable to treat I,5–7 as a sequence leading to I,8.[37]

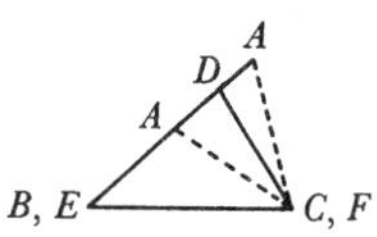

Figure 1.14

Euclid could, of course, prove I,26 by superposition as well. For suppose *BC* is equal to *EF*, angle *ABC* to *DEF*. Then, if *BC* and *EF* are made to coincide (fig. 1.14), *A* will lie somewhere on *ED* (possibly extended). Clearly, if angle *ACB* is equal to *DFE*, *A* and *D* must coincide. On the other hand, if angle *BAC* is equal to *EDF*, *A* and *D* must coincide as well because of I,16. Euclid does not argue in this way. Instead he supposes *DE* to be less than *AB* and makes *BG* equal to *DE* (fig. 1.15). If angle *ACB* is assumed equal to *DFE*, I,4 yields the equality of *GCB* to *DFE* and hence to *ACB*, an absurdity. But if *BAC* is assumed equal to *EDF*, the same proposition yields the equality of *BGC* to *EDF* and hence to *BAC*, contradicting I,16.

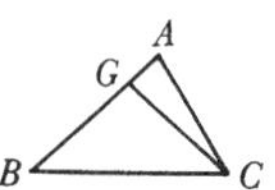

Figure 1.15

Euclid's failure to use superposition in I,26 is somewhat surprising; it constitutes one piece of evidence for a widely held thesis that Euclid found something unsatisfactory in this method of proof.[38] Further evidence is provided by his assumption of postulate 4, which, as Proclus (188.20–189.10) points out, is easily demonstrated using superposition. Suppose that angle *DBC*, *D′B′C′* are right angles with angle *D′B′C′* smaller than *DBC* (fig. 1.16). Then if *A′C′* is placed on *AC* with *B′* on *B*, *D′B′* will fall as in the diagram, and angle *D′B′A′* will be greater than *DBA*. Since by definition 10 angle *DBA* is equal to *DBC*, angle *D′B′A′* is greater than *D′B′C′*, but this inequality is incompatible with definition 10.

On the other hand, Euclid could have avoided the method of superposition in I,8 if he had postponed its proof until he had established the possibility of copying an angle in I,23. For then he could argue for I,8 by supposing angle *FDE* smaller

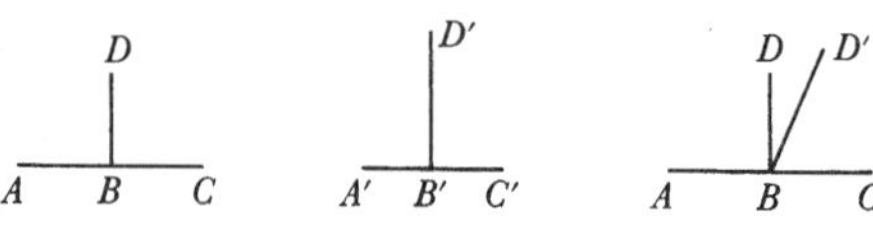

Figure 1.16

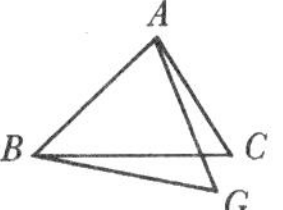

Figure 1.17

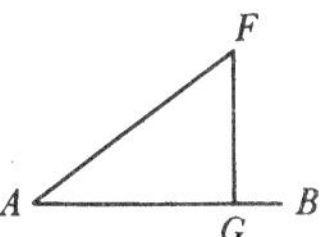

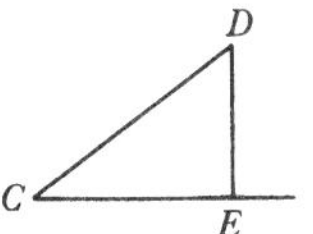

Figure 1.18

than *CAB*, copying *FDE* as *GAB* (fig 1.17) with *GA* equal to *FD*. By I,4, then, *BG* equals *EF*. Hence *BG* equals *BC*, and *AG* equals *AC*, contradicting I,7. Of course, the argument sketched here would be circular unless Euclid's use of I,8 in the proof of I,23 could be circumvented. I have already shown how to avoid the use of 8 in 9, 11, and 12, its only uses prior to 23. To copy the angle *DCE* at the point *A* on the straight line *AB* without using 8, one drops the perpendicular *DE*, cuts off *AG* equal to *CE*, and draws *FG* perpendicular to *AB* at *G* and equal to *DE* (fig. 1.18). Since by postulate 4 all right angles are equal, the angles *FAG*, *DCE* are also equal by I,4.

It is difficult to know what significance to attach to the possibility of avoiding superposition in I,8. Obviously, one cannot maintain both that Euclid was aware of the possibility and that he was dissatisfied with superposition as a method of proof, unless one gives some other explanation for his use of the method in proving 8. Since the alternative proof of I,8 is not very difficult, it seems likely that if Euclid did wish to avoid superposition, his wish was not deep-seated enough to cause him to search very hard for alternatives to it. As for I,4, it is certainly not deducible from Euclid's first principles as they stand. Hilbert assumes a weaker version of I,4 as an axiom and demonstrates its independence from all his other axioms.[39]

Superposition is, of course, a special case of the movement of figures, but there is very little evidence that Euclid found such movement problematic. The *kataskeuē* of I,44 is an interesting example of his attitude. Having set out a straight line *AB*, a triangle *C*, and an angle *D*, Euclid begins the *kataskeuē* with the instructions, "Let the parallelogram *BEFG* be constructed equal to the triangle *C*, in the angle *EBG* which is equal to *D*; let it be placed so that *BE* is in a straight line with *AB*. . . ." Euclid relies on I,42 here, but in that proposition he constructs the parallelogram with one of the sides identical to half of one side of the given triangle. Since Euclid needs to have the parallelogram in a different place, he simply puts it there. He could, of course, have followed a different procedure depending only on constructions at hand. For example, he could have extended *AB* to *E* with *BE* equal to one of the sides of *C* and constructed a duplicate of *C* on *BE*. The crucial question of interpretation is whether Euclid intends his reader to supply the missing steps, or whether he takes for granted the possibility of placing the figure in the required way. *Prima facie* the latter alternative is more plausible simply because Euclid does seem to want to make explicit all of the steps involved in his constructions in book I.[40] In any case Euclid's failure to give the more detailed construction is further evidence that he did not attach very much significance to avoiding the movement of figures, if he attached any at all.

The final evidence on the question of movement to be introduced at this point enables me to complete the description of the constructions involved in the solution of I,45 by considering the first three propositions of book I. In a geometry like Euclid's in which constructions are fundamental, the idea of reducing some constructions to simpler ones would seem to be essential. For what point would there be in just taking for granted constructions such as I,42, 44, and 45? The detailed description of these three constructions presupposes the constructions given by postulates 1 and 2 and propositions 10, 23, and 31. Given I,27, 31 is easily reduced to 23; but the proof of 27 itself depends ultimately on the constructions of I,3 and 11. However, Euclid would need no other constructions than the ones just mentioned in order to carry out his justification of I,45. In other words, he needs only the ability to join points with straight lines, extend straight lines, cut off segments equal to given ones, bisect straight lines, erect perpendiculars, and copy angles. Proclus (333.5) says that this last construction was the discovery of the fifth-century mathematician Oinopides, whom Proclus (283.6–7) also associates with the dropping of a perpendicular. It is possible then that Oinopides' construction for I,23 was the one I gave above rather than Euclid's, as Heath (vol. I, p. 295) suggests. Whoever invented Euclid's construction must have thought the drawing of circles with fixed center and radius a simpler construction than the copying of an angle. Once postulate 3 has been introduced enabling one to draw such a circle, it is not a very great step to seeing that none the constructions of I,9–12 needs to be taken as primitive. The natural reduction of these constructions would seem to require the possibility of erecting an equilateral triangle on a given straight line (I,1), a construction which is easily reduced to postulates 1 and 3.

Thus one is left with the three constructional postulates and I,3 as fundamental constructions. I,3 might seem to be directly reducible to the third postulate. For, given a straight line *AB* and a shorter one *CD*, one could place *CD* with one end at *A* and draw the circle with center *A* and distance *CD*; the intersection *E* of the circle with *AB* would determine a straight line *AE* equal to *CD*. Instead of following this procedure, Euclid inserts proposition 2, which shows how to produce a copy of *CD* with one end at an arbitrary point *A*. He connects *AC*, constructs an equilateral triangle *AFC* on it,

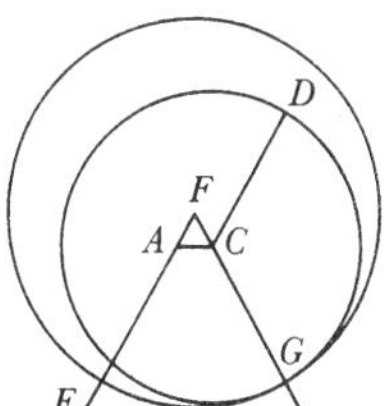

Figure 1.19

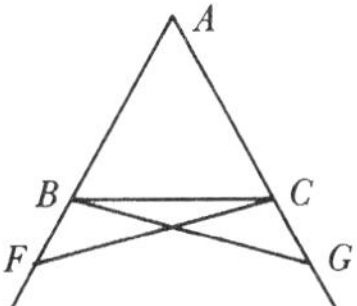

Figure 1.20

and extends *FC* and *FA* (fig. 1.19). He draws the circle with center *C* and distance *CD*, intersecting *FC* extended in *G*, and then the circle with center *F* and distance *FG*, intersecting *FA* extended in *E*. Since *CD*, *CG* and *FC*, *FA* and *FE*, *FG* are equal, so are *CD*, *AE*.

This striking elimination of the construction of I,3 might very well be interpreted as further evidence of Euclid's disdain for moving figures.[41] However, I believe that proposition 2 is best explained in terms of Euclid's concrete conception of geometric objects. To move an object from one position to another is, among other things, to cause it to be no longer in the first position. In the case of proposition 44, moving the parallelogram is acceptable to Euclid because its first position is irrelevant to the proof; the same can be said of Euclid's use of superposition in I,4 and 8. On the other hand, in proposition 2 it is important to maintain the position of the given straight line and, hence, to copy rather than move it. This importance cannot be seen in proposition 2 or 3, but becomes obvious in the first application of 3 in 5. Here Euclid marks off *AG* equal to *AF* and uses proposition I,4 with respect to the triangles *ABG*, *AFC* (fig. 1.20). Obviously, if marking off *AG* equal to *AF* involved displacing *AF*, the construction would eliminate one of the triangles to which I,4 is to be applied. The point made here carries over to a construction like copying an angle at a point on a straight line. It does no good to move the given angle to the point if the angle is also needed in its original position, as it usually is. (See, e.g., I,31.)

The argument of the previous paragraph presupposes that in the *Elements* the use of a construction which has been reduced to more elementary constructions is in theory a carrying-out of those more elementary constructions. Although this presupposition seems reasonable, it does not seem possible to carry it over directly to theorems proved by the use of superposition; that is to say, in such cases one cannot assume that the citation of a theorem is shorthand for the reproduction of the proof of the theorem. For example, if in the proof of I,5, instead of applying the result established in I,4 to the triangles *AFC*, *ABG* of fig. 1.20, one were to try to carry out the proof of I,4 with respect to those triangles, one would have to explain how the side *AB* could be moved to coincide with *AC* in such a way that *AF* would remain where it is so that *AG* could coincide with it. Clearly Euclid believed that the proof of I,4 was perfectly general and that it applied to all pairs of triangles, whatever the relative positions of their sides. However, he perhaps did not think about the possibility of using the method of proof of I,4 in the case of overlapping triangles.

In any case, the evidence from book I that Euclid wished to avoid the movement of figures in geometry is not very strong. Only the presence of postulate 4 carries much weight. For the fact that Euclid does not move figures where he might do so, as in the proof of I,26, shows very little. In general, to introduce motion in these cases would break down deductive structure, substituting motion and immediate evidence for the application of previously proved or assumed propositions. In the case of 26, putting one triangle on the other would replace applications of postulate 1 and propositions 3, 4, and 16. Certainly the use of superposition in 4 and 8 is incompatible with the ascription to Euclid of philosophical qualms about movement.

The logical chain leading to I,45 contains the propositions 1–5, 7–11, 13–16, 18–20, 22, 23, 26, 27, 29–31, 33–38, 41–45. I have attempted to show that the structure of book I is largely determined by this chain, which, although not the only possible sequence to I,45, is a natural one. Of the propositions not included in this chain all except the possibly interpolated 40 are used elsewhere in the *Elements*. The least constrained parts of the chain are those contained in the portions of book I for which I have given no logical diagram, namely propositions 1–12 and 21–26. For the first of these two portions we have seen that 1–3 constitute a sequence which must be established before 5. 4 likewise must be established before 5, although it could be placed before 1–3. 6, as the converse of 5, follows it naturally. 13 presupposes 11, and the constructions of 9–12 are so closely related that it is natural to give them together. Euclid chooses to base their proofs on 8, which necessitates the sequence 5, 7, 8. As we have seen, Euclid could have avoided the use of 8 altogether.

21–26 are more problematic. I have pointed out that, in Euclid's presentation, 23 depends on 22, which uses 20 as a precondition. 21 is not used until III,8, but its position is probably determined by its dependence on 20. Similarly, 24 and its partial converse 25 are not used until III,7 and XI,20 respectively, but 24 depends upon 23. 26 is not used until 34 and does not depend on any proposition later than 16.[42] However, 26 could not be moved further back in the book without destroying Euclid's division of it between those propositions which do and those which do not treat of parallels. Presumably Euclid places 26 as late as he does because it is not used until 34, and because it has no close logical relations to other propositions.

I should perhaps mention briefly the final three propositions of book I, the core of which is 47, the Pythagorean theorem. 48 is the converse of 47 and is reduced to it. In 46 Euclid shows how to construct a square with a given straight line as side. He thinks of 46 as preliminary to 47, since he begins the *kataskeuē* of the latter by constructing squares on the three sides of a right triangle. Logically he could just as well have set out a triangle with the three squares. In 47 Euclid assumes on intuitive grounds what he later proves as II,2. In 48 he takes for granted, as he does elsewhere, the unproven but provable assumption that squares are equal if and only if their sides are, the only elementary deductive lapse in book I. I shall discuss the proof of the Pythagorean theorem in chapter 5 in connection with other propositions which resemble it in their deliberate avoidance of the theory of proportion. Here I shall only remark that the sequence 46–48 makes no use of propositions after I,41.

Even if I have succeeded in making the arrangement of the propositions of book I comprehensible, the problem of Euclid's starting points remains. Thus far I have dealt only with postulates 1–3, the basis of all the constructions of book I. These constructions are, as I have argued, analogues of existence assertions in modern mathematics, but the close correlation between existence and constructibility has been seen to be a product of the Euclidean conception of geometric objects rather than of a conscious constructive philosophy of mathematics. This correlation tends to be preserved throughout Greek mathematics, although, of course, the notion of constructibility varies with the means of construction permitted. However, book I of the *Elements* is unique in Greek mathematics in containing an explicit list of permissible constructions. In connection with other texts one has to say simply that the constructions permitted are the constructions actually used.

But, even in book I of the *Elements*, the means used to prove existence go beyond what is explicitly formulated in the first three postulates. In a certain sense one can see that this assertion must be true, since none of the postulates provides means for establishing the existence of points, even though both postulates 1 and 3 require points in order to be applied. The *kataskeuē* of proposition 1 illustrates two common ways in which Euclid gets the points he needs. The points *A* and *B*, which are used as the centers of the two circles, are given by the setting-out of the straight line as the straight line *AB*, i.e. (in this case), as the straight line segment between the points *A* and *B*. Euclid's procedure here would seem to depend on definition 3, which asserts that the extremities of a line are points. Since Euclid sets out a limited straight line, he is entitled to assume the existence of its limiting points.[43] Strictly speaking, however, this assumption is not constructive, since to know that a straight line is limited does not necessarily include knowing the location of its limits. Euclid would seem to be presupposing that when he calls a straight line segment *AB* he is naming two determinate points which limit it.

The point *C* of proposition 1 is produced by the intersection of two lines, the most common way of producing or finding points in all geometry. One of the important differences between Euclid and Hilbert is that Hilbert makes explicit the conditions under which lines intersect at a point, whereas Euclid almost always takes them for granted. The generation of the point *C* is the standard example of Euclid's use of tacit assumptions.[44] The proof of I,44 is an interesting exception to this general rule. There (see fig. 1.4) Euclid argues, "... since the straight line *EB* falls upon the parallels *KE*, *AB*, the angles *KEB*, *EBA* are equal to two right angles. Therefore the angles *HEC*, *EBA* are less than two right angles; and the straight lines produced indefinitely from angles less than two right angles meet; therefore *EH*, *BA* when produced will meet. Let them be produced and meet at *D*."

Presumably thinking of this passage and another like it in II,10, Zeuthen claimed that postulate 5 is intended to play the role of an existence assumption for points.[45] His position is undermined by the absence of any other equally necessary analogous assumptions to handle other cases in which points are generated by intersecting lines and by Euclid's much more usual practice of failing to note in any way the function of even postulate 5 in guaranteeing the existence of certain points. (See, for example the *kataskeuē* of I,37.) A somewhat more accurate characterization of the postulate can be based on its first application in I,29. There Euclid uses the postulate to argue that two lines will meet, without worrying about their meeting in a point. Thus it can be said that the postulate gives a criterion for the meeting of straight lines rather than asserting the existence of their point of intersection. Euclid takes the existence of this point for granted when he needs it, but the fact of meeting is what is crucial in the formulation of the postulate.

Perhaps the most enlightening examples of Euclid's treatment of points are provided by cases like I,5 and 12 in which Euclid begins a construction by simply "taking" a point "at random" on a line or in a region of space. Hilbert's axioms provide for the existence of such points, but Euclid obviously takes their existence for granted. In general Euclid attempts to prove the existence of a point only if the point satisfies a condition uniquely, or, as in the case of the point *C* in I,1, almost uniquely. In other cases Euclid feels free to invoke points not set out in the *ekthesis* as he needs them. In this respect points are anomalous among the objects of Greek elementary plane geometry.

However, in other parts of Greek geometry there are unpostulated procedures for producing objects. One is the generation of a plane or solid figure by the motion of another, e.g., the generation of a sphere by a full rotation of a semicircle about its diameter. Another is the generation of a plane figure by passing a plane through a solid. The most well-known examples of a figure generated in this way in Greek geometry are the conic sections; an example from the *Elements* is provided by the generation of a greatest circle by passing a plane through the center of a sphere (XII,17). Usually, but not always, these methods of generation are described in definitions of the figure generated, as in XI, def. 14. For example, Apollonius introduces the term 'parabola' after describing in the eleventh proposition of his *Conics* how it is generated and what conditions it satisfies. But neither Euclid nor Apollonius ever bothers to postulate or prove the possibility of generating the figures in question. Nor does one find anywhere in the Greek sources a discussion of the need for justifying their generation.

Why is one allowed to assume the possibility of moving a semicircle about its diameter to generate a sphere but must postulate the possibility of rotating a straight line about its endpoint to generate a circle? Equally, why is one entitled to assume the possibility of passing a plane through a cone to generate a conic section but required to postulate the possibility of connecting two points to generate a straight line? It seems unlikely that there are philosophically satisfactory answers to such questions. Perhaps such questions seem more perplexing than they are because one thinks of the constructional postulates of book I as a norm when, in fact, they are unique in Greek mathematics. Their explicit statement is another example of Euclid's concern for and consciousness of the methods employed in proofs in book I.

Postulates 4 and 5 have been discussed by scholars much more extensively than 1–3. We have already seen one difficulty involving postulate 4: the postulate is provable. Another difficulty lies in determining its use. Most commentators have chosen to ignore the role of the postulate in proofs and to assign it a special purpose. Heath's suggestion (vol. I, p. 201) seems to have won as much support as any:

As to the *raison d'être* and the place of Post. 4 one thing is quite certain. It was essential from Euclid's point of view that it should come before Post. 5, since the condition in the latter that a certain pair of angles are together less than two right angles would be useless unless it were first made clear that right angles are angles of a determinate magnitude.

I do not know what Heath has in mind by "Euclid's point of view," but logically Euclid has no need of postulate 4 before 5, since all he needs to apply the latter is a pair of angles less than some pair of right angles. And if Euclid believed he needed postulate 4 before 5 for the reason Heath suggests, it is hard to see why he didn't think he needed it before the definitions of acute and obtuse angles as angles less than and greater than a right angle, respectively.

Explanations like Heath's are and must be conjectural. It seems preferable to look to the places in book I where Euclid actually uses the postulate. Three of them are hardly obvious, and were apparently first noticed by Todhunter (Heath, vol. I, p. 277). In the proofs of I,14, 15, and 28 Euclid argues for the equality of two pairs of angles from the equality of each of the pairs with some two right angles, an inference which clearly does not go through unless postulate 4 or something like it holds. The other two uses of the postulate in 46 and 47 are unmarked by Heiberg or Heath. In 46 Euclid writes, ". . . the angles *BAD*, *ADE* are equal to two right angles. But the angle *BAD* is right; therefore the angle *ADE* is right." 47 contains the most obvious application of postulate 4: ". . . the angle *DBC* is equal to the angle *FBA*—for each is right. . . ." Even though none of these five inferences involves an explicit citation of postulate 4, and none of them is noted by Proclus, they would seem to provide a reasonable enough explanation of its presence.

It is not my intention to add to or even discuss the vast literature on the parallel postulate in geometry in general or in Euclid's geometry.[46] Proclus (191.21–193.9) reports attempts to prove the postulate as early as the second century A.D. Moreover, there is some evidence from book I itself that Euclid ascribes a special status to the postulate and, in particular, postpones using it until he has to. For there are two simplifications which Euclid might have made in book I by proving I,27–32 immediately after I,16. I,17 is obviously a trivial corollary of 32, and the second part of 26 is easily reduced to the first part, again using 32. As book I now stands, the only proposition after 16 used in the derivation of 27–32 is 23, the copying of an angle. But I have already pointed out that 23 can be proved without using any proposition after 11. The hypothesis that Euclid ascribes special significance to the parallel postulate is attractive, especially in the light of its later role in the history of mathematics. But the evidence for the hypothesis is too indirect to be ultimately convincing. Euclid often proves trivial consequences of propositions. Moreover, one of the organizing principles of the *Elements* is clearly subject matter, and so it is not surprising that Euclid places together all propositions involving the concept of parallelism. Finally, the account of the deductive arrangement of book I which I have given seems to provide a sufficient explanation for the place in which the parallel postulate is first invoked. All of these considerations are, of course, compatible with the generally accepted hypothesis that the postulate was introduced by Euclid himself to cover a deductive flaw in the theory of parallels.[47]

Heiberg follows manuscripts other than P in listing what I have called postulate 6 as the last (and interpolated) common notion. It is difficult to say how often this assumption is used, since if two straight lines could enclose a space, there would be counterexamples to many propositions in book I. It seems sensible to restrict attention to the explicit appeals to it. In I,4 (see above, p. 21) Euclid justifies the inference from the coincidence of *B*, *E* and of *C*, *F* to that of *BC*, *EF* with the words "For if *B* coincides with *E* and *F* with *C*, but the base *BC* does not coincide with *EF*, two straight lines will enclose a space, which is impossible." After leaving these words in place in his edition, Heiberg decided that they were "without doubt not genuine," because in a ninth-century Arabic translation of the *Elements* the equivalent of these words is found after the Q.E.D. instead of in the usual position.[48] According to Heiberg, a commentator first added the words and then added postulate 6, which was subsequently transferred to the common notions. I agree with Heiberg that the postulate is unlikely to be genuine, but disagree about the quoted words. The Arabic translation is certainly not sufficient evidence for his claim. For the same kind of reasoning with almost identical wording occurs in XI,3 and 7. The controversy over the postulate and the apparent general tendency to reject it might well have led someone to remove the words from the text of the proof of I,4 and place them at the end as an explanation. Moreover, Proclus undoubtedly thinks the words are genuine, although he almost certainly knows of texts in which the postulate does not occur.[49] The existence of texts in which the words but not the postulate occur makes it seem likely that the postulate was added to justify the words.

According to Proclus (238.25–239.2), Euclid takes the assertion that two straight lines do not enclose a space for granted as "something agreed upon or accepted" (*homologoumenon*). However, after giving a fallacious proof of the assertion,

Proclus claims that Euclid knows or acknowledges (*eidenai*) the assertion when he states the first postulate, which presupposes the uniqueness of the straight line connecting two points. Proclus' claim is vague enough to be incontestable. Euclid undoubtedly does "know" that only one straight line can be drawn through two points, and probably he is relying on this "knowledge" in the proof of I,4. But when Heath (vol. I, p. 232) suggests that Euclid could omit postulate 6 because "the fact it states is included in the meaning of postulate 1," he is being extremely misleading at best. Postulate 1 implies nothing about uniqueness. And if Euclid wished to assert the uniqueness of the straight line, he surely could have done so—for example, by inserting a definite article before 'straight line' in postulate 1. The evidence suggests that Euclid never raised the question whether the impossibility of two straight lines enclosing a space was a logical consequence of his first principles. For him this assertion rests directly on the intuitive notion of a straight line, and, since this notion has been characterized in the definitions, there is no need to prove a directly obvious fact about it.

The proof of I,1 provides a clear example of the use of a common notion in a proof. In general, however, the role of the common notions is far from clear. This obscurity is in part due to the vexed question of their authenticity. Heiberg accepts as genuine only 1–3, 8, and 9, the same ones which are given by Proclus, who calls them axioms (*axiōmata*). Proclus (196.15–198.15) indicates that there was a great deal of controversy over the axioms. Heron put forward only three; others added or suggested adding 6 and postulate 6; Pappus introduced a more precise version of 4, as well as other principles such as "All parts of a plane and of a straight line coincide with one another." It seems likely that the controversies referred to by Proclus affected the manuscripts of the *Elements*, as Heiberg assumes.[50] On the other hand, Proclus' list need not have been Euclid's. For, if Heiberg is right, common notions 4–7 and postulate 6 were interpolated before Proclus' time. Proclus himself admits the existence of manuscripts which vary in their list of "axioms" (198.3–5). Thus it seems best to examine the common notions in some detail.

For this purpose it is convenient to represent them symbolically. Symbolization requires some care, because the modern concepts represented by $=$, $+$, and $-$ do not correspond exactly to the Greek mathematical concepts of equality, addition, and subtraction. Moreover, Euclid uses these concepts and common notions outside of book I in connection with things other than plane geometric objects, namely, solid geometric objects, numbers, and what he calls magnitudes. For the present it will suffice to think of the common notions as applying to these things, even though the exact sense of the concepts involved in such applications has not been made clear. I shall use $\simeq$ to mean 'equals', which in book I means 'equals in length' for straight lines, 'equals in area' for plane figures,[51] and something like 'can be made to coincide with' for rectilineal angles. Similarly, I shall use $\prec$ for 'is less than' and $\succ$ for 'is greater than', treating $x \prec y$ and $y \succ x$ as synonymous. $x \prec y$ can be taken to mean 'x is equal to a proper part of y' or, when addition has been introduced, $\exists z(x \simeq y + z)$. In book I and, with the possible exception of III, 16 and 31, throughout the *Elements*, Euclid uses $\simeq$ and $\prec$ only in connection with what might be called comparable objects, i.e., with pairs of lines, or pairs of plane figures, or pairs of rectilineal angles, or pairs of solid figures, or pairs of numbers. It is simplest to think of the comparability of x and y in this straightforward sense as a condition for the meaningfulness of the expressions $x \simeq y$ and $x \prec y$. Common notion 1 can be expressed as

CN1 $x \simeq y \;\&\; z \simeq y \rightarrow x \simeq z.$

When Euclid performs an addition in book I, he takes the result of combining, or considering as one, two distinct, i.e., nonoverlapping, figures or lines or angles. Elsewhere, e.g., in II,7, he speaks of adding overlapping figures, but it is simplest to construe such addition as the combining of distinct figures equal to the original ones. I shall use the expression $x + y$ to represent the combining of distinct objects equal to x and y, with the understanding that such x and y are comparable. Common notion 2 is then expressed by

CN2 $x \simeq y \;\&\; w \simeq z \rightarrow x + w \simeq y + z.$

Although Euclid subtracts y from x only when y is a proper part of x, it is simplest to think of $x - y$ as the elimination from x of a comparable proper part equal to y and state common notion 3 as

CN3 $x \simeq y \;\&\; w \simeq z \rightarrow x - w \simeq y - z.$

1, 2, and 3 are the least suspect common notions. Simplicius, a careful scholar of the sixth century A.D., says that only these are found "in the old texts."[52] As has already been mentioned, Proclus attributes the reduction of the axioms to three to Heron (first century A.D.). Probably the axioms in question are common notions 1–3. The axioms eliminated may, of course, have been added to the *Elements* after Euclid. The only reason Proclus cites for retaining 8 and 9 is that Euclid uses them in proofs, a reason which would support the retention of other common notions which Proclus rejects. In any case, Simplicius' explicit assertion about common notions 1–3 is good grounds for assigning them special status. Common notion 1 is used in the proofs of 21 propositions in book I, and is stated in full in the proofs of 1, 2, and 13. 2 is used in twelve proofs, 3 in seven, but neither is cited explicitly.[53] By itself explicit citation does not seem to provide much support for the genuineness of a first principle. For such citation is not common enough in the *Elements* to be considered an essential part of Euclid's style, and, as we have seen, postulate 6 may have been introduced to turn a piece of reasoning into a citation.

Common notions 4 and 5 as stated are essentially unusable. A version of 4, which is used in I,17, 21, and 29, may be formulated as

CN4′ $x \simeq y \ \& \ z \prec w \rightarrow x + z \prec y + w.$

Nothing like common notion 5 is used in book I; and 5, unlike 4, is not found in all manuscripts of the *Elements*. Perhaps, then, 5 was added to be symmetrical with 4, which was itself added in an unskillful attempt to cover the inferences just mentioned. The following precise version of 5 is used by Euclid in other parts of the *Elements*:

CN5′ $x \prec y \ \& \ z \simeq w \rightarrow x - z \prec y - w,$

to which one might add another principle used by Euclid, namely,

CN5″ $x \simeq y \ \& \ z \prec w \rightarrow x - z \succ y - w.$

Common notion 6 is used in the proof of I,42; and a more general form of it, "Doubles of equals are equal to each other," is explicitly invoked in the proof of I,47. If one thinks of the double of x as the sum of two distinct objects, each equal to x, this more general form may be expressed

CN6′ $y \simeq y' \ \& \ y \simeq z \ \& \ y' \simeq z' \rightarrow y + z \simeq y' + z'.$

An analogous general form of common notion 7, which may be formulated

CN7′ $y \simeq z \ \& \ y' \simeq z' \ \& \ y + z \simeq y' + z' \rightarrow y \simeq y',$

is stated in I,37 and 38, but the common notion itself is not used in book I. Heiberg supposes that the statements of CN6′ and CN7′ are interpolations, but this supposition seems unlikely since, if they were interpolated, they would probably have been made to fit existing common notions or common notions added to fit them. It seems more likely that these statements are independent of the common notions, that common notion 6 was added to cover the inference in I,42, and that common notion 7 was added to be symmetrical with 6. There are also logical grounds for rejecting the two common notions. As Proclus (196.20–197.5) says, 6 is derivable from 1 and 2. 7 is not derivable from the other common notions, but it is from comparable assumptions made by Euclid in book I and elsewhere; these will be listed below.

It has already been pointed out that Proclus' grounds for accepting the last two common notions are not strong. Tannery and, following him, Heath reject both.[54] Neither is cited explicitly. 8 is allegedly used in the three applications of superposition in the *Elements* (I,4 and 8 and III,24). Euclid's Greek suggests that for him there is no inference from coincidence to equality and hence no need for a justifying principle. He simply writes, "Such-and-such will coincide with so-and-so and will be equal to it." It seems likely that if Euclid formulated common notion 8, he would be more emphatic about its use, at least to the extent of using some word expressing inference such as 'therefore'.

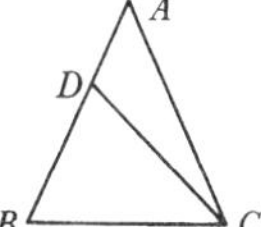

Figure 1.21

In the proof of I,6 Euclid supposes that angle $ABC \simeq$ angle ACB and that $AC \prec AB$ (fig. 1.21). He makes $BD \simeq AC$ and proves triangle DBC is equal to triangle ABC, "the less to the greater, which is absurd." Proclus (254.27–255.3) asserts that these words are an invocation of common notion 9.[55] Tannery rejects this claim on the grounds that Euclid does not say "the part to the whole," as would be expected in a citation of 9.[56] Tannery seems to me right here. The judgment that triangle BDC is smaller than ABC is a direct and intuitive one, made obvious by the diagram. Someone thought to justify it by adding a common notion. In this case it was impossible to make the verbal formulation correspond to Euclid's words without producing a trivial assertion such as "The greater is greater than the less." Moreover, the addition of 9 is no real help, since the judgment that triangle DBC is a part of triangle ABC is as intuitive as the judgment that it is smaller, and 9 involves new concepts, 'part' and 'whole', for which Euclid has no axioms or postulates.

The evidence from book I supports what Simplicius' description of the old manuscripts suggests: only the first three common notions are genuine Euclid. Tannery tried to establish the spuriousness of these three as well, but his arguments are weak and have not been generally accepted.[57] It seems reasonable to treat the first three as genuine. They represent basic facts about equality, addition, and subtraction which Euclid acknowledges using and which he considers worth making explicit in advance. He uses other related facts not stated in advance, sometimes signaling his use of these facts, sometimes not. Sometimes the facts in question can be derived from his explicit assumptions, sometimes not; but in no case does Euclid show any concern for the question of derivability. He uses whatever facts he needs and presumably thinks of them all as obvious. There is then an important difference between Euclid's postulates and his common notions. The postulates are presumably intended to be complete in the sense of containing all of the geometric constructions and assumptions presupposed in Euclid's plane geometry. The common notions, however, are not intended to contain all the assumptions about equality, addition, and subtraction used in book I, but simply the most prominent ones.

Before leaving the topic of the common notions I would like to indicate the principles governing $\simeq$, $\prec$, $+$, and $-$ which Euclid actually uses. One example will show that there is a certain leeway for interpretation on this subject. In I,20 Euclid proves that, in any triangle ABC, $BC \prec AB + AC$. (fig. 1.22). He extends BA to D, making $AD \simeq AC$. Hence, by I,5, angle $ADC \simeq$ angle ACD. (i) Therefore angle $ADC \prec$ angle BCD, and hence, by I,19, $BC \prec DB$. But since $AD \simeq AC$, (ii) $BC \prec AB + AC$. Heiberg and Heath cite common notion 9 in conjunction with the inference (i), but it is clearly insufficient. One could cover the whole inference with $x \simeq y \rightarrow x \prec y + z$, but it seems reasonable to invoke an analogue of 9 ($x \prec x + y$) and a law permitting the substitution of equals for equals in inequalities. Inference (ii) could also be taken care of by one complicated principle. But it is also justified by the substitutivity of equals for equals and $x \simeq y \rightarrow x + z \simeq y + z$. This last assertion is, however, derivable from CN2 and the reflexivity of equality, a fact which Euclid would undoubtedly have considered too trivial to state explicitly. There is then no unique list of "common notions" taken for granted by Euclid, but the following list is sufficient to cover all the standard inferences which he does make. The list should also make clear the character of Euclid's common, i.e., not specifically geometric, assumptions. For convenience I have divided the list into groups.

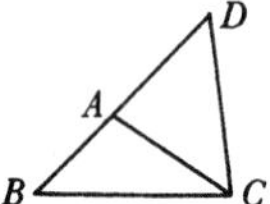

Figure 1.22

Axioms of equality

CN1 $x \simeq y \,\&\, z \simeq y \rightarrow x \simeq z$

Ia (reflexivity)

$$x \simeq x$$

These two assertions are known to be sufficient for deriving all truths containing only $\simeq$ and logical symbols.

Axioms for $\simeq$, $+$, and $-$

CN2 $x \simeq y \;\&\; z \simeq w \rightarrow x + z \simeq y + w$

Ib (commutativity)

$$x + y \simeq y + x$$

Ic (associativity)

$$(x + y) + z \simeq x + (y + z)$$

CN3 $x \simeq y \;\&\; z \simeq w \rightarrow x - z \simeq y - w$

Id $x \simeq (x + y) - y$

Ie $(x + y) - z \simeq (x - z) + y$

If $x - (y + z) \simeq (x - y) - z$

The axioms here added to Euclid's common notions are expressions of (i) the intuitive conception of addition as concatenation of distinct objects, (ii) the intuitive conception of subtraction as the taking away of a proper part, and (iii) the fact that addition and subtraction so conceived are inverse to one another.

Axioms involving $\prec$

Ig (transitivity)

$$x \prec y \;\&\; y \prec z \rightarrow x \prec z$$

Ih (trichotomy)

$$x \prec y \vee x \simeq y \vee y \prec x$$

$$x \prec y \rightarrow \neg (x \simeq y \vee y \prec x)$$

Ii (substitutivity in inequalities)

$$x \simeq y \rightarrow (x \prec z \rightarrow y \prec z)$$

Ij $x \prec x + y$

Ik $x - y \prec x$

CN4′ $x \simeq y \;\&\; z \prec w \rightarrow x + z \prec y + w$

CN5′ $x \prec y \;\&\; z \simeq w \rightarrow x - z \prec y - w$

CN5″ $x \simeq y \;\&\; z \prec w \rightarrow x - z \succ y - w$

The derivation of all principles used by Euclid from these axioms requires the addition of one existential assumption:

Il $x \prec y \rightarrow \exists z(z + x \simeq y)$

This last assumption should be thought of as a simplifying device for the derivation of principles used by Euclid rather than a general assumption made by him. For in actual applications the existence of such a z will be established on the basis of considerations relevant to the particular subject at hand: for example, geometric considerations in the case of geometry, arithmetic ones in the case of arithmetic. In the same way, the question whether expressions such as $x + y$ and $x - y$ are meaningful for particular x and y will be answered on the basis of such considerations. The actual list of axioms is not so important as their general character. For, judged simply in terms of frequency of use, such principles play a role in Euclid's reasoning as large as or larger than his geometric postulates. Clearly the principles transcend the common notions found in all the manuscripts of the *Elements*. Since their application is sometimes transparent, it seems unlikely that Euclid even attempted to formulate all the assumptions on which his reasoning is based, a clear divergence between the Euclidean and the modern notion of axiomatization.

The correlation between terms defined and terms used in book I is certainly closer than the correlation between stated principles and principles used, but it is not perfect either. Some of the terms Euclid defines do not occur in the propositions of the *Elements* at all. A rhomboid is subsequently called a parallelogram, although no definition of 'parallelogram' has been given. An oblong is later called a right-angled parallelogram. Rhombuses and trapezia are not treated under any name. And although Euclid's definitions of 'trilateral', 'quadrilateral', and 'multilateral' suggest a nomenclature for polygons derived from the number of their sides, Euclid subsequently uses a terminology based on the number of angles, 'triangle', 'pentagon', 'hexagon', etc., where the suffix '-gon' comes from *gōnia*, meaning 'angle'. The exception to this general rule is 'quadrilateral', which Euclid does use occasionally, but the reason for the exception is clear. Even before Euclid the word 'tetragon' means 'square', and Euclid simply continues tradition.

Euclid's respect for tradition is the standard explanation for the presence in the *Elements* of definitions which do not harmonize with the subsequent deductive development.[58] Certainly the presence of these definitions is a good example of the logically discomfiting, if relatively insignificant way in which the work of predecessors enters into the *Elements*. On the other hand, it would be a mistake to construe the definitions as nothing but a traditional list of terms. Some of the definitions, notably 10–12, 15–17, 20, 21, and 23, play an essential role in proofs. A typical example is provided by the use of the definition of a circle in the proof of proposition 1. Moreover, in Greek mathematics generally, definitions seem to be more significant foundationally than other kinds of first principles. After the first book Euclid gives further definitions but no new postulates or common notions, even when he turns to new domains, such as solid geometry, arithmetic, or the theory of magnitudes; and in most other Greek mathematical works the only first principles are, or at least are called, definitions. Hence it is worthwhile to say a few words about the character of Euclid's definitions in book I.

For this purpose it is useful to recall the modern doctrine of definition in mathematics which was referred to briefly in the previous section. A definition is always of a symbol or expression and not of a thing. It is legitimate, although pointless, to define a squircle as a square circle. The requirement on a definition is only that it make possible the elimination of any occurrence of the defined symbol in favor of previously defined expressions and hence, ultimately, in favor of expressions used without definition, the so-called primitive terms. Any theory must, of course, have such primitive terms. In the fragment of Hilbert's *Grundlagen* presented in the previous section, the explicit primitive terms are $\mathscr{L}$ and $\mathscr{B}$, but the use of two styles of variables amounts to the use of 'point' and 'straight line' as primitives. In the presentation the eliminability of defined terms such as 'segment' was illustrated by writing axioms without using them. Whatever Hilbert does with defined terms can be accomplished without them.

If one examines the first principles of book I in the light of the modern theory of definition, one is struck not only by the large number of defined terms but also by the large number of undefined ones. In the postulates the following kinds of nonlogical terms occur:

(i) verbs describing mathematical activity: draw, extend, describe;
(ii) expressions of relative position: from ... to ..., continuously, in a straight line, fall between, interior, side, meet;
(iii) expressions of size comparsion: equal, less;
(iv) expressions of extent: finite, *ad infinitum*;
(v) general terms designating geometric properties and objects: straight, line, point, circle, center, distance, right angle, angle.

None of the terms in the first four groups are defined, but all of those in group (v) except 'distance' are. 'Distance' is not, however, a significant exception, since it is clearly to be understood as a synonym for something like 'straight line from the center to the circumference'. Certainly any claim that 'distance' is a primitive term implicitly characterized in postulate 3 would be untenable, because 'distance' can be interpreted in any way in postulate 3, for example as 'diameter' or 'circumference' or 'straight line ten times as long as the radius'. It seems fair to say then that there are no undefined objects in Euclid's geometry. Of course, other objects are defined in the definitions, but, except for the abstractions 'length' and 'breadth', the new primitives introduced fall under the categories (i)–(iv).

Clearly, primitive terms which occur only in the definitions of others cannot be said to have an implicit characterization. They are simply taken for granted. The same would seem to be true of the primitive terms which occur in the postulates and common notions, since in general they occur only once, and more occurrences are needed to give an interesting sense to a term. A possible exception to this general claim is 'equal', which does occur several times. Certainly in the proof of proposition 1 'equal' could be replaced throughout by the letter E without affecting the validity of the argument. Although there is no conclusive mathematical or logical argument against treating 'equal' as at least in part implicitly defined by Euclid's first principles, the fact that Euclid fails to state all the properties of equality which he employs renders the assumption that 'equal' is an exception among Euclid's primitive terms implausible.

In any case the general situation with respect to the primitive terms in book I of the *Elements* is clear. Euclid takes it for granted that terms in the categories (i)–(iv) are understood. Using these terms as a basis, Euclid defines all geometric objects and goes on to specify certain operations performable on them and certain relations holding among them. Then, on the basis of these definitions and specifications, he goes on to determine other operations which can be performed and other relations which hold. The definitions of the most elementary geometric objects—points, lines, straight lines, angles, etc.—are mathematically useless and are never invoked in the subsequent development. On the other hand, the definitions of other objects—the kinds of angles, the kinds of figures, parallel straight lines—in terms of these more elementary ones are frequently invoked. There is, however, no good evidence that the difference between these two sorts of definition was felt by the Greeks.

I have already discussed the extent to which Euclid maintains a constructive point of view with respect to geometric operations. The notion of constructivity can be applied to relations as well as operations. The simplest account of a constructive relation is as one which is decidable, i.e., as one for which there is a uniform way of deciding whether or not it holds of a given object or objects. The precise account of a decidable arithmetic relation has played an important role in modern logic. One way of making the concept precise is to specify certain relations as decidable and to specify certain procedures of reduction to these relations as showing other relations to be decidable. Euclid's policy of introducing an operation by showing it to be reducible to the operations of postulates 1–3 is analogous to this way of treating relations.

But Euclid does not appear to do anything similar in connection with relations. For example, he makes no attempt to show that one can decide whether or not two straight lines are parallel before he treats parallels.

It is possible to argue that Euclid has no reason to adopt a constructive methodology for relations explicitly because all the relations he employs in elementary geometry are patently decidable. For example, the question of the parallelism of two straight lines is reducible to the question of the equality of two angles, which in turn is reducible to a question of coincidability of straight lines, coincidability for Euclid being an immediately decidable relation. In fact, Euclid does not use any relation which is not reducible to coincidability between lines until he treats ratios in book V. However, he cannot be said to follow a constructive methodology for these relations because in the case of relations like parallelism he introduces the relation with a nonconstructive sense (never meeting, even if extended *ad infinitum*) and then assumes or proves propositions showing that this relation can be determined to hold in a constructive way (post. 5, I,27). Since the propositions presuppose the meaningfulness of a nonconstructively defined term, Euclid's constructivism would be circular. The sensible conclusion seems to be that, whatever tendency Euclid has toward using constructibility as a proof of meaningfulness for operations, he has none toward using decidability as a proof of meaningfulness for relations.

1.3 Geometry and Algebra: Book II, Propositions 1–7

In the next chapter I take up Euclidean arithmetic, books VII–IX. Before doing so I want to say a few words about book II, which is a focal point for the discussion of a crucial question in the interpretation of the *Elements*. At this point I want only to indicate the character of the question and to lay the groundwork for discussion in subsequent chapters by looking at the initial seven propositions in book II, the first of which I quote without its standard *sumperasma*.

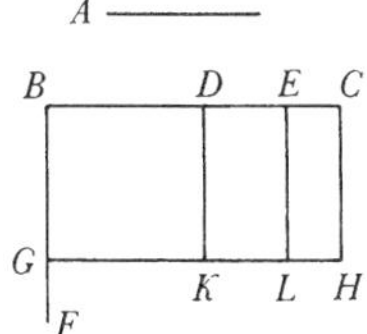

Figure 1.23

II,1 If there are two straight lines, and one of them is cut into any number of segments whatever, the rectangle contained by the two straight lines is equal to the rectangles contained by the uncut straight line and each of the segments.

Let *A*, *BC* be two straight lines, and let *BC* be cut at random (*hōs etuchen*) at the points *D*, *E*; I say that the rectangle contained by *A*, *BC* is equal to the rectangle contained by *A*, *BD*, that contained by *A*, *DE* and that contained by *A*, *EC* [fig. 1.23]. For let *BF* be drawn from *B* at right angles to *BC*, let *BG* be made equal to *A*, through *G* let *GH* be drawn parallel to *BC*, and through *D*, *E*, *C* let *DK*, *EL*, *CH* be drawn parallel to *BG*. Then *BH* is equal to *BK*, *DL*, *EH*. Now *BH* is the rectangle *A*, *BC*, for it is contained by *GB*, *BC*, and *BG* is equal to A; BK is the rectangle A, BD, for it is contained by GB, BD, and BG is equal to A; and DL is the rectangle A, DE, for DK, that is BG, is equal to A. Similarly also EH is the rectangle A, EC. Therefore the rectangle A, BC is equal to the rectangle A, BD, the rectangle A, DE and the rectangle A, EC.

It should be clear that, once the construction is described, II,1 becomes a geometrically trivial proposition. (Moreover, once one gets a sense of the methods of book II, most of the constructions can be figured out directly from the propositions themselves.) If $\mathbf{O}(x, y)$ is used to designate a rectangle with arbitrary straight lines equal to x and y as adjacent sides,[59] II,1 can be expressed in abbreviated form as

If A and $B_1B_2 \ldots B_{n+1}$ are straight lines,

$$\mathbf{O}(A, B_1B_2) + \mathbf{O}(A, B_2B_3) + \ldots + \mathbf{O}(A, B_nB_{n+1}) \simeq \mathbf{O}(A, B_1B_{n+1}).$$

It has long been recognized that II,1 can also be expressed algebraically, if one represents A as x and B_iB_{i+1} as y_i and a rectangle as the product of its sides, and disregards, as Euclid sometimes does, the relative positions of B_1B_2, B_2B_3, etc. II,1 then becomes simply an expression for the distributivity of multiplication over addition:

II,1a $\quad xy_1 + xy_2 + \ldots + xy_n = x(y_1 + y_2 + \ldots + y_n).$

The possibility of such algebraic representations is one of the bases for calling most of the material in book II and other portions of Greek mathematics resembling it geometric algebra. This term, which I shall adopt simply as a label not as a description, was apparently introduced by Zeuthen,[60] who, perhaps more than anyone else, is responsible for what I shall call the algebraic interpretation of Greek mathematics. I do not find it easy to characterize this interpretation, because it seems to me never to have been precisely formulated. Zeuthen himself reached his conclusions on the basis of an analysis of Apollonius' *Conica* according to which the reasoning in it is little different from the treatment of conic sections in analytic geometry. The foundation of the *Conica* is, for Zeuthen, geometric algebra of the kind found in book II, a proposition like II,1 being nothing but the Greek formulation of II,1a, and the straight lines and areas in it representing arbitrary quantities, including numbers. Zeuthen explains the geometric formulation by reference to the discovery of incommensurables or irrationals, and the desire for exactness rather than approximation in the representation of quantities as well as for secure, rigorous proofs.

The algebraic interpretation has a number of advantages, the most important of which is that it renders intelligible to the modern reader parts of Greek mathematics which are so complex geometrically that their development in a purely geometric way seems almost impossible. On the other hand, Zeuthen's theories of the historical development of Greek algebra are necessarily conjectural. Probably, in the absence of a more substantive historical foundation, Zeuthen's interpretation of the mathematical texts themselves would have remained nothing but an interesting hypothesis. It now seems to be generally believed that this foundation is provided by the discovery of an arithmetic algebra in Babylonian problem texts from as early as 1700 B.C. Careful investigation of the Babylonian materials led Neugebauer to the conclusion that "the entire substantive content on which Greek mathematics builds is contained in Babylonian mathematics" and, more particularly, that the characteristic technique of geometric algebra "becomes immediately comprehensible with respect to both its orientation and its methods of solution if one takes it as nothing but the translation of Babylonian methods into the language of geometric algebra."[61] Babylonian mathematics then provides Zeuthen's enlightening mathematical analysis with the historical foundation which it otherwise lacks.

The scope of this book precludes discussion of all the issues raised by the Zeuthen-Neugebauer interpretation of geometric algebra. However, it may be possible to get clear on some issues. A precondition for doing so is an understanding of what is claimed in calling a piece of Greek mathematics algebraic. The development of modern algebra can be viewed as the culmination of the structural conception of mathematics. But that conception is very explicit in the vocabulary of algebra, in which terms like 'group' and 'field' designating arbitrary structures satisfying given formal conditions are prominent. Normally too, at least at the more elementary level, the algebraist will emphasize the diversity of possible mathematical embodiments of a group or a field or other algebraic domain. It is quite clearly out of the question that one could establish an algebraic interpretation of much of Greek mathematics on the basis of vocabulary or explicit procedures. Rather, such an interpretation depends upon reading texts like book II as saying something other than what they appear to be saying. I have already argued that the structural conception of mathematics, which is the core of algebra, is essentially foreign to Euclid. To invoke this argument in the present discussion would perhaps be to beg the question, but the argument surely does provide *prima facie* grounds for doubting the viability of the algebraic interpretation.

However, it should be pointed out that the notion of algebra which seems to be relevant to the issues raised here is not so much the abstract theories of modern algebra as their forerunner, the application of standard algebraic formulas and procedures to the solution of quantitative problems. Zeuthen has no difficulty in showing how important Greek mathematical results can be construed as geometric embodiments of such algebraic solutions. However, he knows perfectly well that there is no such thing as a Greek algebraic equation in our sense. There are only geometric theorems and problems in the manner of II,1. Nor are there any such equations in Babylonian mathematics, most of which consists of step by step solutions to particular numerical problems on the basis of a general procedure. It is presumably for this reason that Neugebauer speaks of a translation of Babylonian *methods* into the language of geometric algebra. The situation may be explained in terms of a simplified fictional example. (A historically more accurate one will be discussed in chapter 4.) Roughly, one has three distinct forms of the same mathematical content:

1. (Babylonian) A standard computation of sums such as $3 \cdot 5 + 3 \cdot 6$ by adding 5 to 6 and multiplying the sum by 3;
2. (Greek) Proposition II,1 and the rules embodied in the first three common notions;
3. (Modern) The equation II,1a and rules for computing sums and products and for substituting equals for equals.

3 is for us the natural way to explain the standard form of the computations like 1, and it also provides a consistent interpretation of 2. The important question seems to me to be whether or not 2 makes sufficiently good sense, on its own and in its own context, to render the importation of the algebraic interpretation unnecessary. Unfortunately, questions of context are difficult to answer in the case of book II, because the applications of many of its propositions are themselves problematic. For example, II,1 is never cited explicitly in the *Elements*, and its content is so trivial that commentators have overlooked its tacit use in X,60–65, 97–102. But reference to applications in book X in no way undermines the algebraic interpretation of II,1, since book X itself is usually explicated algebraically. The situation is quite similar with other propositions in book II, which one can only hope to interpret in a thorough way in connection with their applications later in the *Elements*. This fact is the main reason why the present discussion can only be preliminary.

One point which can be made now is that in the case of some of the fundamental propositions of geometric algebra, the geometric interpretation is, at least *prima facie*, sufficiently plausible to render the importation of algebra unnecessary. For example, it seems reasonable to treat the representation of a rectilineal area as a square (II,14) and the Pythagorean theorem (I,47) as purely geometric propositions, even though the former is construable as the solving of $x^2 = a$, i.e., as the taking of a square root, and the latter is a fundamental tool of geometric algebra. To take another example, the construction of a rectangle on a given side a and equal to a given rectilineal area b—in Greek terms, the application (*parabolē*) to a of a rectangle equal to b—is on the algebraic interpretation the solution of $ax = b$, or the division of b by a. According to Zeuthen and those who follow him, this task is solved in I,43–45.[62] However, although it is easy enough to derive the solution of the algebraic problem from these propositions, and although Euclid subsequently does use the construction corresponding to this solution for parallelograms in VI,25 and for rectangles in X,60–65, 97–102, he actually shows only how to construct in a given angle a parallelogram equal to a given rectilineal figure. On the algebraic interpretation the use of the parallelogram must be read as a pointless geometric generalization, and the failure to prove the algebraically most interesting form of I,45 as the leaving out of a trivial consequence despite its fundamental importance. Since Euclid often bothers to prove trivial consequences whether or not they are fundamental, the more reasonable conclusion would seem to be that his primary, if not exclusive, motivation is geometric and that the form in which he proves I,45 in determined by his immediate concern to lay the groundwork for II,14.

After II,1 Euclid proves the following propositions. (I use $\mathbf{T}(x)$ to stand for the square on a straight line equal to x, and $2 \cdot x$ to designate the result of adding together two objects equal to x.)

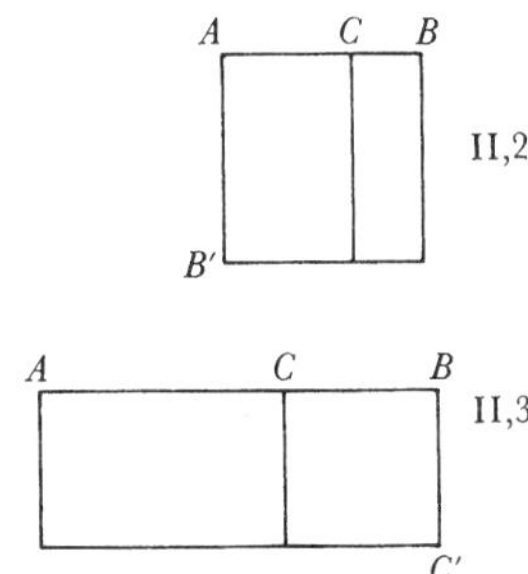

Figure 1.24

II,2 If ACB (fig. 1.24) is a straight line,

$\mathbf{O}(AB, AC) + \mathbf{O}(AB, BC) \simeq \mathbf{T}(AB)$.

II,3 If ACB is a straight line,

$\mathbf{O}(AB, BC) \simeq \mathbf{O}(AC, BC) + \mathbf{T}(BC)$.

II,4 If ACB is a straight line,

$\mathbf{T}(AB) \simeq \mathbf{T}(AC) + \mathbf{T}(CB) + 2 \cdot \mathbf{O}(AC, CB)$.

II,5 If $ACDB$ is a straight line bisected at C,

$\mathbf{O}(AD, DB) + \mathbf{T}(CD) \simeq \mathbf{T}(CB)$.

II,6 If $ACBD$ is a straight line and AB is bisected at C,

$\mathbf{O}(AD, DB) + \mathbf{T}(CB) \simeq \mathbf{T}(CD)$.

II,7 If ACB is a straight line,

$$\mathbf{T}(AB) + \mathbf{T}(BC) \simeq 2 \cdot \mathbf{O}(AB, BC) + \mathbf{T}(AC).$$

The diagrams of II,2 and 3 in fig. 1.24 should be sufficient indications of Euclid's proofs of them. Their algebraic interpretation depends upon how algebraic variables are assigned to lengths. II,2 may be expressed by either of

II,2a $(x + y)x + (x + y)y = (x + y)^2,$ $(AC{:}x;\ BC{:}y),$

II,2b $x(x - y) + xy = x^2,$ $(AB{:}x;\ BC{:}y),$

and II,3 by any of

II,3a $(x + y)y = xy + y^2,$ $(AC{:}x;\ BC{:}y),$
II,3b $xy = (x - y)y + y^2,$ $(AB{:}x;\ BC{:}y),$
II,3c $x(x - y) = y(x - y) + (x - y)^2,$ $(AB{:}x;\ AC{:}y).$

Of course, there is a sense in which the different formulations all say the same thing, but the fact that different algebraic formulas can be expressed by the same geometric proposition is probably the only stylistic advantage of geometric algebra over its modern counterpart. Unfortunately, this advantage is also an obstacle to the evaluation of the algebraic interpretation. For the algebraic relation of one proposition of geometric algebra to another will depend upon how algebraic variables are assigned to line segments. There are, for example, simple algebraic derivations of II,2a and 3a from II,1a. For II,2a one notes that

$$(x + y)^2 = (x + y)(x + y) = [\text{by II,1a}]$$
$$(x + y)x + (x + y)y.$$

Similarly II,3 is a consequence of II,1 because

$$(x + y)y = y(x + y) = yx + yy = xy + y^2.$$

Since Euclid normally takes for granted such geometrically obvious assertions as $\mathbf{T}(x) \simeq \mathbf{O}(x, x)$ and $\mathbf{O}(x, y) \simeq \mathbf{O}(y, x)$, he could have carried out geometrized versions of these arguments. The fact that he does not do so is an indication that he does not perceive the relation between these propositions in the way in which a modern algebraist would. For Euclid each of II,1–3 states an independent geometric fact.[63]

Heron carries out arguments analogous to the ones I have just described for each of II,1–10. These arguments are preserved by Al-Narizi,[64] who also interprets the propositions numerically, assigning numbers to the line segments and corresponding products of those numbers to the rectangles contained by pairs of segments. Thus for II,1 he assigns A 6, BC 10, BD 2, DE 3, EC 5, and points out that $6 \cdot 10 = 2 \cdot 6 + 3 \cdot 6 + 5 \cdot 6$.

The majority of the scholia on book II are similar numerical assignments and computations. Although Heron's arguments and these numerical assignments are more algebraic than book II of Euclid's *Elements*, their late date makes them problematic evidence for the interpretation of book II. Moreover, even Heron's arguments have a strong geometric flavor. He stresses that the arguments do not require a diagram because only one line is set out. However, his reasoning is deductive rather than combinatorial, and in some cases seems to rely on at least an imaginary geometric picture of the situation. For example, in dealing with II,3 he writes,

> The demonstration of this proposition through the demonstration of the first proposition of this book requires no figure. We suppose that we are given two lines *AB*, *BC*, one, *BC*, undivided, the other, *AB*, divided at the point *C*. It is manifest [by II,1] that the rectangle contained by the undivided line and the line *AB* is equal to the sum of the rectangles contained by the undivided line and the parts of the divided line, i.e., the two parts *AC*, *CB*. But the undivided line is equal to *CB*; therefore the rectangle contained by the undivided line and the line *CB* is equal to the square on the line *CB*. Therefore the rectangle contained by the lines *AB*, *BC* is equal to the rectangle contained by the lines *AC*, *CB* and the square on the line *CB*.[65]

As far as I can see, the only substantive difference between Heron's proof and Euclid's is Heron's curious use of *CB* to correspond both to a segment of *AB* and to a line equal to this segment, a line which is separately depicted in Euclid's diagram and proof.

Heron gives a similar reduction of II,4 to II,2 and 3. It may be represented algebraically by expressing II,4 as

II,4a $(x+y)^2 = x^2 + y^2 + 2xy,$

and noting that

$$(x+y)^2 = [\text{by II,2a}]\ (x+y)x + (x+y)y = (y+x)x + (x+y)y = [\text{by II,3a}]\ yx + xx + xy + yy = x^2 + y^2 + 2xy.$$

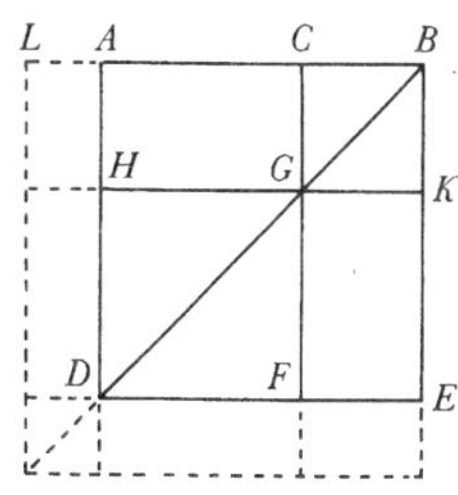

Figure 1.25

Again, there would be no difficulty in carrying out the equivalent of this argument within the framework of the *Elements*. Euclid, however, proceeds as follows. He completes the square *ABED* on *AB*, draws the diagonal *BD*, and then draws *CF* through *C* parallel to *BK* and intersecting *BD* at *G*, and *KH* through *G* parallel to *AB* (fig. 1.25). It is a simple matter to argue that

$\mathbf{T}(BC) \simeq$ square *BG*
$\mathbf{T}(AC) \simeq$ square *HF*
$\mathbf{O}(AC, BC) \simeq$ rectangle *AG* $\simeq$ [by I,43] rectangle *GE*,

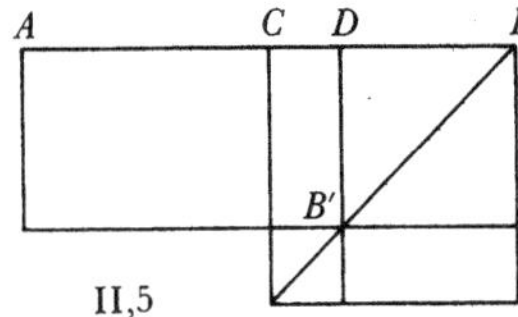

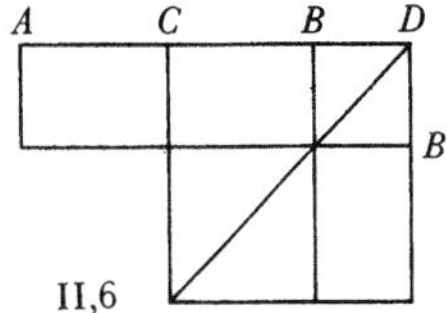

Figure 1.26

so that, since $\mathbf{T}(AB) \simeq$ square AE, the theorem follows. The construction of II,4 is a standard one. Starting with II,7, Euclid, given a square on a straight line ACB, carries it out with the words "Let the figure be drawn." In II,8 and XIII,3 he carries out on a straight line $LACB$ the construction indicated by the diagram including the broken lines with the words "Let the figure be drawn double." Usually the figure itself and knowledge of I,43 suffices to indicate the proof. In such cases I shall simply draw the diagram, possibly with shading to indicate equalities. Figure 1.25 should convey Euclid's proof of II,7, and the two diagrams in fig. 1.26 the proofs of II,5 and 6. As in the case of II,2–4, Euclid could have given more algebraic derivations of II,5–7. Heron derives II,7 immediately from II,3 and 4 in a proof which may be represented algebraically as

$$\begin{aligned}(x+y)^2+y^2 &= [\text{by II,4a}]\ x^2+y^2+2xy+y^2 = \\ &2y^2+2yx+x^2 = 2(y^2+yx)+x^2 = [\text{by II,3a}] \\ &2((y+x)y)+x^2.\end{aligned}$$

Here II,7 is represented by taking AC as x and BC as y. If, however, one takes AB as x and BC as y, one gets the algebraically more interesting

II,7a $\quad x^2+y^2 = 2xy+(x-y)^2,$

which is obviously equivalent to

II,7a′ $\quad (x-y)^2 = x^2+y^2-2xy.$

A direct derivation corresponding to II,7a would be possible using II,3. Since Heron does not give such a derivation, it would seem that even he does not interpret II,7 in what for us would be the algebraically most significant way.

The situation becomes even more complicated in the case of II,5 and 6, because of the additional line segment involved. If one takes AC ($\simeq BC$) as x and CD as y, the two propositions say exactly the same thing algebraically, namely,

II,5a $\quad (x+y)(x-y)+y^2 = x^2,$
II,6a $\quad (y+x)(y-x)+x^2 = y^2.$

It is possible to interpret Heron's proof of II,6 as a realization of this fact, since he proves it by extending DA to E with $AE \simeq$

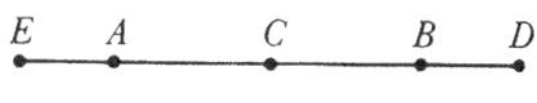

Figure 1.27

DB (fig. 1.27), and arguing that by II,5 $\mathbf{O}(EB, BD) + \mathbf{T}(CB) \simeq \mathbf{T}(CD)$, but $AD \simeq EB$. This argument may be read as a simple interchange of x and y in II,5a. On the other hand, although Heron's proof of II,5 has an algebraic interpretation, its geometric interpretation suggested by Euclid's figure seems more natural. Heron argues that

$$\mathbf{T}(CB) \simeq \text{[by II,2]}\ \mathbf{O}(CB, BD) + \mathbf{O}(BC, CD) \simeq \text{[by II,3]}\ \mathbf{O}(CB, BD) + \mathbf{O}(CD, BD) + \mathbf{T}(CD) \simeq \mathbf{O}(AC, BD) + \mathbf{O}(CD, BD) + \mathbf{T}(CD) \simeq \text{[by II,1]}\ \mathbf{O}(AD, BD) + \mathbf{T}(CD).$$

A modern algebraist might very well treat II,5 and 6 as special cases of II,7 and 4 respectively, a possibility which is seen most clearly by taking AB as x and DB as y, so that II,5 and 6 become

II,5b $(x - y)y + (\frac{1}{2}x - y)^2 = (\frac{1}{2}x)^2$,
II,6b $(x + y)y + (\frac{1}{2}x)^2 = (\frac{1}{2}x + y)^2$.

Particular interest has focused on these two formulations because it is possible to read them as embodiments of the standard procedure, known to the Babylonians, for solving the equations

$ay - y^2 = b$ (or equivalently $x + y = a$, $xy = b$),
$ay + y^2 = b$ (or equivalently $x - y = a$, $xy = b$).

It is simplest to illustrate this fact in terms of II,6b, taking x as a known quantity a and using II,3a to substitute $ay + y^2$ for $(a + y)\,y$ so that II,6b becomes

$$ay + y^2 + (\tfrac{1}{2}a)^2 = (\tfrac{1}{2}a + y)^2.$$

Hence, if $ay + y^2 = b$,

$$(\tfrac{1}{2}a + y)^2 = ay + y^2 + (\tfrac{1}{2}a)^2 = b + (\tfrac{1}{2}a)^2,$$

and

$$y = \sqrt{b + (\tfrac{1}{2}a)^2} - \tfrac{1}{2}a.$$

There is no difficulty in transforming this series of algebraic manipulations or the similar series related to II,5 into a geometric construction in which the computation of y becomes the determination of the point D on AB (extended).[66] Indeed we shall see that Euclid carries out such constructions in a generalized form in VI,28 and 29.

On the other hand, despite the attractiveness of the algebraic interpretation of II,4–7, it should be clear that Euclid's presentation is not exactly what the algebraic interpretation would lead one to expect and that the interpretation is by no means forced upon one. The propositions can just as well be read as geometric assertions about the equality of certain areas useful for the transformation of one area or areas into another. Of course, both the propositions and the transformations have an algebraic representation, but *prima facie* there is no more reason to say the geometric form is an embodiment of the algebraic representation than to say the opposite. It is true that II,5 and 6 read by themselves make little mathematical sense, and that their algebraic interpretation renders them intelligible to the modern reader. However, these propositions are rather clearly lemmas for later applications, and there is no reason to expect them to make sense in isolation. Hence, a final judgment on their meaning depends upon an examination of these applications, a task I will be taking up in later chapters in connection with the whole of the *Elements*.

To conclude this section I would like to sum up and perhaps clarify what seem to me the fundamental issues in the interpretation of geometric algebra. There can be no question of the consistency and coherence of the algebraic interpretation of geometric algebra. Nor can there be any question that geometric algebra plays a role in the *Elements*, the *Conica*, and elsewhere very like the role of algebra in analytic geometry. Sometimes proponents of the algebraic interpretation appear to be proclaiming little more than this fact. But this minimal, correct claim would be compatible with a geometric interpretation, since there is nothing to prevent geometric results from playing the role in descriptive geometry which algebraic ones play in analytic geometry. The claims which seem to me clearly to distinguish the algebraic interpretation from the geometric one are:

1. The lines and areas of geometric algebra represent arbitrary quantities;
2. Geometric algebra is a translation of Babylonian algebraic methods;
3. The "line of thought" in much of Greek mathematics is "at bottom purely algebraic."[67]

The truth of any one of these claims would seem to me sufficient to establish the algebraic interpretation to the exclusion of the geometric one. Ultimately the first of these claims reduces to the view that geometric algebra is intended to be applicable to numbers. Most of the direct evidence for this view is relatively late, the earliest being perhaps the scholia which interpret propositions in book II numerically. These scholia and other evidence make it certain that the possibility of applying geometric algebra to arithmetic problems was an established fact by the first century A.D., but it is difficult to know how much light the later situation throws on the *Elements* themselves. Zeuthen's view that there was a pre-Euclidean arithmetic algebra[68] is in a sense confirmed by our present understanding of Babylonian mathematics; but, in terms of Greek evidence, Zeuthen's view is essentially historical conjecture. Within the *Elements* themselves there are three apparent arithmetic applications of book II, which I will discuss in section 2.3. Here I wish only to suggest the necessity of distinguishing between the recognition and use of arithmetic-geometric analogies and algebraic thought. It seems relatively clear that geometric ideas played a substantial role in early Greek arithmetic thinking which may well have been based entirely on the representation of numbers as plane arrays of units.[69] This way of dealing with numbers would obviously facilitate the recognition of analogies between geometric results and arithmetic ones and would also suggest the possibility of exploiting geometric procedures in arithmetic. Such analogical thinking is to be distinguished from the algebraic approach of combining the treatment of distinct disciplines by abstracting the common features of the objects they deal with. We shall see that Euclid does something like this in book V where he treats magnitudes in general. However, I shall argue that even there he is concerned only with the geometric, which he separates from the arithmetic in keeping with the prevalent Greek opinion that the infinitely divisible or continuous and the discrete are radically different kinds of things.

Taken literally, claim 2 gives one the picture of Greek mathematicians methodically rewriting a given body of mathematics. However, this literal reading involves the difficulty that the Greeks place great emphasis on proof whereas, as far as one can tell, the Babylonians never felt it necessary to justify their procedures. Thus the real notion of Greek translation of Babylonian algebra has to involve both the representation of methods as theorems and problems and the supplying of proofs. This notion of translation is considerably more flexible than the ordinary one invoked in saying, for example, that arithmetic can be translated into set theory. This flexibility becomes particularly problematic in light of the fact that we will presumably never know in what form the Greeks might have come to know Babylonian mathematics, if indeed they did know it in the fifth and fourth centuries B.C. One is left with the choice between a precise hypothesis of methodical translation which the evidence would not seem to justify, and a looser hypothesis which would not seem to be clearly preferable to other alternatives, e.g., the assumption that geometric algebra is the Greek embodiment of a generally shared knowledge[70] or that the Babylonians and the Greeks reached equivalent results independently.

Claim 3 rather obviously needs to be made more precise. In the *Elements* we find transformations of areas into other equal ones by means of the addition and subtraction of lengths and areas and the geometric analogues of multiplication, division, and the extraction of square roots. The fact that similar operations are fundamental in algebra does not seem to me sufficient to settle the question whether the operations in the *Elements* are aptly described as algebraic. The paradigms of algebraic reasoning are abstract structural argument, on the one hand, and the manipulation of equations, on the other. Neither of these paradigms appears to be basically Euclidean. Indeed, however one wishes to describe the results proved in book II, the proofs themselves show no sense of the connection between the propositions involved. This fact suggests strongly that Euclid is approaching his subject by looking at the geometric properties of particular spatial configurations and not by considering abstract relations between quantities or formal relations between expressions. The study of other parts of the *Elements* may reverse this impression, but for the moment there is no good reason to assume that Euclid is reasoning algebraically.

Notes for Chapter 1

Bibliographical Note

The commentary on book I by Proclus is an invaluable source of historical information. The mathematical discussions of the *Elements* tend toward the trivial, and are usually adequately summarized in Heath's notes on particular propositions. In reading the philosophical portions of the commentary, one must bear in mind that Proclus is a fifth-century Neoplatonist.

The deductive structure of books I–IV has been codified in as definitive a way as possible by Neuenschwander in "Die ersten vier Bücher," which also contains what will probably become the standard historical theory based on these books. The tables at the beginning of Neuenschwander's paper (pp. 328–338) saved me from numerous errors, and would, if they had appeared earlier, have saved me a great deal of labor. I should perhaps add that no one interested in deductive structure can rely on the citations in the standard editions of the *Elements*.

The first principles have always been a subject of interpretative discussion. Chapter 14 of Zeuthen's *Histoire* . . . is perhaps the most elaborate attempt to make the principles "acceptable" from a modern point of view. I myself find the general orientation of Seidenberg in "Did Euclid's *Elements* book I develop geometry axiomatically?" more plausible. Von Fritz's "Die APXAI . . ." and "Gleichheit . . ." give a thorough interpretative representation of ancient materials relating to Euclid's first principles, while in "Die Postulate . . ." van der Waerden attempts to locate their fifth-century origins.

I should perhaps also mention two works bearing on book I which have aroused some interest among scholars working in ancient mathematics but of which I have not made much use in this book: Szabò's *Anfänge* . . . and Toth's "Das Parallelproblem . . . ."

Zeuthen's account of geometric algebra is most conveniently read in chapter 1 of *Die Lehre von den Kegelschnitten* . . . and in *Sur l'origine de l'algèbre*. The contents of Babylonian algebra are admirably summarized by Goetsch, who also lists the important collections of primary materials. The algebraic reading of the *Elements* based on Neugebauer's hypothesis of Babylonian origins is vigorously developed by van der Waerden. The article by Gandz is also helpful. Attempts to argue against the algebraic interpretation by Szabò in an appendix to *Anfänge* . . . and by Unguru have not met with much success. The latter has been attacked rather fiercely by van der Waerden ("Defense . . ."), Freudenthal ("What is algebra . . ."), and Weil.

1. For a brief historical account of the decline of descriptive geometry, see Bourbaki, pp. 158–174.

2. I use the following logical notation: $\forall$ for 'for all', $\exists$ for 'at least one', $\neg$ for 'it is not the case that', & for 'and', $\vee$ for 'or', $\rightarrow$ for 'only if', and $\leftrightarrow$ for 'if and only if'. The identity sign $=$ is also considered a logical sign; negations of identities are expressed using $\neq$.

3. Because Hilbert's axioms are for solid geometry, he requires that a lie in the plane of A, B, C. With the elimination of this condition, axiom II,4 becomes an axiom of plane geometry.

4. "But no geometer or philosopher doubted that geometry dealt with actual space and investigated its properties—Pasch, Enriques, Pieri, Klein, they all assert it. . . ." Freudenthal, "Zur Geschichte der Grundlagen der Geometrie," p. 111.

5. *Gesammelte Abhandlungen*, III, p. 291.

6. *Ibid.*

7. Klein, p. 201.

8. Poincaré, pp. 252–253. The first sentence in the quotation refers to the presentation of axiom I,1 in the first edition of the *Grundlagen*:

I,1 Two distinct points A and B always completely determine a straight line. We write $AB = a$ or $BA = a$.

Instead of 'determine' we will also use other expressions, for example, A lies on a, A is a point of a, a passes through A . . . .

9. Translation by E. J. Townsend.

10. See Hilbert's essay "On the infinite."

11. Curry, p. 56.

12. The Archimedean axiom is discussed below in section 3.2.

13. Sometimes this procedure is ruled out of court on the (correct) grounds that not all mathematical questions of the kind in question can be settled by the slavish application of logical rules. To my mind it is more significant that even where such application would work, for example, in proving theorems, it is not used. Slavish application of the logical rules is almost always impractical. This situation may, however, change as computers become more efficient.

14. For a clear indication of the set-theoretic character of Gödel's reasoning see his two essays listed in the bibliography.

15. See, for example, Russell, pp. 5–10.

16. For a demonstration of this assertion see, for example, Eisenhart, pp. 283–292.

17. The possibility of giving a precise definition of truth was established by Tarski.

18. Some of the common systems do determine a unique structure if some word such as 'set' or 'system' is taken as a logical term like ∀ or ¬, and therefore not subjected to alternative interpretations.

19. This section presupposes some familiarity with the contents of book I, the assumptions and propositions of which are listed in Appendix 4.

20. Proclus describes this distinction at 77.7–81.2, and the parts of a proposition at 203.1–210.6.

21. See, for example, the *lēmma* before X,33. (I use '*lēmma*' and '*lēmmata*' to refer to propositions called *lēmmata* by Euclid. I use the English word 'lemma' to describe propositions which are obviously preliminary to more important results, even if Euclid does not call them *lēmmata*.)

22. The word '*diorismos*' is also used to refer to the statement of the conditions required for a problem to be solvable, as in I,22. I use the word 'precondition' to refer to such a *diorismos*.

23. See W. and M. Kneale, pp. 80–81.

24. See my paper "Greek mathematics and Greek logic," pp. 42–43.

25. See Proclus, 207.4–25.

26. This is, of course, the intuitionistic point of view, normally associated with L. E. J. Brouwer. Heyting gives a clear account of the general theory.

27. See Zeuthen, "Die geometrische Construktion . . ."; and, for criticism, Frajese, "Sur la signification . . . ."

28. Seidenberg has attempted to explain the first three postulates as an outgrowth of the use of ropes and pegs in ritual altar construction. See, for example, his "Peg and cord . . . ." The standard secondary sources on compass and straightedge in Greek geometry are Steele and Niebel.

29. See, for example, Gillings, pp. 137–139.

30. Not in his edition, but in "Paralipomena . . . ," p. 50.

31. Neuenschwander, "Die ersten vier Bücher . . . ," p. 331, includes in his tables a dependency of 43 on 31 and 30, because, if the construction indicated in the figure for 43 were carried out, 30 and 31 would be invoked. However, Euclid does not carry out this construction, but says, "Let *ABCD* be a parallelogram, *AC* its diameter, *EH*, *FG* the parallelograms about *AC*, *BK*, *KD* the so-called complements." With this *ekthesis* Euclid introduces a way of looking at parallelograms which becomes fundamental in book II.

32. In fact, Proclus only points out that uniqueness is a consequence of I,30. He does not say that Euclid proves the proposition in order to establish uniqueness. Heath's discussion is somewhat marred by his desire to establish the superiority of Euclid's parallel postulate to Playfair's. I,30 also plays an important role in Toth's attempt to reconstruct a pre-Euclidean theory of parallels; see especially Toth, pp. 278–279 and 292–300.

33. It is not possible for me to discuss the many attempts to show that Euclid was concerned with questions of uniqueness. I do discuss the uniqueness of the straight line between two points on pp. 31–32.

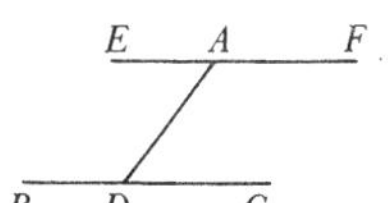

Figure 1.28

34. In 31 Euclid draws *EF* through *A* and parallel to *BC*, by connecting *A* with a random point *D* on *BC* and constructing angle *DAE* equal to *ADC* (fig. 1.28). He could instead drop *AD* perpendicular to *BC* and draw *EF* perpendicular to *AD*, using postulate 4 to get the necessary equalities for applying I,27.

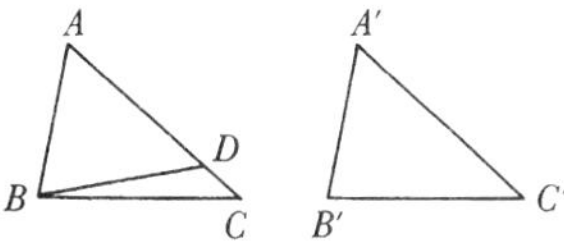

Figure 1.29

35. As Heath (vol. I, pp. 306–307) points out, Euclid does not prove

If *ABC*, *A′B′C′* (fig. 1.29) are two triangles with sides *AB*, *A′B′* and *BC*, *B′C′* and angles *BAC*, *B′A′C′* equal and the angles *ACB*, *A′C′B′* either both less than or both not less than a right angle, angles *ABC*, *A′B′C′* (and hence the other corresponding parts of the two triangles) are equal,

although he does establish a corresponding result for similar triangles in VI,7. I give a proof of the proposition for future reference. Suppose the proposition false, and let angle *A′B′C′* be less than *ABC*. Then, if angle *ABD* is made equal to *A′B′C′*, by I,26 and 4, *BD* will be equal to *B′C′*, and hence to *BC*. But then, by I,5, angle *BDC* is equal to *BCA*, and each is less than a right angle by I,17. Hence angle *B′C′A′* is less than one, and so is its equal, *BDA*. But then the sum of the angles *BDA*, *BDC* is less than two right angles, contradicting I,13.

36. In later books Euclid does take the inference for granted. See, for example, VI,5. However, the compression of arguments by leaving out obvious steps, particularly ones which have been gone through in earlier propositions, is characteristic of the later books.

37. There is no clear reason for the inclusion of the second part of I,5. For Proclus' explanation, see Heath, vol. I, pp. 255, 260.

38. See, for example, Heath, vol. I, pp. 225, 249, Dijksterhuis, vol. I, pp. 144–152, von Fritz, "Die APXAI . . . ," pp. 76–89.

39. The axiom asserts the pairwise equality of the remaining angles of the two triangles described in the antecedent of I,4. Von Fritz ("Gleichheit . . . ," p. 17) suggests that Hilbert's procedure is not open to Euclid because the axiom in question is insufficiently self-evident from the Greek point of view. Since Euclid's parallel postulate has usually been thought to be unself-evident, it is difficult to place much weight on such considerations.

40. This does not seem to be true in the later books, where even logically necessary details are sometimes slid over.

41. See, for example, Dijksterhuis, vol. I, p. 147.

42. In fact, the first case of 26, the only case to be used in book I, depends only on I,4. The second case is first used in III,3.

43. In general, Euclid uses 'straight line' to designate line segments of determinate and known extent. Postulate 2 and propositions I,1, 10, 12, 22, and VI,30, in which he speaks of finite or infinite straight lines, are exceptional.

44. See, for example, Klein, pp. 197, 199; or Eves and Newsom, pp. 43–44.

45. *Histoire . . .*, p. 99.

46. R. Bonola gives a standard account of the mathematical history of the postulate.

47. For the evidence, see Heath, *Mathematics in Aristotle*, pp. 27–30.

48. "Paralipomena . . . ," p. 56.

49. See Proclus' discussion at 238.25–240.10. At 196.21–33 he mentions postulate 6 as something which should not be added to the axioms.

50. See the textual apparatus on the common notions in his edition.

51. Heath's view (vol. I, p. 327) that Euclid sometimes means congruence when he speaks of the equality of rectilineal figures is untenable. See Dijksterhuis, vol. I, p. 195 or von Fritz, "Die APXAI . . . ," p. 71 ff.

52. *Codex Leidensis*, I,1, p. 28.

53. Neuenschwander invokes common notions 1 and 2 to explain inferences involving doubles in I,41, 42, 47 and involving inequalities in I,20. I do not include these alleged applications in my count. The inferences are discussed on pp. 34 and 36.

54. Tannery, "Sur l'authenticité . . . ," p. 167, and Heath, vol. I, pp. 225, 232.

55. Heiberg also cites common notion 9 in connection with this inference and with identical or similar inferences in I,7, 16, 18, 20, 24, and 26. He does not, however, cite it for other similar steps in I,14, 39, 40, and 44, which are noted by Neuenschwander. (Heath cites the common notion only for 16 and 20.)

56. "Sur l'authenticité . . . ," p. 167.

57. Counterarguments to those put forward by Tannery in "Sur l'authenticité . . ." are given by Heath, vol. I, pp. 221–222.

58. See, for example, Heath, vol. I, p. 151.

59. The notation is due to Dijksterhuis (vol. II, p. 2), from whom I have also taken the use of **T** (x) to designate the square on an arbitrary straight line equal to x. **O** represents the first letter of the Greek word for rectangle (*orthogōnion*), **T** the first letter of the word for square (*tetragōnon*). Euclid usually refers to **O** (A, B) as "the [thing contained] by A, B" (*to hupo tōn A, B*) and **T** (A) as "the [thing drawn] on A" (*to apo tēs A*).

60. See *Die Lehre von den Kegelschnitten* . . . , p. 7.

61. "Zur geometrischen Algebra," pp. 258, 252. Compare van der Waerden, p. 124.

62. See Zeuthen, *Histoire* . . . , pp. 36–37, or Heath, vol. I, pp. 346–347.

63. Heath (vol. I, p. 377) suggests that Euclid proves II,2 and 3 and other propositions in book I from scratch rather than deducing them from previously proved propositions to inculcate a general method. Even if this speculation is correct, it shows that the method in question is not what we would normally call algebraic.

64. See *Codex Leidensis*, II,2.

65. *Codex Leidensis*, II,2, p. 16. The translation is of the Latin of Heiberg and Besthorn.

66. See Heath, vol. I, pp. 383–388.

67. For explicit expressions of these views I refer to Zeuthen, *Die Lehre von den Kegelschnitten* . . . , p. 6 for 1 and to the passages cited in note 61 for 2. The words in quotation marks in 3 are from a description of book X by van der Waerden, p. 171.

68. See especially *Sur l'origine de l'algèbre*, pp. 10–29.

69. See, for example, Heath, *A History* . . . , vol. I, pp. 76–84.

70. In "The survival . . . ," p. 530, Neugebauer says that Babylonian mathematics was "common mathematical knowledge all over the ancient Near East." Other articles in which Neugebauer discusses the relation of Babylonian to Greek mathematics are listed in the bibliography.

2 Euclidean Arithmetic

2.1 Book VII

In books VII–IX Euclid develops the subject of arithmetic in almost complete isolation from the remainder of the *Elements*. The position of the arithmetic books is presumably to be explained by the fact that the only applications of arithmetic concepts and results are in book X. However, the contents of the arithmetic books go well beyond what is needed in book X, so that it seems unlikely that Euclid develops arithmetic simply for the applications he makes of it. The dependence of VII–IX on earlier books is even more tenuous. Terms defined for proportions in general in book V are used in connection with numerical proportions, but it seems to be generally agreed that as far as laws of proportionality are concerned, the arithmetic books are intended to be self-contained. There is a problematic use of geometric algebra in IX,15, which will be discussed in section 2.3, but otherwise not even the common notions are explicitly invoked. Thus it is not unreasonable to approach the arithmetic books as a foundational work, the arithmetic analogue of book I.

Because of the independence of these books the modern reader is somewhat surprised to find no specifically arithmetic postulates in the *Elements*. Euclid's only arithmetic first principles are the 23 definitions at the beginning of book VII. The first of these defines the unit in a mathematically useless way; the second defines a number as a "multitude composed of units" (*to ek monadōn sugkeimenon plēthos*). An obvious consequence of these two definitions is that a unit is not a number. Nor is there for Euclid such a thing as a multitude composed of a single unit. Hence, he sometimes repeats a proof for numbers and the unit or distinguishes cases only because of the distinction between a unit and a number.[1] Sometimes, however, Euclid ignores this distinction.[2] I shall often follow this latter practice, using the term 'positive integer' to refer to all multitudes of units, including "multitudes" consisting of one unit.

Euclid's definition of a number reflects an important feature of Greek arithmetic which distinguishes it from the modern theory of positive integers. The modern analogue of the Euclidean unit is a single thing, the positive integer 1. The other positive integers are obtained by repeated applications to 1 of a successor operation σ, which intuitively is the addition of 1 but which is completely characterized for arithmetic purposes by the axioms

P1 $\sigma(k) \neq 1$,
P2 $\sigma(j) = \sigma(k) \rightarrow j = k$.

Given such an operation, one can define 2 to be the successor of 1, 3 the successor of 2, and so on, each defined object being a unique positive integer. In Greek arithmetic there are indefinitely many units and indefinitely many ways of combining them into multitudes. Clearly then, there is no unique 2 or 3; any pair of units is a 2, for example. Moreover, one runs into difficulty in trying to say in a noncircular way what a given positive integer is. I have just used the word 'pair' in characterizing a 2, but obviously there is no real difference between talking about a pair of units and two units. In the late nineteenth century Frege showed that such a characterization need not be considered trivial, since to say a multitude contains two things can be construed as saying that it contains nothing but an x and a y with $x \neq y$.[3] There is no good reason to suppose that the Greeks were aware of the possibility of anything like Frege's analysis of the notion of pair. Perhaps they were also unaware of the difficulty of saying what a pair is, but it is interesting that in Euclid's arithmetic particular numbers play virtually no role and are never explicitly characterized. Euclid does not prove that $2 + 2 = 4$ or that a 2 and a 2 combined yield a 4, nor does he even have the apparatus for doing so. Such facts, insofar as they are used in the *Elements*, are used without proof.

In the modern theory of positive integers one defines certain arithmetic operations (functions) and relations and proves properties of them. The commonest way of defining an operation is by specifying its value for 1 and then its value for $\sigma(k)$ in terms of its value for k.[4] For example, addition is determined by the formulas $l + 1 = \sigma(l)$ and $l + \sigma(k) = \sigma(l + k)$, multiplication by $l \cdot 1 = l$ and $l \cdot \sigma(k) = l \cdot k + l$. Arithmetic relations can be introduced in a similar way, but ordinarily they are defined in the same manner as in geometry. Euclid defines only one operation on numbers, multiplication. To represent the definition (VII, def. 16) I use $\simeq$ to designate the relation of equinumerosity and $+$ to designate the combining of disjoint multitudes of units. (In connection with arithmetic, $\prec$ is used to designate the relation 'not containing as many units as' and $-$ the operation of taking away a proper submultitude from a multitude.) The definition says,

$l \cdot k$ is $k_1 + k_2 + \ldots + k_l$, where, for $1 \leq i < j \leq l$,
$k_i \simeq k_j \simeq k$.

From a technical point of view this definition has a number of curious features. First of all, l is used both as an object language variable ranging over positive integers and as a metalanguage variable or subscript. Such usage is impossible within first order logic but not in an extension to higher order logic. I shall take such an extension for granted because explaining it would take more space than is justified by the illumination to be gained for Greek mathematics.[5] The important point is that the extension incorporates within itself all of elementary arithmetic. In other words, the "logic" used in the proof of arithmetical propositions already contains those propositions among its own laws. This formal situation corresponds exactly to the informal situation in the *Elements*. As will become clear in this chapter, because Euclid takes for granted such ideas as "adding one number to itself as many times as there are units in the other" (VII, def. 16), he presupposes the fundamental arithmetic notions of counting and adding.

The introduction of an operation symbol such as $\cdot$ requires a proof that the application of the operation always yields an object, i.e., a proof that $l \cdot k$ exists. The last phrase in Euclid's definition of multiplication, "and some number comes-to-be," suggests some concern for the question of existence, since it sounds as though Euclid is leaving open the possibility that k added to itself l times might yield nothing. In fact, he never considers this possibility in developing arithmetic. He simply "takes" whatever numbers he needs. In addition, in books VII–IX Euclid does not use locutions like 'construct' and 'erect' which are so prevalent in his geometry. The few arithmetic problems Euclid solves are always phrased in terms of finding numbers satisfying a condition rather than generating them. And, having found the numbers sought, Euclid does not conclude with a 'Q.E.F.', as in a geometric problem, but with a 'Q.E.D.'[6] Thus there seems to be a significant difference between Euclid's notions of plane geometric objects and arithmetic ones. Units can be compared to points which are taken as needed. But numbers too are taken as needed, and even when they are produced by a series of operations, they are thought of as found rather than generated. Despite this difference, however, Euclid's arithmetic is hardly less "active" than his geometry. Proofs still involve the manipulation of objects, dividing them into their parts, producing others equal to given ones, and so on. Of course, in arithmetic it is the combinatorial rather than spatial aspects of the operations which are of fundamental importance.

The simplest way to characterize Euclid's attitude toward existence is to say that he assumes there are infinitely many units any finite selection of which constitutes a number. This assumption can be represented formally by using variables o

and p to range over units, i, j, k, l, m, and n to range over positive integers, and $\in$ to designate the relation between a unit and a positive integer to which it belongs. The assumption is then expressed by

VIIa $\exists o \exists k \forall p (p \in k \leftrightarrow p = o)$,
$\forall k \exists p \exists l (p \notin k \; \& \; p \in l \; \& \; \forall o (o \in k \rightarrow o \in l))$,
$\forall o_1 \ldots \forall o_n \exists k \forall o (o \in k \leftrightarrow o = o_1 \vee \ldots \vee o = o_n)$.

The first formula asserts the existence of a positive integer 1, i.e., of a multiplicity containing a single unit. For any positive integer k, the second formula guarantees the existence of a positive integer l containing all the units of k plus at least one more. The third formula says that any n units constitute a positive integer.

VIIa suffices to guarantee the existence of k multiplied by l. Since this product is not unique, it is technically illegitimate to use the operation symbol $\cdot$ in representing it. Euclid in fact has no analogue of such a symbol. If he wants to use the product of k and l in a proof, he always sets the product equal to a particular number m. However, the train of his reasoning is usually much more easily understood by using the ordinary symbol for multiplication, and so I shall continue to write $l \cdot k$ to stand for "the" result of multiplying k by l, i.e., adding together l k's. It is also convenient to use the expression $\Sigma^n l_i$ for the sum of n integers $l_1, l_2, \ldots, l_n$ and $\Sigma^n l_i (l_i \simeq l)$ for the sum of n integers equal to l, i.e., for the product $n \cdot l$.

Perhaps the fundamental undefined notion in Euclid's arithmetic is the notion of measurement. Intuitively, one positive integer measures a second when it divides the second evenly, i.e., when the second positive integer can be segregated into some number of parts each equal to the first. The notion of measurement can, of course, be applied geometrically as well, and Euclid does do so. Indeed, measurement is as fundamental a notion in Euclid's mathematics in general as, for example, meeting and coinciding are in his plane geometry. For arithmetic I shall use k/l to symbolize 'k measures l' and assume that every number measures itself.[7] Euclid uses the notion of measurement to say what is meant by one positive integer being a part, parts, or multiple of another:

VII, def. 3 $\text{PART}(k, l) \leftrightarrow k \prec l \; \& \; k/l$;
VII, def. 4 $\text{PARTS}(k, l) \leftrightarrow k \prec l \; \& \; \neg \, \text{PART}(k, l)$;
VII, def. 5 k is a multiple of $l \leftrightarrow l \prec k$ and l/k.

A trivial consequence of definitions 3 and 4 is

VII, 4 $k \prec l \rightarrow \text{PART}(k, l) \vee \text{PARTS}(k, l)$.

Euclid does not prove this proposition in the trivial way suggested here. Instead, on the assumption that k is less than l, he argues,

For l, k are either prime to one another or not. First, let l, k be prime to one another. Then, if k be divided into the units in it, each unit of those in k will be some part of l; so that k is parts of l.

Next let l, k not be prime to one another; then k either measures or does not measure l. If now k measures l, k is a part of l. But if not, let the greatest common measure m of l, k be taken (VII,2), and let k be divided into numbers equal to m, namely m_1, m_2, m_3. Now since m measures l, m is a part of l. But m is equal to each of the numbers m_1, m_2, m_3. Therefore, each of the numbers m_1, m_2, m_3 is also a part of l; so that k is parts of l.

It is easy to derive from the contents of this proof a procedure for determining what particular part or parts a given number is of a greater. The notion of particular parts is in fact essential to Euclid's arithmetic, although he nowhere defines it. He already presupposes it in defining four numbers to be proportional when the first is the *same* part, parts, or multiple of the second as the third is of the fourth. Zeuthen argued that the true sense of expressions such as 'part' and 'same part' is to be found in the proof of VII,4 and not in the definitions.[8] Zeuthen's hypothesis has the great advantage of closing a logical gap in the arithmetic books and giving an explanation for the form of the proof of VII,4. However, it is not at all clear why, if Zeuthen's hypothesis is correct, Euclid should have proceeded as he does rather than defining the greatest common measure of two numbers and their ratio in terms of the number of times the common measure measures each.[9] I shall not, therefore, adopt Zeuthen's hypothesis, but rather try to give what seems to me a more historically accurate account of Euclid's approach to arithmetic.

Essentially what Euclid takes for granted is not simply the relation of measurement, but the arithmetic relation of measuring a certain number of times. An expression which he uses for this relation is 'k measures l according to the units in m', i.e., 'k measures l m times'. (See, for example, VII,16.) I shall use $\mathcal{M}(k, l, m)$ to abbreviate this expression. Using it one can define

$k/l \leftrightarrow \exists m \mathcal{M}(k, l, m)$;
m-PART$(k, l) \leftrightarrow k \prec l \;\&\; \mathcal{M}(k, l, m)$;
m-n-PARTS$(k, l) \leftrightarrow k \prec l \;\&\; \exists j(\mathcal{M}(j, k, m) \;\&\; \mathcal{M}(j, l, n))$;
m-MULTIPLE$(k, l) \leftrightarrow m$-PART(l, k).

It is a consequence of the analysis given here that $\mathscr{M}(k, l, m)$, m-PART (k, l), and m-MULTIPLE (l, k) all say the same thing when $k \prec l$. Euclid also takes for granted the equivalence of $\mathscr{M}(k, l, m)$ and $m \cdot k \simeq l$.[10] In a modern formal treatment of Euclid's arithmetic one of these notions would be made primitive and proofs formulated in terms of it. Euclid, however, moves from one to the other without explicitly establishing any logical relationships between them. In addition he makes use of tacit assumptions which are here formulated in terms of $\mathscr{M}$:

VIIb $k \simeq k' \ \& \ l \simeq l' \ \& \ m \simeq m' \rightarrow$
$(\mathscr{M}(k, l, m) \rightarrow \mathscr{M}(k', l', m'))$;

VIIc $k \simeq k' \ \& \ l \simeq l' \ \& \ \mathscr{M}(k, l, m) \ \& \ \mathscr{M}(k', l', m') \rightarrow$
$m \simeq m'$;

VIId $k \simeq k' \ \& \ m \simeq m' \ \& \ \mathscr{M}(k, l, m) \ \& \ \mathscr{M}(k', l', m') \rightarrow$
$l \simeq l'$;

VIIe $l \simeq l' \ \& \ m \simeq m' \ \& \ \mathscr{M}(k, l, m) \ \& \ \mathscr{M}(k', l', m') \rightarrow$
$k \simeq k'$.

These principles should probably not be thought of as fundamental assumptions, since they can be derived in a Euclidean way from the general assumptions listed on pp. 36–37. This fact can be seen most easily by noting that $\mathscr{M}(k, l, m)$ is equivalent to $m \cdot k \simeq l$ or $l \simeq \Sigma^m k_i (k_i \simeq k)$. Hence VIIb and d are tantamount to the assertion that equals added to equals an equal number of times result in equals, a principle which is proved by iterated applications of common notion 2 or, more formally, by induction. It is simplest to prove VIIc and e indirectly. For VIIc is tantamount to the assertion that if equals are added to equals an unequal number of times, the result of adding them fewer times will be less than the result of adding them more times, an assertion which is easily proved using Ij. Similarly, VIIe can be thought of as the assertion that if $k \prec k'$, $m \cdot k \prec m \cdot k'$, which is derivable using Ij and l. The same sorts of derivations would be possible for the analogues of VIIb–e stated for m-n-PARTS.

The point of the preceding discussion is not to suggest that Euclid might have been aware of the possibility of such derivations, but rather to bring out the similarity between the assumptions underlying Euclid's arithmetic and those underlying his geometry. If one makes allowance for the difference between numbers and geometric objects and hence for the somewhat different sense attaching to such notions as addition and equality, his arithmetic assumptions are basically generalizations of his geometric ones. This relationship is perhaps made most clear in the case of Euclid's assumptions about measurement in the arithmetic books:

VIIf (i) $k/l \;\&\; l/m \rightarrow k/m$;
(ii) $k/l \rightarrow k \preceq l$;
(iii) $k/l \;\&\; k/m \rightarrow k/(l+m)$;
(iv) $k/l \;\&\; k/m \rightarrow k/(l-m)$.

Since k/l simply says that for some n, $l \simeq \Sigma^n k_i (k_i \simeq k)$, these assumptions are consequences of

(i) $l \simeq \sum^{n} k_i(k_i \simeq k) \;\&\; m \simeq \sum^{n'} l_i(l_i \simeq l) \rightarrow$
$$m \simeq \sum^{n \cdot n'} k_i(k_i \simeq k),$$

(ii) $k \preceq \sum^{n} k_i(k_i \simeq k)$,

(iii) $\sum^{n} k_i(k_i \simeq k) + \sum^{n'} k_i(k_i \simeq k) \simeq \sum^{n+n'} k_i(k_i \simeq k)$,

(iv) $\sum^{n} k_i(k_i \simeq k) - \sum^{n'} k_i(k_i \simeq k) \simeq \sum^{n-n'} k_i(k_i \simeq k)$.

Of these (i) and (iii) are expressions of the irrelevance of order in addition, (ii) is a generalized form of Ij, and (iv) is provable using Id.

The general nature of the combinatorial assumptions used by Euclid should be clear. However, it is convenient to list here the remaining ones used in the arithmetic books before proceeding. The first of these has already been mentioned.

VIIg $k \prec l \rightarrow (m \cdot k \prec m \cdot l \;\&\; k \cdot m \prec l \cdot m)$;

VIIh $\sum^{n} k_i + \sum^{n} l_i \simeq \sum^{n} (k_i + l_i)$;

VIIh′ $k \cdot \sum^{n} l_i \simeq \sum^{n} (k \cdot l_i)$;

VIIh″ $\sum_{i}^{n}\left(\sum_{j}^{m} k_{ji}\right) \simeq \sum_{j}^{m}\left(\sum_{i}^{n} k_{ji}\right)$;

VIIi $\sum^{n} k_i - \sum^{n} l_i \simeq \sum^{n} (k_i - l_i)$.

VIIh, h′, and h″ are again expressions of the irrelevance of order in addition; VIIi says that the difference of two sums of an equal number of summands is the sum of the differences between the summands (provided, of course, that the differences are defined).

Before turning to the analysis of Euclid's development of arithmetic, it is necessary to clear up two difficulties involving definition 21. There would seem to be two ways of rendering the definiens of PROPORTIONAL (k, l, m, n) in symbols:

(i) $\exists i\,(i\text{-PART}(k, l) \;\&\; i\text{-PART}(m, n)) \vee$
$\exists i \exists j\,(i\text{-}j\text{-PARTS}(k, l) \;\&\; i\text{-}j\text{-PARTS}(m, n)) \vee$
$\exists i\,(i\text{-MULTIPLE}(k, l) \;\&\; i\text{-MULTIPLE}(m, n))$.

(ii) $\forall i\,(i\text{-PART}(k, l) \rightarrow i\text{-PART}(m, n)) \;\&$
$\forall i \forall j\,(i\text{-}j\text{-PARTS}(k, l) \rightarrow i\text{-}j\text{-PARTS}(m, n)) \;\&$
$\forall i\,(i\text{-MULTIPLE}(k, l) \rightarrow i\text{-MULTIPLE}(m, n))$.

Prima facie (i) might appear to be closer to the definition itself. On the other hand, (ii) renders the kind of expression Euclid uses in typical applications of the definition. For example, in VII,11, the first application of definition 21, Euclid invokes it with the words "Since as AB is to CD, so is AE to CF, whatever part or parts AB is of CD, the same part or the same parts is AE of CF also." Taken literally, however, neither definition is in accord with ordinary intuitions about proportionality. For, according to (i), $\neg$ PROPORTIONAL (8, 5, 10, 6), since 8 is not a part, parts, or multiple of 5, and also $\neg$ PROPORTIONAL (3, 3, 4, 4), since 3 is neither a part, parts, or multiple of itself. For the same reasons, according to (ii), PROPORTIONAL (8, 5, 8, 6) and PROPORTIONAL (3, 3, 3, 4).

Euclid handles the first difficulty by assuming, when it is convenient, that if PROPORTIONAL (k, l, m, n), then $k \prec l$ (and so $m \prec n$). Intuitively this assumption is justified, because if PROPORTIONAL (k, l, m, n), PROPORTIONAL (l, k, n, m). Of course, if one makes use of this intuitive assumption, one need not take account of the clause referring to multiples in definition 21. In fact, Euclid does not take this clause into account when he proves laws of proportionality, as is illustrated in the quotation from VII,11 in the preceding paragraph. As far as proportionalities involving equal numbers are concerned, Euclid seems to exclude them from his definition deliberately. However, it would be impossible to maintain this exclusion without complicating the development of arithmetic. And indeed, Euclid does admit such proportionalities when he needs them, e.g., in IX,36. The simplest way out of both of these difficulties would be to define m-n-PARTS* (k, l) without the restriction that k is less than l, and with the explicit proviso that any of k, l, m, n may be 1. Then one could define PROPORTIONAL (k, l, m, n) by either

(i′) $\exists i \exists j$ (i-j-PARTS* (k, l) & i-j-PARTS* (m, n)),

or

(ii′) $\forall i \forall j$ (i-j-PARTS* $(k, l) \rightarrow$ i-j-PARTS* (m, n)).

Of these two definitions, (i′) would seem to be preferable. For, according to (ii′), $\neg$ PROPORTIONAL (6, 8, 3, 4), because 6-8-PARTS* (6, 8), but $\neg$ 6-8-PARTS* (3, 4), since 4 does not have 8th parts. On the other hand, if (i′) is used, some of Euclid's proofs fail because apparently trivial assumptions are no longer trivial. For example, in the proof of VII,14 Euclid takes for granted that

VIIj PROPORTIONAL (k, l, m, n) & PROPORTIONAL $(k, l, i, j) \rightarrow$ PROPORTIONAL (i, j, m, n).

However, there is no direct way to derive

$\exists k' \, \exists l' \, (k'\text{-}l'\text{-PARTS}^*(i, j) \; \& \; k'\text{-}l'\text{-PARTS}^*(m, n))$

from

$\exists i' \, \exists j' \, (i'\text{-}j'\text{-PARTS}^*(k, l) \; \& \; i'\text{-}j'\text{-PARTS}^*(m, n))$

and

$\exists m' \, \exists n' \, (m'\text{-}n'\text{-PARTS}^*(k, l) \; \& \; m'\text{-}n'\text{-PARTS}^*(i, j))$.[11]

The simplest way to overcome these difficulties would be to prove the equivalence of (i′) and (ii′), and the simplest way to do this would be to establish a unique expression for the parts* that one number is of another. On Zeuthen's hypothesis, Euclid carries out this task in VII,4. I am more inclined to think that Euclid simply takes the existence of such a unique expression for granted. If he does so, then for him VIIj would be an immediate consequence of the definition of proportionality in the form (i′). Other consequences of this definition which Euclid assumes without proof are

VIIk (i) $\text{PROPORTIONAL}(k, l, m, n) \rightarrow \text{PROPORTIONAL}(m, n, k, l)$,

(ii) $\text{PROPORTIONAL}(k, l, m, n) \rightarrow (k \simeq m \leftrightarrow l \simeq n)$,

(iii) $k \simeq m \rightarrow (\text{PROPORTIONAL}(k, l, m, n) \leftrightarrow l \simeq n)$,

(iv) $l \simeq n \rightarrow (\text{PROPORTIONAL}(k, l, m, n) \leftrightarrow k \simeq m)$.

Euclid uses other locutions for "k, l, m, n are proportional," notably "as k is to l so is m to n" and "k, l, m, n are in the same ratio." But he rarely speaks of the ratio of k to l or of equalities between ratios. In general a ratio is a relation between two objects and not itself an object or even a pair of objects, as in the modern foundational definition of fractions as pairs of integers. Of course, theorems about arithmetic proportionality can be translated into theorems about fractions, but such translation is not historically accurate with respect to Euclid's *Elements*. Hereafter I shall write $(k, l) = (m, n)$ instead of $\text{PROPORTIONAL}(k, l, m, n)$, because the former notation is shorter.[12] But it must be remembered that $(k, l) = (m, n)$ expresses a four-place relation between positive integers and not an identity between fractions or ratios. That proportionality is a four-place relation is made clearest in book V, where Euclid proves as proposition 11 an analogue of VIIj for proportionalities involving arbitrary magnitudes. If proportionalities expressed identities between objects, this proposition would be a trivial instance of common notion 1, and hence not proved by Euclid.

Euclid proves the fundamental properties of numerical proportionalities in VII,11–14. These propositions may be expressed symbolically as

11 $(k, l) = (m, n) \rightarrow (k - m, l - n) = (k, m)$,

12 $(k_1, l_1) = (k_2, l_2) = \ldots = (k_n, l_n) \rightarrow$
$(k_1, l_1) = (k_1 + \ldots + k_n, l_1 + \ldots + l_n)$,

13 $(k, l) = (m, n) \rightarrow (k, m) = (l, n)$,

14 $(k_1, k_2) = (l_1, l_2)$ & $(k_2, k_3) = (l_2, l_3)$ & ... &
$(k_{n-1}, k_n) = (l_{n-1}, l_n) \rightarrow (k_1, k_n) = (l_1, l_n)$.

In stating 13 and 14 Euclid uses the terms 'alternately' (*enallax*) and '*ex equali*' (*di' isou*) which are defined in book V, defs. 12 and 17, and used in analogues of 13 and 14 for arbitrary magnitudes (V,16 and 22). Presumably Euclid believes that the definitions in book V make clear enough what the terms mean in the case of numerical proportionalities. However, he must not think that the propositions of book V apply to such proportionalities, for, if he did, he would not have to prove any of 11–14 since their analogues have already been proved in book V.

14 follows easily from 13. I give Euclid's presentation of the proposition:

If there be arbitrarily many (*hoposoioun*) numbers and others equal to them in multitude, which taken in pairs (*sunduo*) are in the same ratio, they will also be in the same ratio *ex equali*.

Let there be arbitrarily many numbers *A*, *B*, *C* and others equal to them in multitude *D*, *E*, *F* which taken in pairs are in the same ratio, so that as *A* is to *B* so is *D* to *E*, and as *B* is to *C* so is *E* to *F* [fig. 2.1]. I say that *ex equali* as *A* is to *C* so also is *D* to *F*. For since as *A* is to *B* so is *D* to *E*, therefore alternately as *A* is to *D* so is *B* to *E*. Again, since as *B* is to *C* so is *E* to *F*, therefore alternately as *B* is to *E* so is *C* to *F*. But as *B* is to *E* so is *A* to *D*; therefore also as *A* is to *D* so is *C* to *F*. Therefore, alternately as *A* is to *C* so is *D* to *F*. Q.E.D.

Figure 2.1

In form this proof is quite like a standard geometric one except that it lacks a *kataskeuē* and a *sumperasma*. There is no *kataskeuē* because none is needed for this particular argument. One notes also that the diagram plays no real role in this proof except possibly as a mnemonic device for fixing the meaning of the letters. The nonfunctionality of the diagrams in arithmetic is an indication of the greater abstractness of the subject as compared with geometry. However, although in the present case the reasoning is purely logical, in many propositions combinatorial ideas and geometric analogies will be seen to be functioning.

The absence of a *sumperasma* is typical of the arithmetic books.[13] There would not seem to be any philosophical or mathematical reason for its absence since Euclid's arithmetic proofs are as general as his others. The problem of generality is, however, much more explicit in the arithmetic books. In 14 Euclid wishes to prove something for an arbitrary number of positive integers, but he has no subscript notation. He therefore proves 14 for "arbitrarily many numbers *A*, *B*, *C*," i.e., he proves it for three positive integers with the expectation that the truth of the assertion for four, five, six, etc., positive integers will be clear. In general, the difference between a subscript notation using dots and Euclid's is purely stylistic. I shall use the former because it does make the intent of generality more explicit. In this notation the proof of 14 would run as follows. Let $(k_i, k_{i+1}) = (l_i, l_{i+1})$, for $1 \leq i < n$.[14] Then by VII,13, $(k_i, l_i) = (k_{i+1}, l_{i+1})$, and, by repeated applications of VIIk(i) and j, $(k_1, l_1) = (k_n, l_n)$. Hence, by VII,13 again, $(k_1, k_n) = (l_1, l_n)$. Here one simply has to "see" that the argument applies to every particular n.

However, it is a simple matter to turn the informal use of subscripts in a proof into an application of arithmetic induction. One does this in the case of VII,14 by first proving the proposition for $n = 3$ as in the *Elements* and then proceeding as follows. Suppose (inductive hypothesis) that the theorem is true for $n \geq 3$, and let $(k_i, k_{i+1}) = (l_i, l_{i+1})$, for $1 \leq i \leq n$. Then trivially $(k_i, k_{i+1}) = (l_i, l_{i+1})$ for $1 \leq i < n$, and, by inductive hypothesis, $(k_1, k_n) = (l_1, l_n)$, or (VII,13 and k(i)) $(k_n, l_n) = (k_1, l_1)$. But since, by assumption, $(k_n, k_{n+1}) = (l_n, l_{n+1})$, by VII,13 again, $(k_n, l_n) = (k_{n+1}, l_{n+1})$. Therefore, by VIIj, $(k_1, l_1) = (k_{n+1}, l_{n+1})$, and, by VII,13, $(k_1, k_{n+1}) = (l_1, l_{n+1})$. This proof is more explicit than the previous two in showing how the theorem applies to any number of positive integers. For example, if one wants to apply the proof for $n = 5$, one has a proof for $n = 3$, and the inductive part of the proof just given shows how to get a proof for $n = 4$ from one for $n = 3$, and, again, how to get a proof for $n = 5$ from one for $n = 4$.

Although there is only a notational difference between Euclid's proof of 14 and my proof using subscripts and only a short step separates the latter from an inductive proof, it would hardly seem appropriate to descry in Euclid's proof a knowledge of arithmetic induction.[15] On the other hand, there are other proofs in the arithmetic books which do suggest a conception of the idea of induction. For sometimes Euclid proves separate propositions for the cases $n = 2$ and $n = 3$, carrying out the second by a reduction to the first in a way which could be repeated for $n = 4$, $n = 5$, and so on. For example, having shown how to find the greatest common measure (GCD) of two numbers in VII,2, he shows in VII,3 that the GCD of three numbers k_1, k_2, k_3 is $\text{GCD}(\text{GCD}(k_1, k_2), k_3)$. It would seem perverse to deny Euclid's intention is to convey a general result, especially in the light of his later use of the procedure for finding the GCD in a proposition formulated for arbitrarily many numbers (VII,33).

However, it is important to realize that in every instance where Euclid proceeds in this quasi-inductive way, the implicit induction is on the number of terms involved in a construction or assertion and not on the integers themselves. He does not, for example, prove the commutativity of multiplication (VII,16) by induction, but by a general example. This point is important because arithmetic induction applied directly to the integers is a fundamental principle of modern arithmetic. Indeed, for arithmetic purposes the positive integers are adequately characterized by the axioms P1 and 2 stated on p. 59 and the principle of mathematical induction:

P3 For any property $\mathscr{P}$, $\mathscr{P}(1)$ & $\forall k(\mathscr{P}(k) \rightarrow \mathscr{P}(\sigma(k)) \rightarrow \forall k \mathscr{P}(k)$,

i.e., all positive integers have any property which 1 has and which $k + 1$ has if k does. The truth of P3 is simply an expression of the conception of positive integers as generated from 1 by repeated applications of the successor operation. And, given this conception, the specific character of 1 and the successor operation are irrelevant. They are any object and operation satisfying P1–3. In other words, the structural conception of mathematics applies to arithmetic as well as geometry. For Euclid a unit is clearly not an arbitrary thing, but is somehow connected with our designation of objects as single or one. More importantly, numbers are not characterized as generated from units in a serial order. They are simply finite aggregates of units. Of course, there are important relations between these aggregates, but the relation of successor is not one which plays an important role in the *Elements*. Thus it can be said that the integers themselves are not conceived in the structural way conducive to the use of induction, but that there is inductive reasoning about collections or sequences of positive integers. This difference might be compared to the difference in formal theories between the use of induction in the object language and its use in the metalanguage.

I return to the topic of arithmetic proportionality. If one sets $n = 2$, propositions 11, 12, and 13 are simple reformulations of 7 and 8, 5 and 6, and 9 and 10, respectively. These propositions may be expressed as follows:

a. If

$$l \simeq \sum^{n} k_i (k_i \simeq k) \;\&\; l' \simeq \sum^{n} k'_i (k'_i \simeq k'),$$

then

$$\text{(VII,5)} \quad l + l' \simeq \sum^{n} k_i (k_i \simeq k + k'),$$

$$\text{(VII,7)} \quad l - l' \simeq \sum^{n} k_i (k_i \simeq k - k'),$$

$$\text{(VII,9) (i)} \quad (k' \simeq \sum^{m} k_i (k_i \simeq k) \rightarrow l' \simeq \sum^{m} l_i (l_i \simeq l)) \;\&$$

$$\text{(ii)} \quad (\exists j (k' \simeq \sum^{m'} j_i (j_i \simeq j) \;\&\; k \simeq \sum^{m} j_i (j_i \simeq j)) \rightarrow$$

$$\exists j' (l' \simeq \sum^{m'} j'_i (j'_i \simeq j') \;\&\; l \simeq \sum^{m} j'_i (j'_i \simeq j')));$$

b. If

$$l \simeq \sum^{n} j_i (j_i \simeq j) \;\&\; k \simeq \sum^{m} j_i (j_i \simeq j) \;\&$$

$$l' \simeq \sum^{n} j'_i (j'_i \simeq j') \;\&\; k' \simeq \sum^{m} j'_i (j'_i \simeq j'),$$

then

$$\text{(VII,6)} \quad l + l' \simeq \sum^{n} j_i (j_i \simeq j + j') \;\&$$

$$k + k' \simeq \sum^{m} j_i (j_i \simeq j + j'),$$

$$\text{(VII,8)} \quad l - l' \simeq \sum^{n} j_i (j_i \simeq j - j') \;\&$$

$$k - k' \simeq \sum^{m} j_i (j_i \simeq j - j'),$$

$$\text{(VII,10) (i)} \quad (k' \simeq \sum^{n'} k_i (k_i \simeq k) \rightarrow l' \simeq \sum^{n'} l_i (l_i \simeq l)) \;\&$$

$$\text{(ii)} \quad (\exists j^* (k' \simeq \sum^{n'} j_i (j_i \simeq j^*) \;\&$$

$$k \simeq \sum^{m'} j_i (j_i \simeq j^*)) \rightarrow$$

$$\exists j^* (l' \simeq \sum^{n'} j_i (j_i \simeq j^*) \;\&$$

$$l \simeq \sum^{m'} j_i (j_i \simeq j^*))).$$

These propositions are used only to prove 11–13, and they depend on no explicit arithmetic assumptions or prior propositions. In fact, most of them are quite direct consequences of the combinatorial assumptions I have already formulated. For example, 5 and 6 follow directly from VIIh. From the modern point of view, 7 and 8 follow equally directly from VIIi; but because of Euclid's concrete conception of subtraction as the taking away of a proper part and of addition as concatenation,

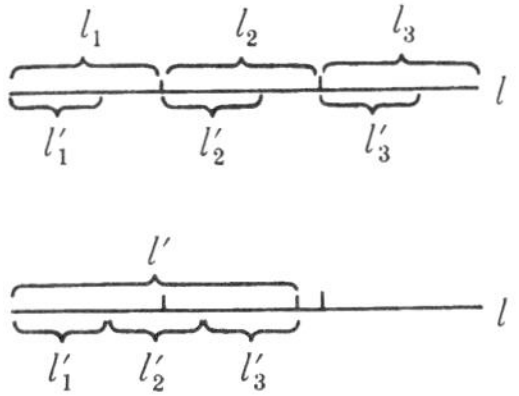

Figure 2.2

he is forced to proceed in a more roundabout way. For Euclid a proof of VII,7 corresponding to the application of VIIi would involve dividing l into actual nth parts $l_1, \ldots, l_n$, each equal to k, and l' into nth parts $l'_1, \ldots, l'_n$, each equal to k' (fig. 2.2). But there is no way to do this without either making l' discontinuous or failing to make each l'_i a subpart of l_i. Hence Euclid proceeds by letting $k - k'$ be an nth part of j and proving that $j \simeq l - l'$. He sets $j \simeq \Sigma^n k_i (k_i \simeq k - k')$ so that, since $(k - k') + k' \simeq k, j + l' \simeq \Sigma^n k_i (k_i \simeq k)$, by VII,5, i.e., $j + l' \simeq l$ and $j \simeq l - l'$. To get around a similar difficulty in VII,8 Euclid constructs new j_i^*, j_i^{**}, with j_i^* a proper part of j_i^{**}, $j_i^* \simeq j, j_i^{**} \simeq j'$, so that $j_i^* - j_i^{**}$ is always defined and is equal to what we would call $j_i - j'_i$. Euclid then argues that

$$k \simeq \sum^m j_i^* (j_i^* \simeq j), \quad k' \simeq \sum^m j_i^{**} (j_i^{**} \simeq j'),$$
$$l \simeq \sum^n j_i^* (j_i^* \simeq j), \quad l' \simeq \sum^n j_i^{**} (j_i^{**} \simeq j'),$$

and in effect applies VIIi.

Euclid's arguments for VII,9 and 10 are more easily understood if their consequents are represented more directly as

VII,9 (i) (m-PART $(k, k') \rightarrow m$-PART (l, l')) &
(ii) (m-m'-PARTS $(k, k') \rightarrow m$-m'-PARTS (l, l'));

VII,10 (i) (n'-PART $(k, k') \rightarrow n'$-PART (l, l')) &
(ii) (m'-n'-PARTS $(k, k') \rightarrow m'$-n'-PARTS (l, l')).

To prove 9, Euclid argues that if

m-PART (k, k')	m-m'-PARTS (k, k'),
m-PART (k_i, k'_i)	m-m'-PARTS (k_i, k'_i),

so that, by V,5 or 6 generalized to n summands,

m-PART $(k_1 + \ldots + k_n, k'_1 + \ldots + k'_n)$	m-m'-PARTS $(k_1 + \ldots + k_n, k'_1 + \ldots + k'_n)$,

i.e.,

m-PART (l, l')	m-m'-PARTS (l, l').

Euclid proves 9 for the case $n = 2$ so that, in a sense, he does not need the generalization to any number of summands. However, there is no reason to doubt his awareness of the need, in the logical sense, for such a generalization. For he uses 5 and 6 in the same way for the proof of 12, which is explicitly stated for arbitrarily many summands. It is curious, however, that Euclid does not prove 5 and 6 in the more general form which could be applied directly in 9 and 12, and also in 10 to which the same remarks apply.

Euclid's argument for 10 founders on the difficulties in his account of part and parts to which I have already referred. On the basis of VII,9 Euclid asserts:

$$(n'\text{-PART}(j_i, j_i') \rightarrow n'\text{-PART}(l, l')) \;\&\; (m'\text{-}n'\text{-PARTS}(j_i, j_i') \rightarrow m'\text{-}n'\text{-PARTS}(l, l')).$$

He then infers, presumably on the ground that k is the sum of the j_i's, k' of the j_i''s, that

$$(n'\text{-PART}(k, k') \rightarrow n'\text{-PART}(l, l')) \;\&\; (m'\text{-}n'\text{-PARTS}(k, k') \rightarrow m'\text{-}n'\text{-PARTS}(l, l')).$$

Heiberg brackets as a Theonine interpolation a purported justification of this inference which shows more clearly its problematic character. The justification invokes the following consequence of the generalized form of VII,5 and 6:

$$(n'\text{-PART}(j_i, j_i') \rightarrow n'\text{-PART}(k, k')) \;\&\; (m'\text{-}n'\text{-PARTS}(j_i, j_i') \rightarrow m'\text{-}n'\text{-PARTS}(k, k')).$$

The resulting argument is patently fallacious, since it is of the form '$(p \rightarrow q)$ & $(p \rightarrow r)$; therefore $q \rightarrow r$'. To get a correct argument one would need to establish $q \rightarrow p$, i.e.,

(a) $(n'\text{-PART}(k, k') \rightarrow n'\text{-PART}(j_i, j_i'))$ &
(b) $(m'\text{-}n'\text{-PARTS}(k, k') \rightarrow m'\text{-}n'\text{-PARTS}(j_i, j_i'))$.

Of these, (a) is provable because the relation of part is necessarily expressed in least terms. But (b) is in fact false on Euclid's definition, since, for example, 4-6-PARTS(8, 12) but not 4-6-PARTS(2, 3).

To sum up the preceding discussion of Euclid's presentation of the theory of arithmetic proportionality I give a schematic representation of the deductive structure of VII,5–14 in fig. 2.3. As I have already indicated, 11–13 are simple reformulations of the propositions from which they are derived, and 5 and 6 are tacitly generalized for the proofs of 9, 10, and 12. Finally, 5–10 have no further use in the *Elements*, and they depend upon no prior explicit assumptions or results.

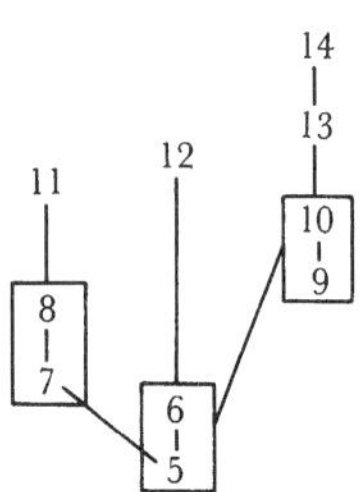

Figure 2.3

In modern treatments of the foundations of analysis it is customary to define equality for fractions by the identity of the cross products, i.e.,

$$\frac{m}{n} \text{ equals } \frac{k}{l} \text{ if and only if } m \cdot l = n \cdot k.$$

The analogue for ratios is

VII,19 $(m, n) = (k, l) \leftrightarrow m \cdot l \simeq n \cdot k$,

which Euclid derives from

VII,17 $(i \cdot m, i \cdot n) = (m, n)$,
VII,18 $(m \cdot i, n \cdot i) = (m, n)$.

For by these two propositions and VIIj and k the following assertions are equivalent: $(m, n) = (k, l)$; $(m \cdot k, n \cdot k) = (m \cdot k, m \cdot l)$; $m \cdot l \simeq n \cdot k$. Euclid derives VII,18 from VII,17 and

VII,16 $m \cdot n \simeq n \cdot m$.

Although 17 is simply a special case of 12, Euclid proves it by pointing out that since $\mathcal{M}(m, i \cdot m, i)$ and $\mathcal{M}(o, i, i)$,

$(o, i) = (m, i \cdot m)$.

Similarly,

$(o, i) = (n, i \cdot n)$.

Therefore, $(m, i \cdot m) = (n, i \cdot n)$ or, by VII,13, $(m, n) = (i \cdot m, i \cdot n)$. It is to be noticed that all but the last step of this argument depends upon unstated ideas about measurement and proportionality.

16, of course, expresses the commutativity of multiplication. For Euclid the commutativity of addition is a trivial fact resting on the conception of addition as concatenation. However, the commutativity of multiplication is not a trivial fact because it is not immediately obvious that the concatenation of m collections of n units each produces a result equal to the concatentation of n collections of m units each. (Of course, commutativity would be obvious if multiplication were identified with the forming of a rectangle.) In modern treatments of arithmetic, commutativity is proved by induction using the definition of multiplication. Euclid uses proportionality, but in a somewhat curious way. He could have argued, as in the proof of 17, that $(o, n) = (m, n \cdot m)$ and $(o, m) = (n, m \cdot n)$, i.e., by VII,13, $(o, n) = (m, m \cdot n)$, so that, by VIIj and k, $m \cdot n \simeq n \cdot m$. Instead of proceeding in this way Euclid uses

VII,15 $\mathcal{M}(o, k, i)$ & $\mathcal{M}(j, k', i) \rightarrow$
$\forall l (\mathcal{M}(o, j, l) \rightarrow \mathcal{M}(k, k', l))$,

and argues that since $\mathcal{M}(n, m \cdot n, m)$ and $\mathcal{M}(o, m, m)$,

o measures n the same number of times that m measures $m \cdot n$.

But also, $\mathcal{M}(m, n \cdot m, n)$ and $\mathcal{M}(o, n, n)$, or

o measures n the same number of times that m measures $n \cdot m$.

Therefore m measures $m \cdot n$ and $n \cdot m$ the same number of times, and they are equal. The only difference between these two proofs of VII,16 which is not transparently terminological is Euclid's use of VII,15 in place of VII,13. But VII,15 is simply

VII,13 or 9 with the first number of the proportion made a unit, as is seen when 15 is reformulated in the language of parts:

VII,15′ i-PART(o, k) & i-PART$(j, k') \rightarrow$
$\forall l(l$-PART$(o, j) \rightarrow l$-PART$(k, k'))$.

Euclid's proof of 15 directly parallels his proof of 9 except that he uses VII,12 instead of 5 and 6. His reasons for introducing 15 and invoking it in the proof of 16 are not at all clear. He cites 15 three other times.[16] In IX,11 he moves from (i) $(o, j) = (k, l)$ to (ii) "o measures j the same number of times as k measures l," and then to (iii) "o measures k the same number of times as j measures l," applying 15. Clearly Euclid could have used VII,13 to get (ii′) $(o, k) = (j, l)$ and inferred (iii) directly. The other two applications of 15 are in purely verbal lemmas for VII,39,

VII,37 $i|j \rightarrow \exists k(i$-PART$(k, j))$,
VII,38 $\exists k(i$-PART$(k, j)) \rightarrow i|j$.

Euclid's reasons for proving these lemmas are not at all clear, since in VII,20 and elsewhere he moves immediately from $\mathscr{M}(i, j, k)$ to $\mathscr{M}(k, j, i)$, the latter of which is obviously equivalent to i-PART(k, j). In any case, Euclid argues that if (i) $\mathscr{M}(i, j, k)$, then, since $\mathscr{M}(o, k, k)$, (ii) o measures k the same number of times as i measures j; and, by 15, (iii) o measures i the same number of times as k measures j; or (iv) o is the same part of i as k is of j. But (v) i-PART(o, i), so that (vi) i-PART(k, j). In proving 38, Euclid simply argues in the reverse order from (vi) to (ii) and infers $i|j$ directly; VII,15 is used to infer (ii) from (iii). Clearly Euclid could have used 13 in place of 15 in each proof, replacing (ii) and (iii) with $(o, k) = (i, j)$ and $(o, i) = (k, j)$, respectively. The only apparent explanation of Euclid's inclusion of VII,15 is an unwillingness to perform alternation on proportions involving units when it has been proved for numbers. However, this explanation seems unsatisfactory since in the proof of 15 itself Euclid applies VII,12 to such proportionalities, although 12 too is stated only for numbers.

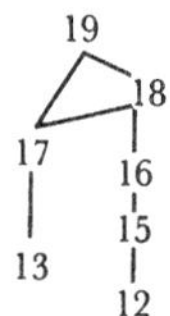

Figure 2.4

The deductive structure of 15–19 is indicated in fig. 2.4. All of 11–19 are basic propositions of Euclid's arithmetic, and are used explicitly and tacitly throughout the arithmetic books. Since, in general, their use is not problematic, I shall not usually even remark on their applications.

In order to deal with the rest of book VII, it is convenient to start with VII,33, the problem of finding least numbers in the same ratio as arbitrarily many given numbers. Euclid, of course, only defines proportionality for two pairs of numbers, but it is easy enough to extend his definition to two ordered n-tuples by defining

$$(k_1, \ldots, k_n) = (l_1, \ldots, l_n) \leftrightarrow (k_i, k_{i+1}) = (l_i, l_{i+1}),$$

for $1 \leq i < n$. It is also convenient to define what it is for $l_1, \ldots, l_n$ to be the least numbers in the ratio of $k_1, \ldots, k_n$ by

$$\begin{aligned} &k_1, \ldots, k_n\text{-LEAST}(l_1, \ldots, l_n) \leftrightarrow \\ &\quad (k_1, \ldots, k_n) = (l_1, \ldots, l_n)\ \& \\ &\quad \forall j_1 \ldots \forall j_n ((j_1, \ldots, j_n) = (l_1, \ldots, l_n) \rightarrow l_1 \preceq j_1). \end{aligned}$$

(I shall write $\text{LEAST}(l_1, \ldots, l_n)$ for $l_1, \ldots, l_n\text{-LEAST}(l_1, \ldots, l_n)$.) The adequacy of this definition is established by

$$(k_1, \ldots, k_n) = (l_1, \ldots, l_n) \rightarrow (k_i \prec l_i \rightarrow k_j \prec l_j).$$

The proof of this proposition follows from features of proportionality which Euclid takes for granted and the fact that, by VII,14 and 13, $(k_i, l_i) = (k_j, l_j)$.

VII,33 may now be stated

VII,33 To find $l_1, \ldots, l_n$ such that $k_1, \ldots, k_n\text{-LEAST}(l_1, \ldots, l_n)$.

Euclid divides the proof into two cases. If $\text{PRIME}(k_1, \ldots, k_n)$, i.e., if they have only the unit as common measure (VII, def. 13), then according to VII,21, they are the least in their ratio. If $\neg\,\text{PRIME}(k_1, \ldots, k_n)$, Euclid finds the greatest common measure k of $k_1, \ldots, k_n$, using VII,2 and 3, and proves that

$$k_1, \ldots, k_n\text{-LEAST}\left(\frac{k_1}{k}, \ldots, \frac{k_n}{k}\right).$$

VII,21 depends on

VII,20 $k_1, \ldots, k_n\text{-LEAST}(l_1, \ldots, l_n) \rightarrow \exists j \mathscr{M}(l_i, k_i, j)$.

22 is the converse of 21; and the three propositions 20–22 constitute another part of the foundation of Euclid's arithmetic. Although he proves these propositions for pairs of numbers, he uses them in a general form; and his proofs are easily generalized. The proof of 20 is indirect. Euclid assumes that $m'\text{-}m\text{-PARTS}(l_i, k_i)$, for some i, so that the same is true for every i. He then argues that

$$\frac{l_1}{m'}, \ldots, \frac{l_n}{m'}$$

are lesser numbers in the same ratio of $l_1, \ldots, l_n$. More exactly, he sets $l_i \simeq \Sigma^{m'} k_j^* (k_j^* \simeq k_i')$ with $m\text{-PART}(k_i', k_i)$ and argues by VII,12 that

$$(k_i', k_{i+1}') = \left(\sum^{m'} k_j^* (k_j^* \simeq k_i'), \sum^{m'} k_j^* (k_j^* \simeq k_{i+1}')\right) = (l_i, l_{i+1});$$

which is impossible, since $k_i' \prec l_i$. In what could be the only application of VII,4 but could equally well be an application of the definition of 'parts', Euclid infers that PART (l_i, k_i), and hence that $\exists j \mathcal{M} (l_i, k_i, j)$.

The proof of 22 is simple, and turns on the fact that $\frac{l_1}{m}, \ldots, \frac{l_n}{m}$ are in the same ratio as $l_1, \ldots, l_n$. In the proof Euclid constructs $\frac{l_i}{m}$ by means of a stylized argument which I shall not repeat again. He says, "As many times as m measures l_i, so many units let there be in $j_i \left(\simeq \frac{l_i}{m}\right)$. Since m measures l_i according to the units in j_i, therefore m by multiplying j_i has made l_i." (Sometimes Euclid concludes that j_i by multiplying m has made l_i.) He is then in a position to apply VII,17 to get the result that $j_1, \ldots, j_n$ are in the same ratio as $l_1, \ldots, l_n$.

For 21 Euclid supposes that PRIME $(l_1, \ldots, l_n)$ and that $(k_1, \ldots, k_n) = (l_1, \ldots, l_n)$ with $k_i \prec l_i$. He then invokes VII,20 to assert the existence of an m such that (i) $\mathcal{M} (k_i, l_i, m)$; hence (ii) $\mathcal{M} (m, l_i, k_i)$, which is impossible since PRIME $(l_1, \ldots, l_n)$. The inference from (i) to (ii) is repeated in VII,24 and 33. Heath (vol. II, p. 323) assumes that this is a matter of applying the definition of multiplication to (i) to get $m \cdot k_i \simeq l_i$, then VII, 16 to get $k_i \cdot m \simeq l_i$, and then the definition of multiplication again to get (ii). Heiberg suggests with more or less equal plausibility that Euclid infers from (i) that the unit measures m as often as k_i measures l_i; then by VII,15 that the unit measures k_i as often as m measures l_i; and then (ii). Hereafter I shall simply cite VII,16 for these inferences.

Euclid's application of VII,20 to infer (i) presupposes that LEAST $(k_1, \ldots, k_n)$, although Euclid says only that the k_i are less than the l_i. He proceeds in the same way elsewhere, for example, in VII,33. In effect then, Euclid argues from

(a) PRIME $(l_1, \ldots, l_n)$ & $\neg$LEAST $(l_1, \ldots, l_n)$

to

(b) $\exists l_1' \ldots \exists l_n' (l_1' \prec l_1 \ \& \ldots \& \ l_n' \prec l_n \ \&$
$(l_1, \ldots, l_n') = (l_1, \ldots, l_n))$,

and then to

(c) $\exists k_1 \ldots \exists k_n (k_1 \prec l_1 \ \& \ldots \& \ k_n \prec l_n \ \&$
$l_1, \ldots, l_n$-LEAST $(k_1, \ldots, k_n))$.

Here (b) is a reformulation of (a), but the existential claim it makes is not explicitly constructive. Only in VII,33 does one learn how to find numbers satisfying (b) (and also, of course, (c)). Naturally, there is always a trivial procedure for finding them, namely searching through all the numbers $l'_1, \ldots, l'_n$ less than $l_1, \ldots, l_n$, respectively, until less (or least) ones in the same ratio are found. In general, in arithmetic as elementary as Euclid's it is always possible to replace an assertion of the existence of a number or numbers of a certain kind with a rudimentary constructive procedure for determining the number or numbers. Hence, one will not find in Euclidean arithmetic any essentially nonconstructive arguments or assertions. On the other hand, as VII,21 indicates, Euclid does not seem to have been particularly concerned to bring out the constructive character of his arithmetic.

The step from (b) to (c) is from a modern point of view an application of the least number principle, which, stated for single positive integers, says that if there exists a positive integer having a property $\mathscr{P}$, there exists a least one having it. The least number principle is a fundamental structural principle of modern arithmetic. One needs only a few elementary assumptions to prove its equivalence to the principle of arithmetic induction.[17] Euclid's use of the least number principle is usually as casual as in the proof of VII,21. He is explicit only when he proves in VII,31 that every composite (i.e., nonprime) positive integer is divisible by a prime. He argues that if k is not a prime, there is a number k_1 dividing it. If k_1 is a prime, the theorem is true of k; but if it isn't, k_1, and hence (by VIIf(i)) k, is divisible by a number k_2. If k_2 is prime, the theorem is true of k; but if not, there is a k_3 dividing k_2. Euclid now says,

> The investigation coming to be in this way, there will be found some prime number which will measure the number before it, which will also measure k. For if it is not found, infinitely many numbers of which each is less than the other will measure k, which is impossible in numbers.

The manuscripts which Heiberg calls Theonine contain prior to this proof a shorter one in which the least number l measuring k is taken, and it is argued that l must be prime because any number measuring it will also measure k and be less than l. Although this proof is more explicit in the application of the least number principle, the first one contains its probable justification: If there is a number satisfying a condition but no least, there will be an infinite descending chain of numbers, which is impossible. In symbols this assertion is

$$\exists k\, \mathscr{P}(k) \rightarrow \neg\, \forall l\,(\mathscr{P}(l) \rightarrow \exists m\,(m \prec l\ \&\ \mathscr{P}(m))),$$

which is clearly logically equivalent to the following formulation of the least number principle:

$\exists k\mathscr{P}(k) \rightarrow \exists l(\mathscr{P}(l) \ \& \ \forall m(m \prec l \rightarrow \neg \mathscr{P}(m)))$.

Thus there can be no doubt that Euclid not only presupposes but also makes explicit use of an equivalent of the principle of mathematical induction. However, unlike the principle of induction, the least number principle or the denial of infinitely descending chains does not seem to depend upon a genuinely structural conception of the positive integers. For these principles can be understood solely in terms of the Euclidean conception of numbers as finite concatenations of units. Given any number k, a number k_1 less than it can be construed as a portion of k, a number less than k_1 as a portion of k_1 and hence of k, and so on. But obviously, since k cannot be divided past its units, there cannot be an infinite descent of this kind. In other words, if k has a property, there must be a least positive integer having it, because the finitely many units in k embody all the positive integers less than k.

For the proof of VII,33 it is necessary to show how to find $\mathrm{GCD}(k_1, \ldots, k_n)$ when $\neg$ PRIME $(k_1, \ldots, k_n)$. Although Euclid treats this topic at the beginning of book VII, he uses the results only in 4 and 33. In VII,2 he gives the now-standard procedure for finding the greatest common measure of two numbers not prime to one another. The method involves so-called alternate subtraction (*antanairesis*). Given two numbers k_1, l_1, with $k_1 \prec l_1$, one checks whether k_1/l_1. If so $k_1 \simeq \mathrm{GCD}(k_1, l_1)$. If not, one checks whether the lesser k_2 of k_1 and $l_1 - k_1$ divides the greater l_2. If so, one argues in a way to be described shortly that $k_2 \simeq \mathrm{GCD}(k_2, l_2)$. Otherwise, one continues the procedure, checking in the general case whether the lesser k_{i+1} of k_i and $l_i - k_i$ divides the greater l_{i+1}. Since (i) $l_{i+1} \prec l_i$ and there are no infinite descents, (ii) if m/k and $m/(l-k)$, m/l (VIIf(iii)), and (iii) measurement is transitive (VIIf(i)), it is clear that this procedure will eventually find a common measure, possibly 1. Euclid proves in VII,1 that if it is 1, PRIME (k_1, l_1). His proof of this proposition and the rest of his proof of VII,2 may be described by pointing out that, since any measure of two numbers measures their difference (VIIf(iv)), and measurement is transitive, any measure of k_1 and l_1 will measure the measure found by the procedure just described. Hence the latter must be $\mathrm{GCD}(k_1, l_1)$; and if it is 1, k_1 and l_1 must have no common measure but 1.

At the end of the proof of VII,2 Euclid adds

VII,2, corollary $\quad m/k \ \& \ m/l \rightarrow m/\mathrm{GCD}(k, l)$, which he uses in VII,3 in showing how to find the greatest common measure of three numbers k_1, k_2, k_3. He lets GCD $(k_1, k_2) \simeq l$ and GCD $(l, k_3) \simeq m$. Clearly (VIIf(i)) m measures each of k_1, k_2, k_3. But if n is a greater measure of them, by the corollary it would have to measure l and also m, contradicting VIIf(ii).

Euclid's proofs of VII,1–3 involve a proliferation of cases for two reasons: first, he does not allow the unit to be the greatest common measure of relatively prime numbers; second, he distinguishes between the situation where the lesser given number divides the greater so that no subtractions are performed and the situation where there is at least one subtraction. If Euclid's distinctions are not made, the proof of VII,3 is easily turned into an inductive derivation of

$$\text{GCD}(l_1, \ldots, l_{n+1}) \simeq \text{GCD}(\text{GCD}(l_1, \ldots, l_n), l_{n+1}) \;\&\; (k/l_1 \;\&\; \ldots \;\&\; k/l_{n+1} \rightarrow k/\text{GCD}(l_1, \ldots, l_{n+1})).$$ [18]

Some such generalization is needed for VII,33, the proof of which is completed by showing

$$\text{GCD}(k_1, \ldots, k_n) = k \rightarrow k_1, \ldots, k_n\text{-LEAST}\left(\frac{k_1}{k}, \ldots, \frac{k_n}{k}\right).$$

For Euclid $\frac{k_i}{k}$ is the number l_i according to which k measures k_i. He asserts, presumably on the basis of the definition of proportionality and alternation, that $(l_1, \ldots, l_n) = (k_1, \ldots, k_n)$, and supposes that some other numbers $j_1, \ldots, j_n$ are least in the ratio of $k_1, \ldots, k_n$. Then by VII,20 there is an m such that $\mathcal{M}(j_i, k_i, m)$ or $\mathcal{M}(m, k_i, j_i)$, and $j_i \cdot m \simeq k_i \simeq l_i \cdot k$. Therefore, by VII,19, $(j_i, l_i) = (k, m)$ and $m \succ k$, which is impossible, since $k \simeq \text{GCD}(k_1, \ldots, k_n)$.

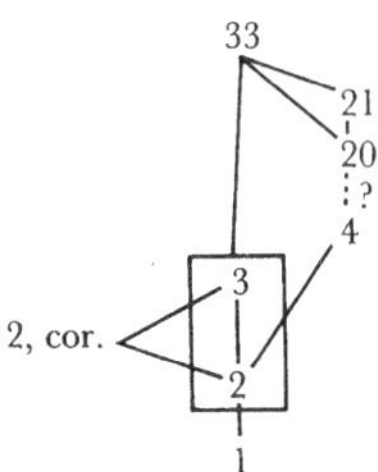

Figure 2.5

The structure of the proof of 33 is indicated in fig. 2.5, where dependencies on VII,5–19 are left out of account. Since 1–4 are not used elsewhere in the *Elements*, their position is very hard to explain, unless one accepts Zeuthen's hypothesis about the true significance of 4.

Euclid uses 33 to carry out three further constructions: the finding of the least number measured by k_1 and k_2 (LCM (k_1, k_2)) (VII,34); the finding of LCM (k_1, k_2, k_3) (VII,36); and the finding of the least number having k_1th, ..., k_nth parts (lcm $(k_1, \ldots, k_n)$) (VII,39). Euclid's proof of 34 is essentially a demonstration that

$$\text{If } k_1, k_2\text{-LEAST}(l_1, l_2), \text{LCM}(k_1, k_2) \simeq k_1 \cdot l_2.$$

The demonstration is complicated because Euclid distinguishes two cases according to whether or not PRIME (k_1, k_2). Combining them, his argument may be represented as follows. If k_1, k_2-LEAST (l_1, l_2), then, by VII,19, $k_1 \cdot l_2 \simeq k_2 \cdot l_1$, and clearly $k_1 \cdot l_2$ is a multiple of k_1 and k_2. If m is a lesser common multiple with $m \simeq i_1 \cdot k_1 \simeq i_2 \cdot k_2$, then, by VII,19, $(k_1, k_2) = (i_2, i_1) = (l_1, l_2)$ and, by VII,20, l_2/i_1. But also $(l_2, i_1) = (k_1 \cdot l_2, k_1 \cdot i_1)$, or, since $k_1 \cdot i_1 \simeq m$, $(l_2, i_1) = (k_1 \cdot l_2, m)$ and $k_1 \cdot l_2/m$; which is impossible, since $m \prec k_1 \cdot l_2$.

In VII,35 Euclid establishes a lemma for 36, which he also uses in VIII,4. The lemma asserts that LCM (k_1, k_2) divides any greater multiple m of k_1 and k_2, and is justified on the grounds that, if it does not, there are i and j such that $m \simeq i \cdot \text{LCM}(k_1, k_2) + j$ & $j \prec \text{LCM}(k_1, k_2)$; but then k_1 and k_2 measure LCM (k_1, k_2), and they also measure m; therefore, by VIIf(iv), they measure j, which is impossible, since $j \prec \text{LCM}(k_1, k_2)$. This proof obviously depends upon nothing but the meaning of 'least common multiple' and assumptions about measurement. 35 itself is an analogue of the corollary to VII,2,[19] as is shown by its use in the proof of 36, which is easily generalized to a proof of

$$\text{LCM}(k_1, \ldots, k_{n+1}) \simeq \text{LCM}(\text{LCM}(k_1, \ldots, k_n), k_{n+1}) \ \& \ (k_1/m \ \& \ \ldots \ \& \ k_{n+1}/m \rightarrow \text{LCM}(k_1, \ldots, k_{n+1})/m).$$

If one sets LCM $(k_1) \simeq k_1$, VII,34 and 35 establish this result for $n = 1$. Suppose it holds for $n \geq 1$ and that LCM $(k_1, \ldots, k_n) \simeq k$ and LCM $(k, k_{n+1}) \simeq l$. Clearly each of $k_1, \ldots, k_{n+1}$ divides l. But if they divide another number m, k/m by inductive hypothesis and l/m by VII,35, so that $l \prec m$, i.e.,

$$l \simeq \text{LCM}(k_1, \ldots, k_{n+1})$$

and l measures any other multiple of $k_1, \ldots, k_{n+1}$.

Euclid uses this generalized form of VII,34–36 in proving 39, which is in fact a terminological variant of them asserting that lcm $(k_1, \ldots, k_n) \simeq \text{LCM}(k_1, \ldots, k_n)$. To prove 39 Euclid establishes the equivalence of 'l has kth parts' and 'k measures l' in 37 and 38, which have already been discussed on p. 74. Hence he can argue that LCM $(k_1, \ldots, k_n)$ has $k_1, \ldots, k_n$th parts and that no number less than LCM $(k_1, \ldots, k_n)$ can have such parts. Euclid's reasons for establishing this terminological variant are unclear, since he never uses it.[20] Figure 2.6 shows the deductive structure of 33–39, giving all the uses of 1–3 and 33–39. A line below a number indicates that the corresponding proposition depends on nothing but definitions and tacit assumptions. In the diagram, applications of VII,5–19 and the questionable use of 4 in 20 are left out of account.

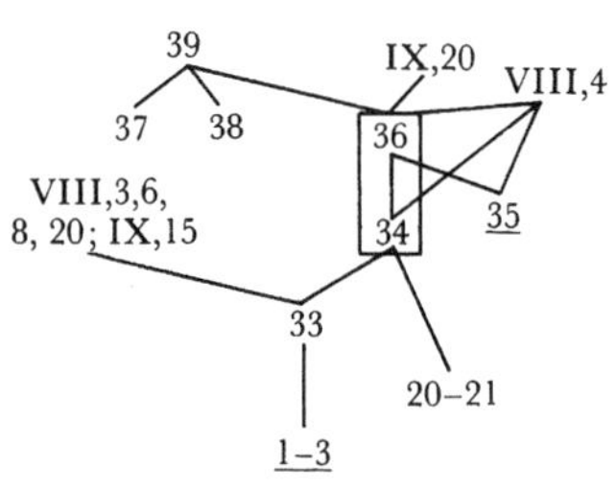

Figure 2.6

Of the remaining propositions in book VII, namely 23–32, I have discussed only 31, "Any composite number is measured by some prime number," of which 32, "Any number is prime or is measured by some prime number," is a trivial reformulation, as Euclid's reduction of the latter to the former shows. Euclid cites 31 explicitly in proving IX,13: "E is not prime. Therefore, it is composite. But any composite number is measured by some prime number; therefore E is measured by some prime number." However, in IX,20, the only other application of 31 or 32, Euclid apparently uses 32, since he moves directly from "E is not prime" to "E is measured by some prime number." This kind of trivial stylistic variation would seem to have no mathematical significance.

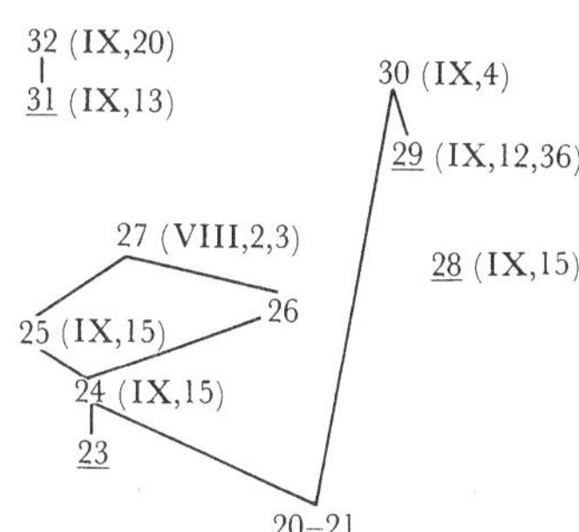

Figure 2.7

The deductive structure of 23–32 is shown in fig. 2.7; the only applications of those propositions in the *Elements* are also indicated. Since most of these propositions are used once or twice in books VIII and IX, it seems likely that they are elementary lemmas inserted in book VII with an eye to their later use. I shall briefly describe 23–30 and their proofs.

VII,23 $\text{PRIME}(k, l) \ \& \ j/k \rightarrow \text{PRIME}(j, l)$

is true because, by VIIf(i), any measure of j and l would also measure k. Euclid also proves

VII,24 $\text{PRIME}(k, l) \ \& \ \text{PRIME}(m, l) \rightarrow \text{PRIME}(k \cdot m, l)$

indirectly on the grounds that if j is a common measure of $k \cdot m$ and l with $i \cdot j \simeq k \cdot m$, $(j, k) = (m, i)$; but, by 23, $\text{PRIME}(j, k)$ so that, by VII,21 and 20, j/m, contradicting $\text{PRIME}(m, l)$. Two obvious consequences of 24 are

VII,25 $\text{PRIME}(k, l) \rightarrow \text{PRIME}(k^2, l)$,

VII,26 $\text{PRIME}(k, l) \ \& \ \text{PRIME}(k, m) \ \& \ \text{PRIME}(j, l) \ \& \ \text{PRIME}(j, m) \rightarrow \text{PRIME}(k \cdot j, l \cdot m)$,

from which it follows that

VII,27 $\text{PRIME}(k, l) \rightarrow \text{PRIME}(k^n, l^n)$.

For $n = 2$ applications of 25 give

$\text{PRIME}(k^2, l)$, $\text{PRIME}(k^2, l^2)$, and also $\text{PRIME}(k, l^2)$.

And if for $n \geq 2$

$\text{PRIME}(k, l^n)$, $\text{PRIME}(k^n, l)$, $\text{PRIME}(k^n, l^n)$,

then application of VII,26 yields $\text{PRIME}(k^{n+1}, l^{n+1})$, and application of VII,24 yields $\text{PRIME}(k, l^{n+1})$ and $\text{PRIME}(k^{n+1}, l)$.

In all manuscripts Euclid states 27 for $n = 2$ and $n = 3$, adding "This always happens with the extremes," a remark which is bracketed by Heiberg. Euclid proves the two cases explicitly mentioned, but in his proof says nothing about extremes. Nor is it immediately clear what he means by "extremes." However, in the only possible applications of 27 Euclid considers proportionalities of the form

$$(k^n, k^{n-1} \cdot l) = (k^{n-1} \cdot l, k^{n-2} \cdot l^2) = \therefore \ldots = (k \cdot l^{n-1}, l^n)$$

with PRIME (k, l). He infers PRIME (k^n, l^n) from PRIME (k, l), presumably using VII,27. Since the first and last terms of any proportion are called extremes, it seems all but certain that the reference to extremes in VII,27 anticipates this application. Moreover, even if Heiberg is right in considering the reference an interpolation, there is no reason to interpret 27 any differently from the other cases in which Euclid gives a quasi-inductive proof by doing the case $n = 2$ and reducing the case $n = 3$ to it. I should perhaps add that for Euclid the induction is not on an exponent n, but on the number of terms in the product $(k \cdot (k \cdot \ldots \cdot (k \cdot (k \cdot k)) \ldots))$.

Euclid reduces

VII,28 PRIME $(k, l) \rightarrow$ PRIME $(k + l, k)$ & PRIME $(k + l, l)$,
PRIME $(k + l, k) \vee$ PRIME $(k + l, l) \rightarrow$ PRIME (k, l)

to VIIf(iv) and (iii) respectively.

VII,29 PRIME (k) & $\neg\, k/l \rightarrow$ PRIME (k, l)

follows from the fact that a measure of k and l distinct from k is a proper measure of k. Euclid uses 29 in the proof of

VII,30 PRIME (k) & $k/l \cdot m \rightarrow k/l \vee k/m$.

He assumes the antecedent and $\neg\, k/l$, and infers by 29 that PRIME (k, l). Hence, LEAST (k, l), and, if $j \cdot k \simeq l \cdot m$, $(k, l) = (m, j)$ and k/m.[21]

30 and 31 provide all the necessary materials for a proof of the so-called fundamental theorem of arithmetic:

Every composite number is a product of primes in one way only.

That every composite number is a product of primes is clear from VII,31. For suppose k is the least composite number which is not a product of primes. By VII,31 there is a prime l and number m such that $k \simeq l \cdot m$. But $m \prec k$, so that m is either a prime or a product of primes; in either case k is a product of primes. To show that the same primes occur in every factorization of a composite number, it suffices to establish the following generalization of VII,30:

$$\text{PRIME}\,(k)\ \&\ k/\prod^{n+1} l_i \rightarrow \exists i(1 \leq i \leq n + 1\ \&\ k/l_i).$$

VII,30 establishes this result for the case $n = 1$. Suppose it is true for $n \geq 1$ and $k/\Pi^{n+2}l_i$, i.e., $k/(l_{n+2}) \cdot (\Pi^{n+1}l_i)$. By VII,30

k/l_{n+2} or $k/\Pi^{n+1}l_i$. In the latter case k measures one of $l_1, \ldots, l_{n+1}$. Thus any prime dividing a product of primes divides—and hence is identical with—one of the primes, i.e., the same primes occur in every factorization of any composite number. In order to show that the same primes occur equally often it suffices to point out that if

$$k \cdot \prod^{n} l_i (l_i \simeq l) \simeq k' \cdot \prod^{n'} l'_i (l'_i \simeq l)$$

with l a prime and $n' < n$, then l measures k'. For under those circumstances

$$k' \simeq k \cdot \prod^{n-n'} l_i (l_i \simeq l),$$

and, since $l/\Pi^{n-n'} l_i (l_i \simeq l)$, l/k'.

Because of the central role of prime factorization in modern arithmetic, it is tempting to read VII,30 and 31 as the Greek embodiment of the fundamental theorem of arithmetic.[22] For although the notion of a product of arbitrarily many factors does not occur in the *Elements*, the ideas in the proof of the theorem would certainly seem to fall within Euclid's range. On the other hand, he certainly could have come closer to expressing the theorem by proving, for example, that every composite number is measured by a plurality of primes or that a prime which divides a product of three numbers divides one of the numbers. In addition, the fundamental theorem would have simplified the proofs of other propositions in the *Elements* such as IX,36. The sensible way to describe this situation would seem to be to say that although the *Elements* contains the materials for proving the fundamental theorem, it contains neither the theorem nor the equivalent of it.

In any case, the discussion in this chapter leads to the conclusion that book VII has the overall structure indicated in fig. 2.8. Propositions 1–10 are the foundational material which is used only in book VII itself; 11–22 are fundamental results applied throughout the arithmetic books. The applications of the remaining propositions are more limited. They are divided into two groups on the basis of topic: 23–32 being concerned with primes and relative primes, 33–39 with expressing ratios in least terms and finding least common multiples. One sees from fig. 2.8 that the structure of VII is not as linear as that of book I.

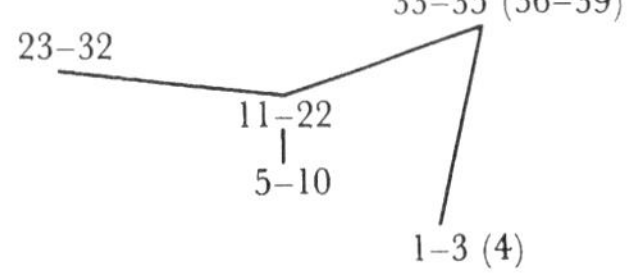

Figure 2.8

2.2 Books VIII and IX

The logical structure of books VIII and IX is puzzling, to say the least. Certain groups of propositions cohere, but the internal and external arrangement of the groups is not always perspicuous, and the details of argumentation sometimes leave much to be desired. Perhaps the central notion of the two books is continuous proportionality, which Euclid never defines. In VIII,4 Euclid uses the term 'continuously proportional' (*hexēs analogon*) in such a way as to suggest the definition

$$k_1, \ldots, k_n, l_1, \ldots, l_n\text{-CPROP}(m_1, \ldots, m_{n+1}) \leftrightarrow (m_i, m_{i+1}) = (k_i, l_i),$$

but elsewhere he only uses the expression when $(k_i, l_i) = (k_j\ l_j)$, so that one could define

$$k, l\text{-CPROP}(m_1, \ldots, m_{n+1}) \leftrightarrow (m_i, m_{i+1}) = (k, l),$$

and

$$\text{CPROP}(m_1, \ldots, m_{n+1}) \leftrightarrow m_1, m_2\text{-CPROP}(m_1, \ldots, m_{n+1}).$$

Heiberg and Heath suggest that the term 'continuously proportional' is used improperly in VIII,4. However, since this usage is just a general form of the standard one, I shall retain all three defined expressions.

It is convenient to begin by formulating what from the modern point of view are the essential facts about continuous proportionality. Suppose one wishes to find $m_1, \ldots, m_{n+1}$ such that $k_1, \ldots, k_n, l_1, \ldots, l_n$-CPROP $(m_1, \ldots, m_{n+1})$. The simplest way is to take $m_i \simeq k_i \cdot \ldots \cdot k_n \cdot l_1 \cdot \ldots \cdot l_{i-1}$. For it is easy to prove, using VII,16–18 and the associativity of multiplication, that

$$(k_i, l_i) = (k_i \cdot \ldots \cdot k_n \cdot l_1 \cdot \ldots \cdot l_{i-1}, k_{i+1} \cdot \ldots \cdot k_n \cdot l_1 \ldots \cdot l_i),$$

so that

VIIIA(i) If

$$m_i \simeq k_i \cdot \ldots \cdot k_n \cdot l_1 \cdot \ldots \cdot l_{i-1}$$

for $1 \leq i \leq n + 1$, then

$$k_1, \ldots, k_n, l_1, \ldots, l_n\text{-CPROP}(m_1, \ldots, m_{n+1}).$$

If one thinks of exponentiation as a special case of multiplication, one also has:

VIIIA(ii) If $m_i \simeq k^{(n+1)-i} \cdot l^{i-1}$ for $1 \leq i \leq n + 1$, then

$$k, l\text{-CPROP}(m_1, \ldots, m_{n+1}),$$

or

VIIIA(iii) $\text{CPROP}(k^n, k^{n-1} \cdot l, \ldots, k \cdot l^{n-1}, l^n)$.

The special case of this proposition when k is 1 is

VIIIA(iv) $\text{CPROP}(1, l, l^2, \ldots, l^n)$.

Clearly one can also prove

VIIIA(v) $\text{CPROP}(1, m_1, \ldots, m_n) \rightarrow m_i \simeq m_1{}^i$.

For if $\text{CPROP}(1, m_1, \ldots, m_n)$, since $m_1 \simeq m_1 \cdot 1$, $m_{i+1} \simeq m_1 \cdot m_i$, and the result follows by induction on n.

In VIII,2 Euclid solves the problem of finding the least $m_1, \ldots, m_{n+1}$ such that k', l'-$\text{CPROP}(m_1, \ldots, m_{n+1})$. The content of his solution is contained in

VIIIB $\text{CPROP}(m_1, \ldots, m_{n+1})$ & $\text{LEAST}(m_1, \ldots, m_{n+1}) \leftrightarrow$
$\exists k \exists l(\text{LEAST}(k, l)$ & $m_i \simeq k^{(n+1)-i} \cdot l^{i-1})$,

which, together with VII,20, implies

VIIIC $\text{CPROP}(m_1, \ldots, m_{n+1}) \leftrightarrow$
$\exists j \exists k \exists l(m_i \simeq j \cdot k^{(n+1)-i} \cdot l^{i-1}$ & $\text{LEAST}(k, l))$.

Using VIIIB one solves the problem of VIII,2 by applying VII,33 to find k, l such that k', l'-$\text{LEAST}(k, l)$ and constructing $m_i \simeq k^{(n+1)-i} \cdot l^{i-1}$. It is simplest to prove VIIIB in two parts.

(a) Suppose $\text{LEAST}(k, l)$ and $m_i \simeq k^{(n+1)-i} \cdot l^{i-1}$. $\text{CPROP}(m_1, \ldots, m_{n+1})$ by VIIIA(ii) or (iii). Suppose $(m_1, \ldots, m_{n+1}) = (m_1', \ldots, m_{n+1}')$. Then $(m_1', m_{n+1}') = (m_1, m_{n+1}) = (k^n, l^n)$. But, since $\text{LEAST}(k, l)$, $\text{PRIME}(k, l)$ by VII,22, $\text{PRIME}(k^n, l^n)$ by VII,27 in its general form, and $\text{LEAST}(k^n, l^n)$, i.e., $\text{LEAST}(m_1, m_{n+1})$; hence $m_1 \lesssim m_1'$.

(b) Suppose now $\text{CPROP}(m_1, \ldots, m_{n+1})$ & $\text{LEAST}(m_1, \ldots, m_{n+1})$. Take k, l least in the ratio of m_1, m_2, and construct $m_i' \simeq k^{(n+1)-i} \cdot l^{i-1}$. Then $(m_1, \ldots, m_{n+1}) = (m_1', \ldots, m_{n+1}')$ and, by part (a), $\text{LEAST}(m_1', \ldots, m_{n+1}')$; hence $m_i' \simeq m_i$.

The argument just given is basically Euclidean, but the formulation is not. Euclid proves

VIII,1,3 $\text{CPROP}(m_1, \ldots, m_{n+1}) \rightarrow$
$(\text{LEAST}(m_1, \ldots, m_{n+1}) \leftrightarrow \text{PRIME}(m_1, m_{n+1}))$.

The proof of VIII,1 is a simplified version of the proof (a), in which Euclid argues that if $\text{CPROP}(m_1, \ldots, m_{n+1})$ and $(m_1, \ldots, m_{n+1}) = (m_1', \ldots, m_{n+1}')$, then, since $(m_1, m_{n+1}) = (m_1', m_{n+1}')$, if $\text{PRIME}(m_1, m_{n+1})$, $m_1 \lesssim m_1'$ so that $\text{LEAST}(m_1, \ldots, m_{n+1})$.[23] Euclid then solves VIII,2, in the manner suggested, for $n = 3$, first constructing k^2, $k \cdot l$, l^2 and then $k \cdot k^2$, $k \cdot k \cdot l$, $l \cdot k \cdot l$, $l \cdot l^2$. He carries out the argument for VIIIA(i) to show first that the three terms and then that the four terms are continuously proportional, adding as a corollary that when three (four) terms are in continuous proportion the extremes are squares (cubes). For VIII,3 Euclid proceeds as in (b) of the foregoing paragraph for the case $n = 3$, but for him this procedure means sketching in outline the proof of VIII,2. It seems fair to say that the *Elements* contain the information of VIIIA(iii) and VIIIB, but that the *protaseis* of the relevant propositions, VIII,1–3, do not by themselves convey the information. Nevertheless, Euclid invokes the information when he needs it. For example, in IX,15, given that CPROP (m_1, m_2, m_3) & LEAST (m_1, m_2, m_3), he takes m_1, m_2-LEAST (k, l) and asserts that $m_1 \simeq k^2$, $m_2 \simeq k \cdot l$ and $m_3 \simeq l^2$. This kind of argument is not strictly speaking a deduction of one proposition from another; but the legitimacy of the procedure is clear enough, and also understandable in the absence of a notation for expressing VIIIB concisely.

It is perhaps worthwhile to mention briefly an alternative way of solving VIII,2. Euclid might have taken k'^n, $k'^{n-1} \cdot l'$, $\ldots$, $k' \cdot l'^{n-1}$, l'^n, and argued that they are continuously proportional in the ratio of k', l', and then taken the least $n + 1$ terms in the same ratio. This procedure would not yield the information embodied in Euclid's proof nor even what is expressed in the corollary without further argumentation, argumentation which in modern presentations would involve the fundamental theorem of arithmetic. A somewhat analogous situation arises in connection with VIII,4, the problem of finding the least $m_1, \ldots, m_{n+1}$ such that $k_1, \ldots, k_n, l_1, \ldots, l_n$-CPROP $(m_1, \ldots, m_{n+1})$, when LEAST (k_i, l_i). In the arithmetic books this construction is employed only in the unused VIII,5, where Euclid does not worry about LEAST (k_i, l_i). Otherwise, its only possible application in the *Elements* is in X,12, where Euclid treats the case $n = 3$ without worrying about either LEAST (k_i, l_i) or LEAST (m_1, m_2, m_3). Clearly knowledge of VIIIA(i) would suffice for this step in X,12. Indeed with such knowledge Euclid could prove VIII,4 by setting $m'_i \simeq k_i \cdot \ldots \cdot k_n \cdot l_1 \cdot \ldots \cdot l_{i-1}$, and then using VII,33 to find the least $m_1, \ldots, m_{n+1}$ in the ratio of $m'_1, \ldots, m'_{n+1}$. Instead, he does the case $n = 3$, first constructing $m'_2 \simeq$ LCM (k_2, l_1) with $m'_2 \simeq j'_1 \cdot l_1 \simeq j'_2 \cdot k_2$, and setting $m'_1 \simeq j'_1 \cdot k_1$ and $m'_3 \simeq j'_2 \cdot l_2$, without indicating that m'_1, m'_2, m'_3 solve the problem for k_1, k_2, l_1, l_2. Thereafter he distinguishes the case in which m'_3 is LCM (m'_3, k_3) and the case in which it is not. The cases are combined simply by setting $m_3 \simeq$ LCM (m'_3, k_3) with $m_3 \simeq j_1 \cdot m'_3 \simeq j_2 \cdot k_3$, and $m_1 \simeq j_1 \cdot m'_1$, $m_2 \simeq j_1 \cdot m'_2$, and $m_4 \simeq j_2 \cdot l_3$. I shall not here describe the proof that these m_1, m_2, m_3, m_4 solve the problem of VIII,4, nor shall I describe the general procedure to be extracted from Euclid's particular case. For it should be clear that Euclid's construction and proof are much more complicated than what the use of VIIIA(i) and VII,33 in the manner just described would require. Moreover, I can see no sense in which Euclid's presentation conveys useful information obscured by the shorter proof. Indeed, from our point of view, the shorter proof would seem more informative, since VIIIA(i) makes clear that if

$k_1, \ldots, k_n, l_1, \ldots, l_n\text{-CPROP}(m_1, \ldots, m_{n+1})$,

then

$(m_1, m_{n+1}) = (k_1 \cdot \ldots \cdot k_n, l_1 \cdot \ldots \cdot l_n)$.

Hence the shorter construction for VIII,4 could be interpreted as the standard method, apparently also Greek,[24] for multiplying fractions by reducing the ratio between the products of numerators and denominators to lowest terms. Euclid's proof shows no sense of a connection between VIII,4 and the multiplication of fractions or multiplication of any sort. Hence, it also shows no sense of the fact that VIII,2 is the special case of VIII,4 when $(k_i, l_i) = (k_j, l_j)$. Indeed, the only evidence that Euclid recognizes the fact is his use of the same term, 'continuously proportional', in the two propositions; but, if he does recognize it, he fails to exploit it.

The concept of multiplication is obviously relevant to

VIII,5 Plane numbers have to one another the ratio compounded [of the ratios] of their sides.

The notion of plane numbers and their sides shows clearly the influence of geometric ways of thinking on Euclid's arithmetic. A plane number is simply a product $k \cdot l$ of which k and l are the sides (VII, def. 17). Similarly, a solid number $k \cdot l \cdot m$ has k, l, and m as its sides (def. 18). It is from this geometric conception of numbers that the locutions 'square' (def. 19) and 'cube' (def. 20) make their way into arithmetic. Euclid himself does not define compounding of ratios; but some manuscripts contain at the beginning of book VI a definition which, although garbled, shows a clear sense of the connection between compounding and multiplication:

A ratio is said to be compounded of ratios when the sizes (*pelikotētes*) of the ratios multiplied together make something.

The scholiasts and Eutocius (82–86) discuss this definition in terms of numerical ratios. They understand the size of a ratio to be the number (in a non-Euclidean sense of number according to which the unit may be divided) corresponding to a ratio as $2\frac{1}{4}$ corresponds to (9, 4). The composition of (4, 3) and (3, 2) is obtained by multiplying $(1 + \frac{1}{3})$ and $(1 + \frac{1}{2})$ to get $1 + \frac{1}{2} + \frac{1}{3} + \frac{1}{6}$, or 2, the number corresponding to the double ratio. In the *Elements* there is no trace of this definition. The notion of compounding which Euclid uses may be defined by

$$(m_1, m_2) \text{ is the compound of } (k_1, k_2) \text{ and } (l_1, l_2) \leftrightarrow \exists m((m_1, m) = (k_1, k_2) \;\&\; (m, m_2) = (l_1, l_2)).$$

Clearly, if (m_1, m_2) is the compound of (k_1, k_2) and (l_1, l_2), then

$\frac{m_1}{m_2}$ is an expression for the product of $\frac{k_1}{k_2}$ and $\frac{l_1}{l_2}$. Hence it is possible to say that compounding ratios is an analogue of multiplying fractions. The serious question for interpretation is whether compounding should be viewed as a representation of multiplying, i.e., as a device for representing the multiplication of fractions in the language of proportionality. I shall be arguing that compounding should not be viewed in this way, but shall postpone this argument until later chapters covering propositions in which Euclid employs the notion of compounding.

VIII,5 itself has an easy proof. For given plane numbers $k \cdot l$ and $j \cdot m$, it is clear by VII,16 and 17 that $l \cdot j$ satisfies the condition that $(k \cdot l, l \cdot j) = (k, j)$ and $(l \cdot j, j \cdot m) = (l, m)$. Euclid, however, does not proceed in exactly this way. Rather he uses VIII,4 to find the least numbers m_1, m_2, m_3 satisfying $(k, j) = (m_1, m_2)$ and $(l, m) = (m_2, m_3)$, and then argues by VII,16, 17, and 14 that $(k \cdot l, j \cdot m) = (m_1, m_3)$. There does not appear to be any advantage to Euclid's procedure. For example, m_1 and m_3 are not necessarily the least numbers in the ratio of $k \cdot l$ and $j \cdot m$; thus, for the numbers $2 \cdot 5$ and $3 \cdot 6$, Euclid's procedure yields $(2 \cdot 5, 3 \cdot 6) = (10, 18)$. On the other hand, this procedure bears a very close relation to the proof of its geometrical analogue VI,23 according to which parallelograms $ABCD$, $A'B'C'D'$ with equal angles at B and B' are to one another in the ratio compounded of $(AB, A'B')$ and $(BC, B'C')$. (See below, pp. 154–155.) If one identifies AB with k, BC with l, $A'B'$ with j, and $B'C'$ with m, and also identifies the parallelograms with the corresponding products, Euclid can be said to prove VI,23 by finding straight lines m_1, m_2, m_3 with $(k, j) = (m_1, m_2)$ and $(l, m) = (m_2, m_3)$ and then arguing for exactly the same proportionalities as in VIII,5, namely, $(k \cdot l, l \cdot j) = (k, j)$, $(l \cdot j, j \cdot m) = (l, m)$, and $(k \cdot l, j \cdot m) = (m_1, m_3)$. In discussing VI,23 in more detail I shall argue that this proposition is best understood in a purely geometric way with no reference to fractions and multiplication. Since VIII,5 appears to be derivative from VI,23, it provides no evidence for a connection between compounding and multiplication. More generally, Euclid's failure to exploit VIIIA(i) except in the more particular form VIIIA(ii) or (iii) suggests that he does not grasp the notion of continuous proportionality in its most general form.

VIIIC yields easy proofs of

VIII,6 $\text{CPROP}(m_1, \ldots, m_{n+1})$ & $\neg\, m_1/m_2 \rightarrow \neg\, m_i/m_j$ for $1 \leq i < j \leq n + 1$,

and

VIII,8 $\text{CPROP}(m, m_1, \ldots, m_n, m') \ \& \ (m, m') = (j, j') \rightarrow$
$\exists j_1 \ldots \exists j_n \ \text{CPROP}(j, j_1, \ldots, j_n, j')$.

Euclid applies 8 to the case where $n = 1$ in VIII,24 and IX,1 and 2, and where $n = 2$ in IX,3–6. He uses VIII,6 for $n = 2$ in VIII,14 and $n = 3$ in VIII,15, but in both cases in the weaker form

VIII,7 $\text{CPROP}(m_1, \ldots, m_{n+1}) \ \& \ m_1/m_{n+1} \rightarrow m_1/m_2$.

Using VIIIC, one can argue for 6 by pointing out that if $\text{CPROP}(m_1, \ldots, m_{n+1})$ and $j \simeq i + j'$, then m_i and m_j are the same multiple of $k^{(n+1)-i} \cdot l^{(i-1)}$ and $k^{(n+1)-(i+j')} \cdot l^{(i+j')-1}$, respectively, so that $(m_i, m_j) = (k^{j'}, l^{j'})$; since $\neg\, m_1/m_2$, k cannot be a unit, and, since $\text{PRIME}(k^{j'}, l^{j'})$, $\neg\, m_i/m_j$. For VIII,8 one can argue that since m and m' are the same multiple of some k^{n+1} and l^{n+1}, respectively, where $\text{LEAST}(k^{n+1}, l^{n+1})$, j and j' must be equal multiples of k^{n+1} and l^{n+1} as well, and the result follows. Euclid's proofs are very like these,[25] but not as explicit about the form of the continuous proportionalities. He is able to avoid this explicitness by citing VIII,3. The proof of 8 essentially contains the information of VIIIC, but at no place in the *Elements* does Euclid show a specific understanding of it.

In VIII,9 and 10 Euclid proves weaker forms of VIIIA (iii) and VIIIB:

VIII,9 $\text{CPROP}(m, m_1, \ldots, m_n, m') \ \& \ \text{LEAST}(m, m') \rightarrow$
$\exists k_1 \ldots \exists k_n \exists l_1 \ldots \exists l_n (\text{CPROP}(o, k_1, \ldots, k_n, m) \ \&$
$\text{CPROP}(o, l_1, \ldots, l_n, m'))$;

VIII,10 $\text{CPROP}(o, k_1, \ldots, k_n, k) \ \& \ \text{CPROP}(o, l_1, \ldots, l_n, l)$

$\rightarrow \exists m_1 \ldots \exists m_n \text{CPROP}(k, m_1, \ldots, m_n, l)$.

For us, VIII,10 is a consequence of VIIIA (v), according to which the antecedent of VIII,10 says that $k \simeq k_1^{n+1}$ and $l \simeq l_1^{n+1}$, and VIIIA (iii). Euclid does the case $n = 2$, repeating much of the argument of VIII,2 to show that $k_2 \simeq k_1^2$, $k \simeq k_1^3$, $l_2 \simeq l_1^2$, $l \simeq l_1^3$, and that $\text{CPROP}(k_1^3, k_1^2 \cdot l_1, k_1 \cdot l_1^2, l_1^3)$. For VIII,9 Euclid infers from the antecedent that $\text{LEAST}(m, m_1, \ldots, m_n, m')$. The desired result now follows from VIIIB and VIIIA (iv), but Euclid again carries out much of the argument of VIII,2. In the case of these two propositions, the repetition of the argument of VIII,2 is not due to a failure to state VIIIB explicitly so much as to a failure to treat 1 as a number and multiplication by 1 as a special case of multiplication. The proofs of VIII,9 and 10 can be said to show a knowledge of VIIIA (iv) and (v); thus the propositions themselves can be said to contain the information that

$$\text{VIIIB}' \quad \text{CPROP}(m, m_1, \ldots, m_{n-1}, m') \;\&\; \text{LEAST}(m, m_1, \ldots, m_{n-1}, m') \rightarrow \exists k \exists l (m \simeq k^n \;\&\; m' \simeq l^n);$$

$$\text{VIIIA (iii)}' \quad \exists m_1 \ldots \exists m_{n-1}\, \text{CPROP}(k^n, m_1, \ldots, m_{n-1}, l^n).$$

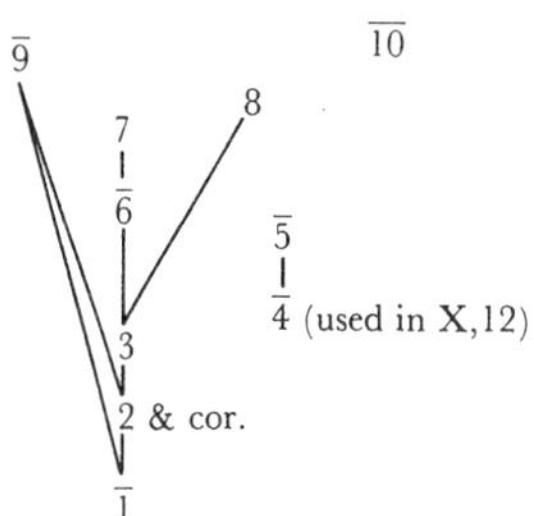

Figure 2.9

The deductive structure of the first ten propositions of book VIII is set out in fig. 2.9, where a line above a proposition indicates that all its uses in the arithmetic books are shown in the figure. Since 5, 9, and 10 are never used, and 4 is not employed until book X, their position here is not really explicable. Diagrammatically the remaining propositions in the sequence have a reasonable arrangement, although, as we have seen, there is a much simpler expression of their essential content in modern notation.

On the other hand, the importance attached to these propositions by the Greeks is not entirely clear. Van der Waerden (pp. 111–112) points to their close connection with Pythagorean harmonics. Zeuthen centers a theory of the history of Greek arithmetic around the fact that the content of VIII,1–10 in its modern formulation can be seen to provide the answer to the question whether the nth root of $\frac{k}{l}$ is rational.[26] For, if $\frac{k}{l}$ is in least terms,

$$\sqrt[n]{\frac{k}{l}} = \frac{\sqrt[n]{k}}{\sqrt[n]{l}} = \frac{j}{m}$$

if and only if

$$\frac{j^n}{m^n} = \frac{k}{l}$$

if and only if $\exists i_1 \ldots \exists i_{n-1}\, \text{CPROP}(k, i_1, \ldots, i_{n-1}, l)$ if and only if k and l are nth powers. Although Zeuthen's interpretation of VIII,1–10 is without doubt mathematically cogent, there is really no direct evidence to support it or the historical theory Zeuthen bases on it.

Indeed, no explanation of VIII,1–10 is likely to prove fully satisfactory, since explicit evidence outside the *Elements* is lacking, and the uses to which Euclid puts the propositions in the *Elements* are themselves problematic. It is these applications which will be my concern. For this purpose it is convenient to look at the major uses of VIII,8 in IX,1–6. These propositions can be interpreted as characterizations of the m for which $\exists l(k^2 \cdot m \simeq l^2)$ and $\exists l(k^3 \cdot m \simeq l^3)$. Since $k^n \cdot m \simeq l^n$ if and only if

$$m = \frac{l^n}{k^n}$$

if and only if

$$\sqrt[n]{m} = \frac{l}{k},$$

these propositions can also be interpreted as answers to the question of when an integer has a rational square or cube root. From a modern point of view these questions are special cases of the general problem which Zeuthen reads into VIII,1–10, but there is much more evidence that the Greeks were interested in these particular questions, especially the question whether one could have $k^2 \cdot 2 \simeq l^2$ or $k^3 \cdot 2 \simeq l^3$. For the first of these would provide a numerical expression for the ratio of the side of a square to its diagonal, the second a numerical expression for the ratio of the side of a cube to the side of a cube double it in volume.

Euclid proves

IX,4,5 $\exists l\,(k^3 \cdot m \simeq l^3) \leftrightarrow \text{CUBE}\,(m)$;

but instead of proving

IX,1′,2′ $\exists l\,(k^2 \cdot m \simeq l^2) \leftrightarrow \text{SQUARE}\,(m)$,

he establishes

IX,1,2 $\text{SQUARE}\,(k \cdot m) \leftrightarrow \text{SIMPLANE}\,(k, m)$,

where k and m are said to be similar plane numbers if they are products $k_1 \cdot k_2$ and $m_1 \cdot m_2$ with $(k_1, k_2) = (m_1, m_2)$ (VII, def. 22, in which similar solid numbers are defined analogously). IX,1,2 is false when 'CUBE' is substituted for 'SQUARE' and 'SIMSOLID' for 'SIMPLANE', since, e.g., $2 \cdot 2 \cdot 3$ and $4 \cdot 4 \cdot 6$ are similar solid numbers whose product is not a cube, and the product of 2 and 4 is a cube although they are not similar solid numbers. To get a "three-dimensional" analogue of IX,1,2, one could prove that the product of three numbers of which the first and second as well as the second and third are similar solid numbers is a cube; however, this result bears no significant relation to IX,4,5.

In a moment I will offer an explanation of how the idea of similar numbers and IX,1,2 might have found their way into the *Elements* out of the attempt to prove IX,1′,2′. Before doing so, I want to look at possible proofs of the latter and of IX,4,5. The general theorem that $\exists l\,(k^n \cdot m \simeq l^n)$ if and only if m is an nth power is easily inferred from the fundamental theorem of arithmetic, which implies that every prime in the prime factorization of an nth power occurs a multiple of n times. Euclid would also have a quick proof of the half of this general theorem which asserts that m is an nth power if $k^n \cdot m$ is. For if $k^n \cdot m \simeq l^n$, then $(o, m) = (k^n, l^n)$, and it is easy to argue by VIIIA(iii) and VIII,8 extended to units that n numbers fall between o and m in continued proportion so that, by VIIIA(v), m is an nth power.

The other half of the general theorem asserts that $k^n \cdot m^n$ is an nth power or, more simply, $k^n \cdot m^n \simeq (k \cdot m)^n$, an assertion which can be proved by a complicated combinatorial argument based on the definition of multiplication. Euclid could bypass these complexities in the particular cases that interest him. For $n = 2$ Euclid could argue that since, by VIIIA(iii), $(k^2, k \cdot m) = (k \cdot m, m^2)$, $k^2 \cdot m^2 \simeq (k \cdot m)^2$, i.e., $k^2 \cdot m^2$ is a square. Euclid instead proves as a separate proposition

VIII,11a $\exists l\,(k^2, l) = (l, m^2)$,

to which he adds

VIII,11b (k^2, m^2) is the duplicate of (k, m).

Euclid also proves in VIII,12 that

VIII,12 $\exists l_1 \exists l_2$ CPROP (k^3, l_1, l_2, m^3) &
(k^3, m^3) is the triplicate of (k, m).

'Duplicate' (*diplasiōn*) and 'triplicate' (*triplasiōn*) are defined for ratios involving arbitrary magnitudes in the ninth and tenth definitions of book V: (x, y) is the duplicate of (z, w) if and only if

$$\exists v\,((x, v) = (v, y) = (z, w)),$$

and the triplicate of it if and only if

$$\exists v_1 \exists v_2\,((x, v_1) = (v_1, v_2) = (v_2, y) = (z, w)).$$

Clearly then, the second halves of 11 and 12 are immediate consequences of VIIIA(iii) and VII,17. More importantly, it is clear that the duplicate of (k, m) is simply the compound of (k, m) and (k, m), and, assuming an obvious notion of compounding three ratios, the triplicate of (k, m) is the compound of (k, m), (k, m), and (k, m). Thus finding duplicates and triplicates of ratios corresponds to squaring and cubing fractions, and VIII,11 (12) is an immediate consequence of VIII,5 (a direct extension of VIII,5). Euclid proves VIII,11 and 12 from scratch, constructing the numbers $k \cdot m$, $k^2 \cdot m$, $k \cdot m^2$, and arguing, as before, that k-m-CPROP $(k^2, k \cdot m, m^2)$ and k-m-CPROP $(k^3, k^2 \cdot m, k \cdot m^2, m^3)$. His failure to exploit VIII,5 is a good indication that he does not construe compounding as a representation of multiplying, even though he knows that

$(k \cdot l, j \cdot m)$ is the compound of (k, j) and (l, m). Since there is no other candidate for the representation of the multiplication of fractions, it also seems unlikely that Euclid construes duplicating and triplicating as squaring and cubing, even though he knows that (k^2, m^2) and (k^3, m^3) are respectively the duplicate and triplicate of (k, m).

In any case, with VIII,11 Euclid has at his disposal an easy proof that the product of two squares is a square. On the other hand Euclid encounters more difficulty with the assertion that m is a square if $k^2 \cdot m$ is. Because

(i) $k^2 \cdot m \simeq l^2 \leftrightarrow (k^2, l) = (l, m)$,

one could use the least number principle and argue that there can be no least square l^2 which is the product of a square k^2 and a nonsquare m. For if l^2 is such a square, there are, by (i) and VIIIC, k_1 and m_1 such that LEAST(k_1, m_1) and, for some j, $k^2 \simeq j \cdot k_1^2$, $l \simeq j \cdot k_1 \cdot m_1$, and $m \simeq j \cdot m_1^2$. Since k^2 is less than l^2, j must be a square; therefore, m, being the product of two squares, must itself be a square.

Euclid does not proceed in this way; instead he uses (i) to reduce the assertion to be proved to

VIII,22 CPROP$(k^2, l, m) \rightarrow$ SQUARE(m),

which he purports to derive directly from

VIII,20 CPROP$(k, l, m) \rightarrow$ SIMPLANE(k, m).

Euclid argues for this proposition by taking k-l-LEAST(k', l') so that there are j and j' with $k \simeq j \cdot k'$, $l \simeq j \cdot l' \simeq j' \cdot k'$, and $m \simeq j' \cdot l'$, and $(j', j) = (m, l) = (l', k')$ and $(j', l') = (j, k')$, so that k and m are similar plane numbers.[27] This argument goes through with k^2 substituted for k, and establishes

CPROP$(k^2, l, m) \rightarrow$ SIMPLANE(k^2, m).

To complete the proof of VIII,22 one needs a principle which Euclid takes for granted, namely

VIIID SIMPLANE$(k^2, m) \rightarrow$ SQUARE(m).

This principle is true, but not obviously so. Euclid presumably takes it for granted because of a misleading aspect of the analogy between geometric figures and numbers. When a rectilineal figure is given, its sides are also given; but in general the sides of a plane or solid number are not given when it is. When 12 is given as a plane number, its sides could be 1 and 12, 2 and 6, or 3 and 4. Therefore, although it is possible to infer immediately from "*A* has equal sides, and *A* and *B* are similar figures" that *B* has equal sides, it is not possible to make an analogous immediate inference in the case of numbers. Two plane or solid numbers might be similar under certain factorizations into sides but not under others. More importantly, there might be a factorization into sides of one of two similar numbers with no similar factorization of the other. For example, 6 and 24 are similar plane numbers because $(2, 3) = (4, 6)$, but 6 does not have sides in the ratio of $(3, 8)$. Clearly, therefore, Euclid ought to prove VIIID rather than assume it. Similar remarks apply to Euclid's inference of VIII,23 from VIII,21, in which he takes for granted

VIIIE $\text{SIMSOLID}(k^3, m) \rightarrow \text{CUBE}(m)$.

The proofs of VIIID and E all involve the same kind of considerations raised on pp. 91–92 in connection with IX,1′,2′ and IX,4,5. I shall therefore sketch a general argument establishing

$n\text{SIM}(k^n, m) \rightarrow m$ is an nth power,

where n-similarity is defined by

$$n\text{SIM}(k, m) \leftrightarrow \exists k_1 \ldots \exists k_n \exists m_1 \ldots \exists m_n (k \simeq k_1 \cdot \ldots \cdot k_n \ \& \\ m \simeq m_1 \cdot \ldots \cdot m_n \ \& \ (k_1, \ldots, k_n) = (m_1, \ldots, m_n)).$$

Let $k^n \simeq \Pi^n k_i$, $m \simeq \Pi^n m_i$ with $(k_1, \ldots, k_n) = (m_1, \ldots, m_n)$. Find $k_1, \ldots, k_n$-LEAST$(l_1, \ldots, l_n)$, so that for some j and l, $k_i \simeq j \cdot l_i$ and $m_i \simeq l \cdot l_i$, and $k^n \simeq \Pi^n j \cdot l_i$, $m \simeq \Pi^n l \cdot l_i$. Combinatorial arguments give $k^n \simeq j^n \cdot \Pi^n l_i$, and $m \simeq l^n \cdot \Pi^n l_i$. Therefore, since k^n is an nth power, $\Pi^n l_i$ also is; and since l^n and $\Pi^n l_i$ are nth powers, so is their product m.

It does not seem unreasonable to suppose that the attempt to prove IX,2′ led to the concept of similar plane numbers and to VIII,20. But VIII,20 and VII,19 suffice to prove the stronger IX,1, while IX,2 is a consequence of the converse of VIII,20, namely,

VIII,18 $\text{SIMPLANE}(k, m) \rightarrow \exists l((k, l) = (l, m))$ & (k, m) is the duplicate of the ratio of the corresponding sides.

In 19 Euclid proves the analogous result for similar solid numbers. The word 'corresponding' (*homologos*) is defined for ratios generally in definition 11 of book V: When $(x, y) = (z, w)$, then x and z correspond and so do y and w. VIII,18 and 19 are simply more general versions of 11 and 12 which could be derived simply from VIII,5 and its extension to solid numbers, if Euclid saw the connection of duplicating and triplicating with compounding. Euclid, however, chooses to go back to the basics of book VII, taking, in the case of 18, k as $k_1 \cdot k_2$, m as $m_1 \cdot m_2$, and l as $m_1 \cdot k_2$, and demonstrating that k_1, m_1-CPROP$(k_1 \cdot k_2, m_1 \cdot k_2, m_1 \cdot m_2)$.

Thus Euclid could derive IX,1,2 from VIII,18 and 20, which depend upon nothing in book VIII prior to 18.[28] However, Euclid does not adopt this procedure, perhaps because it has no parallel in the case of cubes. For, there is no simple analogue for cubes to the assertion that CPROP (k, l, m) if and only if $k \cdot m \simeq l^2$. Euclid is able to give parallel proofs for IX,1,2 and IX,4,5 by establishing as a lemma for the latter

IX,3 CUBE $((k^3)^2)$.

He carries out constructions establishing CPROP (o, k, k^2, k^3) and $(o, k^3) = (k^3, (k^3)^2)$ and applies VIII,8 (with o treated as a number) and 23. The proofs of IX,1,2 and 4,5 can be represented in parallel columns. Since

$(k^2, k \cdot m) = (k, m)$,	$((k^3)^2, k^3 \cdot m) = (k^3, m)$,
therefore	therefore
(i) SIMPLANE (k, m)	CUBE (m)
if and only if (VIII,18,20)	if and only if (VIII,19,23)
(ii) $\exists l$CPROP (k, l, m)	$\exists l_1 \exists l_2$CPROP (k^3, l_1, l_2, m)
if and only if (VIII,8)	if and only if (VIII,8)
(iii) $\exists l'$CPROP $(k^2, l', k \cdot m)$	$\exists l'_1 \exists l'_2$CPROP $((k^3)^2, l'_1, l'_2, k^3 \cdot m)$
if and only if (VIII,18,22)	if and only if (VIII,19,23)
(iv) SQUARE $(k \cdot m)$.	CUBE $(k^3 \cdot m)$.

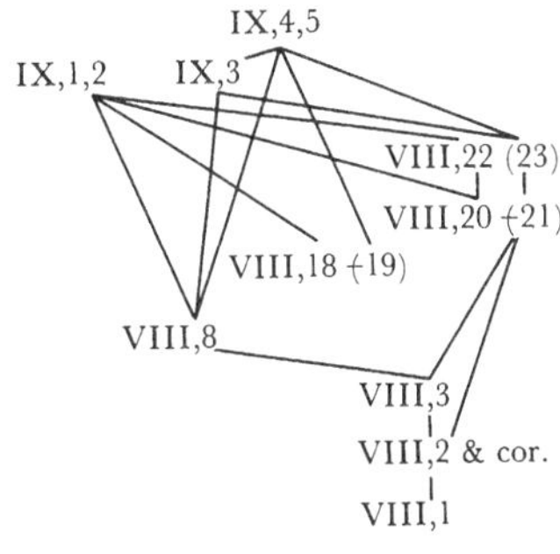

Figure 2.10

Figure 2.10 shows the deductive structure of the proofs of IX,1,2 and 4,5. In this diagram "solid" theorems are cited in parentheses after their "plane" analogues; a line joining them indicates that the latter is used in the proof of the former. One sees that, except for 7, all the propositions from VIII,1–10 which find any use in the arithmetic books beyond 1–10 are represented. These propositions are, to be sure, more general than their applications, but such generality is a common feature of the *Elements*.

Propositions VIII,24–27 appear merely to involve filling out details in a theory already developed, but Euclid's choice of details is somewhat perplexing. VIII,25 ($(k^3, l) = (j^3, m^3)$ $\rightarrow$ CUBE (l)) makes it possible to move directly from step (i) to step (iv) in the proof of IX,4; and, in fact, Euclid's proof of 25 simply reproduces those steps. Similarly, once it is inferred that two square numbers are similar plane numbers, steps (i) to (iv) of the proof of IX,1 become Euclid's proof of 24, the analogue for squares of 25. To prove in 26 (27) that similar plane (solid) numbers k, l are in the ratio of square (cube) numbers, Euclid argues that, by VIII,18 (19), CPROP (k, m, l) (CPROP (k, m_1, m_2, l)), for some m (m_1, m_2); hence, by VIII,2, corollary, the extremes of the least numbers in the ratio of of k, m, l $(k, m_1, m_2, l)$are squares (cubes) and in the same ratio as k, l. The proof of IX,2 amounts to a proof of the converse of 26; and, if an application of VIII,21 is substituted for the application of VIII,23 in the proof of IX,5, the result is a proof of the converse of 27. Euclid's failure to prove these converses is somewhat surprising, not only because of his standard practice of proving converses, but also because he appears to apply the converse of 26 in the derivation of IX,10.[29] None of 24–27 is ever used in the arithmetic books, although 26 and its converse are mentioned as having been proved in a puzzling *lēmma* after X,9, and 24 is employed in X,29 and 30.

IX,6 is the converse of IX,3 and derived, like it, from VIII,8, 19, and 23, and applied in IX,10. The presence of IX,7 seems inexplicable for three reasons: it is never used; it is a direct consequence of definitions; and its content seems to be taken for granted by Euclid in the proof of VIII,21 when he says, "Therefore N by multiplying the product of H, K has made A; therefore A is solid ...." I am inclined to group IX,8–10 with the propositions already discussed and to separate them from those which follow, because they are the last uses of the material between VIII,3 and IX,6[30] and are not subsequently used; and, in addition, 11 marks the beginning of a series of propositions in which VII,23–32 are given their only applications. IX,8–10 can also be seen as a kind of appendage to the material already established. For Euclid proves that if CPROP $(o, m_1, \ldots, m_n)$, then (i) if m_1 is a square (cube), so is each m_i, and (ii) if m_1 is not square (cube), SQUARE $(m_i) \leftrightarrow 2/i$ (CUBE $(m_i) \leftrightarrow 3/i$). He argues explicitly that m_2 is a square, m_3 a cube, and uses VIII,22 and 23 to infer that m_{i+2} (m_{i+3}) is a square (cube), if m_i is. Result (i), for squares, follows from this result and VIII,22. To argue similarly for cubes Euclid needs that m_2 is a cube, if m_1 is; to establish this he gives an explicit argument that $m_2 \simeq m_1^2$, and applies IX,3.

In IX,10 Euclid does the case $n = 6$ and argues that

$$\neg\,\text{SQUARE}\,(m_1) \rightarrow \neg\,\text{SQUARE}\,(m_3) \qquad \neg\,\text{CUBE}\,(m_1) \rightarrow \neg\,\text{CUBE}\,(m_4).$$

He proceeds indirectly, asserting that since, by IX,8,

$$\text{SQUARE}\,(m_2) \qquad \text{CUBE}\,(m_3)$$

and

$$(m_1, m_2) = (m_2, m_3) \qquad (m_2, m_3) = (m_3, m_4),$$

therefore

$$^{*}\text{SQUARE}(m_3) \rightarrow \text{SQUARE}(m_1). \qquad ^{*}\text{CUBE}(m_4) \rightarrow \text{CUBE}(m_2),$$

but, by IX,6,

$$\text{CUBE}(m_2) \rightarrow \text{CUBE}(m_1).$$

In the inferences marked with an asterisk Euclid argues from $(i, j^2) = (k^2, l^2)$ $((i, j^3) = (k^3, l^3))$ to "i is a square (cube)." These inferences can be made using VIII,24 (25) if one first infers $(j^2, i) = (l^2, k^2)$ $((j^3, i) = (l^3, k^3))$. However, in the case of squares, most manuscripts, including P, contain the words "i and j^2 have to one another the ratio which a square number has to a square number, so that i and j^2 are similar plane numbers; and j^2 is a square; therefore i is a square," which would, of course, involve an application of the unproven converse of VIII,26. The parallelism between the two cases in the proof suggests that Euclid has the converse of 27 in mind in the second one even though he does not mention that SIMSOLID (m_3, m_2).

If one adds VIII,24–27 and IX,6–10 to the deductive development already depicted, the result is the diagram shown in fig. 2.11. As the lines above the proposition numbers indicate, none of the propositions after VIII,8 represented here have applications in the arithmetic books not depicted in the diagram.

The only propositions from VIII,11 to IX,10 which do not appear in this structure are the first seven, VIII,11–17. None of these are applied in the arithmetic books, although 11 is used in X,9. As has already been indicated, 11 and 12 may represent steps toward proving IX,1′, 2′ and 4,5, which became otiose with the introduction of the concept of similarity to establish VIII,22 and 23. Euclid appears to go out of his

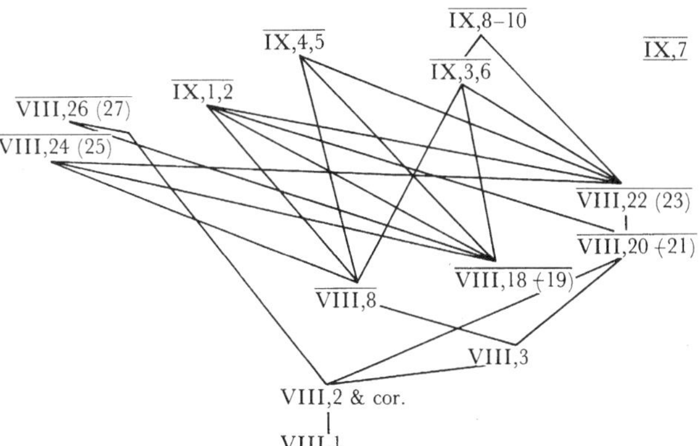

Figure 2.11

way to avoid citing 11 and 12 in the proofs of VIII,24 and 25 and of IX,2, 4–6. In each of these, rather than apply 11 or 12 directly to square or cube numbers, he argues that the numbers are similar and applies VIII,18 or 19. Probably the special case was lost sight of when the more general one was proved.

Propositions 13–17 are very perplexing. 16 and 17 are trivial logical equivalents of 14 and 15, which assert that $k^n/l^n \leftrightarrow k/l$, for $n = 2$ and $n = 3$. These equivalences in turn follow from VIIIA(ii), VIII,7, and the transitivity of measurement. For since k, l-CPROP$(k^n, k^{n-1} \cdot l, \ldots, k \cdot l^{n-1}, l^n)$, if k/l, each term in the continuous proportion measures the next and, by transitivity, k^n/l^n; on the other hand, if k^n/l^n, then, by VIII,7, $k^n/k^{n-1} \cdot l$, and k/l. In place of VIIIA(ii) Euclid refers to the constructions of 11 and 12. In VIII,13 Euclid again carries out these constructions and argues in the standard way for the conclusion that $m_1^2, \ldots, m_n^2$ and $m_1^3, \ldots, m_n^3$ are continuously proportional if $m_1, \ldots, m_n$ are. Whatever the purpose of 13–17—and it is tempting to suppose that they represent further unavailing efforts in the direction of IX,1′,2′ and 4,5—it seems reasonable to treat them together with 11 and 12. As fig. 2.12 indicates, their presence accounts for the presence of VIII,7, which, in turn, depends upon VIII,6. If this diagram is added to the one for the rest of VIII,1–IX,10, all the propositions among VIII,1–10 which are applied in the arithmetic books are accounted for. Indeed, all of their applications are accounted for except for a reference to the construction of VIII,2 in IX,15, which will be discussed in the next section. Most of the foundational propositions VII,11–22, 33 are used fairly heavily in the proofs of VIII,1–IX,10, although 11, 12, 15, and 19 are not used at all.[31] VII,34 and 35 are used in the proof of VIII,4, but except for the use of the generalized form of VII,27 in the proofs of VIII,2 and 3, there are no uses of VII,23–32 in VIII,1–IX,10. All of the uses of these propositions are represented in the tripartite diagram of fig. 2.13.

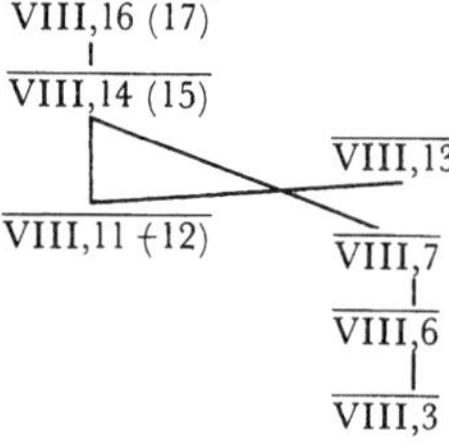

Figure 2.12

I shall discuss IX,11–20 in a somewhat different order from Euclid's, leaving 15 until the next section. In 16–19 Euclid investigates the conditions under which there is an m such that $(k, l) = (j, m)$. Since $(k, l) = (j, m)$ if and only if $k \cdot m \simeq l \cdot j$, it is clear that there is such an m if and only if $k/l \cdot j$, in which case m is the number by which k measures $l \cdot j$. Euclid treats the case in which $l \simeq j$ separately in IX,18 and divides both 18 and 19 into subcases: in 18 according to whether or not PRIME(k, l), in 19 according to whether or not PRIME(k, l) & CPROP(k, l, j).[32] To handle the cases in which PRIME(k, l) or PRIME(k, l) & CPROP(k, l, j) he proves 16 and 17 as lemmas.

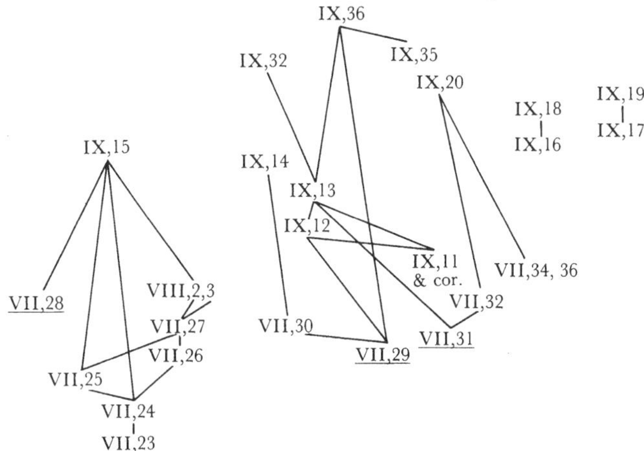

Figure 2.13

The first of these is a special case of the second, according to which

IX,17 $\text{CPROP}(m_1, \ldots, m_n)$ & $\text{PRIME}(m_1, m_n) \rightarrow$
$\neg\, \exists m((m_1, m_2) = (m_n, m))$.

Euclid argues indirectly that if the proposition were false, one would have, by VII,13, $(m_1, m_n) = (m_2, m)$, and by VII,21 and 20, m_1/m_2. Hence, by the transitivity of measurement, m_1/m_n, contradicting $\text{PRIME}(m_1, m_n)$.

IX,18 and 19 show that the standard division of Greek mathematical propositions into theorems and problems is an incomplete schematism. For these two propositions are formulated as "investigations," as we might call them, into the possibility of finding a third or fourth number proportional to given numbers. Clearly the *protasis* of such propositions could play no role in other derivations; rather the contents of the proof of the *protasis* must be applied.

Heath (vol. II, p. 403) appears to identify IX,14 with the fundamental theorem of arithmetic. He gives no reasons, and it is very difficult to say what he has in mind. The argument given by Euclid would show that a number which is the product of primes $k_1, \ldots, k_n$ is not divisible by any prime not equal to one of the k_i. But for the fundamental theorem one would still need proofs that every number is a product of primes and that the same primes cannot occur a different number of times in different factorizations of the same integer. Moreover, in IX,14 Euclid speaks only of the least number divisible by primes $k_1, \ldots, k_n$ without asserting or implying that this number is the product of $k_1, \ldots, k_n$; and it seems likely, although unverifiable, that he thinks of $k_1, \ldots, k_n$ as different primes. Thus IX,14 is considerably weaker than the fundamental theorem, and, since Euclid never uses it, his reasons for proving it must remain unknown. The proof of IX,14 is simple. If k is the least number measured by $k_1, \ldots, k_n$ and, for some prime l distinct from the k_i, $k \simeq l \cdot m$, then, by VII,30, since PRIME (k_i, l), k_i/m, contradicting the fact that k is the least number measured by the k_i.

IX,11 and 12 are clearly lemmas for

IX,13 $\text{CPROP}(o, m_1, \ldots, m_n) \mathbin{\&} \text{PRIME}(m_1) \mathbin{\&} m/m_n \rightarrow \exists i(1 \leq i \leq n \mathbin{\&} m \simeq m_i)$,

an obvious consequence of VIIIA(v) and the fundamental theorem of arithmetic. The two lemmas say

IX,11, corollary $\text{CPROP}(o, m_1, \ldots, m_n) \rightarrow \mathcal{M}(m_j, m_i, m_{i-j})$;

IX,12 $\text{CPROP}(o, m_1, \ldots, m_n) \mathbin{\&} \text{PRIME}(m) \mathbin{\&} m/m_n \rightarrow m/m_1$.

I have cited the corollary to IX,11 rather than the proposition itself, because the corollary gives the constructive content of the proof, and it is this content which is used in 12 and 13. In this case, then, Euclid's proof is of the corollary which is stated more abstractly in the *protasis*. The proof itself is simply an application of VII,14, according to which $(o, m_j) = (m_{i-j}, m_i)$.

The proof of IX,12 amounts to a proof of

IX,12′ $\text{CPROP}(o, m_1, \ldots, m_n) \mathbin{\&} \text{PRIME}(m, m_1) \mathbin{\&} m/m_i \rightarrow m/m_{i-1}$.

The essentials of the proof are that if the antecedent of this assertion holds and $m_i \simeq k \cdot m$, $m_1 \cdot m_{i-1} \simeq k \cdot m$,[33] and m/m_{i-1}, by VII,30. To complete the proof of IX,12 one need only note that if PRIME (m) and $\neg$ PRIME (m, m_1), then m/m_1. Euclid proves 12 for $n = 4$ and does a step-by-step reduction from m_4 to m_3 to m_2 to m_1. Instead of invoking VII,30 he carries out the argument for it. Finally, he proceeds in a confused, indirect way; he infers PRIME (m, m_1) from $\neg(m/m_1)$, but after he has reached the conclusion m/m_1, which should be the end of his proof, he continues as if the hypothesis he had derived an inconsistency from were PRIME (m, m_1).

IX,13 receives the same kind of proof, but a more elaborate one, which is perhaps best understood in terms of the lemma

$$\text{CPROP}(o, m_1, \ldots, m_n) \mathbin{\&} \text{PRIME}(m_1) \mathbin{\&} m/m_i \mathbin{\&} \neg\exists_j(m \simeq m_j) \rightarrow \exists k(k/m_{i-1} \mathbin{\&} \neg\exists j(k \simeq m_j)).$$

The lemma obviously implies that if IX,13 were false there would be a k dividing m_1 but not equal to it, contradicting the fact that m_1 is prime. Euclid begins what amounts to the proof of this lemma by pointing out that m cannot be prime, since, if it were, it would, by IX,12, divide and hence be equal to m_1. Therefore, there is a prime l such that l/m and, by the transitivity of measurement, l/m_i; again by IX,12, l divides and is equal to m_1, so that m_1/m. However, since m/m_i, there is a k such that $k \cdot m \simeq m_i \simeq m_1 \cdot m_{i-1}$, and $(m_1, m) = (k, m_{i-1})$, so that k/m_{i-1}. If k were some m_j, then, since by IX,11, corollary $m_i \simeq m_j \cdot m_{i-j}$, m would be m_{i-j}, contrary to hypothesis.

Euclid uses IX,13 in the proof of IX,36, the last proposition of the arithmetic books, in which he considers for the only time perfect or complete numbers, a perfect number being the sum of all positive integers less than it which measure it (VII, def. 23). $6(= 1 + 2 + 3)$, $28(= 1 + 2 + 4 + 7 + 14)$, and $496(= 1 + 2 + 4 + 8 + 16 + 31 + 62 + 124 + 248)$ are the first three perfect numbers. If the variable i^* is used to range over the nonnegative integers, these numbers can be represented in the form

$$2^0 + 2^1 + \ldots + 2^n + 2^0 \cdot \sum^n 2^{i^*} + \ldots + 2^{n-1} \cdot \sum^n 2^{i^*},$$

or more simply in the form $2^n \cdot \Sigma^n 2^{i^*}$. Clearly, not every number of this form is perfect, since, e.g., $2^3 \cdot \Sigma^3 2^{i^*} = 120$, but the sum of the proper divisors of 120 is 240. Euclid proves that if $\Sigma^n 2^{i^*}$ is a prime, then $2^n \cdot \Sigma^n 2^{i^*}$ is perfect.[34] In fact every even number is perfect if and only if it satisfies the Euclidean condition; it is not known whether there are any odd perfect numbers.

The proof of IX,36 may be divided into three parts which respectively establish

36a $\quad i^* \leq n \rightarrow (2^{i^*}/2^n \cdot l \;\&\; 2^{i^*} \cdot l/2^n \cdot l)$;

36b $\quad \text{PRIME}(l) \;\&\; j/2^n \cdot l \rightarrow$
$\qquad \exists i^* (i^* \leq n \;\&\; (j = 2^{i^*} \vee j = 2^{i^*} \cdot l))$;

36c $\quad 2^n \cdot \sum^n 2^{i^*} = \sum^n 2^{i^*} + \left(\sum^{n-1} (2^{i^*} \cdot \sum^n 2^{i^*}) \right)$.

36a and b say that the only factors of $2^n \cdot l$ with l prime are the numbers of the forms 2^{i^*} and $2^{i^*} \cdot l$ with $i^* \leq n$. 36c, which is a direct consequence of the fact that $2^n = 2^0 + \Sigma^{n-1} 2^{i^*}$, says that if $l = \Sigma^n 2^{i^*}$, then $2^n \cdot l$ is the sum of the numbers 2^{i^*}, for $i^* \leq n$, and $2^{i^*} \cdot l$, for $i^* < n$. Euclid, of course, does not speak of powers of 2 or of products of powers of 2 and l, but of numbers in continuous double proportion starting from a unit or from l. His formulation of IX,36 may be expressed

IX,36 $\quad m_1 \simeq o \;\&\; m_2 \simeq 2 \;\&\; \text{CPROP}(m_1, \ldots, m_{n+1}) \;\&$
$\qquad m \simeq \sum^{n+1} m_i \rightarrow (\text{PRIME}(m) \rightarrow \text{PERFECT}(m_{n+1} \cdot m))$.

Euclid's proof of this proposition can be represented by assuming its antecedent and 1-2-CPROP $(l_1, \ldots, l_{n+1})$. With these assumptions 36a–c may be expressed

36a′ $m_i/m_{n+1} \cdot l_1 \;\&\; l_i/m_{n+1} \cdot l_1$;

36b′ PRIME $(l_1) \;\&\; j/m_{n+1} \cdot l_1 \rightarrow$
$\exists i(i \leq n + 1 \;\&\; (j \simeq m_i \vee j \simeq l_i))$;

36c′ $l_1 \simeq m \rightarrow m_{n+1} \cdot l_1 \simeq l_1 + \sum^n l_i$.

36a′ follows from a lemma according to which

$$m_{n+1} \cdot l_1 \simeq m_{(n+2)-i} \cdot l_i \simeq m_i \cdot l_{(n+2)-i},$$

i.e.,

$$2^{n^*} \cdot l = 2^{n^*-i^*} \cdot (2^{i^*} \cdot l) = 2^{i^*} \cdot (2^{n^*-i^*} \cdot l),$$

and the obvious facts that $m_i/m_i \cdot l_{(n+2)-i}$ and $l_i/m_{(n+2)-i} \cdot l_i$. The lemma itself is easily inferred from $(m_1, m_2) = (l_1, l_2) = (1, 2)$, using VII,14 and 19.

For 36b′ Euclid assumes the antecedent and lets $j \cdot k \simeq m_{n+1} \cdot l_1$, i.e., $(j, m_{n+1}) = (l_1, k)$. If j/m_{n+1}, j is an m_i, by IX,13. But if $\neg j/m_{n+1}$, $\neg l_1/k$; since PRIME (l_1), PRIME (l_1, k), by VII,29; and, by VII,21 and 20, k/m_{n+1}; hence, by IX,13 again, k is an m_i, so that $j \cdot m_i \simeq m_{n+1} \cdot l_1 \simeq m_i \cdot l_{(n+2)-i}$, by the lemma for 36a′; hence $j \simeq l_{(n+2)-i}$.

Since, by the lemma for 36a′, $m_{n+1} \cdot l_1 \simeq l_{n+1} \cdot m_1 \simeq l_{n+1}$, to establish 36c′ it suffices to show that if $l_1 \simeq m$, then $l_{n+1} \simeq l_1 + \Sigma^n l_i$, or $\Sigma^n l_i \simeq l_{n+1} - l_1$. Since $l_2 - l_1 \simeq l_1$, this fact follows from a more general lemma proved by Euclid which enables one to compute the sum of the first n members of any series of $n + 1$ or more numbers in continuous proportion, namely,

IX,35 CPROP $(k_1, \ldots, k_{n+1}) \rightarrow$
$(k_2 - k_1, k_1) = (k_{n+1} - k_1, \sum^n k_i)$.

In the proof of this proposition Euclid makes his only application of VII,11, according to which

$$(k_1, k_i) = (k_2, k_{i+1}) = (k_2 - k_1, k_{i+1} - k_i),$$

so that, by VII,13,

$$(k_2 - k_1, k_1) = (k_{i+1} - k_i, k_i)$$

and, by VII,12,

$$(k_2 - k_1, k_1) = \left(\sum^n (k_{i+1} - k_i), \sum^n k_i\right);$$

but elementary combinatorial arguments make it clear that

$\sum^{n}(k_{i+1} - k_i) \simeq k_{n+1} - k_1.$

This completes the proof of IX,36, which is obviously a culminating point in the arithmetic books.[35] Propositions IX,13 and 35 are rather clearly lemmas for 36; and 13 explains the presence of IX,11 and 12. IX,16–19 are an isolated body of propositions, the proofs of which are extremely cumbersome. IX,14 too stands in isolation, and so does one of the monuments of Greek arithmetic, IX,20, the assertion of the infinity of the primes. Euclid refutes the hypothesis that $k_1, \ldots, k_n$ are all the primes by taking $\mathrm{LCM}(k_1, \ldots, k_n) + o \simeq l$; l then is either a new prime or divisible by a prime k which cannot be equal to any of the k_i, since if it were, it would divide both l and $\mathrm{LCM}(k_1, \ldots, k_n)$ and, hence, their difference—the indivisible unit.

There is a dramatic change after IX,20. IX,21–34 constitute a very elementary body of material, which, except for one application of IX,13 in the proof of 32, depends on no previously proved propositions. Becker[36] attempted to explain this anomaly by arguing that 21–34 are a remnant of an early treatment of arithmetic which culminated in a proof of IX,36 different from Euclid's. His suggestion remains one of the most persuasive historical hypotheses based on the *Elements*. The explicit foundations of 21–34 are the definitions 6–11 of book VII:

VII, def. 6 $\quad \text{EVEN}(k) \leftrightarrow \exists l \exists m(l \simeq m \;\&\; l + m \simeq k)$;

VII, def. 7 $\quad \text{ODD}(k) \leftrightarrow \neg\,\text{EVEN}(k) \leftrightarrow \exists l(\text{EVEN}(l) \;\&\; (k \simeq l + o \vee l \simeq k + o))$;

VII, def. 8 $\quad \text{EVENEVEN}(k) \leftrightarrow \exists l \exists m(\text{EVEN}(l) \;\&\; \text{EVEN}(m) \;\&\; \mathcal{M}(l, k, m))$;

VII, def. 9 $\quad \text{EVENODD}(k) \leftrightarrow \exists l \exists m(\text{EVEN}(l) \;\&\; \text{ODD}(m) \;\&\; \mathcal{M}(l, k, m))$;

VII, def. 11 $\quad \text{ODDODD}(k) \leftrightarrow \exists l \exists m(\text{ODD}(l) \;\&\; \text{ODD}(m) \;\&\; \mathcal{M}(l, k, m))$.

Definition 7 obviously involves the hypothesis that a collection of units which is divisible into two equal parts ceases to be so when a unit is added to or taken away from it, and one which is not so divisible becomes so by the addition or elimination of a unit. The manuscripts include a corresponding definition of ODDEVEN; but the definition is incompatible with IX,33 in which Euclid speaks of a number being even-times odd only.[37]

The true foundation of 21–34 is made clear by Euclid's proof of 21, the assertion that the sum of n even numbers is even. Euclid does the case $n = 4$, and argues as follows: (i) since each of $k_1, \ldots, k_n$ is even, it has a half part; (ii) therefore $\Sigma^n k_i$ has a half part; (iii) therefore $\Sigma^n k_i$ is even. In step (iii) definition 6 is explicitly quoted, and obviously the same definition is at work in step (i). Step (ii) rests on some such law as VIIh and common notion 2; but in this case, if such combinatorial laws are not cited, it is hard to see that there is any real proof at all. For, given the definitions, 21 is nothing but an instance of a general combinatorial law. Becker suggested that the original proof for 21 was accompanied by the manipulation of pebbles. Clearly, Euclid's straight lines could be used just as well. The crucial point for my purposes is that without such manipulations or the laws they embody, the proof of 21 is empty.

The remainder of 21–34 can be discussed briefly. In 22 and 23 Euclid proves that the sum of evenly many odd numbers is even, that of oddly many odd numbers odd. He reduces 22 to 21, using the law

$$\sum^{n}(k_i - o_i) + \sum^{n} o_i \simeq \sum^{n} k_i$$

and the definition of odd number, and 23 to 22 using the law

$$\sum^{n} k_i \simeq \sum^{n-1} k_i + k_n - o + o.$$

In 24–27 Euclid establishes the parity of the difference of two numbers of given parity. Had he employed his usual assumption that $(x - y) + y \simeq x$, he could have reduced these laws of subtraction to his laws for addition plus the easily proved fact that the sum of two numbers of different parity is odd. Instead he proves that the difference of two even numbers is even in the same way that he proves 21, and then reduces the other cases of subtraction to this one.

In 28 and 29 Euclid proves that an odd number multiplying a number of given parity produces a number of the same parity by reduction to 21 and 23 and the fact that $n \cdot k$ contains n k's. Obviously Euclid could have established the more general result that the multiplication of an even number always produces an even number. In IX,30 Euclid shows that an odd number which measures an even number measures its half. He lets $n \cdot k \simeq l$ with k odd and l even; he then invokes IX,23 to show that n must be even and concludes that since k measures l an even number of times, it measures $\frac{1}{2}l$. The use of 23 here is curious because Euclid could have applied IX,29 more simply. More curious still is the last inference, which is the heart of the proof but receives no explanation. The reason none is given may be that if l is represented by an even number n of collections of k units each, it is obvious that $\frac{1}{2}l$ is represented by $\frac{1}{2}n$ such collections.[38] Euclid makes the same kind of inference in IX,33. To show that k is even-times odd only if $\frac{1}{2}k$ is odd, he first points out that such a k is even-times odd, since it is equal to $2 \cdot \frac{1}{2}k$. He then refutes the possibility of k's being even-times even on the grounds that, if it were, "the half of it will be measured by an even number though it is odd." Euclid is here taking for granted that the product of two even numbers is even, the more general form of 28, which depends on 21; he is also making tacit application of 29, since he doesn't bother to rule out the possibility of ODDODD (k). Euclid also takes for granted the general form of 28 in deriving 31 in which he proves that PRIME $(k, 2 \cdot l)$, if ODD (k) and PRIME (k, l). Proceeding indirectly, he lets j be the common measure of k and $2 \cdot l$ and infers immediately that j must be odd, since k is, so that, by 30, j/l, contradicting PRIME (k, l).

In IX,32 and 34 Euclid uses the notion of a number being doubled from a dyad, i.e., being a power of 2. He shows in 32 that such numbers are even-times even only on the grounds that if k is such a number, there exist $k_1, \ldots, k_n$ such that CPROP $(o, 2, k_1, \ldots, k_n, k)$, and, by IX,13, only the even numbers $2, k_1, \ldots, k_n$ divide k. In 34 Euclid considers k which are neither doubled from a dyad nor have an odd half and shows that they are both even-times even and even-times odd. The former is clear since $k \simeq 2 \cdot \frac{1}{2}k$. For the latter Euclid argues indirectly. "For if we bisect k, then bisect its half, and do this continually, we shall come upon some odd number which will measure k according to an even number. For if not we shall come upon a dyad, which is contrary to the hypothesis [that k is not doubled from the dyad]."

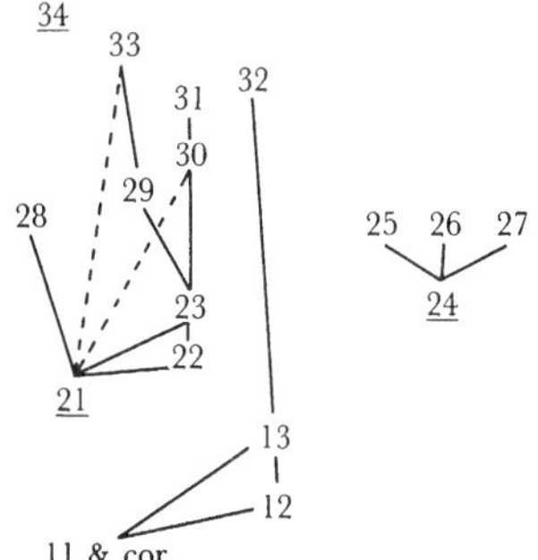

Figure 2.14

The logical structure of 21–34, including all dependencies from books VIII and IX, is indicated in fig. 2.14. The use of IX,13 is striking, especially since, as Becker pointed out, 32 is easily derived from 31, according to which an odd number, since it is prime to 2, must be prime to 2^2, to 2^3, and so on. Substitution of this derivation for Euclid's makes 21–23, 28–34 a cohesive deductive unit with a fairly clear purpose: the classification of composite integers according to the parity of their factors. For it is easy enough to derive from these propositions that all and only odd composite integers are odd-times odd only and that even composite integers are even-times even only, even-times odd only or both even-times even and even-times odd, according to whether they are doubled from a dyad, are double an odd number, or neither. As for 24–27, it is a rather striking fact that the only subsequent use made by Euclid of any of 21–34 is a remark based on 24 and 26 in a *lēmma* after X,28. It is conceivable that this remark is the explanation for the inclusion of 24–27.

Becker also showed that IX,36 could be proved using 21–34 and the kind of considerations involved in their demonstrations. I shall reproduce the proof suggested by Becker with some variation designed to increase the role of 21–34. 36a is easily proved since it is clear that both 2^{i^*} and $2^{i^*} \cdot l$ measure $2^{i^*} \cdot l$ and therefore measure $2 \cdot 2^{i^*} \cdot l$, $2 \cdot 2 \cdot 2^{i^*} \cdot l$, and so on, up to $2^n \cdot l$. The proof of 36b is more complicated and depends ultimately on all of IX,21–23, 28–34 (provided Becker's alternative derivation of 32 is substituted for Euclid's). The proof is divided into cases according to whether or not j is odd. If j is odd, then since $2^n \cdot l$ is always even, if $j/2^n \cdot l, j$ divides $\frac{1}{2}(2^n \cdot l)$, i.e., $2^{n-1} \cdot l$, by 30, $\frac{1}{2}(2^{n-1} \cdot l)$, and so on, so that j/l; but since l is prime, $j \simeq l \simeq 2^0 \cdot l$. If j is even, j is either doubled from a dyad or even-times odd. In the first case $j \simeq 2^i$, and all that has to be shown is that $i \leq n$; but if $i > n$, one can argue that if $2^i/2^n \cdot l$, $\frac{1}{2}(2^i)/\frac{1}{2}(2^n \cdot l)$, i.e., $2^{i-1}/2^{n-1} \cdot l$, and so on down to $2^{i-n}/l$, contradicting PRIME (l). For the case where EVENODD (j), one can set $j \simeq j_2 \cdot j_1$ with $j_3 \cdot j_2 \cdot j_1 \simeq 2^n \cdot l$ and j_1 odd. Using IX,30 one can argue that j_1 divides and hence is equal to l. Since then $j_3 \cdot j_2 \cdot l \simeq 2^n \cdot l, j_3 \cdot j_2 \simeq 2^n$, and, because 2^n is even-times even only, j_2 must be as well, so that j_2 must be 2^i, for some $i \leq n$.

Becker proves 36c by giving an entirely combinatorial argument for $2^n \cdot l \simeq l + \Sigma^{n-1} 2^i \cdot l$. The argument should be clear from fig. 2.15, in which each section of the configuration is supposed to contain the number of units indicated, so that the whole represents $2^5 \cdot l$ as $2^4 \cdot l + 2^3 \cdot l + 2^2 \cdot l + 2 \cdot l + l + l$. Such a combinatorial argument could well be convincing, and the picture it suggests of early mathematicians giving proofs by means of arrays of pebbles has certain attractions. Whether or not Becker is correct in his historical hypothesis about the exact character of the pre-Euclidean embodiment of 21–36, the view that the *Elements* contain a revision of a proof of 36 based on 21–34 seems almost certainly correct. On the other hand, it is hard to see why Euclid would replace with a more complicated demonstration, a proof using combinatorial principles of the same kind which he employs elsewhere, or why, having made the replacement, he would retain propositions which the replacement had rendered useless. Perhaps, then, 21–34 are a relatively independent section of the arithmetic books aimed at classifying composite numbers in terms of the parity of their factors and proving an elementary result presupposed by book X; while 36, together with its lemma, 35, may be placed at the end of the arithmetic books because Euclid attaches some special significance to perfect numbers. Personally I am inclined to Becker's theory of the origin of 21–36, but it seems to me to raise as many questions as it solves.

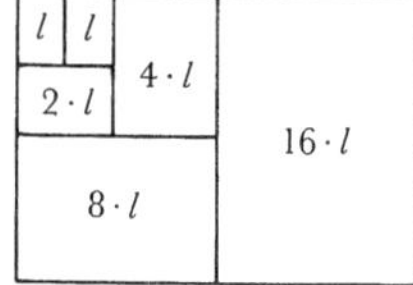

Figure 2.15

2.3 Arithmetic and Algebra; Applications of Arithmetic in Book X

In section 1.3 I tried to give a general characterization of geometric algebra. The discussion of the arithmetic books should have made clear the extent to which Euclidean arithmetic involves geometric modes of thinking and the kind of concrete combinatorial reasoning that is common to both Euclid's arithmetic and his geometry. It is striking, however, that although Euclid's arithmetic thought is often governed by geometric analogies, nothing in books VII–IX which has been discussed involves an actual transference of a geometric truth into arithmetic. In particular, although such notions as those of plane and square numbers seem to invite the use of geometric algebra, we have seen no cases in which it has been used. There are, however, two possible arithmetic applications of geometric algebra in the *Elements* which I now want to consider.

The first is the proof of

IX,15 $\text{CPROP}(k, l, m)$ & $\text{LEAST}(k, l, m) \rightarrow \text{PRIME}(k + l, m)$ & $\text{PRIME}(l + m, k)$ & $\text{PRIME}(k + m, l)$.

Figure 2.13 assigns to this proposition a perhaps undeserved prominence, but it is not unreasonable to suppose that Euclid proved at least VII,28 with an eye to its role in IX,15. Itard[39] points out a possible reason for the inclusion of 15 in the *Elements*: it yields a simple proof that no integers satisfy the golden section, i.e., there are no i and j such that $(i, j) = (j, i + j)$ or $j^2 \simeq i \cdot (i + j)$. For suppose there were and that $i, j, i + j$-$\text{LEAST}(k, l, m)$. Then, for some n, $i \simeq n \cdot k$, $j \simeq n \cdot l$, and $i + j \simeq n \cdot m$. Hence $n \cdot m \simeq n \cdot k + n \cdot l \simeq n \cdot (k + l)$, so that $m \simeq k + l$, contradicting IX,15, according to which $\text{PRIME}(k + l, m)$. It will be seen in chapters 5 and 7 that the golden section, called "division into extreme and mean ratio" by the Greeks, plays an extremely important role in the *Elements*. If Euclid is interested in the golden section in IX,15, he does nothing to make his interest clear to the reader.

However, for my purposes the proof of IX,15 is of more interest than the possible significance of the proposition itself. By VIIIB, one has that $k \simeq i^2$, $l \simeq i \cdot j$, $m \simeq j^2$, for some i, j with $\text{LEAST}(i, j)$. Successive applications of VII,22, 28, 24, and 25 yield that $\text{PRIME}((i + j) \cdot i, j^2)$. "But the product of $i + j$, i is the square on i together with the product of i, j." Therefore, $\text{PRIME}(i^2 + i \cdot j, j)$, i.e., $\text{PRIME}(k + l, m)$. One proves similarly that $\text{PRIME}(l + m, k)$. Finally, since by VII,28 $\text{PRIME}(i + j, i)$ and $\text{PRIME}(i + j, j)$, one has by VII,26 $\text{PRIME}((i + j)^2, i \cdot j)$. "But the squares on i, j together with twice the product of i, j are equal to the square on $i + j$." Therefore,

$$\text{PRIME}(i^2 + j^2 + 2 \cdot i \cdot j, i \cdot j);$$

and two applications of VII,28 yield that

PRIME $(i^2 + j^2, i \cdot j)$,

i.e., PRIME $(k + m, l)$.

The two quotations from Euclid assert the equalities which we would express by $(i + j) \cdot i = i^2 + i \cdot j$ and $(i + j)^2 = i^2 + j^2 + 2 \cdot i \cdot j$; given the commutativity of addition, these formulas are special cases of the algebraic identities II,3a and II,4a which can be read out of II,3 and 4. Moreover, the language which Euclid uses here and elsewhere in arithmetic is very similar to the language of geometric algebra, the principal difference being the use of the neuter article in connection with geometric figures and of the masculine one in connection with numbers. *To apo tēs AB* (*tetragōnon*) is the square on the straight line *AB*, *ho apo tou AB* (*tetragōnos arithmos*) the square of the number *AB*. Similarly, *to hupo tōn AB*, *BΓ* (*periechomenon orthogōnion*) is the rectangle contained by *AB*, *BΓ*, *ho ek tōn AB*, *BΓ* (*genomenos arithmos*) is the product of *AB*, *BΓ* (although Euclid sometimes refers to this product as *ho hupo tōn AB*, *BΓ*, and he defines a square number as one "contained by" two equal numbers (VII, def. 19)).

Before discussing the significance of the proof of IX,15 I would like to look at Euclid's other apparent application of geometric algebra to numbers. In book X, after proposition 28, Euclid proves two *lēmmata* showing how to find four pairs of square numbers such that their sum or difference is or is not a square number. He uses IX,24 or 26 to take i and j such that $i - j$ is even. He then invokes an arithmetic analogue of II,6 to assert what we might express by

$$i \cdot j + \left(\frac{i - j}{2}\right)^2 = \left(\frac{i + j}{2}\right)^2,$$[40]

and sets $k \simeq i \cdot j$, $l \simeq \frac{i - j}{2}$, and $m \simeq \frac{i + j}{2}$. One could in fact show that all possible integral solutions to $x^2 + y^2 = z^2$ can be represented in the form $x = \sqrt{i \cdot j}$, $y = \frac{i - j}{2}$, $z = \frac{i + j}{2}$, where i and j are similar plane numbers.[41] Euclid contents himself with applying IX,1 and 2 to infer

If i and j are similar plane numbers, k and l^2 are square numbers the sum of which is a square, m^2 and l^2 are square numbers the difference of which is a square; and if i and j are not similar plane numbers, m^2 and l^2 are square numbers the difference of which is not a square.

The construction of square numbers the sum of which is not a square is more complicated. In Euclid's presentation, it depends in part on the relative positions of straight lines representing the numbers. I give a more abstract version of his argument which establishes by *reductio* that

If i and j are similar plane numbers, k and $(l - 1)^2$ are square numbers the sum of which is not square.

If $k + (l - 1)^2 \simeq n^2$, then $n^2 \preceq (m - 1)^2$, because, if $n^2 \succ (m - 1)^2, m^2 \succ n^2 \succ (m - 1)^2, m \succ n \succ m - 1$, "and the unit is divided." Euclid again invokes the arithmetic analogue of II,6 to justify

$$(i - 2\cdot(m - n))\cdot j + \left(\frac{i - 2\cdot(m - n) - j}{2}\right)^2 \simeq \left(\frac{i - 2\cdot(m - n) + j}{2}\right)^2.$$

But since

$$\frac{i - 2\cdot(m - n) \pm j}{2} \simeq \frac{i \pm j}{2} - (m - n),$$

therefore

$$(i - 2\cdot(m - n))\cdot j + (l - (m - n))^2 \simeq (i - 2\cdot(m - n))\cdot j + \left(\frac{i - j}{2} - (m - n)\right)^2 \simeq \left(\frac{i + j}{2} - (m - n)\right)^2 \simeq (m - (m - n))^2 \simeq n^2 \simeq i\cdot j + (l - 1)^2;$$

but this is impossible, since $i - 2\cdot(m - n) \prec i$ and $l - (m - n) \preceq l - 1$.[42]

The application of propositions from book II in arithmetic would constitute strong evidence for the algebraic interpretation of book II, although some reason would have to be given for the relative isolatedness of these applications. Unfortunately, the presence of such applications would also represent a striking drop in rigor in the *Elements*, since the inferences in question would be based entirely on a geometric analogy without arithmetic foundation. However, we have seen that the geometric analogy functions in arithmetic quite independently of any algebraic considerations. The same may be true in the present case. The diagrams for II,3, 4, and 6 in figs. 1.24–1.26 transform directly into representations, using discrete units, of the arithmetic versions of these propositions. But, whereas Euclid has a procedure for manipulating and arguing about geometric figures, he is unwilling to use arguments explicitly based on arrays of units in arithmetic. And abstract arguments for the arithmetic analogues of propositions of geometric algebra would be at least as hollow as the proof of IX,21. For

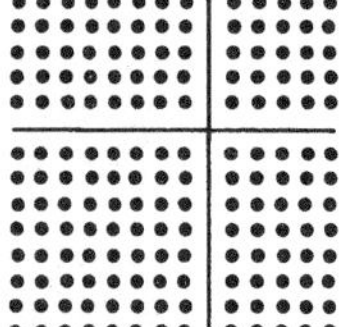
Figure 2.16

example, the geometric equality of the "complements of the parallelograms about the diameter" becomes in the arithmetic of arrays of units the transparent equinumerosity of two arrays on m rows of n units (fig. 2.16). This equinumerosity corresponds to the assertion that multiplication is commutative (VII,16); but without a concrete representation of the relevant figure, the equinumerosity is insufficient to establish an arithmetic analogue of a proposition like II,4 within Euclidean conceptual limits. For within those limits there would seem to be no way to establish that a square number is always divisible into two square numbers and two other equal ones. Euclid chooses simply to assert that such equalities hold.

The point I am making can perhaps be clarified by recalling the discussion of II,3,4, and 6 in section 1.3. Euclid's proofs of these propositions involve standard algebraic-geometric constructions and argument. The general tenor of the arithmetic books makes it clear that Euclid does not wish to base arithmetic on such considerations. However, it was shown that these three propositions can be reduced to II,1 using only substitution of equals for equals and the geometric version of the commutativity of multiplication, $\mathbf{O}(x,y) \simeq \mathbf{O}(y,x)$. The same kind of reduction would work in the same way for the arithmetic analogues of the three propositions. But the arithmetic version of II,1 which is needed for this purpose is simply $n\cdot(k + l) \simeq n\cdot k + n\cdot l$, which, by the definition of multiplication, is equivalent to the special case of VIIh when the summands k_i and l_i are equal. Thus the algebraic identities assumed by Euclid in arithmetic depend only upon a law of addition which he takes for granted throughout the *Elements* and the commutativity of multiplication. This fact does not, of course, mean that the assumed identities are arithmetically obvious, but only that their proof would consist entirely of steps of a kind which Euclid customarily passes over.[43]

As indicated at the end of the last chapter, there is conclusive evidence that geometric algebra was interpreted and used arithmetically in later antiquity. The evidence that it was used in the same way in pre-Euclidean times is sparse but intriguing.[44] However, the two isolated apparent arithmetic applications of geometric algebra in the *Elements* seem insufficient to undermine the claim that Euclid's foundational program involves the separation of arithmetic from geometric algebra. There remains the question whether the line of thought in Euclidean arithmetic is algebraic. It is, of course, true that numbers which are continuously proportional in the usual sense are the terms of a geometric series; and from a modern point of view the results proved by Euclid about continuously proportional numbers are very naturally interpreted as results about such series. In this sense, at least, it can be said that many of Euclid's arithmetic results are algebraic. But they do not seem to be proved in an algebraic way. For in his arguments Euclid constantly reverts to the fundamental combinatorial facts about measurement. He does not develop a set of laws which he exploits for the manipulation of continuous proportionalities in the way in which a modern algebraist manipulates geometric series. Thus, even if from a modern point of view some of Euclid's results are algebraic, his line of thought appears to be a combinatorial one based on the conception of numbers as collections of units.

Before leaving the topic of Euclidean arithmetic, it is convenient to discuss briefly all the applications of arithmetic propositions in book X. We have already seen that in the *lēmmata* after X,28 Euclid uses four otherwise unused arithmetic propositions: IX,24 and 26 in a relatively trivial way, IX,1 and 2 in an essential way. This use of IX,1 and 2 is noteworthy, because it gives the concept of similar numbers a fundamental role in book X and means that much of the deductive structure of VII and VIII makes a contribution toward X. In X,29 (and 30) Euclid, having carried out the construction of the *lēmmata*, takes for granted that if $k^2 - l^2(k^2 + l^2)$ is not a square, then l^2 and $k^2 - l^2(k^2 + l^2$ and $l^2)$ do not have the ratio of a square number to a square number—a direct consequence of the otherwise unused VIII,24. Euclid does essentially the same thing in X,48–53 and 85–90, sometimes even citing the *lēmmata* as if they directly specified the construction of numbers not in the ratio of a square number to a square number.

The other applications of arithmetic results in book X all presuppose that it makes sense to speak of pairs of geometric objects having the same ratio as pairs of numbers and that the standard laws of proportionality can be applied to proportions which might be written as $(x, y) = (k, l)$. In particular, in X,5–8 Euclid uses the conception of numbers as concatenations of units to argue that two objects are commensurable (have a common measure) if and only if they stand in the ratio of a number to a number. It is simplest to take this foundational proposition and its presuppositions for granted now; discussion of it will be taken up in section 3.2. In X,12 Euclid shows

X,12 $\quad (x, y) = (k_1, l_1) \;\&\; (z, y) = (l_2, k_2) \rightarrow$
$\qquad \exists k \exists l((x, z) = (k, l))$.

He apparently applies VIII,4 to construct numbers k, m, l such that k_1, k_2, l_1, l_2-CPROP (k, m, l), and then argues that since $(k, m) = (k_1, l_1) = (x, y)$ and $(m, l) = (k_2, l_2) = (y, z)$, therefore $(x, z) = (k, l)$. Euclid clearly could have carried out the construction of k, m, l directly without relying on a previous proposition, but, with the exception of the *lēmmata* before X,29, which are very specifically related to book X, he seems to want to restrict purely arithmetic argument to the arithmetic books. X,12 is fundamental to the project of book X. It establishes that the things which are commensurable with a given object form a set which is closed and connected with respect to the relation of commensurability.

The crucial arithmetic content of book X relates to the commensurability of straight lines and of the squares on straight lines. In X,9 Euclid in effect establishes

$$(x, y) = (k, l) \leftrightarrow (\mathbf{T}(x), \mathbf{T}(y)) = (k^2, l^2),$$

but he expresses what he proves as

$$\text{X,9} \quad \exists k \exists l((x, y) = (k, l)) \leftrightarrow \exists k \exists l((\mathbf{T}(x), \mathbf{T}(y)) = (k^2, l^2)).$$

He uses a geometric result (VI,20, cor.) to assert that $\mathbf{T}(x)$ is to $\mathbf{T}(y)$ in the duplicate of the ratio of x to y and quotes VIII,11 for the fact that k^2 is to l^2 in the duplicate of the ratio of k to l; he then infers the desired conclusion, taking for granted that ratios are equal if and only their duplicates are—an assumption to be discussed in section 3.2. Although in itself X,9 is unproblematic, applications of it to establish that straight lines x and y are incommensurable require an argument that

$$\neg \exists k \exists l((\mathbf{T}(x), \mathbf{T}(y)) = (k^2, l^2)),$$

i.e., that $\mathbf{T}(x)$ and $\mathbf{T}(y)$ do not have to one another the ratio of a square number to a square number. After X,9 there is a *lēmma* which quotes VIII,26 and its converse as propositions which have been proved in the arithmetic books. The upshot of the *lēmma* is the assertion of the logical equivalent of the converse,

$$\neg \text{SIMPLANE}(j, m) \rightarrow \neg \exists k \exists l((j, m) = (k^2, l^2)).$$

This assertion is invoked in X,10 to enable one to find numbers which do not have the ratio of a square number to a square number. This application of the *lēmma* shows its significance. It provides a completely general criterion for deciding whether or not two numbers have the ratio of a square number to a square number.

X,9 and 10 and the *lēmma* between them are perplexing for a number of reasons.[45] The use of the converse of VIII,26 is paralleled in the proof of IX,10; but, as we have seen, in the arithmetic books Euclid always substitutes a more roundabout application of VIII,18 for the simpler use of VIII,11. Why he should change his practice in X,9 is not clear. In any case, 9 and 10 are fundamental propositions in book X. In conjunction with 12, they make it possible to categorize all straight lines x, according to whether or not, for a given straight line r, x is commensurable with r or x is incommensurable with r but $\mathbf{T}(x)$ is commensurable with $\mathbf{T}(r)$ or x is incommensurable with r and $\mathbf{T}(x)$ is incommensurable with $\mathbf{T}(r)$.

Although one can hardly claim that books VII–IX are assembled only for the sake of their subsequent applications, it is clear that much of the content of VII and VIII plays a role in the deductive structure leading to X. The relatively casual uses of VIII,4 and IX,24 and 26 are also significant, since they provide an explanation for the presence of these otherwise isolated propositions. On the other hand, Euclid obviously proves much more arithmetic than he needs for book X and presents some of what he does need in a much more general form than he needs.

Notes for Chapter 2

Bibliographical Note

The combination of logico-mathematical analysis and historical hypothesis which has dominated twentieth-century discussions of the arithmetic books gets its real beginning with Zeuthen, whose theories are presented most fully in "Sur la constitution . . ." and *Sur l'origine* . . . . Van der Waerden, using pretty much the same techniques as Zeuthen, reached quite different conclusions in "Die Arithmetik . . . ." His analyses, which I have found extremely helpful, are summarized in *Science Awakening*. This tradition of research is admirably summarized and continued by Knorr, whose book includes a useful bibliography.

In the last twenty years there have been two books devoted to Euclid's arithmetic. Itard's translation and commentary includes many useful remarks on the mathematical content of the propositions, but does not take into account the scholarly tradition. (See the review by Schmidt.) Taisbak's monograph is somewhat idiosyncratic and can only be read as a whole. Reading it in this way is very rewarding; on most issues of detail I find Taisbak convincing. Finally I should perhaps mention Malmendier's axiomatization of book VII, which takes as primitives the notions of unit, number, addition, and the four-place relation expressed by 'same part'. Malmendier's axiomatization has the advantage of not making Euclid's more formal arithmetic dependent on an informal arithmetic, but this advantage is obtained at the price of making VII,5 and 9 axioms. I would describe Malmendier's work as a modern version of Euclid rather than a historical reconstruction.

1. Compare, for example, VII,9 and 15.

2. See, for example, the use of VII,12 in the proof of VII,15.

3. See Frege, chapter 4.

4. Operation symbols introduced in this way can be eliminated like symbols defined in the ordinary way only in sufficiently strong theories. See Hilbert and Bernays, vol. I, p. 286ff. and Beth, pp. 113–131.

5. For such an extension see Robbin, p. 132ff.

6. In the case of one arithmetic problem, VII,3, the Theonine manuscripts have 'Q.E.F.'.

7. Euclid is inconsistent on this question. In the proof of VII,1 he asserts explicitly that a certain number measures itself, but he defines a prime (VII, def. 12) as a number which is measured only by the unit.

8. See, for example, "Sur la constitution . . . ," p. 410ff. Quite similar hypotheses are adopted by Dijksterhuis and Itard. Schmidt's review of the latter contains a concise discussion of the issues involved.

9. See also Taisbak, pp. 31–32.

10. "For, since *A* by multiplying *B* has made *C*, therefore *B* measures *C* according to the units in *A*." (VII,16) "*F* also measures *D* according to the units in *E*; therefore *E* by multiplying *F* has made *D*." (VII,24)

11. See also Taisbak, pp. 37–38, 111–112.

12. The notation is due to Dijksterhuis, *Archimedes*, p. 51, who, however, represents it as the expression of an equality between two ratios.

13. A *sumperasma* is found only in VII,4, 31, 32, and VIII,14. IX,35 has a kind of half *sumperasma* which repeats the consequent of the *protasis*.

14. Hereafter I indicate limitations in the range of a variable only in cases where it might not be completely obvious.

15. My discussion of arithmetic induction is indebted to Freudenthal's "Zur Geschichte der vollständige Induktion."

16. For other possible uses of VII,15 see p. 76.

17. For the various forms of induction and their logical relations see Kleene, section 40.

18. Heath (vol. II, pp. 302–303) suggests a different way of generalizing VII, 2, and 3. ". . . We can find the GCD of pairs, then the GCD of pairs of these, and so on, until only two numbers are left and we find the GCD of these." He appears to credit this suggestion to Heron on the basis of Al-Narizi's *Commentary* . . . . However, all Heron says is that some generalization can be made "because it has now been proved that any number measuring other numbers measures their greatest common measure." This remark is probably compatible with any way of making the generalization.

19. Indeed, since VII,35 is related to the proof of VII,34 as the corollary of VII,2 is to the proof of VII,2, it is striking that Euclid doesn't make 35 a corollary of 34. However, while 35 does have a simple proof, there does not seem to be any way of proving the corollary of VII,2 except by reproducing the process of finding GCD (k_1, k_2) and arguing that the number produced is divided by any measure of k_1 and k_2. Euclid prefers to state as a corollary to VII,2 what is really an observation on its proof.

20. Itard (p. 128) suggests that 37–39 are a vestige of an earlier arithmetic treatise somehow connected with the use of auxiliary numbers in Egyptian calculation. The suggestion rests on treating auxiliary

numbers as least common denominators. For arguments against treating them in this way see Neugebauer, *Vorgriechische Mathematik*, p. 137ff. or van der Waerden, pp. 26–27.

21. As Taisbak (pp. 79–80) points out, if the application of VII,29 in this proof is eliminated, the proof establishes what is sometimes called Gauss's theorem: PRIME (k, l) & $k/l \cdot m \rightarrow k/m$.

22. For discussions of the relationship between the fundamental theorem and Euclidean arithmetic see Dijksterhuis, vol. II, pp. 160–162 and Knorr, "Problems . . . ."

23. Here, as in several other propositions in the arithmetic books, Euclid proceeds in a superficially indirect way which has no point.

24. Vogel, *Beitrage* . . . , pp. 436–440, describes the rather sparse evidence about Greek procedures for multiplying fractions.

25. Again he uses a superficially indirect argument in the proof of 6, perhaps to be able to "pick" the terms m_i and m_j.

26. See, for example, "Sur la constitution . . . ," pp. 414–417. Zeuthen ascribes the main features of books VII and VIII to Plato's younger contemporary Theaetetus; van der Waerden ascribes VIII to the somewhat earlier Archytas, and VII to the fifth-century Pythagoreans.

27. In the mss. Euclid's proof of 20 is slightly flawed. See Heath, vol. II, p. 376.

28. Such proofs are given in scholia 2 and 4 to book IX.

29. Heron added the converses of 26 and 27 to book VIII. See the *Commentary* . . . of Al-Narizi, 194.27–195.22.

30. Heiberg cites IX,8 in the proofs of IX,12 and 13. For the reasons he is probably wrong, see note 33 below.

31. There are no explicit uses of VII,16 either; but their absence is due to the fact that Euclid takes the commutativity of multiplication for granted in VIII and IX. See, for example, the proof of VIII,4.

32. I am here describing the "Theonine" texts. Heiberg follows P, in which the proof of 19 contains a fallacious argument that PRIME (k, m) & $\neg$ CPROP (k, l, m) implies $\neg \exists j((k, l) = (m, j))$. See Heath, vol. II, p. 411, where, however, the counterexample is printed incorrectly. A correct one is $k = 4$, $l = 8$, and $m = 9$.

33. Euclid carries out this inference for different values of i five times in propositions 12 and 13. Heiberg refers two to IX,11, corollary, one to IX,11, and two to IX,8. Euclid's language suggests that he is always using the corollary. He first justifies $m_4 \simeq m_1 \cdot m_3$ with "since m_1 measures m_4 according to the units in m_3, therefore m_1 by multiplying m_3 has made m_4." He shortens the reference to the corollary in justifying $m_3 \simeq m_1 \cdot m_2$: "But further, by the preceding [proposition], m_1 by multiplying m_2 has made m_3." Although the third time he leaves the reference implicit, saying only, "But further m_1 by multiplying itself has made m_2," there is no reason to follow Heiberg in thinking that the intended justification has changed, from the corollary, to IX,8. Indeed, Heiberg himself assigns the justification "But further m_1 by multiplying m_2 has made m_3" to the corollary.

34. Taisbak ("Perfect numbers ...") has shown how recognition of the truth of IX,36 could easily arise from the Egyptian technique of multiplication by successive doubling.

35. It is perhaps worthwhile to correlate this presentation of the proof of IX,36 with Euclid's. He starts out with the assumption of the antecedent of 36 as formulated in the text plus PRIME (m), CPROP $(m, l_2, \ldots, l_n)$, and $(m, l_2) = (m_1, m_2)$. He then argues for a special case of the lemma for 36a′, namely $m_{n+1} \cdot m \simeq m_2 \cdot l_n$, so that CPROP $(m, l_2, \ldots, l_n, m_{n+1} \cdot m)$. He then proves 36c′, using IX,35 as indicated in the text, and simply asserts 36a′ without proof. This mere assertion is curious, since 36a′ does turn on the lemma, and Euclid essentially proves the lemma when he completes the proof of 36 by establishing 36b′ as in the text.

36. In "Die Lehre . . . ."

37. On this subject see Heath, vol. II, pp. 282–284.

38. A scholiast (408.2–8) argues as follows. If $\mathcal{M}(k, l, n)$, then $\mathcal{M}(n, l, k)$; but $(n, l) = (\frac{1}{2}n, \frac{1}{2}l)$; so $\mathcal{M}(\frac{1}{2}n, \frac{1}{2}l, k)$ and $\mathcal{M}(k, \frac{1}{2}l, \frac{1}{2}n)$.

39. Page 181.

40. For the formulation of II,6 given in section 1.3 (If $ACBD$ is a straight line and AB is bisected at C, then $\mathbf{O}(AD, DB) + \mathbf{T}(CB) \simeq \mathbf{T}(CD)$), this formula involves taking AD as i, DB as j, so that CB is $\frac{i-j}{2}$ and CD is $j + \frac{i-j}{2}$, which is easily seen to be equal to $\frac{i+j}{2}$.

41. In modern textbooks the integral solutions to $x^2 + y^2 = z^2$ are usually characterized by

$$x = k(m^2 - n^2), \quad y = k \cdot 2mn, \quad z = k(m^2 + n^2),$$

with PRIME (m, n) and one of m and n even. To show the equivalence of this modern condition and the Euclidean one it is simplest to begin by pointing out that, by VIII,18 and 20 and VIIIC,

$$\text{SIMPLANE}(i_1, i_2) \leftrightarrow \exists l \exists l_1 \exists l_2 (\text{LEAST}(l_1, l_2) \;\&\; i_1 \simeq l \cdot l_1^2 \;\&\; i_2 \simeq l \cdot l_2^2).$$

To see that a solution satisfying the modern condition satisfies the Euclidean one it is only necessary to set $i = k(m+n)^2$ and $j = k(m-n)^2$.

The proof that integers satisfying the Euclidean condition satisfy the modern one is more complex. Suppose i and j are similar plane numbers with $i = k \cdot m^2$ and $j = k \cdot n^2$, and LEAST (m, n). Then, either (a) one of m and n is even, or (b) they are both odd. In the first case, since

$$\sqrt{ij} = kmn, \quad \frac{i-j}{2} = \frac{k(m^2 - n^2)}{2}, \quad \frac{i+j}{2} = \frac{k(m^2 + n^2)}{2},$$

k cannot be odd, since, if it were, $k(m^2 - n^2)$ would be odd and $\frac{i-j}{2}$ would not be an integer. Hence, in case (a) the Euclidean solution is expressible in the form

$$\frac{k}{2} \cdot 2mn, \quad \frac{k}{2}(m^2 - n^2), \quad \frac{k}{2}(m^2 + n^2).$$

In case (b) m and n are of the form $2m' + 1$ and $2n' + 1$, so that

$$m^2 - n^2 = (m + n)(m - n) = 2(m' + n' + 1)2(m' - n').$$

Since $m' + n'$ and $m' - n'$ have the same parity, exactly one of $m' + n' + 1$ and $m' - n'$ is even. Moreover, any prime dividing $m' + n' + 1$ and $m' - n'$ divides their sum and difference, i.e., divides the relatively prime m and n; therefore $m' + n' + 1$ and $m' - n'$ are relatively prime. In addition,

$$(m' + n' + 1)^2 - (m' - n')^2 = 4m'n' + 2m' + 2n' + 1 = (2m' + 1)(2n' + 1) = mn,$$

and

$$(m' + n' + 1)^2 + (m' - n')^2 = 2m'^2 + 2n'^2 + 2m' + 2n' + 1 = \frac{4m'^2 + 4m' + 1 + 4n'^2 + 4n' + 1}{2} = \frac{(2m' + 1)^2 + (2n' + 1)^2}{2} = \frac{m^2 + n^2}{2}.$$

Thus, if one sets $m^* = (m' + n' + 1)$ and $n^* = (m' - n')$, the Euclidean solution is expressible as $k(m^{*2} - n^{*2})$, $k \cdot 2m^*n^*$, $k(m^{*2} + n^{*2})$ with m^* and n^* relatively prime and one of them even.

42. Euclid treats the two cases $n \simeq m - 1$ and $n \prec m - 1$ separately. He also takes for granted, as I have done in the text, that $2 \cdot (m - n) \prec i - j$. To see that this must be the case, assume $i - j \preceq 2 \cdot (m - n)$, i.e., $\frac{i - j}{2} \preceq m - n$, or $n \preceq m - \frac{i - j}{2} \simeq \frac{i + j}{2} - \frac{i - j}{2} \simeq \frac{2 \cdot j}{2} \simeq j$. Hence $i \cdot j + (l - 1)^2 \simeq n^2 \preceq j^2$, which is impossible, because $i \succ j$ so that $i \cdot j \succ j^2$.

43. Heiberg printed as Appendix IV to the *Scholia* the "Arithmetic proofs of the propositions proved by means of lines in the second book of the *Elements*," written in the fourteenth century by Barlaam. The title is descriptive. Barlaam establishes by combinatorial argument the arithmetic analogues of those propositions of book II which have such analogues. The proofs are not derived in any obvious way from Euclid's or Heron's, although Barlaam does exploit deductive consequences in the manner of Heron. The foundation of the proofs is clearly Euclid's tacit arithmetic assumptions. One assumption which Barlaam exploits particularly often is

$$\mathcal{M}(m, k_i, l_i) \rightarrow \mathcal{M}(m, \sum^{n} k_i, \sum^{n} l_i),$$

which is just an alternative formulation of VIIh′.

44. See, for example, Heath, vol. I, pp. 385, 398–401. A more speculative account is given by Zeuthen in *Sur l'origine . . .*, pp. 10–29.

45. For Heiberg and Heath's suspicions of the *lēmma* and proposition 10, see Heath, vol. III, pp. 31–33. See also p. 278 of the present work. Proposition 9 has played an especially important role in reconstructions of pre-Euclidean mathematics, because scholium 62 on it credits Theaetetus with its discovery. See the works by Zeuthen, van der Waerden, and Knorr mentioned in the bibliographical note, and also Dijksterhuis's commentary on X,9 (vol. II, pp. 173–178).

3 Magnitudes in Proportion

3.1 Book V

In the previous chapter I tried to make clear the differences between Greek ratios and fractions. The fundamental one is that whereas fractions are objects, ratios are not. An equality $(k, l) = (j, m)$ is really an assertion that a four-place relation holds among k, l, j, m. It is, of course, possible to deny that $(k, l) = (j, m)$, but in the arithmetic books Euclid does not even provide the apparatus for asserting $(k, l) > (j, m)$. The corresponding apparatus for fractions is essential for incorporating the real numbers into the system of fractions. The need for the reals is customarily explained by the lack of a rational solution for certain equations, e.g., $x^2 = 2$—a fact which might be expressed in a more Euclidean way by saying that there is no number x satisfying $(k, x) = (x, 2 \cdot k)$.

The orderly introduction of the reals into the system of fractions is one of the major accomplishments of the foundational work of the nineteenth century. There are a number of different ways of making the introduction, all of them equivalent for mathematical and philosophical purposes. I shall briefly describe a modified version of the method of Dedekind, restricted to positive fractions and real numbers.[1] One begins from the system of positive fractions, taking for granted the ordering relation of being less than and the operations of addition, subtraction, multiplication, etc. The example of the previous paragraph makes clear the key to a satisfactory introduction of the reals. For, if $\sqrt{2}$ is the positive solution to $x^2 = 2$, then one should have

$$\forall k \forall l \left(\frac{k}{l} < \sqrt{2} \vee \frac{k}{l} > \sqrt{2}\right),$$

$$\forall k \forall l \forall j \forall m \left(\left(\frac{k}{l} < \sqrt{2} \;\&\; \frac{j}{m} \leq \frac{k}{l} \rightarrow \frac{j}{m} < \sqrt{2}\right) \;\&\; \left(\frac{k}{l} > \sqrt{2} \;\&\; \frac{j}{m} \geq \frac{k}{l} \rightarrow \frac{j}{m} > \sqrt{2}\right)\right);$$

or, in other words, that $\sqrt{2}$ partitions the positive fractions into two parts, one containing all the fractions less than $\sqrt{2}$, the other all those greater than $\sqrt{2}$. For simplicity one can identify the reals with such partitions, or, for even greater simplicity, with their lower parts. In addition, the rational reals can be accommodated by including partitions made by a fraction, i.e., those the upper part of which contains a least fraction. Finally, since there is no real greater than every fraction and none less than every fraction, one does not want the empty set or the set of all reals to be a real. Hence one defines

Rdef. 3 $\text{REAL}(x) \leftrightarrow \exists k \exists l\left(\frac{k}{l} \in x\right) \;\&\; \exists k \exists l\left(\frac{k}{l} \notin x\right) \;\&$
$$\forall k \forall l \forall j \forall m\left(\frac{k}{l} \in x \;\&\; \frac{j}{m} \leq \frac{k}{l} \rightarrow \frac{j}{m} \in x\right) \;\&$$
$$\forall k \forall l\left(\frac{k}{l} \in x \rightarrow \exists j \exists m\left(\frac{j}{m} \in x \;\&\; \frac{k}{l} < \frac{j}{m}\right)\right);$$

Rdef. 7 $x > y \leftrightarrow \exists k \exists l\left(\frac{k}{l} \in x \;\&\; \frac{k}{l} \notin y\right);$

Rdef. 5 $x = y \leftrightarrow \forall k \forall l\left(\frac{k}{l} \in x \leftrightarrow \frac{k}{l} \in y\right).$

These definitions, the last of which is unnecessary because it is a set-theoretic truth, are numbered to make them correspond with their Euclidean analogues in book V.

The introduction of the reals is the most elementary example of the move from the constructive to the nonconstructive in mathematics. Reals are infinite classes; and nonconstructive or infinitistic methods are sometimes required to establish their fundamental properties. Two examples will suffice to illustrate the point. One, the law of trichotomy, is simple:

R1 $x > y \vee x = y \vee y > x.$

By definition this is equivalent to

$$\exists k \exists l\left(\frac{k}{l} \in x \;\&\; \frac{k}{l} \notin y\right) \vee \forall k \forall l\left(\frac{k}{l} \in x \leftrightarrow \frac{k}{l} \in y\right) \vee$$
$$\exists k \exists l\left(\frac{k}{l} \in y \;\&\; \frac{k}{l} \notin x\right),$$

which is logically equivalent to

$$\forall k \forall l\left(\frac{k}{l} \in x \leftrightarrow \frac{k}{l} \in y\right) \vee \exists k \exists l \neg\left(\frac{k}{l} \in x \leftrightarrow \frac{k}{l} \in y\right),$$

an instance of the logical law of the excluded middle,

$$\forall x \mathscr{P}(x) \vee \exists x \neg \mathscr{P}(x).$$

Hence R1 is logically true in standard logic. However, it is not true under the constructive interpretation of the quantifiers. For under this interpretation an assertion of the form $\exists x \mathscr{P}(x)$ means that an example of a $\mathscr{P}$ can be produced, i.e., constructed. On the other hand, an assertion of the form $\forall x \mathscr{P}(x)$ can be made constructively only if there is a known general law establishing that every particular x is a $\mathscr{P}$. Thus, on the constructive interpretation, R1 asserts that, for any two reals, either a law can be proved establishing that any fraction in one is in the other or an example of a fraction in one but not the other can be produced, by no means an obvious truth. R1 is true in ordinary mathematics because there the reals and the class of fractions are treated as fixed totalities, determinate in all their mathematical properties. Given this approach, one of the alternatives $\frac{k}{l} \in x$ or $\frac{k}{l} \notin x$ must hold for any $\frac{k}{l}$ and x; and if the first alternative does not hold for all fractions, the second must hold for at least one, whether or not an example can be produced.

It is important to realize that not all forms of the law of the excluded middle are constructively meaningless. In particular, for elementary assertions of equality between geometric objects or numbers, one is entitled to assume, as Euclid does, that either they or their negations are true; hence one is also entitled to infer the equality of two such objects from the impossibility of their inequality. Because trichotomy for fractions is constructively true, it is possible to show in a constructive way that at most one of the alternatives $x > y$, $x = y$, $y > x$ must hold for any reals x and y.[2]

I take as a second example of a nonconstructive assertion about real numbers the least upper bound theorem: if there is a real number x not less than any of the reals in a nonempty set S of reals, there is a least such real number y. A constructive proof of this theorem would show how, given x and S, one determines y—at least to the extent of telling whether or not any given $\frac{k}{l}$ is in y. The ordinary proof takes as y the set of all fractions contained in at least one member of S (the union of S). I shall not rehearse the proof that this set is the least upper bound of S. The crucial point is that specifying y as the union provides no general way of determining membership in y. For, to decide whether or not $\frac{k}{l}$ is in y, one would have to decide whether or not $\frac{k}{l}$ is in a member of a member of S; but since S may be infinite, each member of S is infinite, and there need be no way of determining membership in S, there may very well be no way of determining membership in y. Hence, the least upper bound theorem asserts the existence of a least upper bound, but its proof does not give a procedure for constructing one. Thus the theorem is nonconstructive; and so, in fact, are most of the basic theorems of real number theory and the calculus.

If my account of Euclidean arithmetic is correct, there can be no doubt that the *Elements* do not contain the basis for the development of an equivalent of real number theory. For numerical ratios do not form a system of objects ordered by a relation of being less than; indeed, ratios are not objects at all. Hence there can be no question of interpolating reals into such a system or of defining reals as sets of numerical ratios. On the other hand, there can be no doubt that the Greeks were aware of problems analogous to those that are now solved by the theory of real numbers; for example, they realized that there was no numerical ratio corresponding to the relation of the side of a square to its diagonal. The definitive Greek way of dealing with this situation is given in book V, to which I now turn.

The starting point for this approach is allowing what are called magnitudes (*megethē*) to occur in proportionalities, and defining when a first magnitude is in the same ratio to a second as a third is to a fourth and when a first has a greater ratio to a second than a third has to a fourth. There has been some disagreement among scholars concerning the exact nature of magnitudes. I shall be defending the view that the most appropriate interpretation of magnitudes in the *Elements* involves construing them as abstractions from geometric objects which leave out of account all properties of those objects except quantity: i.e., length for lines, area for plane figures, volume for solids, size, however characterized, for angles. Magnitudes are in this respect quite like units and collections of units (numbers), which can be thought of as abstractions from objects which leave out of account all properties except self-identity and numerosity. There are indeed many similarities between Euclid's treatment of numbers and his treatment of magnitudes, the most important of which concerns multiplication. The definitions of equality and inequality for ratios presuppose that magnitudes can be multiplied and the results of multiplication compared. It is clear from book V that m-fold multiplication of a magnitude $x (m \cdot x)$ is thought of in the same way as arithmetic multiplication, namely, as the concatenation or addition of m distinct magnitudes each equal to x. However, since Euclid does not have the notion of number to work with in book V, he consistently uses locutions like "x is the same multiple of y as z is of w" rather than invoking more explicitly arithmetic notions.[3]

Modern commentators have not found anything particularly problematic about Euclid's treatment of multiplication in either arithmetic or the theory of magnitudes, perhaps because, in the analogous modern theories, multiplication by an integer introduces no particular problem. However, in the case of geometric objects Euclid has laid down, at least tacitly, specific notions of constructibility; and it is possible to ask whether or not the notion of multiplication makes constructive sense for a given kind of geometric object. For a straight line AA_1 one can think of n-fold multiplication as the extension of AA_1 to $AA_1A_2 \dots A_n$ with each A_iA_{i+1} equal to AA_1. For a rectilineal plane figure one can imagine its transformation into a rectangle of given height and base AA_1 and then the construction of another rectangle of the same height with base $AA_1A_2 \dots A_n$. In other cases, however, the situation is quite different. Multiplication of an angle eventually produces angles not thought to be angles by Euclid, namely, angles greater than 180°. In the case of circles and certain solids, the theorems in book XII which make possible the production of a figure which is a multiple of a given one depend for their proof on the possibility of taking such multiples. One could, of course, treat the n-fold multiplication of a figure as the construction of n figures equal to it, but it seems more reasonable to give up the attempt to interpret the operation of multiplication constructively. For Euclid makes no attempt to indicate how the operation is to be performed. In his diagrams magnitudes are represented by straight lines, but his vocabulary suggests that these lines have no more geometric significance than their counterparts in arithmetic diagrams. Magnitudes, like numbers, are objects of which arbitrary multiples can simply be taken, i.e., conceived.

There is a similar point to be made in connection with another operation which Euclid assumes to be possible in the proof of V,5, namely, the taking of an mth part of a given magnitude. This assumption may be expressed

Va $\forall x \exists y (x \simeq m \cdot y)$.

This assumption is, of course, false for numbers; and, from a constructive point of view, taking parts is even less satisfactory than taking multiples. For it is known that no construction with straightedge and compass will produce the third part of a circular arc or rectilineal angle. It is possible to revise Euclid's proof of V,5 so that Va is not invoked.[4] However, there is no reason to do so if one admits that the constructive point of view is not functioning in book V. I shall be returning to this question of constructivity later in this chapter. I now wish to look at the foundational aspects of book V.

In general the tacit foundation of book V is quite like the foundation of Euclidean arithmetic.[5] Corresponding to the tacit assumption of infinitely many units in the arithmetic books, there is an assumption, which makes multiplication possible, of the existence, for any magnitude, of arbitrarily many magnitudes equal to it. Euclid also uses terms like 'part' and 'measure' in much the same way as in arithmetic and makes the same kind of assumptions about them, assumptions which are easily expressed by simple changes from numerical variables to variables ranging over magnitudes. Here I shall only state

Vb (i) $x \simeq y \leftrightarrow m \cdot x \simeq m \cdot y$;
(ii) $x \prec y \leftrightarrow m \cdot x \prec m \cdot y$;
(iii) $m \prec n \leftrightarrow m \cdot x \prec n \cdot x$.

One also finds the same kind of combinatorial assumptions about addition and subtraction in the arithmetic books and in book V. On the other hand, in book V Euclid does attempt to prove analogues of some of his arithmetic assumptions, e.g.,

V,1 (cf. VIIh′) $\sum^{n} m \cdot x_i \simeq m \cdot \sum^{n} x_i$.

Euclid treats the case $n = m = 2$, but the general argument may be represented as follows. Let

$$m \cdot x_i \simeq \sum_{j}^{m} x_{i_j}.$$

Then

$$\sum_{i}^{n} m \cdot x_i \simeq \sum_{i}^{n}\left(\sum_{j}^{m} x_{i_j}\right) \simeq \sum_{j}^{m}\left(\sum_{i}^{n} x_{i_j}\right) \simeq m \cdot \sum_{i}^{n} x_i.$$

In interchanging the two sum signs Euclid takes for granted VIIh″ for magnitudes, which is simply another form of what he is trying to prove. Euclid also proves an analogue of VIIi ($m \cdot x - m \cdot y \simeq m \cdot (x - y)$). The core of his argument is a reduction to V,1 which might be expressed as follows: Suppose, using Va, that $m \cdot x - m \cdot y \simeq m \cdot z$, so that $m \cdot x \simeq m \cdot z + m \cdot y$; then, by V,1, $m \cdot x \simeq m \cdot (z + y)$, i.e., by Vb, $x \simeq z + y$, or $z \simeq x - y$. Euclid's argument is not quite so straightforward, because he does not make full use of numerical ideas in treating multiples. He states the analogue of VIIi as

V,5 If x_1 is the same multiple of x_2 as y_1 is of y_2, then $x_1 - y_1$ is the same multiple of $x_2 - y_2$ as x_1 is of x_2,

and argues as follows. Let (by Va) $x_1 - y_1$ be the same multiple of z as y_1 is of y_2. Then y_1 is the same multiple of y_2 as x_1 is of $y_2 + z$. (Here the implicit argument is that if $x_1 - y_1 \simeq m \cdot z$ and $y_1 \simeq m \cdot y_2$, $x_1 \simeq m \cdot y_2 + m \cdot z \simeq$ (by V,1) $m \cdot (y_2 + z)$.) Hence x_1 is the same multiple of x_2 and of $y_2 + z$. Therefore (by Vb), $x_2 \simeq y_2 + z$, i.e., $x_2 - y_2 \simeq z$, and y_1 is the same multiple of y_2 that $x_1 - y_1$ is of $x_2 - y_2$; thus if the antecedent of V,5 is true, so is the consequent.

Such minimizing of the numerical aspect of multiplication is noticeable in three other early theorems of book V. Instead of proving

V,2′ (cf. VIIh) $y \simeq m \cdot x \;\&\; z \simeq n \cdot x \rightarrow y + z \simeq (m + n) \cdot x$

(i.e., $m \cdot x + n \cdot x \simeq (m + n) \cdot x$),

V,3′ $y \simeq m \cdot x \;\&\; z \simeq n \cdot y \rightarrow z \simeq (n \cdot m) \cdot x$

(i.e., $n \cdot (m \cdot x) \simeq (n \cdot m) \cdot x$),

V,6′ $y \simeq m \cdot x \;\&\; z \simeq n \cdot x \rightarrow y - z \simeq (m - n) \cdot x$

(i.e., $m \cdot x - n \cdot x \simeq (m - n) \cdot x$),

Euclid proves the weaker, less explicit

V,2 $y_1 \simeq m \cdot x_1 \;\&\; y_2 \simeq m \cdot x_2 \;\&\; z_1 \simeq n \cdot x_1 \;\&\; z_2 \simeq n \cdot x_2 \rightarrow \exists k(y_1 + z_1 \simeq k \cdot x_1 \;\&\; y_2 + z_2 \simeq k \cdot x_2)$,

V,3 $y_1 \simeq m \cdot x_1 \;\&\; y_2 \simeq m \cdot x_2 \;\&\; z_1 \simeq n \cdot y_1 \;\&\; z_2 \simeq n \cdot y_2 \rightarrow \exists k(z_1 \simeq k \cdot x_1 \;\&\; z_2 \simeq k \cdot x_2)$,

V,6 $y_1 \simeq m \cdot x_1 \;\&\; y_2 \simeq m \cdot x_2 \;\&\; z_1 \simeq n \cdot x_1 \;\&\; z_2 \simeq n \cdot x_2 \rightarrow \exists k(y_1 - z_1 \simeq k \cdot x_1 \;\&\; y_2 - z_2 \simeq k \cdot x_2)$.

In proving V,2 Euclid never even introduces particular multiples. Rather, he proceeds as follows:

For since y_1 is the same multiple of x_1 that y_2 is of x_2, therefore as many magnitudes as there are in y_1 equal to x_1 so many are there in y_2 equal to x_2. For the same reason also, as many as there are in z_1 equal to x_1, so many are there in z_2 equal to x_2. *Therefore as many as there are in the whole $y_1 + z_1$ equal to x_1 so many are there in the whole $y_2 + z_2$ equal to x_2. Therefore whatever multiple $y_1 + z_1$ is of x_1 that multiple also is $y_2 + z_2$ of x_2.

It is difficult to see that there is any proof here at all since the crucial inference, marked with an asterisk, seems to depend upon VIIh for magnitudes, the stronger form of V,2. Of course, VIIh is obvious, given the intuitive conception of addition as concatenation; but then so is V,2.

Euclid proves V,3 for the case $n = 2$, making it possible to use V,2. For if the antecedent of V,3 is true and $z_1 \simeq y_1' + y_1''$ and $z_2 \simeq y_2' + y_2''$ with $y_1' \simeq y_1'' \simeq y_1 \simeq m \cdot x_1$ and $y_2' \simeq y_2'' \simeq y_2 \simeq m \cdot x_2$, then, by V,2 with $n = m$, z_1 and z_2 are equimultiples of x_1 and x_2. It is clear enough that Euclid could have proved V,2 for arbitrarily many summands.[6] The proof of V,6 is very confusing. If Euclid had proved V,2 in an explicit numerical form, he could have argued, on analogy with his proof of V,5, that if $m \cdot x - n \cdot x \simeq (m - n) \cdot z$, then $m \cdot x \simeq (m - n) \cdot z + n \cdot x \simeq$ (by V,2′) $(m - n) \cdot x + n \cdot x$, so that $(m - n) \cdot x \simeq (m - n) \cdot z$, and $x \simeq z$. Because he has foregone a numerical treatment of multiplication but wants to reduce V,6 to V,2, Euclid assumes that if the antecedent of 6 is true, $y_1 - z_1$ is some multiple $k \cdot x_1$ of x_1. He then argues that $y_2 - z_2$ is the same multiple of x_2, on the grounds that since $y_1 \simeq (y_1 - z_1) + z_1$, y_1 is the same multiple of x_1 that $z_2 + k \cdot x_2$ is of x_2, by V,2;[7] hence y_2 is the same multiple of x_2 that $z_2 + k \cdot x_2$ is of x_2, so that $y_2 \simeq z_2 + k \cdot x_2$, and $y_2 - z_2 \simeq k \cdot x_2$. This argument is sound; but a proof of the assumption that $y_1 - z_1$ is some multiple of x_1 would seem to require proving that $y_1 - z_1 \simeq (m - n) \cdot x_1$—the stronger form of V,6 itself.

V,1–3, 5, and 6 do not depend on anything except Euclid's understanding of the nature of magnitudes and operations on them. The other propositions of book V depend ultimately on one or both of the fundamental definitions

V, def. 5 $(x, y) = (z, w) \leftrightarrow$
$\forall m \forall n((m \cdot x \succ n \cdot y \rightarrow m \cdot z \succ n \cdot w)$ &
$(m \cdot x \simeq n \cdot y \rightarrow m \cdot z \simeq n \cdot w)$ &
$(m \cdot x \prec n \cdot y \rightarrow m \cdot z \prec n \cdot w))$,

V, def. 7 $(x, y) > (z, w) \leftrightarrow$
$\exists m \exists n (m \cdot x \succ n \cdot y \ \& \ \neg (m \cdot z \succ n \cdot w))$.

Hereafter I shall abbreviate $(x \succ y \rightarrow z \succ w)$ & $(x \simeq y \rightarrow z \simeq w)$ & $(x \prec y \rightarrow z \prec w)$ by COMP(x, y, z, w) so that the *definiens* of definition 5 would be written

$\forall m \forall n$COMP$(m \cdot x, n \cdot y, m \cdot z, n \cdot w)$.

As an example of the use of this definition, I give the proof of

V,4 $(x, y) = (z, w) \rightarrow (m \cdot x, n \cdot y) = (m \cdot z, n \cdot w)$.

By V,3 any multiples $k \cdot (m \cdot x)$, $l \cdot (n \cdot y)$, $k \cdot (m \cdot z)$, $l \cdot (n \cdot w)$ of $m \cdot x$, $n \cdot y$, $m \cdot z$, $n \cdot w$ are multiples $i \cdot x$, $j \cdot y$, $i \cdot z$, $j \cdot w$ of x, y, z, w; hence if the antecedent of V,4 is true,

COMP$(k \cdot (m \cdot x), l \cdot (n \cdot y), k \cdot (m \cdot z), l \cdot (n \cdot w))$.

V,4 is used later in book V, as are V,1 and 2. However, the only application of V,3 is made in the proof of V,4; and 5 and 6 are never used in the *Elements*. Figure 3.1 gives the deductive structure of these six propositions. (In it and succeeding diagrams the overbar continues to indicate that a proposition is not used later in the book under discussion.)

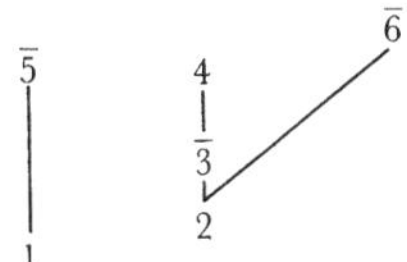

Figure 3.1

The relationship of Euclid's definitions 5 and 7 to Dedekind's definitions for real numbers can be seen by construing magnitudes as positive real numbers, positive integers as integral reals, and ratios (x, y) as quotients $\frac{x}{y}$, while also taking for granted the ordinary mathematical operations on real numbers. Then cross multiplication in definitions 5 and 7 yields

$$5' \quad \frac{x}{y} = \frac{z}{w} \leftrightarrow \forall m \forall n \left(\left(\frac{x}{y} > \frac{n}{m} \rightarrow \frac{z}{w} > \frac{n}{m} \right) \& \left(\frac{x}{y} = \frac{n}{m} \rightarrow \frac{z}{w} = \frac{n}{m} \right) \& \left(\frac{x}{y} < \frac{n}{m} \rightarrow \frac{z}{w} < \frac{n}{m} \right) \right),$$

$$7' \quad \frac{x}{y} > \frac{z}{w} \leftrightarrow \exists m \exists n \left(\frac{x}{y} > \frac{n}{m} \; \& \; \neg \left(\frac{z}{w} > \frac{n}{m} \right) \right).$$

However, if x_1 and x_2 are positive reals, so is $\frac{x_1}{x_2}$; and any real x can be represented as a quotient of two reals, e.g., as $\frac{x}{1}$. Hence, 5′ and 7′ can be read as definitions of $=$ and $>$ for arbitrary positive real numbers. To complete the comparison with Dedekind's definitions one need only write $x > \frac{n}{m}$ and $x < \frac{n}{m}$ as $\neg \frac{n}{m} \in x$ and $\frac{n}{m} \in x$, respectively, and note that the clause $x = \frac{n}{m} \rightarrow y = \frac{n}{m}$ is vacuously true if quotients of positive integers are thought of as fractions, and x and y as ranging over sets of fractions. 5′ and 7′ may now be written as

$$5'' \quad x = y \leftrightarrow \forall m \forall n \left(\left(\frac{n}{m} \in x \rightarrow \frac{n}{m} \in y \right) \& \left(\neg \left(\frac{n}{m} \in x \right) \rightarrow \neg \left(\frac{n}{m} \in y \right) \right) \right),$$

$$7'' \quad x > y \leftrightarrow \exists m \exists n \left(\frac{n}{m} \in x \; \& \; \neg \left(\frac{n}{m} \in y \right) \right);$$

these formulations are obviously equivalent to Rdefs. 5 and 7. Thus it can be said that the *Elements* contain equivalents of the modern definitions of $=$ and $>$ for positive real numbers, but no equivalent of the definition of 'real number'. Euclid's definition of 'ratio' is mathematically useless:

V, def. 3. A ratio is a sort of relation in respect of size between two homogeneous magnitudes. (*Logos esti duo megethōn homogenōn hē kata pēlikotēta poia schesis.*)

Philosophically the difference between the modern and the ancient treatment of irrationality is crucial. For the modern definition of the reals presupposes the densely ordered system of the rationals and determines a new system of objects, each of infinite complexity. The only objects of book V are magnitudes capable of being multiplied and of being compared to one another in respect of size. Thus the fundamental definitions of book V do not in themselves involve nonconstructive or infinitistic mathematical ideas. On the other hand, certain "obvious" facts can only be proved nonconstructively. The most elementary of these is trichotomy for ratios:

(i) $(x, y) > (z, w) \vee (x, y) = (z, w) \vee (z, w) > (x, y)$,

which, by definition, is equivalent to the logical truth[8]

(i′) $\forall m \forall n \text{COMP}(m \cdot x, n \cdot y, m \cdot z, n \cdot w) \vee$
$\exists m \exists n \neg \text{COMP}(m \cdot x, n \cdot y, m \cdot z, n \cdot w)$,

or

Either all pairs of integers m, n satisfy COMP $(m \cdot x, n \cdot y, m \cdot z, n \cdot w)$ or some pair does not.

Under a constructive interpretation, then, (i′) asserts

Either there is a law establishing COMP $(m \cdot x, n \cdot y, m \cdot z, n \cdot w)$ for arbitrary m and n or particular m and n such that $\neg$COMP $(m \cdot x, n \cdot y, m \cdot z, n \cdot w)$ can be found.

There is no reason to believe that this assertion is true, even if x, y, z, w are taken to be straight lines so that it is immediately decidable whether or not COMP $(m \cdot x, n \cdot y, m \cdot z, n \cdot w)$ for any particular m and n.

Becker pointed out that Euclid never uses the law (i)[9]—a fact which Becker took to be further evidence of Euclid's adherence to a constructive point of view. However, Euclid uses another nonconstructive assumption,

Vc $\exists w((x, y) = (z, w))$,

in book V and in book XII. Vc is stronger than (i) in the sense that there are domains, e.g., the positive integers, in which (i) is true and Vc false but (i) can be derived from Vc.[10] At present I want only to point out the necessity, from a constructive point of view, of distinguishing between two basic uses of (i): the one to infer an inequality from the denial of a proportionality; the other to infer a proportionality from the denial of an inequality. In other words, one must distinguish between the two laws

(ia) $\neg((x, y) = (z, w)) \rightarrow$
$((x, y) > (z, w) \vee (z, w) > (x, y))$

and

(ib) $\neg((x, y) > (z, w) \vee (z, w) > (x, y)) \rightarrow (x, y) = (z, w)$.

Only the first of these is nonconstructive.[11] In section 3.2 I will show that Euclid's use of Vc in book V could in fact be replaced with an application of (ib) and thereby show it to be very unlikely that the absence of citations of (i) reflects a concern for constructivity.

Book V is generally thought to be the work of Eudoxus because of two scholia, the more extensive of which (*Scholia*, 282.13–20) says,

> This book is said to be by Eudoxus of Knidos, the mathematician who was born in Plato's time; it has also been ascribed to Euclid; but this is not a false ascription; for nothing prevents it from being someone else's as far as discovery is concerned; but it is agreed by all that as far as the arrangement of it with respect to fundamentals and the logical relation (*akolouthia*) of the things thus arranged to the other [propositions], it is Euclid's.

There is no reason to doubt the scholiast's ascription of the content of Book V to Eudoxus. But what is one to make of his description of the role of Euclid? Euclid might have simply inserted a treatise by Eudoxus into the *Elements*, perhaps adding a few definitions. Or he might have made fairly substantive changes. For example, Eudoxus might have proceeded more arithmetically in the early propositions or perhaps even have taken the combinatorial laws proved in these propositions for granted. Again, he might have used the propositions V,5 and 6 either in proofs which Euclid revised or in connection with propositions which Euclid eliminated entirely. Although I see no clear way to decide among these possibilities, my discussion of the deductive structure of book V will include some reference to alternative proofs available to Euclid. For discussion of this structure it is best to begin with the analogue of VII,13:

V,16 $(x, y) = (z, w) \rightarrow (x, z) = (y, w)$.

Given that $(x, y) = (z, w)$, Euclid has to prove that COMP $(m \cdot x, n \cdot z, m \cdot y, n \cdot w)$. This conclusion is implied by the following propositions which one expects to be true:

V,15 $(x, y) = (m \cdot x, m \cdot y)$,
V,14 $(x, y) = (z, w) \rightarrow$ COMP (x, z, y, w),
V,11 $(x, y) = (z, w) \mathbin{\&} (z, w) = (u, v) \rightarrow (x, y) = (u, v)$,
VA $(x, y) = (z, w) \rightarrow (z, w) = (x, y)$.

For, by V,15, $(x, y) = (m \cdot x, m \cdot y)$ and $(z, w) = (n \cdot z, n \cdot w)$. Applications of VA and V,11 give $(m \cdot x, m \cdot y) = (n \cdot z, n \cdot w)$, so that, by 14, COMP $(m \cdot x, n \cdot z, m \cdot y, n \cdot w)$.

Of the laws used in this proof Euclid takes only VA for granted, presumably as a trivial consequence of definition 5. V,11 is an equally trivial consequence of this definition, since it says $\text{COMP}(m \cdot x, n \cdot y, m \cdot z, n \cdot w)$ & $\text{COMP}(m \cdot z, n \cdot w, m \cdot u, n \cdot v) \rightarrow \text{COMP}(m \cdot x, n \cdot y, m \cdot u, n \cdot v)$. Euclid's proof of 11 amounts to no more than bringing out this definitional expansion. As the example of VA and V,11 indicates, there is no clear general characterization distinguishing trivial laws of proportionality which Euclid is willing to take for granted from those which he thinks require proof. He proves

V,7 $\quad x \simeq y \rightarrow (x, z) = (y, z) \;\&\; (z, x) = (z, y),$

although on definitional expansion it says merely

$$x \simeq y \rightarrow \text{COMP}(m \cdot x, n \cdot z, m \cdot y, n \cdot z) \;\& \\ \text{COMP}(m \cdot z, n \cdot x, m \cdot z, n \cdot y),$$

which, given Euclid's tacit acceptance of Vb(i), is surely obvious. On the other hand, he uses the law of inversion,

$$(x, y) = (z, w) \rightarrow (y, x) = (w, z),$$

without a satisfactory proof, although in the manuscripts this law is inserted as a corollary to V,4 or 7, where it patently does not belong.[12]

Because multiplication is iterated addition, V,15 says that

$$(x, y) = \left(\sum^{m} x_i (x_i \simeq x), \sum^{m} y_i (y_i \simeq y)\right).$$

The equality of the x_i to x and of the y_i to y means by V,7 that $(x, y) = (x_i, y_i)$. Hence V,15 reduces to

V,12 $\quad (x_1, y_1) = (x_2, y_2) = \ldots = (x_m, y_m) \rightarrow$
$$(x_1, y_1) = \left(\sum^{m} x_i, \sum^{m} y_i\right).$$

Euclid proves this proposition by pointing out that if the antecedent is true, $\text{COMP}(k \cdot x_1, n \cdot y_1, k \cdot x_i, n \cdot y_i)$; he then cites V,1 to justify the conclusion that $\text{COMP}\left(k \cdot x_1, n \cdot y_1, k \cdot \sum^{m} x_i, n \cdot \sum^{m} y_i\right)$. It is plausible to explain this inference as dependent upon an intermediate step where it is inferred by way of generalizations of common notion 1 and of a consequence of CN4′ (namely, $x \prec y \;\&\; z \prec w \rightarrow x + z \prec y + w$) that

$$\text{COMP}\left(k \cdot x_1, n \cdot y_1, \sum^{m} k \cdot x_i, \sum^{m} n \cdot y_i\right).$$

Euclid divides the proof of V,14 into three cases, depending upon whether (a) $x \succ z$, (b) $x \simeq z$, or (c) $x \prec z$. He does only case (a), which he derives from the following propositions:

$$\begin{array}{ll} \text{V,8} & x \succ y \rightarrow (x, z) > (y, z) \,\&\, (z, y) > (z, x), \\ \text{V,13} & (x, y) = (z, w) \,\&\, (z, w) > (u, v) \rightarrow (x, y) > (u, v), \\ \text{V,10} & ((x, z) > (y, z) \vee (z, y) > (z, x)) \rightarrow x \succ y. \end{array}$$

For, by V,8, if $x \succ z$, $(x, y) > (z, y)$. Hence, if the antecedent of V,14 is true, $(z, w) > (z, y)$, by V,13 and VA, and, by V,10, $y \succ w$. Of the propositions used here, V,13 is trivial, since on application of VA and definitional expansion it is a consequence of

$$m \cdot z \succ n \cdot w \,\&\, \neg (m \cdot u \succ n \cdot v) \,\&\, \text{COMP}\,(m \cdot z, n \cdot w, m \cdot x, n \cdot y) \\ \rightarrow m \cdot x \succ n \cdot y \,\&\, \neg (m \cdot u \succ n \cdot v).$$[13]

V,7, 8, and 10 form a group with the converse of V,8,

$$\text{V,9} \quad ((x, z) = (y, z) \vee (z, x) = (z, y)) \rightarrow x \simeq y.$$

This proposition is never explicitly used in book V, although it probably would play a role in the proof of case (b) of V,14.[14] Although 9 and 10 are directly inferable from definitions 5 and 7,[15] Euclid attempts indirect reductions of them to 7 and 8, taking for granted that

$$\begin{array}{ll} \text{VB} & \neg((x, y) = (z, w) \,\&\, (x, y) > (z, w)), \\ \text{VC} & \neg((x, y) > (z, w) \,\&\, (z, w) > (x, y)). \end{array}$$

The first of these is an easy consequence of definitions 5 and 7. For us the second is also an easy consequence of these definitions; since, if it were false, there would be m, n, j, l such that $m \cdot x \succ n \cdot y$, $m \cdot z \preceq n \cdot w$, $j \cdot x \preceq k \cdot y$, and $j \cdot z \succ k \cdot w$; hence $k \cdot (m \cdot x) \succ k \cdot (n \cdot y), n \cdot (k \cdot y) \succeq n \cdot (j \cdot x), k \cdot (m \cdot z) \preceq k \cdot (n \cdot w)$, and $n \cdot (k \cdot w) \prec n \cdot (j \cdot z)$. For us there is no difficulty in concluding that $k \cdot m \succ n \cdot j$ and $k \cdot m \prec n \cdot j$, which is impossible; but to carry out this inference in Euclidean form would require quite complex combinatorial argument. One could avoid such argument by invoking VIIh″ for magnitudes to infer $k \cdot (n \cdot y) \simeq n \cdot (k \cdot y)$ and $k \cdot (n \cdot w) \simeq n \cdot (k \cdot w)$ so that

$$\text{(i)} \quad k \cdot (m \cdot x) \succ n \cdot (j \cdot x) \,\&\, k \cdot (m \cdot z) \prec n \cdot (j \cdot z).$$

But clearly $(x, x) = (z, z)$, so that two applications of V,4 give $(m \cdot x, j \cdot x) = (m \cdot z, j \cdot z)$ and $(k \cdot (m \cdot x), n \cdot (j \cdot x)) = (k \cdot (m \cdot z), n \cdot (j \cdot z))$, which is incompatible with (i). This argument depends upon two elementary laws which might be stated

$$\begin{array}{ll} \text{VD} & (x, x) = (y, y), \\ \text{VE} & (x, y) = (z, w) \,\&\, x \succ y \rightarrow z \succ w. \end{array}$$

The second of these is used by Euclid in V,25.

V,8 is in a sense the fundamental proposition of book V. It has an unusually complex proof which I shall discuss in the next section. For now I shall merely point out that the only

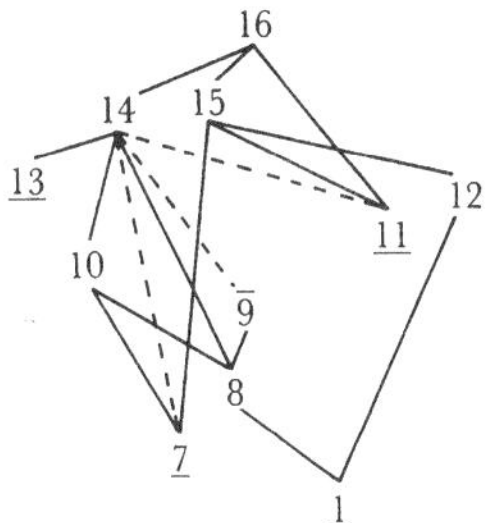

Figure 3.2

proposition presupposed by 8 is 1. The diagram of the deductive structure of the proof of 16 (fig. 3.2) shows that it alone accounts for the presence of 1 and 7–15.

After 16 the deductive structure of book V becomes somewhat more perplexing. Consider, for example, the analogue of VII,14:

V,22 $(x_1, x_2) = (y_1, y_2) \;\&\; \ldots \;\&\; (x_{n-1}, x_n) = (y_{n-1}, y_n) \rightarrow (x_1, x_n) = (y_1, y_n)$.

This has a simple proof like that of its analogue, in which alternation (V,16) is used to infer $(x_i, y_i) = (x_{i+1}, y_{i+1})$ so that $(x_1, y_1) = (x_n, y_n)$ and, again using alternation, $(x_1, x_n) = (y_1, y_n)$. Euclid gives a longer proof of 22, doing the case $n = 3$ in an easily generalizable way. He reduces 22 to V,4 and

V,20 $(x_1, x_2) = (y_1, y_2) \;\&\; (x_2, x_3) = (y_2, y_3) \rightarrow \text{COMP}\,(x_1, x_3, y_1, y_3)$.

For, if the antecedent of 22 holds, then, by V,4, $(m \cdot x_1, k \cdot x_2) = (m \cdot y_1, k \cdot y_2)$ and $(k \cdot x_2, l \cdot x_3) = (k \cdot y_2, l \cdot y_3)$, and so, by V,20, $\text{COMP}\,(m \cdot x_1, l \cdot x_3, m \cdot y_1, l \cdot y_3)$.

In proving 20 Euclid does the case $x_1 \succ x_3$ and argues that $(x_1, x_2) > (x_3, x_2)$, by V,8, and $(x_3, x_2) = (y_3, y_2)$ (inversion); hence $(y_1, y_2) > (y_3, y_2)$, and, by V,10, $y_1 \succ y_3$. This last inference is quite trivial, but it is not covered explicitly by propositions in book V. From V,13 one can get $(y_1, y_2) > (x_3, x_2)$ and $(y_2, y_3) > (x_2, x_1)$, but in order to infer $y_1 \succ y_3$ from either of these one needs

VF $(x, y) > (z, w) \;\&\; (z, w) = (u, v) \rightarrow (x, y) > (u, v)$.

VF like V,13, is a direct consequence of definitions 5 and 7. Since V,13 is proved, the failure to establish VF is presumably an oversight.

Euclid's avoidance of the shorter proof of 22 is commonly explained by the attention he pays to the homogeneity of magnitudes in proportions. A scholiast (286.18–22) explains the use of the word 'homogeneous' in definition 3:

> He says 'homogeneous' because inhomogeneous things cannot have a ratio to one another, neither line to surface nor plane to solid; but to a line a line has a ratio, to a surface a surface, and to a plane a plane.

Quite clearly, the proportion $(x, y) = (z, w)$ might be true even if x and z were inhomogeneous. And certain laws, such as V,22, do not require homogeneity of all terms involved; but others, such as V,16, do. In enunciating propositions Euclid does not make explicit which magnitudes must be taken to be homogeneous, which not, probably because the propositions themselves make this clear enough. In any case Euclid's failure to use 16 in his proof of 22 makes it very probable that he or Eudoxus[16] wished to avoid proofs depending upon propositions in which homogeneity restrictions are stronger than those of the proposition being proved. Euclid's proof of

V,18 $(x, y) = (z, w) \rightarrow (x + y, y) = (z + w, w)$

provides further confirmation of this suggestion. He foregoes an elementary constructive proof in which one applies alternation and then V,12 to the antecedent to get $(x + y, z + w) = (y, w)$, from which the consequent results by another alternation. Instead he uses a non-constructive proof in which Vc is used to reduce V,18 to

V,17 $(x, y) = (z, w) \rightarrow (x - y, y) = (z - w, w)$.

Euclid supposes V,18 false and asserts that there is a v not equal to w such that $(x + y, y) = (z + w, v)$. If $v \prec w$, then by V,17, VA, and V,11 $((z + w) - v, v) = (x, y) = (z, w)$; but since $(z + w) - v \succ z$, $v \succ w$, by V,14, contradicting $v \prec w$. (The case $v \succ w$, which Euclid does not do, would be handled in the same way.)

Euclid uses V,17 and alternation to prove

V,19 $(x, y) = (z, w) \rightarrow (x - z, y - w) = (x, y)$,

but this proposition obviously presupposes that all four magnitudes involved are homogeneous. Clearly Euclid could also derive V,17 from 19, but such a derivation would represent a violation of the policy of making minimal assumptions about homogeneity. Moreover, using Vc, Euclid could have reduced 19 to 12 in just the way he reduces 18 to 17.[17] Euclid gives a fallacious but correctible derivation of 17 from V,1 and 2. He takes multiples $m \cdot (x - y)$, $m \cdot y$, $m \cdot (z - w)$, $m \cdot w$, $n \cdot y$, $n \cdot w$, and argues by V,1 and 2 that $m \cdot (x - y) + m \cdot y$ and $m \cdot (z - w) + m \cdot w$ are equimultiples of x and z, and that $m \cdot y + n \cdot y$ and $m \cdot w + n \cdot w$ are equimultiples of y and w. Hence, if $(x, y) = (z, w)$,

(i) $\text{COMP}\,(m \cdot (x - y) + m \cdot y, m \cdot y + n \cdot y,$
$\quad m \cdot (z - w) + m \cdot w, m \cdot w + n \cdot w)$.

Since Euclid wishes to prove that $\text{COMP}\,(m \cdot (x - y), n \cdot y, m \cdot (z - w), n \cdot w)$, he should turn to the three cases $m \cdot (x - y) \succ n \cdot y$, $m \cdot (x - y) \simeq n \cdot y$, $m \cdot (x - y) \prec n \cdot y$. Instead he turns to the analogous cases for $m \cdot (x - y) + m \cdot y$ and $m \cdot y + n \cdot y$, doing the first case. He supposes that (ii) $m \cdot (x - y) + m \cdot y > m \cdot y + n \cdot y$ and infers, using (i), that (iii) $m \cdot (z - w) + m \cdot w > m \cdot w + n \cdot w$, so that (iv) $m \cdot (x - y) \succ n \cdot y$ and (v) $m \cdot (z - w)$

$\succ n \cdot w$. Euclid should, of course, have assumed (iv) and derived in succession (ii), (iii), and (v).

Euclid's treatment of 17–19 is a rather clear indication that he is concerned to avoid unnecessary assumptions about homogeneity. On the other hand, he is not consistently successful, as is shown by his proof of

V,23 $(x_1, x_2) = (y_2, y_3) \;\&\; (x_2, x_3) = (y_1, y_2) \rightarrow$
$(x_1, x_3) = (y_1, y_3).$

Euclid derives this unused proposition from

V,21 $(x_1, x_2) = (y_2, y_3) \;\&\; (x_2, x_3) = (y_1, y_2) \rightarrow$
$\text{COMP}\,(x_1, x_3, y_1, y_3),$

the proof of which is exactly like the proof of 20. To reduce 23 to 21 Euclid needs to establish (a) $(m \cdot x_1, m \cdot x_2) = (n \cdot y_2, n \cdot y_3)$ and (b) $(m \cdot x_2, n \cdot x_3) = (m \cdot y_1, n \cdot y_2)$. Euclid establishes (a) by applying VA and V,11 to $(x_1, x_2) = (m \cdot x_1, m \cdot x_2)$ and $(y_2, y_3) = (n \cdot y_2, n \cdot y_3)$, which are direct consequences of V,15. (b) could be derived immediately from V,4, as Euclid derives analogous assertions in V,20.[18] Instead, Euclid uses alternation to infer that $(x_2, y_1) = (x_3, y_2)$, and then, as in the proof of (a), $(m \cdot x_2, m \cdot y_1) = (n \cdot x_3, n \cdot y_2)$; another alternation gives (b). Euclid's reliance on an unnecessary homogeneity assumption in this case is puzzling and must be balanced against his apparent care to avoid such assumptions in 18 and 22. I am inclined to think of 23 as an inexplicable exception rather than as a refutation of the view that Euclid is generally concerned with minimizing homogeneity assumptions.

It remains to mention briefly the last two propositions of book V:

V,24 $(x_1, x_2) = (y_1, y_2) \;\&\; (x_3, x_2) = (y_3, y_2) \rightarrow$
$(x_1 + x_3, x_2) = (y_1 + y_3, y_2),$

V,25 $(x, y) = (z, w) \;\&\; x \succ y \;\&\; x \succ z \rightarrow x + w \succ y + z.$

To prove 24 Euclid inverts the second conjunct of the antecedent and applies 22 to get $(x_1, x_3) = (y_1, y_3)$; hence $(x_1 + x_3, x_3) = (y_1 + y_3, y_3)$, by V,18; and, by 22 again, the consequent follows. V,24 is never used explicitly in the *Elements*, but there are inferences in VI,31 and X,68[19] which are directly covered by

VG $(x_1, x_2) = (y_1, y_2) \;\&\; (x_1, x_3) = (y_1, y_3) \rightarrow$
$(x_1, x_2 + x_3) = (y_1, y_2 + y_3).$

Since this proposition is obviously derivable from 24 using inversion, and Euclid takes inversion for granted, it is possible that V,24 is included in book V with an eye to these applications.

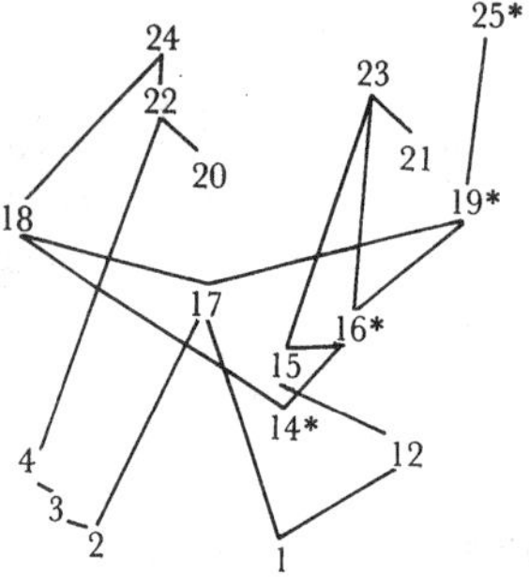

Figure 3.3

Nothing resembling V,25 is ever used in the *Elements*. Heath (vol. II, p. 185) points out that the special case of it when $y = z$ has as a consequence that the "geometric mean", i.e., y, is always smaller than the "arithmetic mean", i.e., $\frac{1}{2}(x + w)$. Since Euclid never defines either of these means, it seems unlikely that this application to them explains the presence of V,25. Whatever the explanation of its presence, V,25 has a simple derivation. For, if the antecedent is true, then, by V,19, $(x, y) = (x - z, y - w)$, and (VE) $x - z \succ y - w$, i.e., $x + w \succ y + z$.

In fig. 3.3 I have indicated the deductive structure of the last part of book V. The numerous applications of V,7–11 and 13 are left out of account. An asterisk beside a number indicates that the proposition in question is only applicable to proportions involving four homogeneous magnitudes. It should be pointed out that the application of 14 in 18 is to a proportion all four terms of which are necessarily homogeneous so that the previously discussed application of 16 in 23 is the only case in which Euclid makes unnecessarily strong homogeneity assumptions.

3.2 Problems in the Interpretation of Book V

In this section I want to discuss a number of topics relating to book V. For convenience I have placed them under three headings: (i) proportion and calculation; (ii) the nature of magnitudes; (iii) the distinctive assumptions of book V. Under each heading I will also be discussing laws of proportion not proved by Euclid.

Proportion and Calculation

Because of the analogy between the theory of proportion and the theory of real numbers, it is tempting to look to the *Elements* for indications of procedures for calculating with ratios, e.g., for producing the sum, difference, product, or quotient of two ratios, or for raising one ratio to some power or finding its nth root. We have already seen how little trace there is of calculational procedures in Euclidean arithmetic. In one sense there is even less in the theory of proportion, since in book V Euclid solves no problems, finds no ratios satisfying given conditions. However, in VI,9–13 Euclid shows how to construct straight lines satisfying certain conditions of proportionality relative to given straight lines, and these constructions would enable one to carry out analogues of calculations with real numbers, analogues which might be justified by reference to book V. For example, one could calculate the "sum" of (x, y) and (z, w) for straight lines x, y, z, w by using VI,12 to find a straight line v such that $(w, z) = (y, v)$ and taking this sum as $(x + v, y)$. V,24 could then be used to show that this sum is unique, i.e., that equal ratios added to equal ratios produce equal ratios.

Obviously a similar procedure could be used for subtraction and could be justified by an analogue of V,24.

Similarly, to "multiply" (x, y) and (z, w), one need only find v such that $(z, w) = (y, v)$ and take the product to be (x, v). V,22 shows that this product is unique, V,23 that the order of factors is irrelevant. Since division is simply multiplication by an inverse ratio, and raising to a power is multiplication of equal ratios, they too are taken care of. The "square root" of (x, y) can be taken by using VI,13 to find v such that $(x, v) = (v, y)$; since (x, y) is the product of (x, v) and (v, y), either of these can represent the square root. Analogous procedures work to find any 2^nth root, but the other roots are not constructible with compass and straightedge. As for uniqueness, Euclid proves no theorem which would establish the uniqueness of the square root of a ratio, but later he takes for granted

VH (i) Equal ratios have equal duplicates and triplicates;
(ii) Equal duplicates and triplicates are duplicates or triplicates of equal ratios.

The first of these is a trivial consequence of V,22. The second follows from

VH′ $\text{CPROP}(x_1, \ldots, x_n)\ \&\ \text{CPROP}(y_1, \ldots, y_n)\ \&$
$(x_1, x_2) > (y_1, y_2) \rightarrow (x_1, x_n) > (y_1, y_n)$,

i.e., greater ratios have greater n-tuplicates. For then if the duplicates or triplicates of (x, y) and (z, w) are equal, one cannot have either $(x, y) > (z, w)$ or $(z, w) > (x, y)$, by VB; a constructive application of trichotomy yields that $(x, y) = (z, w)$. VH′ has a direct proof which I give for the case $n = 3$. In this case the antecedent of VH′ says

$$(x_1, x_2) = (x_2, x_3)\ \&\ (y_1, y_2) = (y_2, y_3)\ \&\ (x_1, x_2) > (y_1, y_2),$$

i.e., there are m and n such that

$$m \cdot x_1 \succ n \cdot x_2\ \&\ m \cdot y_1 \preceq n \cdot y_2\ \&\ m \cdot x_2 \succ n \cdot x_3\ \&\ m \cdot y_2 \preceq n \cdot y_3.$$

But then

$$m \cdot (m \cdot x_1) \succ m \cdot (n \cdot x_2)\ \&\ m \cdot (m \cdot y_1) \prec m \cdot (n \cdot y_2),$$
$$n \cdot (m \cdot x_2) \succ n \cdot (n \cdot x_3)\ \&\ n \cdot (m \cdot y_2) \prec n \cdot (n \cdot y_3).$$

But since, by VIIh″ for magnitudes, $m \cdot (n \cdot x) \simeq n \cdot (m \cdot x)$, there are, by V,3, j and k such that $j \cdot x_1 \succ k \cdot x_3\ \&\ j \cdot y_1 \preceq k \cdot y_3$, i.e., $(x_1, x_3) > (y_1, y_3)$.

It is not, of course, possible to show that Euclid does not have calculations or something like them in mind when he deals with ratios. But he certainly does not make any such concern explicit, and, as we shall see, the way he proceeds in applying ratios suggests anything but a calculational model. In particular, although he uses the compounding of ratios in later books, he shows no clear sense of its relation to duplicating and presumably, therefore, none of its relation to multiplication. A particular law of compounding which is obvious when compounding is construed as multiplication of quotients but of which Euclid shows no knowledge is

VJ If (x, y) is compounded of (z, w) and (u, v), then $x \simeq y \leftrightarrow (z, w) = (v, u)$.

Under the interpretation of ratios as quotients this proposition says that the product of two ratios is 1 if and only if the ratios are reciprocals. Euclid says that z, w, u, v reciprocate (*antipaschein*) when $(z, w) = (v, u)$.[20] As we shall see, VJ would simplify parts of books VI, XI, and XII. VJ also has a simple proof. For if $(z, w) = (x, x')$ and $(u, v) = (x', y)$, then $x \simeq y$ if and only if $(x, x') = (y, x')$ if and only if $(z, w) = (v, u)$.

The Nature of Magnitudes

Euclid's apparent concern for the homogeneity conditions of propositions in book V throws some light on his conception of magnitudes. Presumably Euclid thinks of a proposition of the theory of proportion as a unified formulation of a number of analogous propositions concerning various particular kinds of magnitudes, straight lines, plane figures, etc. rather than as a single proposition about more abstract objects called magnitudes. The latter conception is appropriate to a development of the theory of proportion "for its own sake." However, although Euclid proves some propositions (notably V,23 and 25) which he does not use, and fails to prove some propositions (notably VH) which he does use, it seems reasonably clear that the principal point of book V is to show that the ordinary laws of proportion to be applied in subsequent books can be derived from definitions 5 and 7.

We have seen that, although Euclid presupposes some definitions from book V in the arithmetic books, he proves laws of proportion separately for numbers, and he does not appear to intend to apply any laws from book V to numbers. Moreover, in book V he uses the assumptions Va and c, which are false of numbers. Thus it seems quite certain that, for Euclid, magnitudes do not include numbers. Certainty would seem to me complete if Euclid did not introduce proportionalities involving magnitudes and numbers together in book X. For example, he proves

X,5 z measures x & z measures $y \rightarrow \exists k \exists l ((x, y) = (k, l))$.

He argues that if z measures x k times and y l times, then z

measures x the same number of times as o measures k, and y the same number of times as o measures l, so that $(z, x) = (o, k)$ and $(z, y) = (o, l)$. Inverting, one has $(x, z) = (k, o)$ and *ex equali* $(x, y) = (k, l)$.

Obviously Euclid cannot be using the arithmetic theory of proportion to justify proportionalities such as $(z, x) = (o, k)$, since the arithmetic theory applies only to numbers. But he cannot be using book V either, because he has not proved laws such as "If x measures y the same number of times as z measures w, then $(x, y) = (z, w)$." It is clear that Euclid needs some law or laws connecting proportionality involving magnitudes and proportionality as defined for numbers. The simplest way to do this is probably to expand the notion of magnitude to include numbers and to show that, if x is part or parts of y, then x is the same part or parts of y that z is of w if and only if $(x, y) = (z, w)$ (i.e., if and only if $\forall m \forall n \, \text{COMP}(m \cdot x, n \cdot y, m \cdot z, n \cdot w))$. This equivalence may be expressed as four laws, three of which are easily proved:

VK (i) $y \simeq m \cdot x \;\&\; w \simeq m \cdot z \rightarrow (x, y) = (z, w)$;

(ii) $y \simeq n \cdot u \;\&\; x \simeq m \cdot u \;\&\; w \simeq n \cdot v \;\&\; z \simeq m \cdot v \rightarrow (x, y) = (z, w)$;

(iii) $(x, y) = (z, w) \;\&\; y \simeq m \cdot x \rightarrow w \simeq m \cdot z$.

Proof of (i): If the antecedent is true, then, by V,3, any equimultiples of y and w are equimultiples $l \cdot x$, $l \cdot z$ of x and z; but clearly, by Vb, $\text{COMP}(k \cdot x, l \cdot x, k \cdot z, l \cdot z)$.

Proof of (ii): If the antecedent is true, then, by (i), $(u, y) = (v, w)$, and, by an obvious special case of V,4, $(x, y) = (z, w)$.

Proof of (iii): If the antecedent is true, then, by the same special case of V,4, $(m \cdot x, m \cdot x) = (m \cdot x, y) = (m \cdot z, w)$, and $w \simeq m \cdot z$.

The fourth law might be expressed

$$(x, y) = (z, w) \;\&\; y \simeq n \cdot u \;\&\; x \simeq m \cdot u \rightarrow \exists v (w \simeq n \cdot v \;\&\; z \simeq m \cdot v).$$

This law is easily proved for the w's which have nth parts v. For in that case, if the antecedent holds, $(z, w) = (x, y) = (m \cdot v, w)$, by (ii), and $z \simeq m \cdot v$, by V,9. I shall assume that a magnitude (in the extended sense) w can fail to have an nth part only if it is a number. If w (and therefore z) is a number, it is intuitively clear that a sufficient condition for z to be m nth parts of w is that m and n be least in their ratio. Hence the fourth law may be expressed

VK(iv) $(x, y) = (z, w) \;\&\; y \simeq n \cdot u \;\&\; x \simeq m \cdot u \;\&\; \text{LEAST}(m, n) \rightarrow \exists v (w \simeq n \cdot v \;\&\; z \simeq m \cdot v)$.

To prove (iv) it suffices to show that if z and w are numbers k and l, l has an nth part. Since m is m nth parts of n, if the antecedent of (iv) holds, then, by (ii) $(m, n) = (x, y) = (k, l)$ and COMP $(n \cdot m, m \cdot n, n \cdot k, m \cdot l)$; hence, by VII,16, $n \cdot k \simeq m \cdot l$, and, by VII,19, $(m, n) = (k, l)$ in the arithmetic sense; but then, by VII,20, n/l and l has an nth part.

Euclid's failure to establish a correlation between his two treatments of proportionality before developing the material in book X is probably the greatest foundational flaw in the *Elements*. Zeuthen[21] attempted to show that there really is no flaw on the ground that the *Elements* contain the propositions needed to establish the equivalence of the two definitions of proportionality for numbers. It can, of course, be doubted whether this ground is sufficient to establish Zeuthen's conclusion. It should also be pointed out that this equivalence is insufficient unless one thinks of Euclid's treatment of numbers as in some way a treatment of commensurable magnitudes in general.[22] An alternative way of dealing with the flaw is to suppose that book X is based on an original which made use of a treatment of proportionality not found in the *Elements*. This hypothesis is attractive,[23] but it provides no answer to the question why Euclid overlooked the shortcoming we find in the *Elements*. I suggest that a major reason why he did so lies in his conception of definitions as characterizations of independently understood notions. For us a definition gives a term its sense, so that the same term can be given two definitions only if those definitions are proved equivalent. For Euclid proportionality is an independently understood concept of which he gives two characterizations for two different kinds of objects. The fact that both of those characterizations enable one to prove the ordinary laws of proportionality is a sufficient indication of their correctness. Although Euclid characterizes proportionality when applied to geometric magnitudes and proportionality applied to numbers, he overlooks his failure to characterize it for proportionalities involving numbers and magnitudes together. Since he overlooks this and since I have indicated how the resulting incoherence can be overcome, I shall make no further references to this situation but take for granted Euclid's blending of his two theories of proportion.

The Distinctive Assumptions of Book V

It is clear that the only laws of book V which can be correctly applied to numbers are those which do not depend upon the assumptions Va and c, which are not valid for numbers. I have already mentioned that Heath (vol. II, p. 146) shows how to eliminate the only use of Va. I would now like to show that the use of Vc in V,18 can also be eliminated without making unnecessary homogeneity assumptions. The argument is indirect. One supposes that $(x, y) = (z, w)$ and refutes each of $(x + y, y) > (z + w, w)$ and $(z + w, w) > (x + y, y)$; constructive application of trichotomy yields that $(x, y) = (z, w)$. Since the two refutations are the same, I do only the first. If $(x + y, y) > (z + w, w)$, then there are m and n such that $m \cdot (x + y) \succ n \cdot y$ and $m \cdot (z + w) \preceq n \cdot w$. Using V,1 and subtracting equals from both sides of these inequalities, one gets that $m \cdot x \succ n \cdot y - m \cdot y$ and $m \cdot z \preceq n \cdot w - m \cdot w$; by V,6 there is a k such that $m \cdot x \succ k \cdot y$ and $m \cdot z \preceq k \cdot w$, contradicting $(x, y) = (z, w)$.

The unnecessary use of Vc in book V is one more indication that the constructive point of view does not play a role in book V. Euclid's failure to use trichotomy for ratios in a nonconstructive way is no evidence to the contrary, because there is no case in which he would have to use it nonconstructively. Moreover Vc implies trichotomy for ratios. For if x, y and z, w are two pairs of homogeneous magnitudes, then, by Vc, there is a v such that $(x, y) = (z, v)$. But either $w \succ v$ or $w \simeq v$ or $v \succ w$; in the first case $(x, y) > (z, w)$. by V,8 and 13; in the second $(x, y) = (z, w)$, by V,7 and 11; and in the third $(z, w) > (x, y)$, by V,8 and VF. It seems clear that no Greek ever questioned this "assumption of the existence of a fourth proportional," perhaps because the use was not noticed, but more probably because the existence of such a proportional to three given geometrical objects was considered obvious on the basis of intuitive ideas about continuity.[24]

Euclid's proof of V,8 involves another assumption which, because of the central role of 8, is fundamental to the theory of proportion. The assumption is also used in the proof of X,1, and may be stated

Vd $\quad x \prec y \rightarrow \exists m (m \cdot x \succ y)$.

Euclid's proof of 8 is needlessly complex,[25] but its essentials can be described fairly briefly. The idea is to find, for given magnitudes x, y, z with $x \succ y$, multiples $k \cdot x$, $k \cdot y$, $l \cdot z$ with $k \cdot x \succ l \cdot z$ and $k \cdot y \preceq l \cdot z$. One supposes that $x \simeq y + w$ and lets u be the greater of y and w or y if $y \simeq w$, and v the other of the two. Two applications of Vd give that there are m and n such that (i) $m \cdot v \succ z$ and (ii) $(n + 1) \cdot z \succ m \cdot y \succeq n \cdot z$. V,1 and standard combinatorial laws yield $m \cdot x \succ (n + 1) \cdot z$ because

$$m \cdot x \simeq m \cdot (u + v) \simeq m \cdot u + m \cdot v \succeq m \cdot v + m \cdot y \succ z + n \cdot z \simeq (n + 1) \cdot z.$$

In X,1 Euclid shows that the process of taking away at least a half of a greater magnitude y, at least a half of what is left, and so on, eventually produces a magnitude smaller than a given magnitude x less than y.[26] I shall state and prove this result somewhat formally, representing the repeated subtractions by means of a function satisfying the conditions

$\phi(1,y) \simeq y$ and $\phi(i+1,y) \preceq \frac{1}{2}(\phi(i,y))$.

X,1 can be stated

$$\text{X,1} \quad \forall z(\phi(1,z) \simeq z \;\&\; \phi(i+1,z) \preceq \tfrac{1}{2}(\phi(i,z))) \;\&\; x \prec y \rightarrow \exists n(\phi(n,y) \prec x).$$

The proof of this proposition may be represented as follows. By Vd there is an n such that $n \cdot x \succ y$, where, of course,

$$n \cdot x \simeq \sum^{n} x_i (x_i \simeq x).$$

One can define a function ψ satisfying

$$\psi(1,x) \simeq n \cdot x$$

and

$$\psi(i+1,x) \simeq \psi(i,x) - x_i$$

for $1 \leq i < n$. Clearly,

$$\psi(i+1,x) \succeq \tfrac{1}{2}(\psi(i,x))$$

and

$$\psi(n,x) \simeq x.$$

But also

$$\psi(1,x) \simeq n \cdot x \succ y \simeq \phi(1,y);$$

and if

$$\psi(i,x) \succ \phi(i,y),$$

then

$$\psi(i+1,x) \succeq \tfrac{1}{2}(\psi(i,x)) \succ \tfrac{1}{2}(\phi(i,y)) \succeq \phi(i+1,y),$$

and, by induction up to n,

$$x \simeq \psi(n,x) \succ \phi(n,y).$$

For the use of Vd in X,1 and the first use of it in V,8 Euclid says explicitly that the lesser magnitude if multiplied will sometime be greater than the greater (*to Γ pollaplasiadzomenon estai pote tou AB meidzon*). For inference (ii) of V,8 Euclid simply says, "Let L be taken double of z, M triple of it, and successive multiples increasing by one, until what is taken is a multiple of z and the first that is greater than $m \cdot y$." Heiberg explains inference (i) of V,8 by reference to

V, def. 4 Magnitudes which when multiplied can exceed one another (*ha dunatai pollaplasiadzomena allēlōn huperechein*) are said to have a ratio to one another.

In X,1 he also refers to this definition, but with a 'cf.'. Obviously inference (ii) of V,8, which Heiberg leaves unexplained, is as much dependent on the definition as the other two are. The question is whether any of the inferences are dependent on it.

The attempt to connect the definition with these inferences seems to depend upon reading an expression like 'x and y are magnitudes which when multiplied can exceed one another' as 'some multiple of x exceeds y and some multiple of y exceeds x', so that definition 4 would say

x has a ratio to $y \leftrightarrow \exists m(m \cdot x \succ y)$ & $\exists n(n \cdot y \succ x)$.

It seems likely to me that if Euclid wished to convey the contents of this equivalence, he would have used words like those used in connection with the problematic inferences. The definition might then run, "Magnitudes are said to have a ratio to one another if the lesser when multiplied can exceed the greater." However, even if one accepts the proposed reading of the definition there are reasons to doubt that it is the basis of the inferences. For in none of them does Euclid mention anything about the magnitudes involved having a ratio, and, indeed, ratios are irrelevant to X,1. In the case of (ii) it is legitimate to assume that y and z do have a ratio to one another, but in (i) it can be concluded that v and z have a ratio to one another only on the basis of some such principle as 'If each of two magnitudes has a ratio to a third, then so does their difference.' Euclid might, of course, take such a principle for granted as "obvious," but I am inclined to think that he would not have done so had he felt a need for a justification of the inferences in V,8 and X,1. This interpretation would seem to me especially plausible if, as has often been suggested,[27] definition 4 was somehow intended to exclude infinitesimals; for someone interested in excluding infinitesimals would presumably want to show that there could not be infinitesimal differences between finite magnitudes. One final point to be brought out is the terminological discrepancy between definition 4 and the inferences. In definition 4 Euclid uses the word 'exceed' (*huperechein*) to express $\succ$, but in the inferences he uses 'is greater (*meidzon*) than'. This disparity is especially striking in the context of book V where Euclid appears to be paying particular attention to giving verbal formulations which make deductive dependencies clear. In V,8 itself, in order to bring his vocabulary into line with definition 7, Euclid transforms '$m \cdot y$ is less than $(n+1) \cdot z$' to '$m \cdot y$ does not exceed $(n+1) \cdot z$', and infers from '$z + n \cdot z$ is equal to $(n+1) \cdot z$' and '$m \cdot (y+w)$ is greater than $z + n \cdot z$' that $m \cdot (y+w)$ exceeds $(n+1) \cdot z$.

Thus the textual evidence is against referring V,8 and X,1 to definition 4. Archimedes throws some further light on this question. At the beginning of his work *On the Sphere and Cylinder* I (SCI) he lists as the last of his assumptions (*lambanomena*)

SCI,L5 Of unequal lines, unequal surfaces, and unequal solids, the greater exceeds the less by an amount which added to itself can exceed any given one of those things which are said to be related to one another (*pantos tou protethentos tōn pros allēla legomenōn*).

Apparently this assumption amounts to

If x, y, z are all lines or all planes or all solids,

$$x \prec y \rightarrow \exists n \left(\sum^{n} w_i (w_i \simeq y - x) \succ z \right),$$

or, since there is no reason to distinguish n-fold addition and multiplication by n, and none to be concerned with the particular values of x, y, z so long as they are homogeneous,

SCI,L5 $x \prec y \rightarrow \exists n (n \cdot (y - x) \succ z)$.

In his prefatory letter to the *Quadrature of the Parabola* (QP) Archimedes refers to geometers who use questionable *lēmmata*. He himself takes as a *lēmma* one similar[28] to that used by earlier geometers whose results depending on their *lēmma* have been accepted no less than those not depending on it. Archimedes' *lēmma* is

QP,L Of unequal areas the excess by which the greater exceeds the less, if added to itself, can exceed any given finite area,

apparently SCI,L5 for areas x, y, z. The earlier results referred to by Archimedes are all proved in book XII of the *Elements* and depend directly on a special case of X,1 which may be stated

X,1′ $\forall z(\phi(1, z) \simeq z \ \& \ \phi(i+1, z) \preceq \frac{1}{2}(\phi(i, z))) \ \& \ x \prec y$
$\rightarrow \exists n(\phi(n, y) \prec y - x)$.

X,1′ is derivable from SCI,L5 because SCI,L5 enables one to assert the existence of an n such that $n \cdot (y - x) \succ y$, and then to reproduce the steps of the proof of X,1. X,1′ also implies SCI,L5 because of the easily proved fact that

$$\forall z(\phi(1, z) \simeq z \ \& \ \phi(i+1, z) \simeq \tfrac{1}{2}(\phi(i, z))) \rightarrow 2^{n-1} \cdot (\phi(n, y)) \simeq y).$$

This same fact shows that X,1 implies Vd. Hence Vd and X,1 are equivalent; and so are SCI,L5 and X,1′. In addition, each of the former pair implies each of the latter. Implication in the other direction requires the hypothesis that $y - x \prec y$, i.e., that x is not infinitesimal. For if x is infinitesimal, no multiple of x will exceed y even though a multiple of $y - x$ does. Thus Archimedes' *lēmma* is equivalent to the form of X,1 which Euclid uses in proving the results referred to by Archimedes; but unless one assumes $y - x \prec y$, it is weaker than the principle explicitly proved by Euclid, X,1, and the assumption he uses in proving this principle, Vd.

These logical relations are, of course, insufficient to allow one to infer the exact form of the *lēmma* used by earlier geometers to which Archimedes refers. Since in the context of Archimedes' discussion the word *lēmma* seems to have the sense of 'assumption' rather than of 'preliminary result', it would seem more likely that he is referring to something like Vd than to an explicitly proved proposition. If this suggestion is correct, then it is also likely that Archimedes is not referring to an explicitly formulated assumption. For in no other case does Archimedes formulate an assumption which we know to have been formulated earlier, and it seems improbable that he would have to justify the use of an assumption already made explicitly in the *Elements*. Probably, then, his use of the *lēmma* in SCI was questioned and to justify it he pointed out that a "similar" assumption had already been made tacitly by his predecessors.[29]

It remains to discuss briefly the point of definition 4 or, more precisely, the significance of the phrase 'which when multiplied can exceed one another'. The scholia contain a number of suggestions concerning its significance: (a) that it is a characterization of homogeneity; (b) that it excludes infinite magnitudes; (c) that it shows that magnitudes may be incommensurable. Heath (vol. II, p. 120) apparently accepts all three suggestions, but adds to (b) the exclusion of infinitesimals. (a) seems unlikely, because if the phrase in question were a characterization of homogeneity, Euclid would presumably use it to define 'be homogeneous' rather than 'have a ratio to one another'. Suggestion (b) and especially Heath's extension of it have played an important role in various hypotheses about fifth- and fourth-century Greek mathematics.[30] As far as the *Elements* themselves are concerned, although definition 4 does exclude infinitely large and infinitely small magnitudes from having a ratio to finite ones, it seems improbable that Euclid is using the definition to rule out this possibility. For infinitely small and large magnitudes would be as troublesome in book I as in book V, if they were admitted as a possibility at all. In book I Euclid excludes such magnitudes in the same way in which he excludes infinite collections in the arithmetic books, namely by leaving them out of account. There is no reason why he could not do the same in book V.

The suggestion (c) has in its favor the fact that the whole theory of book V was almost certainly devised to provide a way of dealing with proportionalities involving incommensurables. Definition 4 itself makes no reference to incommensurables, nor does Euclid have any reason to invoke the difference between them and commensurables in book V. Hence one cannot expect to find an explicit characterization or invocation of the difference in definition 4. Furthermore, when the phrase 'x and y when multiplied can exceed one another' is interpreted as 'some multiple of x exceeds y and some multiple of y exceeds x', there is no reference whatsoever to the difference. However, if this phrase is interpreted to mean 'some multiple of x exceeds some multiple of y and some multiple of y exceeds some multiple of x' so that definition 4 is rendered

$$x \text{ and } y \text{ have a ratio} \leftrightarrow \exists m \exists n (m \cdot x \succ n \cdot y) \;\&\; \exists m \exists n (m \cdot y \succ n \cdot x),$$

the connection with the problem of incommensurability becomes much clearer. For one obvious characterization of commensurable magnitudes x and y is that there are m and n such that $m \cdot x \simeq n \cdot y$, i.e., "Magnitudes which when multiplied can equal one another are commensurable." This characterization not only suggests definition 4 as it is now being interpreted, but it also has a close connection with definition 5. For the characterization suggests that proportionality for commensurable magnitudes be defined by

$$(x, y) = (z, w) \leftrightarrow \exists m \exists n (m \cdot x \simeq n \cdot y \;\&\; m \cdot z \simeq n \cdot w)$$

or

$$(x, y) = (z, w) \leftrightarrow \forall m \forall n (m \cdot x \simeq n \cdot y \rightarrow m \cdot z \simeq n \cdot w).$$

We have seen that Euclid moves back and forth between the analogues of these two equivalences in the arithmetic books. It is a short step from either of them, and especially from the second, to definition 5. For although one will not expect incommensurable magnitudes to satisfy either of these equivalences except in a vacuous sense, one will expect them to satisfy $\forall m \forall n (m \cdot x \succ n \cdot y \leftrightarrow m \cdot z \succ n \cdot w)$ if they are proportional; but this condition is exactly the Eudoxean characterization of proportionality for incommensurables. I suggest, then, that V, def. 4 is intended to indicate that proportionality is being defined for more than commensurable magnitudes and that it plays no mathematical role in the *Elements*. Vd then is just another tacit assumption, but in this case one made explicit by Archimedes.

Before leaving the topic of Vd I would like to show that it implies

$$\text{VL}\quad (x,y) > (z,w) \leftrightarrow \text{(a)}\ \exists m\exists n(m\cdot x \succ n\cdot y\ \&\ m\cdot z \preceq n\cdot w)$$
$$\leftrightarrow \text{(b)}\ \exists m\exists n(m\cdot x \succeq n\cdot y\ \&\ m\cdot z \prec n\cdot w)$$
$$\leftrightarrow \text{(c)}\ \exists m\exists n(m\cdot x \succ n\cdot y\ \&\ m\cdot z \prec n\cdot w).$$

Since (a) is by definition equivalent to $(x,y) > (z,w)$ and (c) implies both (a) and (b), it need only be shown that each of (a) and (b) imply (c). However, the two cases are symmetrical, and so it suffices to show that (a) implies (c), or simply that (c) holds if there are m and n such that $m\cdot x \succ n\cdot y$ and $m\cdot z \simeq n\cdot w$. But if this is so, then, by Vd, there is a k such that $k\cdot(m\cdot x - n\cdot y) \succ y$ so that $(k\cdot m)\cdot x - (k\cdot n)\cdot y \succ y$, and $((k\cdot n)+1)\cdot y \simeq (k\cdot n)\cdot y + y \prec (k\cdot m)\cdot x$; but also $(k\cdot m)\cdot z \simeq (k\cdot n)\cdot w \prec ((k\cdot n)+1)\cdot w$.[31] An obvious consequence of VL is

$$\text{VM}\quad (x,y) > (z,w) \leftrightarrow (w,z) > (y,x).$$

The last topic I wish to discuss in this subsection is the relative strength of various assumptions either made by Euclid in book V or obviously related to the book. For this purpose it is necessary to think of magnitudes as the objects of an axiomatic theory in the modern sense. I shall assume that magnitudes constitute an ordered system satisfying the ordinary laws of addition and subtraction and of the taking of multiples, i.e., that magnitudes satisfy the axioms given on pp. 36–37 and their extensions to multiples. It is also necessary to assume that the system contains at least one magnitude and, in order that multiplication be possible, infinitely many magnitudes equal to it. The assumptions I wish to consider are

(i) nonexistence of a greatest magnitude $(\forall x\exists y(y \succ x))$;
(ii) nonexistence of a least magnitude $(\forall x\exists y(x \succ y))$;
(iii) density $(x \prec y \rightarrow \exists z(x \prec z \prec y))$;
(iv) existence of mth parts (Va);
(v) existence of a fourth proportional (Vc);
(vi) the Archimedean condition (Vd);
(vii) continuity (Every cut in the system of magnitudes is made by a magnitude, i.e., if S and S' are disjoint nonempty sets together exhausting the system of magnitudes, and every member of S is less than every member of S', then either S has a greatest member or S' has a least).

(i) is a consequence of the existence of arbitary multiples of any magnitude, (ii) of the existence of mth parts, i.e., of (iv).

(iii) follows from (i) and (ii) because if $w \prec y - x$, then

$x \prec x + w \prec x + (y - x) \simeq y.$

Euclid would almost certainly have thought of (i)–(iii) as obvious truths, at least for geometric objects, and so it seems reasonable to add them to the fundamental assumptions in considering (iv)–(vii).

It is clear that (iv)–(vi) together do not imply continuity, since the positive rationals[32] are a model for (iv)–(vi) in which continuity does not hold. On the other hand, continuity implies each of (iv)–(vi). I sketch the proofs. For the existence of mth parts I first show that continuity implies

(a) $\forall m \exists y (m \cdot y \prec x)$.

This is clearly true for $m = 1$ because there is no least x. Suppose $m \cdot y \prec x$, and consider $x - m \cdot y$. If $(m + 1) \cdot (x - m \cdot y)$ is not less than x, it is either equal to or greater than x. In the former case there is a $z \prec x - m \cdot y$ so that $(m + 1) \cdot z \prec x$; but, if $(m + 1) \cdot (x - m \cdot y) \succ x$, then $m \cdot x \simeq (m + 1) \cdot x - x \succ (m + 1) \cdot (m \cdot y) \simeq m \cdot ((m + 1) \cdot y)$, and $x \succ (m + 1) \cdot y$. Now, let S contain every magnitude y satisfying $m \cdot y \prec x$, and let S' contain all other magnitudes. It is easy to see that S and S' constitute a cut. But S contains no greatest member because, if $m \cdot y \prec x$, there is, by (a), a z such that $m \cdot z \prec x - m \cdot y$, so that $m \cdot (z + y) \simeq m \cdot z + m \cdot y \prec x$. Hence S' contains a least member y satisfying $m \cdot y \succeq x$. It must be the case that $m \cdot y \simeq x$, because, if $m \cdot y \succ x$, there is, by (a), a z such that $m \cdot z \prec m \cdot y - x$, so that $x \prec m \cdot (y - z)$.

To show that continuity implies the Archimedean condition, it suffices to show that if continuity holds, every multiple of a magnitude x cannot be less than a magnitude y. If this were the case, one could define S to be the set of magnitudes w satisfying $\exists m (w \prec m \cdot x)$ and S' to be the complement of this set. It is easy to see that these two sets constitute a cut with S as the lower part. Moreover, S has no greatest member, because if w is in S, $w \prec m \cdot x \prec (m + 1) \cdot x$, for some m, and $m \cdot x$ is in S. Therefore, if continuity holds, S' has a least member z. But, since $z - x \prec z$, $z - x$ is in S, i.e., for some m, $z - x \prec m \cdot x$, or $z \prec m \cdot x + x \simeq (m + 1) \cdot x$, and z is in S, a contradiction.

To show that continuity implies the existence of a fourth proportional, one can take S to be the set of all w satisfying $(z, w) > (x, y)$ and S' to be its complement, i.e., the set of w satisfying $(x, y) \geq (z, w)$. S and S' are clearly a cut with S as lower part, if neither is empty. If $x \simeq y$, neither set is empty because, by (i) and (ii), there are magnitudes greater than and less than z. If $x \prec y$, then $(z, z) > (x, y)$, and, if $m \cdot x \preceq y \prec$

$(m + 1) \cdot x$, then $(x, y) > (z, (m + 1) \cdot z)$. Finally, if $x \succ y$, $(x, y) > (z, z)$, and, if $m \cdot y \preceq x \prec (m + 1) \cdot y$, $\left(z, \frac{1}{m + 1}(z)\right) > (x, y)$. The proof is completed by showing that there is no greatest w such that $(z, w) > (x, y)$ and no least w such that $(x, y) > (z, w)$. Suppose first that $(z, w) > (x, y)$, i.e., that $m \cdot z \succ n \cdot w$ and $m \cdot x \preceq n \cdot y$, for some m and n. Then

$$n \cdot \left(w + \frac{1}{n + 1}(m \cdot z - n \cdot w)\right) \simeq n \cdot w + \frac{n}{n + 1}(m \cdot z - n \cdot w)$$
$$\prec n \cdot w + (m \cdot z - n \cdot w) \simeq m \cdot z,$$

so that

$$\left(z, w + \frac{1}{n + 1}(m \cdot z - n \cdot w)\right) \succ (x, y).$$

On the other hand, if $(x, y) > (z, w)$, there are, by VL, m and n such that $m \cdot x \succeq n \cdot y$ and $m \cdot z \prec n \cdot w$. Hence

$$n \cdot \left(w - \frac{1}{n + 1}(n \cdot w - m \cdot z)\right) \simeq n \cdot w - \frac{n}{n + 1}(n \cdot w - m \cdot z)$$
$$\succ n \cdot w - (n \cdot w - m \cdot z) \simeq m \cdot z,$$

so that

$$(x, y) > \left(z, w - \frac{1}{n + 1}(n \cdot w - m \cdot z)\right).$$

Thus the assumption of continuity implies all of Euclid's important tacit existential assumptions in book V. In addition, since if $(m \cdot x, x) = (x, y)$ y is an mth part of x, it is clear that the assumption of the existence of the fourth proportional and the Archimedean condition would suffice without continuity for Euclid's purposes. However, no further similar reductions are possible, because (iv) and (v) do not imply (vi), and (iv) and (vi) do not imply (v). Hilbert's model to establish the independence of an Archimedean axiom from his other geometric assumptions[33] shows that (iv) and (v) do not imply (vi). An example of a domain in which (iv) and (vi) but not (v) hold is provided by a domain containing the positive rationals, $\frac{m}{n}\pi$, for any $\frac{m}{n}$, and the sums and (positive) differences $\frac{j}{k} + \frac{m}{n}\pi$, $\frac{j}{k} - \frac{m}{n}\pi$, $\frac{m}{n}\pi - \frac{j}{k}$. If the Euclidean operations and relations are defined in the standard way, the resulting model will satisfy (iv) and (vi). However, although it contains 1 and π, it will not contain an x satisfying $(1, \pi) = (\pi, x)$, i.e., it will not contain π^2. For if $\pi^2 = \left|\frac{j}{k} \pm \frac{m}{n}\pi\right|$, π would be an algebraic number $\sqrt{\frac{m^2}{4n^2} \pm \left(\frac{j}{k}\right)} \pm \frac{m}{2n}$, and the circle would be squarable with compass and straightedge.

This last discussion has obviously taken us beyond the conceptual limits of Euclidean mathematics, but the fact that it has done so is indicative of the modern, "abstract" character of book V. Hasse and Scholz[34] call the book the first attempt at a complete axiomatization. A more accurate characterization is perhaps that it is the first attempt to reduce the treatment of an intuitive notion (proportionality) to precise, formally correct definitions (definitions 5 and 7). The same kind of reduction is characteristic of modern foundational work of which Dedekind's characterization of continuity in terms of cuts is a good example. What is missing in book V from a modern point of view is exactly the axiomatic foundation—the existential assumptions and combinatorial laws which underlie the whole book. This shortcoming should not, however, blind one to the foundational achievement of Eudoxus, an achievement which has no parallel until the nineteenth century.

Notes for Chapter 3

Bibliographical Note

Beckmann's monograph is a thorough study of book V from a modern point of view in which the interpretations of others are summarized and discussed. It also includes a useful bibliography. The four "Eudoxos-Studien" of Becker have greatly influenced my own thinking.

1. Dedekind presents his treatment of the reals in "Continuity . . . ." Landau carries out the program in a more formal way in *Foundations . . .*, in which results described here are proved in detail.

2. This assertion is established by showing that the assumption of any two of the alternatives $x > y$, $x = y$, and $x < y$ leads to contradiction. The analogous result for proportions is established as VB and C below on p. 130.

3. The one exception is the fundamental V,8, which is discussed in section 3.2.

4. See Heath, vol. II, p. 146.

5. For a thorough treatment of the foundational aspects of book V see Beckmann.

6. The general form of V,2 could be formulated as

$$y_i \simeq m_i \cdot x_1 \;\&\; z_i \simeq m_i \cdot x_2 \rightarrow \exists k\Big(\sum^{n} y_i \simeq k \cdot x_1 \;\&\; \sum^{n} z_i \simeq k \cdot x_2\Big).$$

7. Since Euclid does not acknowledge multiplication by 1, he states the consequent of V,6 as the assertion that $y_1 - z_1$ and $y_2 - z_2$ are either equal to x_1 and x_2 or equimultiples of them. In the proof he

handles only the case of equality, i.e., of $k = 1$, so that the inference in the text presupposes

$$\text{(i)}\quad y_1 \simeq m\cdot x_1 \,\&\, y_2 \simeq m\cdot x_2 \,\&\, z_1 \simeq x_1 \,\&\, z_2 \simeq x_2 \rightarrow \exists k(y_1 + z_1 \simeq k\cdot x_1 \,\&\, y_2 + z_2 \simeq k\cdot x_2).$$

It seems unlikely that Euclid would refer this inference to V,2, because he would then be allowing multiplication by 1 in that theorem. Probably Euclid thinks of (i) as obvious since elsewhere he takes for granted that $m\cdot x + x \simeq (m+1)\cdot x$. Of course, for the general case of V,6 Euclid would still need to invoke V,2.

8. It is clear that (i) implies (i′). The inference in the other direction depends upon VL, which is discussed in the next section.

9. In "Prinzip . . . ," pp. 374–375.

10. The derivation is given in the next section, p. 139.

11. Roughly speaking, from a constructive point of view (ia) asserts that, if there is no general law to establish that COMP $(m\cdot x, n\cdot y, m\cdot z, n\cdot w)$ for all m and n, m and n can be found such that $\neg$COMP $(m\cdot x, n\cdot y, m\cdot z, n\cdot w)$; (ib) asserts that if such m and n cannot be found, there is such a general law. Clearly, however, whereas to refute the possibility of a general law is not necessarily to be able to produce a counterexample, to show that no counterexample can be produced is to establish a general law.

12. See Heath, vol. II, pp. 144, 149. On pp. 174–175 Heath points out that the insertion of the addition to V,19 $[(x, y) = (z, w) \rightarrow (x, x - y) = (z, z - w)]$ is equally clumsy. Although this corollary is applied frequently in book X, Euclid would probably be willing to take it for granted as an obvious consequence of V,17 and inversion.

13. In the proof of V,13 Euclid appears to commit the fallacy of affirming the consequent, since he does not apply VA but infers $m\cdot x \succ n\cdot y$ directly from $m\cdot z \succ n\cdot w$ & COMP $(m\cdot x, n\cdot y, m\cdot z, n\cdot w)$.

14. Heath (vol. II, p. 163) reproduces Simson's proofs for the other two cases; the derivation of (b) uses V,9 as well as 7 and 11. In V,20 Euclid proceeds in the same way as in 14, doing only one of three cases.

15. See Heath, vol. II, pp. 154–155, 157.

16. In "Homogeneity . . ." I argue that book V is a revision of Eudoxus' work designed to eliminate unnecessary homogeneity restrictions. The argument is an attempt to account for the presence of the unused propositions of book V, namely 5, 6, and 25.

17. I give the argument. If V,19 is false, then, by Vc, there is a v not equal to $y - w$ such that $(z, w) = (x, y) = (x - z, v)$, and, by V,12, $(x, y) = (z, w) = (x - z + z, v + w) = (x, v + w)$; but then, by V,8, $y \simeq v + w$, i.e., $y - w \simeq v$, a contradiction.

18. In two manuscripts this short argument is inserted before the longer, presumably Euclidean one—the only one in P.

19. The same inference would be made in X,69 and 70 if the proofs sketched for them by Euclid were carried out in full. Beckmann (p. 17) also mentions XII,4 and XIII,11 as propositions in which 24 is

employed. The inferences in question presuppose $(x, y) = (z, w) \rightarrow (2 \cdot x, y) = (2 \cdot z, w)$ or, in the case of XII,4, an instance of this law, namely; $(2 \cdot x, x) = (2 \cdot y, y)$. Both of these inferences might be intuitive; the principles governing them can be derived from VD, V,16, V,12, VA, and V,11.

20. Heath renders 'reciprocate' 'are reciprocally proportional'. For Euclid's use of the term, see VI,13 and Heath, vol. II, p. 189.

21. See, for example, "Sur la constitution . . . ," p. 412 and, for more detail, Dijksterhuis, vol. II, pp. 137–139.

22. Zeuthen rather clearly did think of the arithmetic books in this way. The section on VII–IX in *Histoire* . . . is entitled "Grandeurs commensurables et leur traitement numerique."

23. This hypothesis was first fully developed by Becker in "Eine voreudoxische Proportionenlehre . . . ." It is described by van der Waerden, pp. 175–179.

24. This explanation is put forward and developed by Becker in "Warum haben die Griechen . . . ."

25. For Euclid's proof and criticism of it see Heath, vol. II, pp. 149–153.

26. Euclid states X,1 in terms of subtractions of more than a half and remarks at the end of the proof that the proposition is true if exact halves are subtracted, presumably because he applies the theorem to exact halves in XII,16. The proposition is, of course, true as long as there is a lower bound on the ratio of magnitude subtracted to magnitude from which it is subtracted; but the proof of this more general assertion involves combinatorial complications. I have incorporated Euclid's remark into the statement of X,1 and have proceeded analogously in the case of X,1′ to simplify the derivation of Vd from X,1.

27. See, for example, Heath, vol. II, p. 120.

28. After stating the *lēmma* Archimedes says that the earlier geometers used the *lēmma* itself; but after stating the results proved by these geometers, he says that they used a similar *lēmma*.

29. Similar considerations are put forward by Dehn in "Beziehungen . . . ," pp. 19–22.

30. For an example of the attempt to import a concern for infinitesimals into the history of Greek mathematics antecedent to book V, see Hasse and Scholz, p. 8ff., and for criticism of their view, van der Waerden, "Zenon . . . ," especially section 2.

31. VL implies the equivalence of (i) and (i′) on p. 127. Because of definition 5, to show this equivalence it suffices to show that of $(x, y) > (z, w) \vee (z, w) > (x, y)$ to $\exists m \exists n \neg \text{COMP}(m \cdot x, n \cdot y, m \cdot z, n \cdot w)$. The former of these clearly implies the latter. But the latter asserts that either $m \cdot x \succ n \cdot y$ & $m \cdot z \preceq n \cdot w$ or $m \cdot x \simeq n \cdot y$ & $m \cdot z \lessgtr n \cdot w$ or $m \cdot x \prec n \cdot y$ & $m \cdot z \succeq n \cdot w$, for some m and n; VL establishes that each of these alternatives implies an inequality between (x, y) and (z, w).

32. In order to make the models discussed in this section simple, it is best to leave out of account the existence of infinitely many equals of a given magnitude and to give the operation and relation symbols their standard modern sense.

33. *Grundlagen* . . . , pp. 48–50.

34. Page 17.

4 Proportion and the Geometry of Plane Rectilineal Figures

Among the elementary facts of traditional geometry are formulas for computing the areas and volumes of figures on the basis of certain lengths. For example,

The area of a parallelogram is the product of its base and its height;

The area of a triangle is $\frac{1}{2}$ of the product of its base and its height.

Knowledge of such formulas can be traced with reasonable probability to a period long before classical Greek civilization.[1] The fact that such formulas are not to be found in the *Elements* or in other mathematical works of the third century is obviously not due to ignorance. To understand why these formulas are not found in the *Elements* one must first understand what conceptual apparatus would be required in a satisfactory proof of their validity.

Clearly, the first thing needed is a number system with multiplication and a way of assigning numbers to lines and figures. But to assign a number to a line is not to make an abstract correlation; it is to say that the line has a certain length. Hence the formulas for areas presuppose a notion of signed quantities. A straight line is not 9 but 9 centimeters in length; a triangle is not 3 but 3 square meters in area, and so on. Obviously one needs as well an understanding of the relations between signed quantities; for example, one needs to know that the product of a length in meters and a length in meters is an area in square meters. Finally, it seems, one needs to understand notions like length, area, and volume abstractly. A figure isn't an area; it has an area, an area which another figure might have as well.

I have already indicated the general absence of abstract notions in Euclid's geometry. The absence of the other concepts underlying the standard geometric formulas does not need to be documented. The assignment of numbers to geometric objects and, therefore, signed quantities are simply not to be found in the *Elements*. Nevertheless Euclid does prove various analogues of the formulas. As early as I,36 he proves the equality of parallelograms on equal bases and in the same parallels. Had he the notion of height, he could easily prove the equality of parallelograms with equal bases and heights. For, given two such parallelograms, he could construct a copy of one on the base of the other and argue for the equality of copy to original by using I,4 and 34, and for the equality of the two parallelograms to one another by using I,35. Euclid does not define the height of a figure until book VI, and there he defines it inadequately as the perpendicular drawn from the vertex to the base. (VI, def. 4. See Heath, vol. II, p. 189.) Moreover, the only application he makes of the concept in connection with plane geometry concerns triangles and parallelograms "under the same height," where being under the same height is the same as being in the same parallels. This restriction of the notion of height is not crucial, however, because in book VI Euclid follows the procedure—first used in I,43—of placing figures where he can use them.

Since Euclid proves in I,34 that the two triangles formed by the diagonal of a parallelogram are half the parallelogram, he is in a position to derive I,36 for triangles. Again, it would be a simple step to extend this result to triangles with equal bases and heights; however, Euclid does not take it. Similarly, Euclid establishes in I,41 the factor $\frac{1}{2}$ for triangles and parallelograms on the same base and in the same parallels without extending the result to equal bases, let alone to equal heights. These obvious extensions of book I propositions would be equivalent to parts of the formulas for the areas of triangles and parallelograms. They would establish that triangles or parallelograms with equal heights and bases are equal and that a triangle is half the area of a parallelogram with equal base and height. Euclid's failure to establish these extensions suggests that in book I he is not interested in them, but, as has already been argued, in the propositions needed to establish I,45.

Euclid could also have proved in book I a fact which he uses without proof in book VI:

VIA Of two triangles or parallelograms on unequal bases and in the same parallels, the one on the greater base is the greater.

Thus in book I Euclid has all the means needed to show that triangles or parallelograms in the same parallels grow larger as their bases do. He gives this result a precise formulation as

VI,1 If f_1 and f_2 are triangles or parallelograms under the same height with bases b_1 and b_2 respectively, then $(f_1, f_2) = (b_1, b_2)$.

This proposition is indicative of how Euclid uses proportions to express approximations of the formulas for areas. VI,1 does not, of course, give the full strength of the formulas for triangles and parallelograms. For parallelograms the closest approximation to this full strength expressed in terms of proportionality might be "Parallelograms are to one another as the products of their bases and heights," or, since a parallelogram is equal to the rectangle with the same base and height, "Rectangles are to one another as the products of two adjacent sides of each." It is not necessary to concern ourselves further with triangles, since any result for parallelograms is easily extended to triangles using I,34. The trouble with either of the formulations just given is that they use the notion of the product of two straight lines, a notion presupposing most of the conceptual apparatus implicit in the traditional formulas. One might identify the product of two straight lines with the rectangle contained by them. However, this identification turns the proportionality between parallelograms and the products of their bases and heights into a trivial consequence of I,36, and the proportionality between rectangles and the products of two of their adjacent sides into a vacuous assertion of the form $(x, y) = (x, y)$.

In discussing VIII,5 I pointed out that the compounding of ratios is an analogue of the multiplication of fractions, even though Euclid's treatments of compounding and of duplicating suggest no awareness of the connection. Using the notion of compounding one can approximate the formula for the area of a parallelogram with the law

VIB Parallelograms are to one another in the ratio compounded of the ratio of their heights and the ratio of their bases.

Although Euclid does not prove this law, he does prove

VI,23 Equiangular parallelograms have to one another the ratio compounded of [the ratios of] their sides.

As Heath (vol. II, p. 251) points out, VIB is a consequence of VI,23, since any parallelogram is equal to a rectangle of equal base and height, and the height of a rectangle is just one of its sides. It is also not difficult to derive VI,23 from VIB. Despite this equivalence, it seems highly unlikely that VI,23 should be interpreted as in any sense an intentional representation of the formula for the area of a parallelogram. Certainly one would expect Euclid to make more explicit the truth of VIB if he were interested in the formula.

In general the geometric books confirm the impression gained from the arithmetic ones that Euclid does not construe compounding as multiplication. VI,23 itself is, in a sense, evidence of this fact, since the product of the lengths of two sides of a parallelogram does not produce a value of any mathematical significance. Moreover, one can give a purely geometric, natural

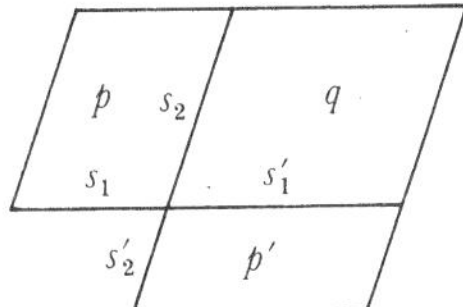

Figure 4.1

account of the idea of compounding ratios. Suppose p and p' (fig. 4.1) are two equiangular parallelograms with adjacent sides s_1, s_2 and s'_1, s'_2, and let them be placed so that s_1, s'_1 and also s_2, s'_2 are in a straight line. (Euclid makes such placements in both VI,14 and 23.) If then q is the parallelogram determined by s_2 and s'_1, one has, by VI,1, $(p, q) = (s_1, s'_1)$ and $(q, p') = (s_2, s'_2)$, and, by definition, p has to p' the ratio compounded of (s_1, s'_1) and (s_2, s'_2).

The only difference between this argument and Euclid's proof of VI,23 is that Euclid first constructs straight lines r, s, t satisfying $(s_1, s'_1) = (r, s)$ and $(s_2, s'_2) = (s, t)$, and then argues that $(p, p') = (r, t)$. The greater complexity of Euclid's proof is perhaps to be explained by a desire to show that (p, p') can be represented as the ratio of two straight lines. The simpler proof gives very little content to the proposition established, although it would give more if compounding were taken as multiplication. On the other hand, the two proportions of the simpler proof show how the idea of compounding might have arisen from the figure, since the two proportions already show how (p, p') is related to (s_1, s'_1) and (s_2, s'_2). All that remains is to give this relation a name.

I am inclined, then, to say that VI,23 is not indicative of an interest in proving an analogue of the ordinary formula for the area of a parallelogram. Rather it appears to be an isolated fact much like VIII,5, its arithmetic parallel. It is difficult to feel certain of this interpretation because VI,23 is never used, while so much else in the *Elements* seems pointed toward some goal or other. However, the alternative interpretations considered here appear to involve the application of too much mathematical apparatus not actually present in the *Elements*. Moreover, the interpretation given here will receive further confirmation from Euclid's treatment of solid figures and from the rest of book VI, to which I now turn.

As the title of this chapter indicates, book VI can be described as the application of the theory of proportion to the plane geometry of straight lines. However, although heavy use is made of the laws of proportionality in book VI, the definition of proportionality on which those laws are based is invoked only twice, namely in the first and last propositions of the book. The last proposition, 33, is the one exception to my characterization of the subject of book VI as rectilineal figures, since it concerns circles. However, 33 is a completely isolated proposition, depending on nothing else in book VI, and is clearly inserted with an eye to its applications in XIII,8–10.

VI,1, by contrast, is fundamental. To prove it Euclid takes triangles ABB_1, ACC_1 with B_1B, CC_1 on one straight line B_1C_1

(fig. 4.2). To prove $(ABB_1, ACC_1) = (BB_1, CC_1)$, it is necessary to show COMP $(m \cdot ABB_1, n \cdot ACC_1, m \cdot BB_1, n \cdot CC_1)$. But since, by I,38, triangles under the same height and with equal bases are equal, equimultiples of ABB_1, BB_1 and of ACC_1, CC_1 are formed by extending the straight line B_1C_1 in either direction to B_mBCC_n, making each of the segments B_1B_2, B_2B_3, $\ldots$, $B_{m-1}B_m$ equal to BB_1 and each of C_1C_2, C_2C_3, $\ldots$, $C_{n-1}C_n$ equal to CC_1, and connecting the B_i's and C_i's with A. Then, by I,38 and VIA, COMP $(AB_mB, ACC_n, B_mB, CC_n)$, i.e., COMP $(m \cdot ABB_1, n \cdot ACC_1, m \cdot BB_1, n \cdot CC_1)$. Euclid then uses I,34 (or 41) and V,15 and 11 to infer the same result for parallelograms.

Becker[2] has pointed out that, within the theory of proportion which he has reconstructed as pre-Eudoxean, VI,1 is an immediate consequence of I,34 and 36 and VIA. Similarly, one could prove VI,1 for commensurable bases, using a theory of proportion analogous to Euclid's treatment of numerical proportionality, a kind of theory many scholars suppose the Greeks to have worked with at some time.[3] Since VI,1 is the only important use of Eudoxus' definition in book VI, it is clear that any theory enabling one to prove VI,1 and standard laws of proportion would suffice as a basis for book VI. The relationship between V and VI is therefore of some interest for theories of the composition of the *Elements*. There are more or less explicit quotations of propositions from book V in VI,1, 2 and 20 and a quotation of the definition of duplicate ratios in VI,19. Such quotations show very little both because of their infrequency and because they could simply be citations of propositions known to be true. In general, but not always, Euclid uses the theory of proportion very explicitly in VI, making transformations step by step. In most cases the steps are easily referred to propositions in V, but there are at least two cases (VI,30 and 31) in which such reference presupposes a tacit application of alternation or inversion. V,17, 19, and 23 are not used at all in book VI, and there are inferences in VI,9 and 22 which apparently presuppose VK(iii) and VH(ii) respectively. These disparities suggest that V was not framed with an eye to specific applications in VI; nor do the proofs in

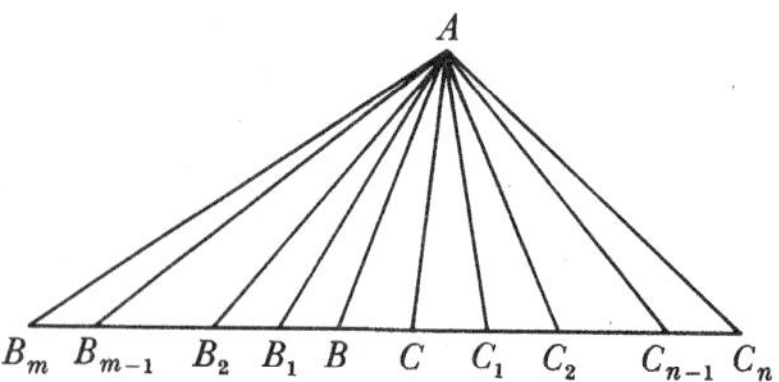

Figure 4.2

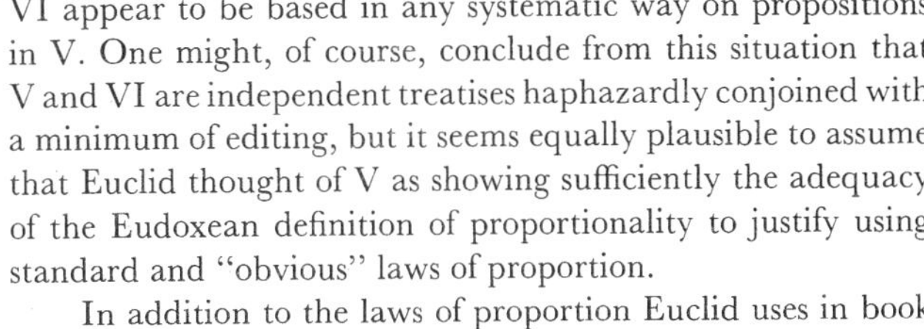

VI appear to be based in any systematic way on propositions in V. One might, of course, conclude from this situation that V and VI are independent treatises haphazardly conjoined with a minimum of editing, but it seems equally plausible to assume that Euclid thought of V as showing sufficiently the adequacy of the Eudoxean definition of proportionality to justify using standard and "obvious" laws of proportion.

In addition to the laws of proportion Euclid uses in book VI most of the content of book I up to proposition 46,[4] but nothing from books II or IV. In VI,33 he uses material from book III, but the only use of book III in the principal part of VI is in proposition 13 where Euclid invokes the first part of III,31, according to which the "angle in a semicircle" (the angle contained by two straight lines drawn to a point on the circumference from the ends of a diameter) is a right angle. This proposition, which is frequently used as an example by Aristotle, has an elementary proof depending upon book I materials only. To prove it Euclid considers an angle BAC in a semicircle with center O (fig. 4.3). He connects BC and AO, and extends BA to D. He then argues that, since the radii of a circle are equal, angle $OBA \simeq$ angle OAB and angle $OAC \simeq$ angle OCA, so that angle $BAC \simeq$ angle OBA + angle $OCA \simeq$ [I,32] angle CAD. But then, by definition, BAC is right. It seems fair to say then that the foundation of book VI is simply the geometry of book I and the theory of proportion.

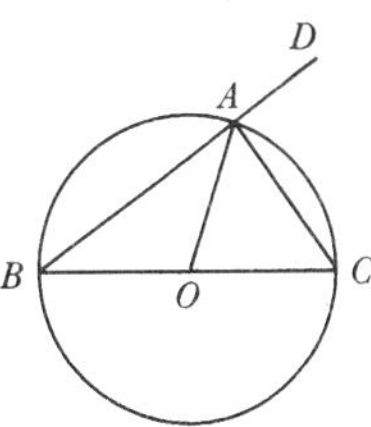

Figure 4.3

In book VI itself a fundamental role is played by the first definition, according to which similar rectilineal figures are those having "their angles severally equal and the sides about the equal angles proportional." Euclid's intention here may again be simply to give a rough sense of an intuitively understood notion. However, the definition does play a mathematical role in the *Elements*, and, from a modern point of view at least, it does so inadequately, because it does not make explicit the need to take the sides about the equal angles in the same order.[5] Thus, according to the definition, the two obviously dissimilar polygons in fig. 4.4 are similar because in them one has the

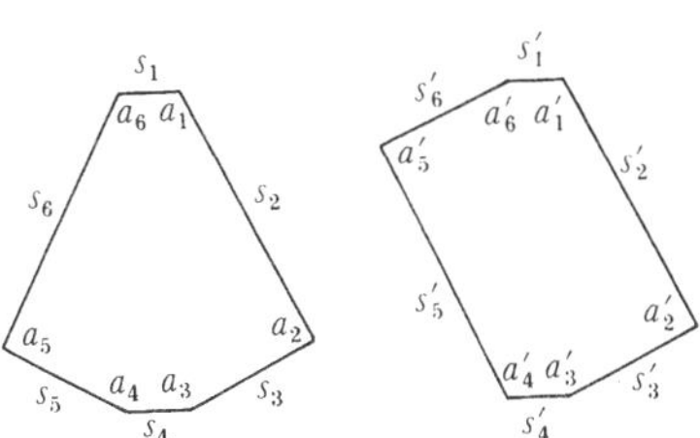

Figure 4.4

angle equalities $a_1 \simeq a_1'$, $a_2 \simeq a_2'$, $a_3 \simeq a_3'$, $a_5 \simeq a_5'$, $a_4 \simeq a_6'$, and $a_6 \simeq a_4'$, and the proportions $(s_1, s_2) = (s_1', s_2')$, $(s_2, s_3) = (s_2', s_3')$, $(s_3, s_4) = (s_3', s_4')$, $(s_5, s_6) = (s_6', s_5')$, $(s_4, s_5) = (s_1', s_6')$, and $(s_6, s_1) = (s_5', s_4')$. To avoid this shortcoming it is necessary to make explicit what Euclid clearly has in mind by saying that p and p' are similar polygons if they are contained by straight lines $s_1, s_2, \ldots, s_n$ and $s_1', s_2', \ldots, s_n'$, respectively, and have the interior angles $a_1, a_2, \ldots, a_n$ and $a_1', a_2', \ldots, a_n'$ with a_i (a_i') contained by s_i and s_{i+1} (s_i' and s_{i+1}'), for $1 \leq i < n$, and a_n (a_n') contained by s_n and s_1 (s_n' and s_1'), and if $a_i \simeq a_i'$, for $1 \leq i \leq n$, and $(s_i, s_{i+1}) = (s_i', s_{i+1}')$, for $1 \leq i < n$. The remaining definitions do not need to be discussed here. However, it is perhaps worthwhile to indicate the meaning of one expression which Euclid uses without definition. Two similar polygons p and p' are similarly situated on the straight lines s and s' if s and s' are corresponding sides of the two polygons, i.e., if s and s' play the role of some s_i and s_i' in the similarity condition for p and p'.

VI is one of the more perplexing books from a logical point of view, both in terms of individual proofs and in terms of overall structure. I shall pass over many of the problems of logical detail, most of which are discussed by Heath; and I shall not attempt to depict the deductive structure of the book as a whole, but rather divide it into a few separately discussible parts, namely (i) 2–8, (ii) 9–13, (iii) 14–17, 19, 20, 23 and (iv) 18, 19–21, 24–30. This listing itself is indicative of the lack of overall coherence in book VI. And outside the list there remain 22 and 31–33, of which I will discuss 22 and 31 at the end of the chapter.

Propositions 4–7 give the basic conditions for the similarity of two triangles t and t' (fig. 4.5) with sides s_1, s_2, s_3 and s_1', s_2', s_3' facing the angles a_2, a_3, a_1 and a_2', a_3', a_1', respectively:

(VI,4) $a_1 \simeq a_1', \quad a_2 \simeq a_2', \quad a_3 \simeq a_3'$;
(VI,5) $(s_1, s_2) = (s_1', s_2'), \quad (s_2, s_3) = (s_2', s_3'),$
$(s_3, s_1) = (s_3', s_1')$;
(VI,6) $a_1 \simeq a_1', \quad (s_1, s_2) = (s_1', s_2')$;
(VI,7) $a_3 \simeq a_3', \quad (s_1, s_2) = (s_1', s_2')$, and both of a_2, a_2' either less than or not less than a right angle.

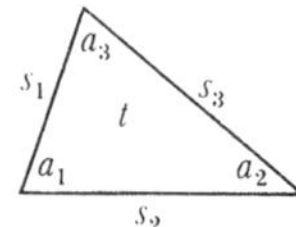

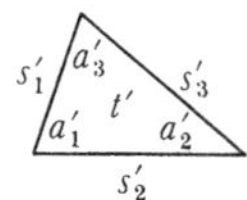

Figure 4.5

In stating these propositions Euclid does not use the term 'similar', but speaks only of angle equalities and proportionalities of sides. The word 'similar' occurs for the first time in the propositions of book VI in

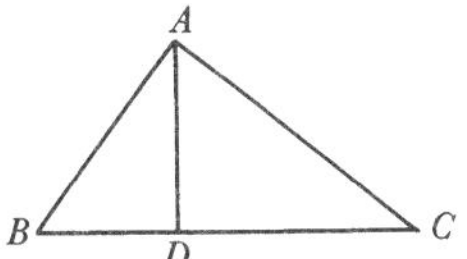

Figure 4.6

VI,8 If in a right-angled triangle *ABC* (fig. 4.6) a perpendicular *AD* is drawn from the right angle at *A* to the base *BC*, then the triangles *ABD*, *ADC* are similar to each other and to *ABC*, and (corollary) (*BD*, *AD*) = (*AD*, *DC*) and (*BC*, *AB*) = (*AB*, *BD*).

Each of 5–8 is reduced to 4. In the case of 8 the reduction is simple because each of the three triangles has a right angle, and each of *ABD*, *ADC* shares another angle in common with *ABC*. Hence, by I,32, the three triangles are equiangular. For 5–7 Euclid constructs a triangle t'' with angles a''_2, a''_3, a''_1 equal to angles a_2, a_3, a_1, respectively, and facing sides s''_1, s''_2, s''_3, where $s''_1 \simeq s'_1$. It is easy to establish using V,11 and 9 that $s'_2 \simeq s''_2$ and $s'_3 \simeq s''_3$ in VI,5 and that $s'_2 \simeq s''_2$ in VI,6 and 7. It follows from the congruence theorems for triangles[6] that t' and t'' are equiangular and hence that t' and t are.

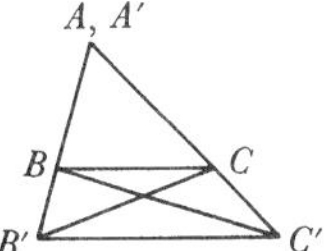

Figure 4.7

Euclid's proof of 4 can perhaps be understood by imagining triangles *ABC*, *A′B′C′* (fig. 4.7) with angles at *A*, *B*, *C* equal to the angles at *A′*, *B′*, *C′*, respectively, and *AB* shorter than *A′B′*. It is easy to argue that if *AB*, *AC* are made to coincide with *A′B′*, *A′C′*, *BC* will fall inside triangle *A′B′C′*, as in the figure, and will in fact be parallel to *B′C′*. In order to show that (*A′B′*, *AB*) = (*A′C′*, *AC*), it suffices to establish that (*BB′*, *AB*) = (*CC′*, *AC*). Euclid in fact proves in VI,2 that if *AB′C′* is a triangle and *B*, *C* lie on *AB′*, *AC′*, respectively, then *BC* is parallel to *B′C′* if and only if (*BB′*, *AB*) = (*CC′*, *AC*). He connects *BC′* and *CB′*. By VI,1, (*BB′C*, *ABC*) = (*BB′*, *AB*) and (*CC′B*, *ABC*) = (*CC′*, *AC*); therefore, (*BB′*, *AB*) = (*CC′*, *AC*) if and only if $BB'C \simeq CC'B$ if and only if (I,37 and 39) *B′C′* and *BC* are parallel. One could now complete the proof of VI,4 by making *AC*, *BC* coincide with *A′C′*, *B′C′*. Euclid avoids the need for two such placements by imagining the two triangles positioned so that *C* and *B′* coincide and *BCC′* is a straight line (fig. 4.8). He extends *BA* and *C′A′*, invoking postulate 5 to show that they will meet at a point *D*. Since, by I,28, *AC*, *A′C* are parallel to *A′D*, *AD*, respectively, VI,2 yields (*DA*, *AB*) = (*C′B′*, *CB*) and (*CB*, *C′B′*) = (*DA′*, *A′C′*). Since $DA \simeq A'B'$ and $DA' \simeq AC$, the desired proportionalities are easily derived.

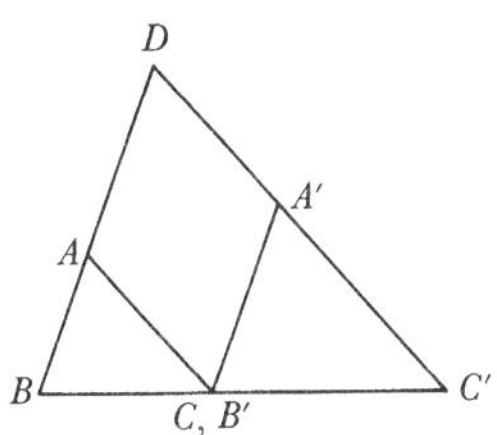

Figure 4.8

Euclid also derives from VI,2 a proposition which Archimedes uses in the *Measurement of a Circle* but which is not used in the *Elements*, namely,

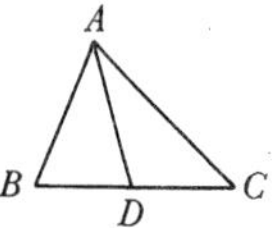

Figure 4.9

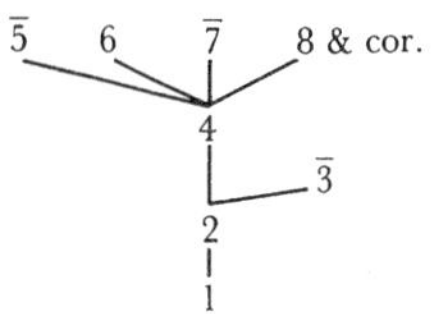

Figure 4.10

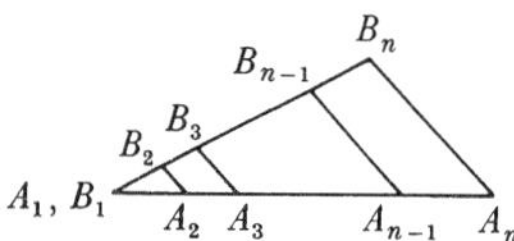

Figure 4.11

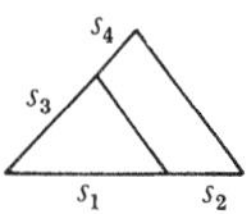

Figure 4.12

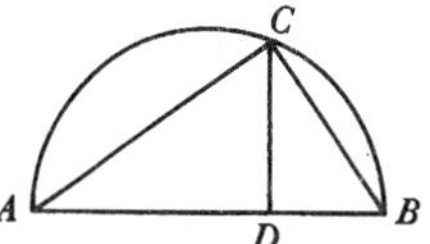

Figure 4.13

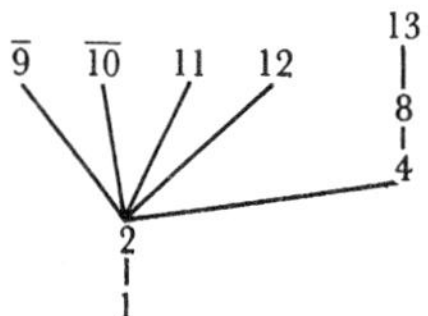

Figure 4.14

VI,3 If ABC (fig. 4.9) is a triangle and D is a point on BC, then AD bisects angle BAC if and only if $(BD, DC) = (AB, AC)$.

The structure of VI,2–8 is represented in fig. 4.10. 7, like 3, is never used in the *Elements*. 5, although not used in book VI, is used in XII,12. It seems likely that Euclid would choose to include all of 4–7 simply to give a complete account of the similarity conditions for triangles.

In 9–13 Euclid uses VI,2 and 8 to solve problems of cutting straight lines in given proportions or determining straight lines satisfying given proportionalities. In 10 he shows how, given a straight line $A_1A_2 \ldots A_n$ (fig. 4.11), another straight line B_1B_n can be cut into segments $B_1B_2, B_2B_3, \ldots, B_{n-1}B_n$ satisfying $(A_iA_{i+1}) = (B_i, B_{i+1})$. He makes A_1 and B_1 coincide and forms the triangle $A_1A_nB_n$. It is easy to show, using VI,2, that the parallels to A_nB_n through each of $A_2, A_3, \ldots, A_{n-1}$ divide B_1B_n in the desired way. VI,9, "to cut off an nth part of B_1B_{n+1}," is really a special case of 10. For given B_1B_{n+1}, one can take an arbitrary straight line A_1A_2, produce its nth multiple A_1A_{n+1}, and proceed as in 10. To prove 9 Euclid does the case $n = 3$ and argues that since $(B_2B_4, B_1B_2) = (A_2A_4, A_1A_2)$ and $A_2A_4 \simeq 2 \cdot A_1A_2$, $B_2B_4 \simeq 2 \cdot B_1B_2$, so that $B_1B_4 \simeq 3 \cdot B_1B_2$. Becker[7] takes this application of VK(iii) as evidence that Euclid's source for 9 was based on a theory of proportion different from the one found in book V. Becker's position is weakened by the fact that VK(iii) is a simple consequence of the Eudoxean definition of proportionality.

In 12 Euclid uses VI,2 to construct a fourth proportional s_4 to three given straight lines s_1, s_2, s_3 (fig. 4.12). He speaks of finding the fourth proportional, just as he speaks in the arithmetic books of finding numbers, but the locution does not seem to have any philosophical significance here since Euclid ends the proof of VI,12 with a 'Q.E.F.'. 11 is the special case of 12 in which the second and third straight lines are identical. Euclid's proofs, indicated in fig. 4.12, are the same for both.

Euclid uses III,31 in VI,13 to show how to find a "mean proportional" between two straight lines s_1 and s_2, i.e., a straight line s such that $(s_1, s) = (s, s_2)$. He represents s_1 and s_2 as segments AD, DB of a straight line ADB, draws the semicircle with ADB as diameter, erects a perpendicular CD to ADB intersecting the circle in C, and connects AC, CB (fig. 4.13). It follows from III,31 and the corollary to VI,8 that CD is the desired mean proportional.

The deductive structure of 9–13 is depicted in fig. 4.14. Although 9 and 10 are not used in book VI, one of them, probably 9, is used in XIII,11, 13, 15, 16, and 18. 10 may represent another case of generalization for the sake of generalization. As was made clear in the first part of section 3.2

(pp. 134–135), propositions 11, 12, and 13 can be thought of as the geometric foundation for calculation with ratios. 9 and 10 could be thought of as serving the same purpose in the special case in which the ratios correspond to fractions. It should be clear that if Euclid viewed the propositions in this way he made no effort to make his point of view known to the reader.

Propositions 14–17 may be stated as follows:

VI,14 If p and p' are equiangular parallelograms with adjacent sides s_1, s_2 and s_1', s_2', then $p \simeq p'$ if and only if $(s_1, s_1') = (s_2', s_2)$.

VI,15 If ABC and $A'B'C'$ are triangles with the angle at A equal to that at A', then $ABC \simeq A'B'C'$ if and only if $(AB, A'B') = (A'C', AC)$.

VI,16 14 is true if p and p' are rectangles.

VI,17 14 is true if p is a rectangle and p' a square.

16 and 17 are trivial consequences of 14, and Euclid's proofs, reducing 16 to 14 and 17 to 16 in an obvious way, show that he recognizes this fact. There are, however, curious explicit citations of 14 where 16 or 17 would do in X,22 and VI,30. 15 is also a trivial consequence of 14 since any triangle ABC is half of the parallelogram determined by AB and BC. Euclid chooses to derive 15 from 1 in exactly the way he derives 14 from 1, so that the deductive structure of 14–17 is as shown in fig. 4.15. 14 itself is a direct consequence of VI,23 and VJ, which, as I pointed out, is obvious on the fractional interpretation of ratios. Euclid seems totally unaware of the connection between 23 and 14. For in proving 14 he carries out the construction of 23, described on p. 155, and argues that since, by VI,1, $(p, q) = (s_1, s_1')$ and $(p', q) = (s_2', s_2)$, $p \simeq p'$ if and only if $(p, q) = (p', q)$ if and only if $(s_1, s_1') = (s_2', s_2)$.

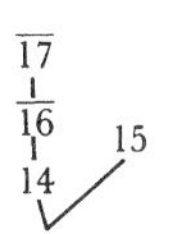

Figure 4.15

17 is used frequently in the later books of the *Elements*, particularly in X and XIII. It makes clear that the finding of a mean proportional x between two straight lines a and b is equivalent to the finding of an x such that $\mathbf{T}(x) \simeq \mathbf{O}(a, b)$, or, in algebraic terms, the solving of $x^2 = ab$. Since I,45 shows how to construct a rectangle equal to a given rectilineal figure, VI,13 is tantamount to a solution of II,14, the problem of constructing a square equal to a given rectilineal figure. Euclid's proof of II,14 appears to be a revision of the proof of VI,13 designed to avoid the theory of proportion and III,31. He constructs the rectangle equal to the given rectilineal figure and represents its consecutive sides as segments AD, DB of a straight line ADB (fig. 4.16). He then carries out the construction of VI,13. With III,31 Euclid could use the Pythagorean theorem and II,4 to argue that $\mathbf{T}(DE) + \mathbf{T}(AD) + \mathbf{T}(DE) + \mathbf{T}(BD)$

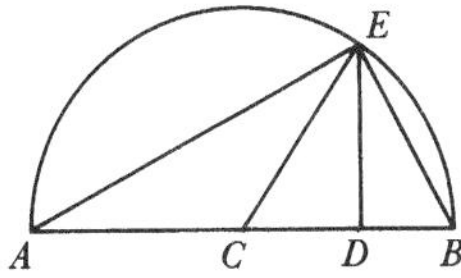

Figure 4.16

$\simeq \mathbf{T}(AE) + \mathbf{T}(EB) \simeq \mathbf{T}(AB) \simeq \mathbf{T}(AD) + \mathbf{T}(BD) + 2 \cdot \mathbf{O}(AD, DB)$, so that $2 \cdot \mathbf{T}(DE) \simeq 2 \cdot \mathbf{O}(AD, DB)$, and $\mathbf{T}(DE) \simeq \mathbf{O}(AD, DB)$. To avoid the need for III,31 Euclid connects the center C of the semicircle with E and argues that $\mathbf{T}(CB) \simeq \mathbf{T}(CE) \simeq \mathbf{T}(DE) + \mathbf{T}(DC)$. Clearly, if the construction is correct, it must be the case that $\mathbf{O}(AD, DB) + \mathbf{T}(DC) \simeq \mathbf{T}(CB)$; but this equation is simply II,5.

There is no compelling reason to classify either II,14 or VI,13 as an algebraic problem. The formulation of II,14 is geometric and has a natural geometric interpretation. VI,13 appears to rest entirely upon the geometric conception of similarity, and it is reasonable to suppose that it was long recognized as the basis of the solution to the problem of II,14.[8] If one supposes that Euclid's proof of II,14 is explained by its position in the *Elements*, the part of the proof which looks most algebraic, namely, the application of II,5, can be read as the invocation of a proposition inserted to be used in 14 and known to be true on independent geometric grounds. This use of II,5 then does not require an algebraic interpretation, since a natural geometric one is available.

Euclid's failure to reduce VI,14 to 23 is one piece of evidence that he does not view compounding of ratios as some form of multiplication. Further evidence is provided by VI,19 and 20, which establish that similar rectilineal figures are to one another in the duplicate of the ratio of the corresponding sides. Euclid proves this result for triangles in 19 and extends it to polygons in 20 by dividing two similar polygons into an equal number of similar triangles "in the same ratio as the polygons" and applying V,12.[9] The relationship between 23 and 19 is slightly more complicated than the one between 23 and 14 because 23 is formulated in terms of parallelograms. However, because a triangle is half of the parallelogram determined by two consecutive sides of the triangle, 23 has as a direct corollary

VI,23t If ABC and $A'B'C'$ are two triangles with the angle at A equal to the angle at A', then $(ABC, A'B'C')$ is the ratio compounded of $(AB, A'B')$ and $(AC, A'C')$.

19 follows immediately from 23t. For if the two triangles of 23t are similar with the obvious correspondences, $(AB, A'B') = (AC, A'C')$, so that the ratio compounded of $(AB, A'B')$ and $(AC, A'C')$ is the duplicate of the ratio of the corresponding sides.

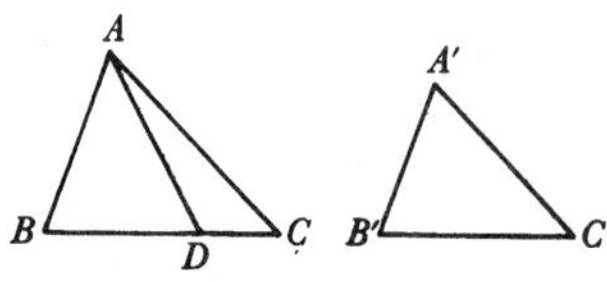

Figure 4.17

To prove 19 Euclid assumes that ABC, $A'B'C'$ (fig. 4.17) are similar triangles with corresponding sides AB, $A'B'$ and BC, $B'C'$, with $BC > B'C'$. He uses VI,11 to determine BD so that $(BC, B'C') = (B'C', BD)$, and connects AD. Since

$(AB, A'B') = (BC, B'C')$, $(AB, A'B') = (B'C', BD)$, and, by VI,15, $ABD \simeq A'B'C'$. Hence, using VI,1, $(BC, BD) = (ABC, ABD) = (ABC, A'B'C')$; but (BC, BD) is by construction the duplicate of $(BC, B'C')$. This proof gives no indication of an awareness of a connection between duplicating and compounding. Even more significant, perhaps, is Euclid's apparent lack of interest in the connection between duplicating and squaring. Of course, it is a trivial consequence of VI,19 or 20 that squares are to one another in the duplicate of the ratio of their sides, so that similar polygons are to one another as the squares on corresponding sides; but Euclid does not bother to prove these facts explicitly. Perhaps the following analysis provides a satisfactory explanation of Euclid's use of duplicate ratios in VI,19. Suppose one wishes to express the ratio between similar triangles ABC, $A'B'C'$, as in fig. 4.17, in terms of straight lines. Since, by VI,1, if D lies on BC, $(ABC, ABD) = (BC, BD)$, this problem reduces to constructing triangle ABD equal to $A'B'C'$ and determining (BC, BD). Since the angles ABC, $A'B'C'$ are equal, one knows by VI,15 that $(AB, A'B') = (B'C', BD)$. But the similarity of ABC, $A'B'C'$ entails $(AB, A'B') = (BC, B'C')$, so that (BC, BD) is the duplicate of $(BC, B'C')$. In this analysis, which leads directly to Euclid's proof, the notion of squaring or forming the square with a given side plays no role. If the analysis is correct, Euclid would appear to be interested in VI,19 only because the problem of determining the ratio between similar polygons reduces directly to it.

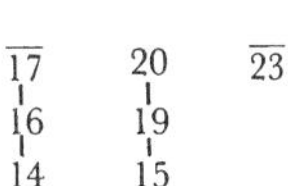

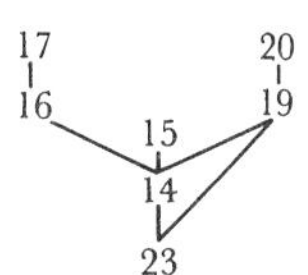

Figure 4.18

The upper part of fig. 4.18 depicts the logical structure of 14–17, 19, 20, and 23, no account being taken of dependencies prior to 1–13. The lower part shows the structure which results from a full exploitation of the relationship between a triangle and a corresponding parallelogram and of the power of the notion of compounding. From a modern point of view the second structure brings out the relation of the mathematical content of these propositions, and in particular brings out the fundamental importance of 23, which is never used in the *Elements*. Another curious feature of the *Elements* is the corollaries to 19 and 20. The latter of these generalizes 20 from polygons, rectilineal figures with more than four sides, to rectilineal figures in general. The corollary to 19 states the equivalent of the corollary of 20 for "figures," which can only mean rectilineal figures.[10] Euclid explicitly employs each of the corollaries in different places and once (XII,1) appears to use VI,20 itself. In these case and in cases in which it is impossible to determine a specific citation, I shall refer to VI,19–20.

In VI,28 and 29 Euclid considers two problems which might seem strange to a person unfamiliar with Greek mathematics. One starts with a straight line AB, a parallelogram p,

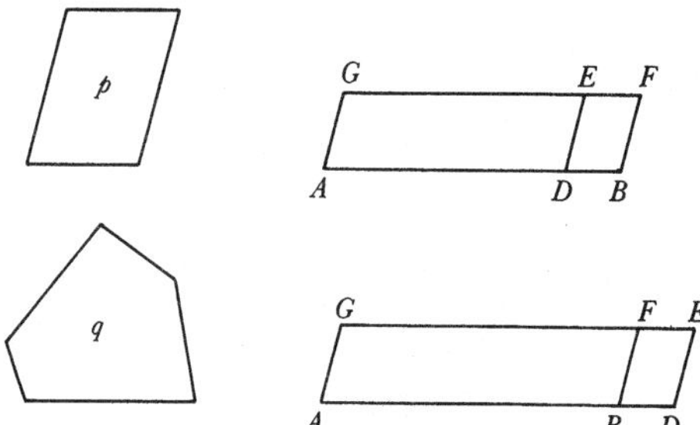

Figure 4.19

and a rectilineal figure q (fig. 4.19). The problem is to find a point D either on AB (VI,28) or on it extended (VI,29) such that if the parallelogram $BDEF$ is made similar to p, the parallelogram $ADEG$ determined by AD and DE will be equal to q. When this has been done Euclid says that there has been applied (*paraballein*) to AB the parallelogram $ADEG$ equal to q and deficient (*elleipon*, VI,28) or exceeding (*huperballon*, VI,29) by a figure similar to p.[11] These problems are easily explained algebraically if one takes p to be a square (as it is in Euclid's applications of VI,28 and 29), and q and AB to be known quantities b and a. The determination of D can then be thought of as the finding of an x (namely BD) such that

VI,28a $ax - x^2 = b$, VI,29a $ax + x^2 = b$,

i.e., as the solution of two forms of quadratic equation.

We would solve VI,29a directly by "completing the square" to get

$$\left(\frac{a}{2}\right)^2 + ax + x^2 = b + \left(\frac{a}{2}\right)^2,$$

so that

$$x + \left(\frac{a}{2}\right) = \pm\sqrt{b + \left(\frac{a}{2}\right)^2}$$

and

$$x = \pm\sqrt{b + \left(\frac{a}{2}\right)^2} - \frac{a}{2}.$$

Because we recognize a positive and a negative square root of $b + \left(\frac{a}{2}\right)^2$, there are for us two solutions of VI,29a, only one of which is positive and hence geometrically significant. Euclid's proof of VI,29 can be interpreted as a geometric generalization of the algebraic procedure for finding this solution. I describe its several steps.

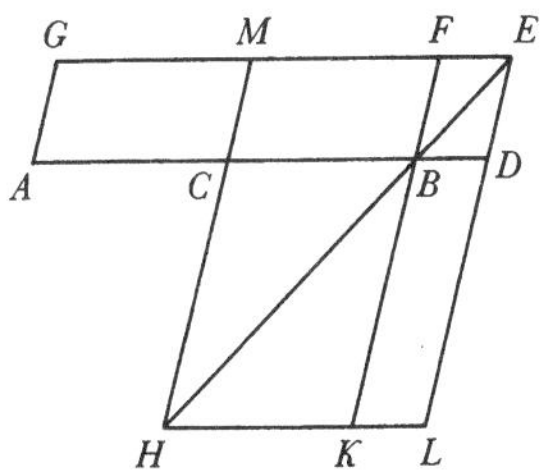

Figure 4.20

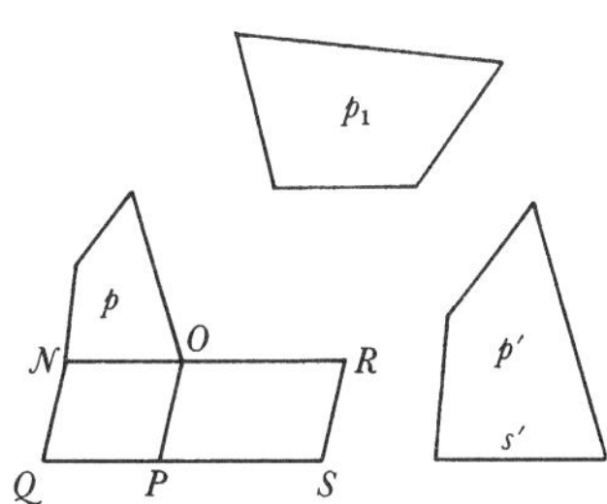

Figure 4.21

(i) The first step, corresponding to the computation of $\left(\frac{a}{2}\right)^2$, is the bisection of AB at C and the construction on BC of a parallelogram $BCHK$ "similar and similarly situated to p" (fig. 4.20). (When p is a square, this is simply the construction of a square on BC.) Euclid shows in VI,18 how to carry out this construction for an arbitrary rectilineal figure p by dividing p into triangles and constructing new triangles equiangular with them, starting from the given straight line.

(ii) The second step, corresponding to the computation of $b + \left(\frac{a}{2}\right)^2$, is the construction of a parallelogram $H'L'E'M'$ similar and similarly situated to p, and equal to $q + BCHK$. For this step Euclid takes for granted the possibility of representing $q + BCHK$ as a single rectilineal figure and applies VI,25, which is formulated for arbitrary rectilineal figures p. (When p is a square, the required construction is II,14 (or VI,13 and the construction of a square).) In 25, to construct a figure p' similar to p and situated on s' as p is situated on NO and also equal to a rectilineal figure p_1 (fig. 4.21), Euclid constructs on NO a parallelogram $NOPQ$ equal to p and on NO extended to R a parallelogram $OPSR$ equal to p_1.[12] Euclid now uses VI,13 to find the mean proportional s' between NO and OR and VI,18 to construct on s' the figure p' similar and similarly situated to p. He invokes the corollary to VI,19 to infer that $(p, p') = (NO, OR)$; but, by VI,1, $(p, p_1) = (NOPQ, OPSR) = (NO, OR)$, so that $p' \simeq p_1$.

(iii) Euclid next argues that $BCHK$ and $H'L'E'M'$ are similar since they are each similar to p, a step which would clearly be unnecessary if p were a square. For this purpose he invokes VI,21 in which he proves that figures are similar if they are similar to the same rectilineal figure. The proof is simply a matter of going through the motions:

For let each of the rectilineal figures A, B be similar to C; I say that A is also similar to B. For since A is similar to C, it is equiangular with it and has the sides about the equal angles proportional. Again, since B is similar to C, it is equiangular with it and has the sides about the equal angles proportional. Therefore each of the figures A, B is equiangular with C and has the sides about the equal angles proportional; therefore A is similar to B.

(iv) The next step is a geometric manipulation to make the analogue of the computation of the value of x possible. Euclid supposes that KH corresponds to $H'L'$, $H'M'$ to HC, and infers without explicit justification that since $H'L'E'M'$ is greater than $HKBC$, $H'L'$, $H'M'$ are respectively greater than HK, HC. He then copies $H'L'E'M'$ as $HLEM$, positioned as in fig. 4.20. In order to complete the figure and establish that *BDEF* is similar to p, he invokes

VI,24,26 If *HLEM* and *HKBC* are parallelograms and *K*, *C* lie on *HL*, *HM* respectively, then *HLEM* and *HKBC* are similar if (24) and only if (26) *B* lies on *HE*.

24 is derived from VI,2 and 4 in an unnecessarily elaborate but elementary way,[13] and 26 is reduced indirectly to 24.

(v) The "computation" is now relatively simple. Since *HLEM* corresponds to $b + \left(\frac{a}{2}\right)^2$ and *HKBC* to $\left(\frac{a}{2}\right)^2$, it is clear that the "gnomon" (II, def. 2) consisting of the figures *MB*, *BE*, *BL* corresponds to b. But since, by I,43, $BL \simeq MB \simeq GC$, $GC + MB + BE$ also corresponds to b; and the problem has been solved. (In the case in which p is a square, *CD* correspounds to $\sqrt{b + \left(\frac{a}{2}\right)^2}$, and *BD* to $\sqrt{b + \left(\frac{a}{2}\right)^2} - \frac{a}{2}$, i.e., to x.)

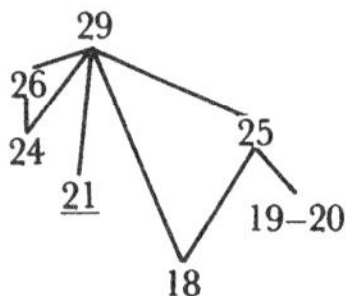

Figure 4.22

The structure of the proof of 29 is indicated in fig. 4.22. All of the propositions used, except 18–20, appear to be proved for the sake of their applications in 29 and in 28,[14] the proof of which has the same structure except for a precondition established in 27. However, despite this same structure, the relation of 28 to its modern algebraic analogue is somewhat more complex than the corresponding relation for 29. We would transform VI,28a into $x^2 - ax = -b$ and proceed in the same way as in the case of VI,29a to get that

$$x = \pm\sqrt{\left(\frac{a}{2}\right)^2 - b} - \frac{a}{2}.$$

Euclid cannot perform an analogue of this first transformation because he has no representation of negative quantities. His procedure can be thought of as a transformation of VI,28a into

$$ax - x^2 - \left(\frac{a}{2}\right)^2 = b - \left(\frac{a}{2}\right)^2$$

and then into

$$x^2 - ax + \left(\frac{a}{2}\right)^2 = \left(\frac{a}{2}\right)^2 - b;$$

but, of course, there is no actual geometric analogue of this second transformation because the same configuration represents both equalities.

The first step of Euclid's proof of VI,28 is again the erection of a parallelogram *BCHK* on the half *BC* of *AB* and similar

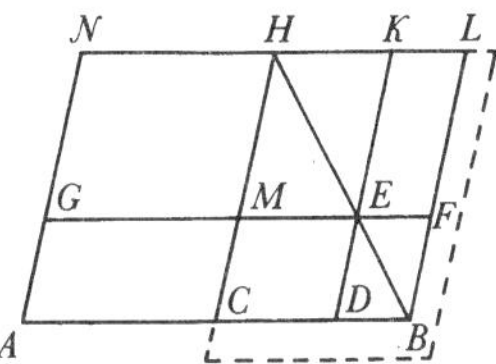

Figure 4.23

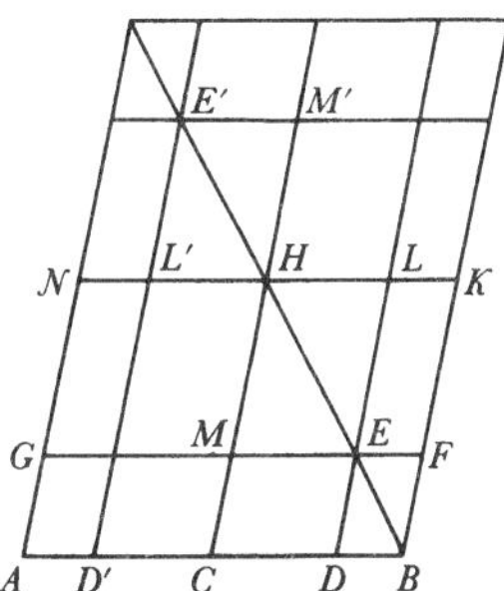

Figure 4.24

and similarly situated to p (fig. 4.23). One sees from the broken lines of the diagram that if q were greater than $BCHK$, there would be no way to carry out the proof. This situation corresponds to the fact that in modern terms the solution to VI,28a might be imaginary, even for positive a and b. For it to be positive and real $\left(\frac{a}{2}\right)^2 - b$ must be nonnegative, i.e., it must be the case that $b \leq \left(\frac{a}{2}\right)^2$, or, in algebraic-geometric terms, $\mathbf{O}(AD, DB) \preceq \mathbf{T}(AC)$. Euclid proves the geometric generalization of this condition in

VI,27 If $ACDB$ (fig. 4.24) is a straight line bisected at C and $CBKH$, $DBFE$ are similar parallelograms, the parallelogram determined by AC, CH is greater than the parallelogram determined by AD, DE.

The proof is simple because $CBKH$ and $DBFE$ can be assumed to be on the same side of $ACDB$ and so about the same diameter. When the figure is completed, one argues easily that $AH \simeq CBKH \succ LF + BE + CE \simeq AE$. The precondition for VI,28 is then that $q \preceq CBKH$. If $q \simeq CBKH$, $ACHN$ solves the problem; if $q \prec CBKH$, then the construction indicated in fig. 4.23 and an argument like that for VI,29 gives the result.

One can see from fig. 4.24 that there is a second solution to VI,28 in which AB is divided at D' with $AD' \simeq BD$. This second solution corresponds to the second positive root of VI,28a. To produce this second solution directly Euclid would have to show that similar parallelograms positioned as $CBKH$ and $HL'E'M'$ are about the same diameter and show that AH is greater than AE'. Although Euclid does not do the first of these, the manuscripts include an argument for the second after the 'Q.E.D.' of 27. For several reasons (recapitulated by Heath, vol. II, p. 260) Heiberg considered this argument to be an interpolation. Even if he was wrong, since Euclid does not deal with the second solution to VI,28, there does not seem to be any way to verify the claim that he was somehow aware of this solution.[15] If Euclid were interested in solving VI,28a, he presumably would be interested in the two solutions. For his own purposes the two solutions are the same. For when Euclid applies VI,28 in book X he is interested only in the way AB is broken at D or D'; but, since $AD' \simeq BD$, the two breaks are equivalent so that Euclid has no reason to consider them both. The fact that he considers the particular solution he does may be explained by reference to the lines to which he applies his book X results in book XIII; for in these the relevant break is at D rather than D'.

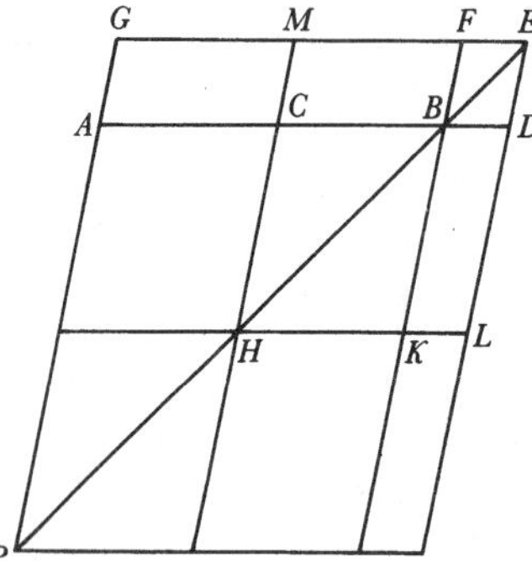

Figure 4.25

The suggestion that Euclid solves only the problems he has a use for may explain why he does nothing explicit corresponding to an obvious third algebraic problem,

VI,29a′ $x^2 - ax = b$.

It is sometimes suggested that Euclid does not do the algebraic-geometric analogue of this problem because he knows it is reducible to 29.[16] For if the parallelogram *AE* of fig. 4.25 solves VI,29, then *EP* is a parallelogram similar to the given one p and exceeds the rectilineal figure q by an appropriate parallelogram, the parallelogram contained by *GF*, the correspondent of a, and *GP*, the correspondent of x. (This geometric construction corresponds to solving VI,29a′ by setting $x - a = y$ so that $b = x^2 - ax = x(x - a) = (a + y)y = ay + y^2$, and VI,29 is applicable.) I am not sure how the appropriateness of the parallelogram would be described in Euclidean terms. However it might be described, the terminological difference between its description and those in VI,28 and 29 would reflect a geometric asymmetry which vanishes when the problems are treated algebraically. This asymmetry and the fact that Euclid has no use for an analogue of VI,29a′ seem to me to provide a much more plausible explanation for the absence of this analogue than does the assumption that Euclid knew he had already provided a solution to the problem in question.

Despite the discrepancies I have mentioned, the algebraic interpretation of VI,28 and 29 is attractive because these propositions have no apparent geometric motivation when they are read in isolation. In this sense they are unlike the problem of constructing a square equal to a given figure, since this problem makes perfectly good sense on its own terms without importation of algebraic ideas. I shall be arguing that VI,28 and 29 can be interpreted geometrically as lemmas for other geometric propositions. Of course, it could also be argued that the proof of these propositions was discovered by an algebraic analysis using 28 or 29 in some form or other. The point of view adopted here is that, if there are no independent grounds for choosing between an algebraic and a geometric interpretation in connection with the *Elements*, the latter is preferable because it does not use concepts which are not explicitly in the work.

VI,28 is used only in book X and will be discussed in section 7.2. VI,29 is used only in VI,30. All applications of either proposition are to cases in which p is a square, so that the geometric generalization to parallelograms is as unnecessary for Euclid's purposes as it is for the solution of quadratic equations. In VI,30 Euclid shows how to divide a straight line into extreme and mean ratio, i.e.,

VI,30 To cut a straight line AB at D so that $(AB, BD) = (BD, AD)$.

The geometric significance of this problem will become clear in connection with the discussion of the regular pentagon in the next chapter. It is now clear that because of VI,17 (Euclid uses 14) the problem reduces to finding D so that $\mathbf{T}(BD) \simeq \mathbf{O}(AB, AD) \simeq \mathbf{O}(AB, AB - BD) \simeq \mathbf{T}(AB) - \mathbf{O}(AB, BD)$, i.e., so that $\mathbf{T}(BD) + \mathbf{O}(AB, BD) \simeq \mathbf{T}(AB)$. However, this problem is simply VI,29 for the case in which p is a square and q is the square on the given straight line AB. This reduction of VI,30 to 29 involves straightforward geometric (or geometric-algebraic) manipulations. Moreover, given II,4 and the obvious fact that $2 \cdot \mathbf{O}(\frac{1}{2}AB, BD) \simeq \mathbf{O}(AB, BD)$, it is easy to see that, if VI,30 is solved, one will have $\mathbf{T}(AB) + \mathbf{T}(\frac{1}{2}AB) \simeq \mathbf{T}(BD) + \mathbf{O}(AB, BD) + \mathbf{T}(\frac{1}{2}AB) \simeq \mathbf{T}(BD + \frac{1}{2}AB)$; and this information leads relatively directly to the construction of 29 for the relevant special case, a construction which it is not difficult to generalize in Euclid's manner. Thus it can be said that if the interpretation of II,4 as a geometric truth is accepted, the desire to prove VI,30 provides the basis for a satisfactory geometric explanation of the discovery and proof of VI,29.

The direct application of VI,29 to 30 leads to a determination of D as a point on an extension of AB so that a little geometric argumentation is needed to determine the required point on AB itself. Euclid does this by erecting the square $ABA'B'$ on AB (fig. 4.26) and applying to $A'B$ the rectangle $AD'EG$ equal to $\mathbf{T}(AB)$ and exceeding by a square $BD'ED$, where D obviously lies on AB and solves the problem. The manuscripts of the *Elements* include an alternative proof of VI,30 which depends on

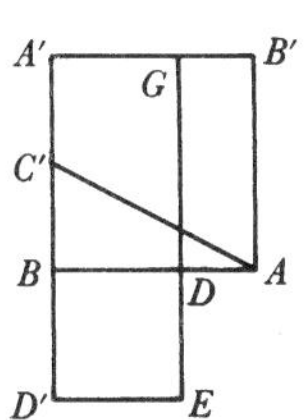

Figure 4.26

II,11 To cut AB at D so that $\mathbf{T}(BD) \simeq \mathbf{O}(AB, AD)$.

Clearly, given VI,17, propositions II,11 and VI,30 are equivalent. If one looks at Euclid's solution to II,11 in isolation it does not seem possible to understand how it was arrived at. But the solution becomes comprehensible as soon as it is recognized as a reworking of the proof of VI,30 to avoid the application of VI,29. As we have seen, the analysis of VI,30 leads to

If $AC'BD'$ is a straight line and AB is bisected at C', then, $\mathbf{T}(AB) + \mathbf{T}(BC') \simeq \mathbf{T}(BD') + \mathbf{O}(AB, BD') + \mathbf{T}(BC') \simeq \mathbf{T}(C'D')$.

According to the converse of the Pythagorean theorem, then $C'D'$ is the hypotenuse of a right triangle with legs AB and BC'. Hence $C'D'$ can be constructed by erecting the square $ABA'B'$

on AB, bisecting $A'B$ at C', and making $C'BD'$ equal to AC'. The proof of II,11 is completed by drawing the square $BD'ED$ to determine D on AB, and showing that $\mathbf{T}(BD') + \mathbf{O}(A'B, BD') + \mathbf{T}(BC') \simeq \mathbf{T}(C'D')$. However, since it is obvious that $\mathbf{T}(BD') + \mathbf{O}(A'B, BD') \simeq \mathbf{O}(A'D', BD')$, the desired equality follows immediately from II,6, which Euclid invokes in proving II,11. Thus II,6, like its counterpart II,5, can be interpreted as a proposition known to be true on independent grounds and proved because it is wanted already in book II. That these grounds are geometric will become clear in the next chapter.

Before leaving VI,28 and 29 I would like to comment briefly on them and on Babylonian procedures corresponding to the solutions of quadratics. The three equations VI,28a, 29a, and 29a′ are the special cases of

VI,28b $ax - cx^2 = b$,
VI,29b $ax + cx^2 = b$,
VI,29b′ $cx^2 - ax = b$,

in which $c = 1$. It is easy to convince oneself that these are the only quadratics which make geometric sense.[17] Moreover, these equations can be reduced to VI,28a, 29a, and 29a′ by dividing through by c; so there is a sense in which Euclid can be interpreted as showing how to solve at least VI,28b and 29b. On the other hand, the geometric representation of cx^2 would be a solid, and, although the extension of VI,28 and 29 to three-dimensional figures is conceivable, there is no satisfactory evidence that it was ever made by the Greeks. Thus a "Euclidean solution" to a general quadratic would most likely have to involve a combination of computation with fractions and geometric construction of a kind not found in the *Elements*.

There are a few Babylonian tablets in which computations naturally explained in terms of single quadratics are carried out. The quadratics are always of the forms VI,29b and b′, never of the form VI,28b.[18] The procedure of solution is not reduction to VI,29a and a′ through division by c, but the computationally simpler, geometrically meaningless transformations:

$$c^2x^2 \pm cax = cb;$$
$$c^2x^2 \pm cax + \left(\frac{a}{2}\right)^2 = cb + \left(\frac{a}{2}\right)^2;$$
$$x = \frac{1}{c}\left(\sqrt{cb + \left(\frac{a}{2}\right)^2} \mp \frac{a}{2}\right).$$

Computations directly interpretable in terms of single quadratics are much less frequent in the Babylonian tablets than computations which can be referred to the pairs of equations

VI,28c $xy = b, \quad x + y = a,$ VI,29c $xy = b, \quad x - y = a,$

which are respectively equivalent to VI,28a and, because of the commutativity of multiplication, to both VI,29a and a′. The bulk of the algebraic tablets in which quadratic methods are involved at all consist of solutions of examples of these pairs or of reductions of other problems to instances of the pairs and then their solution. The methods of reduction are sufficient to solve the general forms of quadratic. The methods of solution are subject to more than one interpretation. However, it seems certain that they are not solved by reduction to a single quadratic any more than single quadratics are solved by reduction to such pairs. In almost every case[19] the solutions take the form

$$x = \sqrt{\left(\frac{a}{2}\right)^2 \pm b} + \frac{a}{2}, \quad y = \pm\left(\sqrt{\left(\frac{a}{2}\right)^2 \pm b} - \frac{a}{2}\right).$$

And in some cases at least, the solution of the pairs VI,28c and 29c appears to involve transformation of them into a pair of quadratics

VI,28c′ $(a - y)y = ay - y^2 = b, \quad x(a - x) = ax - x^2 = b,$
VI,29c′ $(a + y)y = ay + y^2 = b, \quad x(x - a) = x^2 - ax = b,$

and solving each member of the pair separately by completing the square.[20] In this sense, at least, the Babylonians can be said to show explicit knowledge of all three forms of single quadratic VI,28a, 29a, and 29a′.

This brief description perhaps indicates how difficult it is to discuss the precise mathematical relationship between Babylonian and Greek "solutions of quadratic equations" even if one ignores the completely obscure issue of actual mathematical contact between the two cultures. There is direct evidence that the Babylonians worked with what we would call quadratics of the forms VI,29a and a′, but none that they treated VI,28a. On the algebraic interpretation of VI,28 and 29 as solutions of single quadratics there is direct evidence that the Greeks solved VI,28a and 29a, but none that they solved 29a′. One can, of course, claim that Euclid was aware of the solution to this last equation or point to the less direct evidence that the Babylonians worked with VI,28.[21] Alternatively one can interpret VI,28 and 29 and their companions II,5 and 6 as solutions to the pairs of equations VI,28c and 29c by treating *AD* as *y*. *Prima facie* one would expect a different formulation of these propositions if they were such solutions, but because of the flexibility of our algebraic notation for interpreting geometric configurations, the mathematical facts cannot rule out this interpretation. In addition the interpretation fits well with Euclid's use of VI,28 in book X, and increases the likelihood that Euclid was aware of the double solution to VI,28 since the symmetry of x and y in VI,28c makes the interchangeability of their values obvious. Nevertheless, I am not inclined to adopt the "Babylonian" explanation of Euclid's geometric algebra. Given the present state of our knowledge of the two bodies of mathematics and of the contact between the two cultures, the explanation raises as many problems as it solves. I shall therefore continue to try to establish that the "line of thought" in the *Elements* can be satisfactorily understood in terms of its surface geometric character.

The remaining propositions of book VI, namely 22 and 31–33, are not an essential part of the deductive structure of the book as it stands. I have already mentioned that 33 is probably included with an eye to its applications in XIII,8–10. The same sort of thing would seem to be true of 32, which is very explicitly cited in XIII,17.[22] 31 is a generalization of the Pythagorean theorem which is usually attributed to Euclid himself.[23]

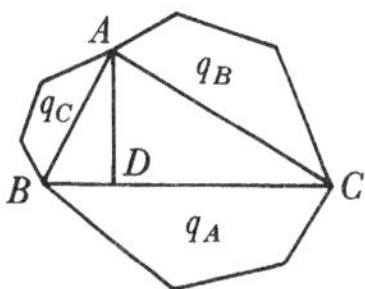

Figure 4.27

VI,31 If ABC (fig. 4.27) is a right triangle with its right angle at A and q_A, q_B, q_C are similar (rectilineal) figures similarly situated on the sides facing the angles at A, B, C respectively, then $q_A \simeq q_B + q_C$.

To prove this result Euclid drops AD perpendicular to BC and invokes VI,8 and the definition of similarity to get $(CB, BA) = (BA, BD)$ and $(CB, CA) = (CA, CD)$. Hence, by VI,19–20, $(q_A, q_C) = (CB, BD)$ and $(q_A, q_B) = (CB, CD)$, and (inversion and V,24) $(q_A, q_C + q_B) = (CB, BD + CD)$. Euclid now infers, using something like VD, that, since $CB \simeq BD + CD$, $q_A \simeq q_C + q_B$.

Heath (vol. I, pp. 350–366) discusses a number of proofs of the Pythagorean theorem and conjectures about which of them might be pre-Euclidean. However the theorem may have been originally established, there seems to be little reason to doubt that Euclid's proof of it as I,47 is derived from the proof of VI,31. The first proportions established in the latter proof imply, because of VI,17, that $\mathbf{T}(BA) \simeq \mathbf{O}(CB, BD)$ and $\mathbf{T}(CA) \simeq \mathbf{O}(CB, CD)$, where clearly $\mathbf{T}(CB) \simeq \mathbf{O}(CB, BD) + \mathbf{O}(CB, CD)$.[24] Therefore, to establish I,47 it is sufficient to prove the two equalities. Since the proof for each is the same, I describe that of the first, $\mathbf{T}(BA) \simeq \mathbf{O}(CB, BD)$. The strongest equality Euclid proves for parallelograms in book I is I,36, which is not directly relevant to the Pythagorean theorem. However, he has the stronger congruence theorems for triangles, and he knows that a triangle is one-half of the parallelogram with the same base and in the same parallels. Connecting AH and EC as in fig. 4.28 produces demonstrably equal triangles

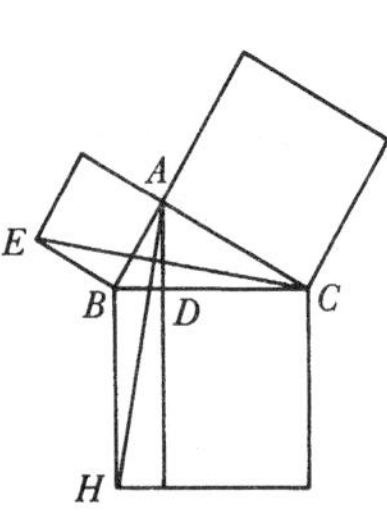

Figure 4.28

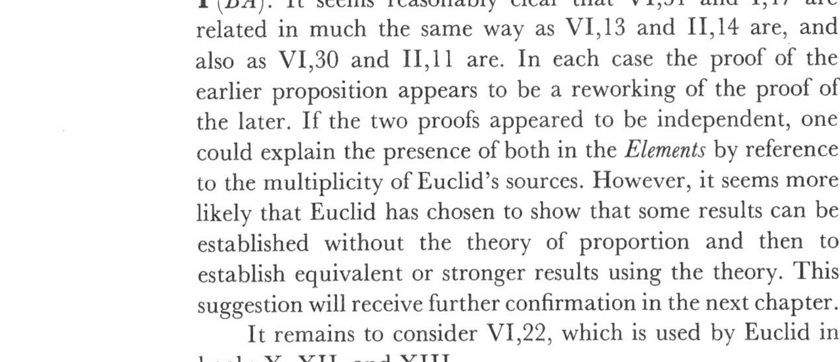

ABH and *EBC*, which are respectively half of **O** (CB, BD) and **T** (BA). It seems reasonably clear that VI,31 and I,47 are related in much the same way as VI,13 and II,14 are, and also as VI,30 and II,11 are. In each case the proof of the earlier proposition appears to be a reworking of the proof of the later. If the two proofs appeared to be independent, one could explain the presence of both in the *Elements* by reference to the multiplicity of Euclid's sources. However, it seems more likely that Euclid has chosen to show that some results can be established without the theory of proportion and then to establish equivalent or stronger results using the theory. This suggestion will receive further confirmation in the next chapter.

It remains to consider VI,22, which is used by Euclid in books X, XII, and XIII.

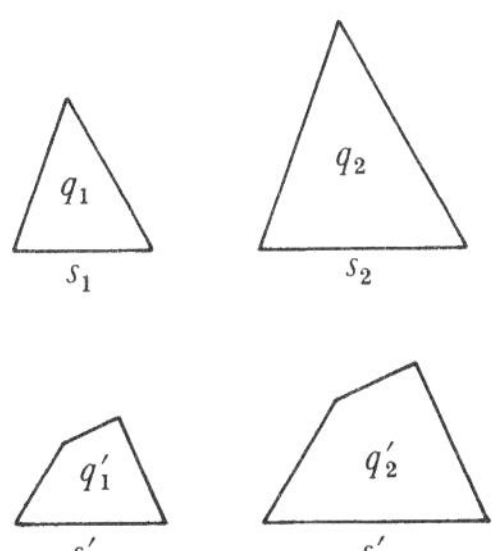

Figure 4.29

VI,22 If q_1, q_2 and q_1', q_2' are two pairs of similar rectilineal figures similarly situated on the straight lines s_1, s_2 and s_1', s_2' (fig. 4.29), then

(i) $(s_1, s_2) = (s_1', s_2') \rightarrow (q_1, q_2) = (q_1', q_2')$,
(ii) $(q_1, q_2) = (q_1', q_2') \rightarrow (s_1, s_2) = (s_1', s_2')$.

This proposition is an immediate consequence of VI,19–20 and VH; but, although Euclid uses VH in the proof of the analogous proposition for parallelepipeds, XI,37, he does not use it here. Instead, for (i) he uses VI,11 to construct straight lines s_3 and s_3' as third proportionals to s_1, s_2 and s_1', s_2' and invokes V,22 to get that $(s_1, s_3) = (s_1', s_3')$, from which the desired result follows using VI,19–20. In effect, then, Euclid goes through a proof of VH(i) for straight lines. For (ii) Euclid argues indirectly that if the assertion were false, there would be a straight line s such that $(s_1, s) = (s_1', s_2')$. Although this s could be constructed, Euclid's vocabulary suggests that he is simply applying the nonconstructive Vc. He says only, "For if s_1 is not to s_2 as s_1' is to s_2', let s_1 be to s as s_1' to s_2'." In any case Euclid uses VI,18 to construct on s the rectilineal figure q similar and similarly situated to q_1 on s_1, and argues that, since by the first part of the proposition $q \simeq q_2, s \simeq s_2$, so that $(s_1, s_2) = (s_1', s_2')$. Euclid justifies the inference to the equality of s and s_2 with the words "But q is also similar and similarly situated to q_2," apparently assuming

VIC Similar and equal rectilineal figures have equal corresponding sides.

This assumption is of a class with

VID Of similar and unequal rectilineal figures, the side of the greater is greater than its correspondent in the lesser figure, which we have seen Euclid uses in VI,28 and 29 (p. 165). Both of these assumptions can, of course, be derived from VH (ii) and VI,19–20. In all of the manuscripts of the *Elements* there is a *lēmma* in which VIC is proved by reduction to the tacit assumption

VIE Of similar rectilineal figures with unequal corresponding sides, the one with the greater side is the greater.

Although VIE can be derived from VH(ii), it also has a more elementary proof. Since similar rectilineal figures can be divided into equally many similar triangles in the same ratio as the original figures, it suffices to prove VIE for triangles. Suppose (fig. 4.30) ABC, $A'B'C'$ are similar triangles with corresponding sides AB, $A'B'$ and BC, $B'C'$ and with $AB \succ A'B'$. Then $BC \succ B'C'$, and if D, E are taken on AB, AC, respectively, with $BD \simeq A'B'$ and $BE \simeq B'C'$, then, if DE is joined, one will have

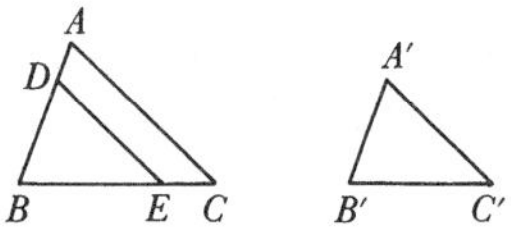

Figure 4.30

$$\text{triangle } A'B'C' \simeq \text{triangle } BDE \prec \text{triangle } ABC.$$

Since VIC is indirectly reducible to VIE, and VID is similarly reducible to VIC and E, there are elementary constructive proofs of these three assertions. However, it seems unlikely that the author of book VI was aware of these or any other proofs. In assuming VIC–E, he was probably thinking of the intuitive rather than the precisely defined notion of similarity.

Notes for Chapter 4

Bibliographical Note

I know of no recent work on book VI as a whole. Bibliographical references for its geometric algebra and Babylonian algebra are given in the bibliographical note for chapter 1. I would like to record here my particular indebtedness to Goetsch's paper on Babylonian algebra.

1. See, for example, Neugebauer, *Vorgriechische Mathematik*, pp. 121–124, 166–171. Neugebauer makes clear the difficulties involved in the interpretation of the primary sources for Babylonian and Egyptian "geometry."

2. In "Eine voreudoxische Proportionenlehre . . . ," p. 326.

3. For example, according to Heath, "The evidence suggests the conclusion that geometry developed itself for some time on the basis of the numerical theory of proportion which was inapplicable to any but commensurable magnitudes . . ." (*A History* . . . , vol. I, p. 155).

4. According to Neuenschwander's table ("Die ersten vier Bücher . . . ," pp. 334–336, 27 of the first 46 propositions of book I are used in VI. This statistic can only be an approximate index of the dependence of VI on I both because some of the alleged uses are tacit and questionable and because some of the propositions in book I are clearly lemmas for others. However, one does not need a more precise analysis in order

to be able to infer from Neuenschwander's statistic that the direct dependence of VI on I is substantial.

5. For discussion of this definition see Heath, vol. II, p. 188 and especially Dijksterhuis, vol. II, pp. 87–88.

6. In proving 7 Euclid carries out the congruence argument given in note 35 of chapter 1.

7. "Eine voreudoxische Proportionenlehre . . . ," p. 329.

8. Proclus (213.3–11) credits the fifth-century mathematician Hippocrates of Chios with the discovery that the construction of a cube double the size of a given one reduces to the problem of finding two mean proportionals between one straight line and another twice its length. It is difficult to believe that the simpler relation between the problems of II,14 and VI,13 was not comprehended at the same time or earlier.

9. Euclid's proof of 20 is unnecessarily complicated. See Heath, vol. II, pp. 238–239.

10. See Heath, vol. II, p. 234, where the complicated manuscript situation is also described. In the "Theonine" manuscripts there is an appropriate substitution of 'triangle' for 'figure' in the corollary to 19, which is stated for arbitrary figures in a second corollary to 20.

11. For the connection of this terminology with Apollonius' names for the conic sections, see Heath, vol. I, pp. 344–345.

12. Neither of these two constructions is exactly covered by earlier propositions, but I,44 and 45 indicate how they are to be done. See p. 45.

13. See Heath, vol. II, pp. 252–253, where, however, the attempt to explain away the elaborateness seems forced.

14. Neuenschwander ("Die stereometrischen Bücher . . . ," pp. 93–97) lists tacit uses of 21 in XI,27 and XII,8; and 26 is used in X,91–96; but 24, 25, and 27 are never applied again.

15. The claim has been made by Zeuthen (*Die Lehre* . . . , pp. 19–20), Heath (vol. II, p. 264), and Dijksterhuis (vol. II, pp. 111–112).

16. See Zeuthen, *Die Lehre* . . . , pp. 18–19 and Dijksterhuis, vol. II, p. 112.

17. See Dijksterhuis, vol. II, pp. 111–112.

18. Gandz, p. 480, attempts to explain the absence of problems representable by means of VI,28b by reference to its double positive root, which, according to Gandz, the Babylonians avoided because they found it "embarassing" and "nonsense."

19. Goetsch, pp. 124–125, gives two examples where the solutions appear to be in a different form.

20. A mechanical procedure applied to the two equations in VI,28c′ would, of course, lead to the same value for x and for y. Goetsch (p. 121) assumes that the correct solution was obtained by a process amounting to treating an expression of the form $z^2 - az + \left(\frac{a}{2}\right)^2$ once as equivalent

to $\left(\frac{a}{2} - z\right)^2$ and once as equivalent to $\left(z - \frac{a}{2}\right)^2$.

21. I should perhaps remark that the evidence is convincing to me, but that Gandz denies that the Babylonians ever used the pairs of equations VI,28c′ and VI,29c′ in the way suggested in the text.

22. VI,32 is false in its Euclidean formulation. See Heath, vol. II, pp. 271–272.

23. See Heath, vol. II, p. 269. The evidence is a remark by Proclus (426.9–18). However, since Proclus is primarily concerned to point out that 31 is a generalization of the Pythagorean theorem and since he considers the *Elements* to be the work of Euclid, little can be read into his saying that Euclid "secured" VI,31 "by the irrefutable proofs of science." And he says nothing stronger.

24. When Euclid asserts this last equality in the proof of I,47, he takes for granted an obvious fact which he subsequently establishes as II,2.

5 The Circle and Its Relation to the Triangle, the Square, and the Regular Pentagon, Hexagon, and *Pentekaidekagon*

5.1 The Circle

Prior to book III the circle functions only as a means for constructing straight lines equal to given ones. No proposition in the first two books contains the word 'circle'. By contrast all but two of the 53 propositions of books III and IV contain the word, a clear indication of Euclid's policy of using subject matter as an organizing principle in the *Elements*. In book III Euclid's subject is the properties of circles, their arcs, and certain straight lines which meet them. In IV he deals with the problem of inscribing a circle in or circumscribing it about certain rectilineal figures and with that of circumscribing or inscribing the figures about or in a circle. I have chosen to divide my discussion of this material in this chapter into two sections which do not correspond exactly to Euclid's division of the two books. In this section I treat most of book III—namely propositions 1–22, 25, and 31–34. In section 5.2 I consider the inscriptions and superscriptions, the material in book III which is closely related to them, and the general character of Euclid's treatment of plane geometry.

Book III is like book VI and unlike book I in being, from the standpoint of logic, very loosely organized. It is convenient to divide the representation of its logical structure into two parts (figs. 5.1 and 2). Figure 5.1 consists of propositions

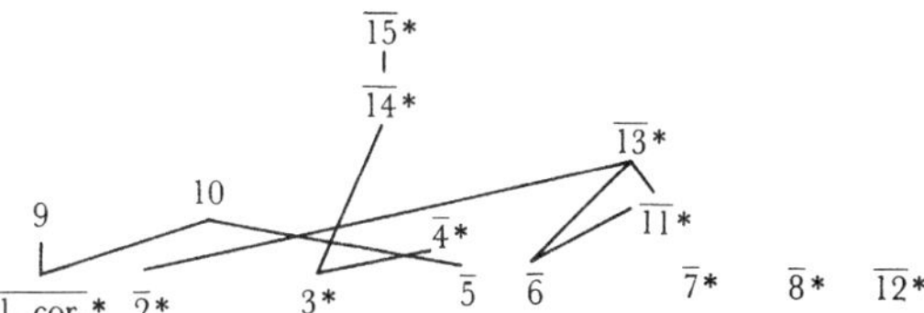

Figure 5.1

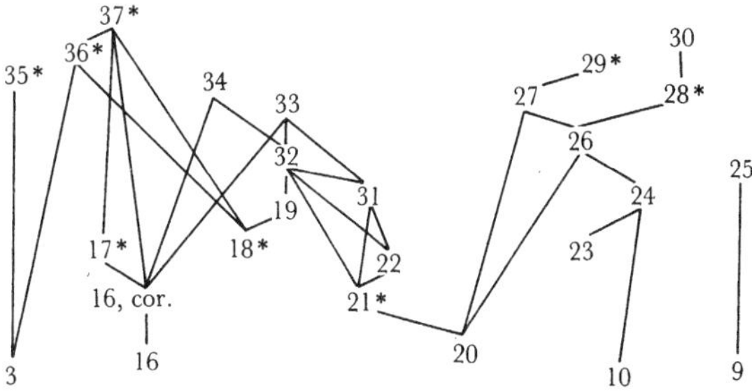

Figure 5.2

1–15, which deal with some of the elementary properties of circles which meet and of straight lines passing through circles. The only proposition among III, 1–15 not represented by a numeral in fig. 5.1 is 1, in which Euclid shows how to find the center of a circle by taking the midpoint of the perpendicular bisector of a chord. This construction is one of the most frequently used in the *Elements*. In figs. 5.1 and 2 its applications are indicated with an asterisk. Besides 1, the only propositions from the first part of book III used in the second part are 3, 9, and 10. Figure 5.2 shows the structure of the second part of book III.

Perhaps the most striking fact brought out by tnese two diagrams is the relatively high degree of independence among the propositions. If the use of III,1 is left out of account, there are fourteen propositions (2, 3, 5–8, 12–16, 18, 20, and 23) which depend on no prior propositions in book III. There are also thirteen propositions (4, 7, 8, 12, 13, 15, 25, 29, 30, 33–35, and 37) which are never used to prove anything in the book, and only four of these (15, 29, 30, and 37) are used anywhere in the *Elements*. Other features of book III are also perplexing. The term 'sector' is introduced in definition 10, but never used in the *Elements*. The proofs of propositions 7 and 8 make use of a concept of distance between intersecting straight lines which is never explained. There are also the curious references to so-called mixed angles (*miktai gōniai*) made by straight and curved lines in definition 7 and propositions 16 and 31; these apparently are remnants of an earlier approach to geometry.[1] As far as dependence on the two previous books is concerned, the situation is again striking. In the last three propositions of book III, Euclid uses I,47 and II,5 and 6. But prior to III,35, although he makes heavy use of I,1–26, he applies subsequent propositions—and therefore the theory of parallels—only in III,14, where he applies I,47, and in III,20, 22, 31, and 32, where he uses I,32.

Perplexity only mounts with more detailed study of Euclid's proofs. In many cases there are alternative, simpler proofs which would increase logical dependence. Heath's notes on book III provide a good survey of the most important suggestions for revision which have been made by commentators and editors. They also indicate many of Euclid's tacit assumptions. A study of the text and these suggestions makes it reasonably clear that there is no way to "rectify" book III short of starting over again and reformulating the whole. In this sense book III is somewhat different from the other books of the *Elements*, in which what we see as shortcomings can be overcome by making explicit tacit assumptions. The kind of assumption which is relevant to book III concerns order and continuity or, more simply, facts of spatial intuition. But in book III these assumptions are sometimes taken for granted and sometimes argued for. For example, Euclid's proof of III,1 presupposes that a straight line and a circle cannot have more than two points in common, but III,2 is an attempt to prove what amounts to this same fact. In light of the general character of Euclidean mathematics, these attempts at proof are rather more surprising than the employment of tacit spatial assumptions. I shall try to explain in the course of discussion why these attempts are made, but in most cases I will not go into the logical shortcomings discussed by Heath.

The definition of a circle (I, def. 15) guarantees the existence and uniqueness of the center of any given circle. Euclid could use this guarantee to bypass proposition 1. For in all cases in which he uses the proposition to find or, as he says, take (*lambanein*) the center *O* of a circle, he could simply say, as he sometimes does, "Let *O* be the center." Proposition 1 is, then, another indication of the constructive character of Euclid's elementary geometry. On the other hand, it is not exactly like the ordinary geometric problem. It does not establish the existence of an object by generating it, but rather finds an object whose existence is guaranteed by definition. The proof of proposition 1 itself depends upon this guarantee. Euclid argues indirectly that the center of a given circle must be the midpoint of the perpendicular bisector of a chord because no other point could be the center. Obviously this argument would not work unless it was already known that some point must be the center.

Proclus (301.22–302.11) uses proposition 1 as an example of a porism (*porismos*, also the Greek word for a corollary), which he distinguishes from a theorem and a problem on the grounds that a porism involves finding, as distinguished from generating or making, as in a problem, and does not just involve theorizing, as in a theorem. In addition to this rather negative characterization, Proclus gives one other example of a porism: the finding of the greatest common measure of two commensurable magnitudes (X,3). It is difficult to generalize from the two examples with any certainty. Proclus also refers to a work of three books of porisms written by Euclid. Unfortunately this work is lost, but Pappus' treatment of the work[2] makes it seem likely that the term 'porism' originally had more mathematical than philosophical significance and was used to refer to a kind of proposition lying outside the scope of elementary mathematics. The *Elements* themselves provide no evidence of an interest in the distinction between porisms and problems. Euclid's use of the word 'find' (*heurein*) does not seem to be restricted to porisms, since VI,11–13 and numerous propositions in book X contain it, without seeming to be porisms in Proclus' sense. Nor does Euclid have a special tag for porisms to go along with 'Q.E.D.' for theorems and 'Q.E.F.' for problems. The distinction, then, would not seem to have played a role in the composition of the *Elements*.

Nevertheless, it is perhaps worthwhile to indicate which propositions in the *Elements* are porisms in Proclus' sense. For this purpose I use the following characterization which seems to formulate what Proclus has in mind: A porism is a proposition with the grammatical form of a problem in which an object is sought that can be proved to exist independently of any construction producing it. This characterization clearly applies to III,1. It also applies to X,3 (and 4), if one takes for granted the existence of arbitrary nth parts of magnitudes (Va). For any common measure z of x and y is an mth part of x and an nth part of y, for some m and n. But, by the least number principle, if x and y are commensurable, there must be a least m' and n' for which some z' is an m'th part of x and an n'th part of y; such a z' is a greatest common measure of x and y. The same kind of argument works for all the propositions of arithmetic which have the grammatical form of problems[3] because they all involve finding least numbers satisfying some condition which is demonstrably satisfied by some number or numbers. Thus, if one takes for granted the existence of arbitrary numbers and the least number principle, Euclid's arithmetic contains only theorems and porisms, but no problems.

I am inclined to think that all other propositions in the *Elements* are theorems or problems. Heath (vol. I, p. 13) mentions as "real porisms" III,25, VI,11–13, X,10, and XIII,18, in addition to Proclus' examples and some of the relevant arithmetic propositions. In III,25 Euclid completes (*prosanagraphein*) the circle of which a given segment is a segment. There does not seem to be any reason to suppose that the existence of a segment of a circle or of an arc of its circumference presupposes the existence of the whole circle; nor does Euclid prove III,25 as if it did. Although VI,11–13 are phrased in terms of finding straight lines proportional in given ways to given ones, they could just as well be phrased in terms of constructing. Moreover, the existence of a mean proportional between two straight lines (VI,13) cannot be proved on the basis of Euclid's existential assumptions.[4] The existence of third and fourth proportionals does, of course, follow from Vc, but so does the existence of the straight line segments which are cut off in VI,9 and 10. VI,9–12 should then be classed together; and, since Vc plays no role in any of their proofs, they are best classified as problems. X,10 depends on VI,13 and therefore cannot be a porism. XIII,18 is an anomaly among the propositions of Greek mathematics and is comparable to IX,18 and 19. In it the sides of the five regular solids, which have already been constructed, are set out (*ekthesthai*) and compared (*sugkrinai*). The setting out involves constructing certain straight lines from the given diameter of the sphere and arguing that the lines are equal to the sides of the regular solids. Comparing the sides is characterizing the ratios among them. This characterization amounts to a theorem, and the setting out is really a *kataskeuē* for the proof of the theorem. XIII,18 is then another proposition in which the proof rather than the *protasis* gives the mathematical content. If it is to be classed as porism, problem, or theorem, it would seem most appropriate to call it the last of these.

Euclid formulates part of the content of the proof of III,1 as a corollary according to which the center of a circle lies on the perpendicular bisector of any chord. In proposition 3 he proves that a straight line through the center of a circle is perpendicular to a chord if and only if it bisects it. For if (fig. 5.3) OC is drawn from the center O, intersecting the chord AB at C, and the figure is completed, the triangles OAC, OBC will have equal sides OA, OB and equal angles OAC, OBC. Hence, by I,8 and 26, the angles OCA, OCB are right, i.e., equal to each other, if and only if $AC \simeq CB$. In proposition 4 Euclid proves that intersecting chords which are not both diameters do not bisect each other. He does the case in which neither is a diameter, arguing that if they did bisect one another the straight line from the center to their point of intersection would, by 3, be perpendicular to both chords, an impossibility.

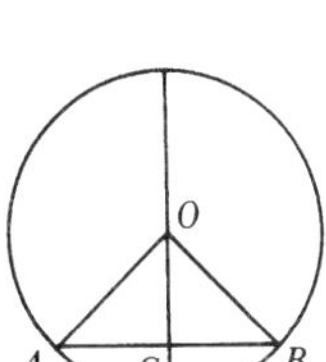

Figure 5.3

Euclid also uses 3 to prove

III,14 Two chords of a circle are equal if and only if they are equidistant from the center,

from which he infers

III,15 Of chords in a circle, the greater is closer to the center,

where the relative distance of chords from the center of a circle is defined in terms of the length of the perpendiculars from the center to the chords (III, defs. 4 and 5). To prove 14 Euclid considers two chords (fig. 5.4) A_1B_1, A_2B_2 in the circle with center O and having OC_1, OC_2 as perpendicular bisectors. If Euclid had the congruence theorem of note 35 of chapter 1, he could complete the argument by using it. Instead he invokes the Pythagorean theorem and argues that since $\mathbf{T}(OC_1) + \mathbf{T}(B_1C_1) \simeq \mathbf{T}(OC_2) + \mathbf{T}(B_2C_2)$, $OC_1 \simeq OC_2 \leftrightarrow B_1C_1 \simeq B_2C_2 \leftrightarrow A_1B_1 \simeq A_2B_2$. For 15 Euclid imagines the same situation with $OC_1 > OC_2$. He finds C_3 on OC_1 so that $OC_3 \simeq OC_2$, and draws the chord $A_3C_3B_3$ perpendicular to OC_1. By III,14,

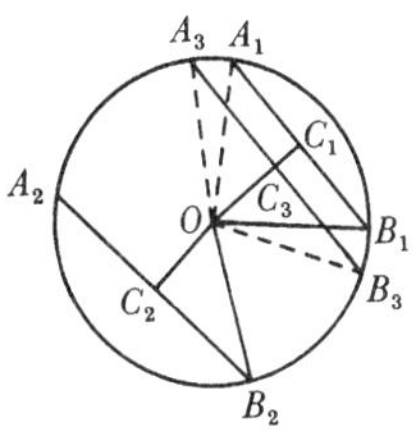

Figure 5.4

$A_3B_3 \simeq A_2B_2$. Euclid connects O with the endpoints of A_1B_1, A_3B_3, and uses the intuitive fact that angle $A_3OB_3 \succ$ angle A_1OB_1 to conclude via I,24 that $A_3B_3 \succ A_1B_1$. (The intuitive assumption could have been avoided by using the Pythagorean theorem, as in 14.) Euclid also proves that a diameter is greater than any chord which is not a diameter on the grounds that the two radii to the ends of such a chord are together equal to a diameter, but, by I,20, greater than the chord. This part of III,15 is cited rather explicitly in XII,17 and functions tacitly as a *diorismos* for IV,1; 14 is invoked in XIII,18, and used tacitly in XII,16. Otherwise the two propositions are not applied.

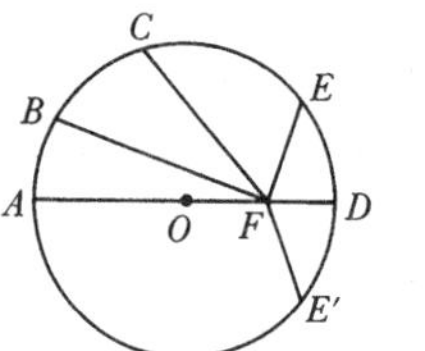

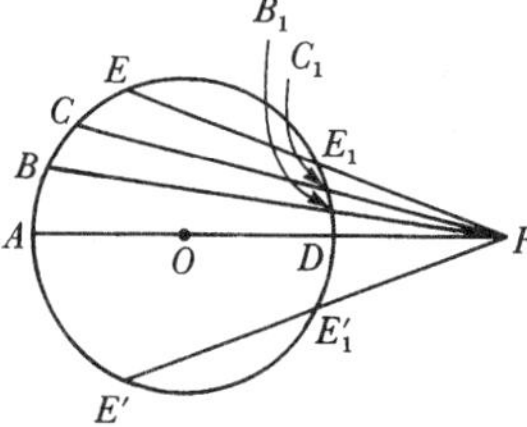

Figure 5.5

In III,7 and 8 Euclid uses without definition another concept of relative distance, namely the distance between two straight lines with a common endpoint, the distance being measured by the angle between the two straight lines. He proves

III,7 If (fig. 5.5, top) O is the center of the circle $ABCD$ and F a point distinct from O on the straight line (diameter) $AOFD$, then the length of straight lines from F to the circumference depends only on distance from AOF and increases as the distance decreases.

III,8 If (fig. 5.5, bottom) O is the center of the circle $ABCD$ and F is a point on the straight line $AODF$, the length of straight lines from F to the concave (convex) circumference depends only on distance from AOF and increases (decreases) as the distance decreases.

Since these propositions are not used in the *Elements* I shall not discuss their proofs. Their presence may be due to their value in applied mathematical contexts. Theodosius makes use of the content of 7 in *Spherica* III,1, Aristarchus of 8 in proposition 3 of *On the Sizes and Distances of Sun and Moon*.

Part of Euclid's formulation of III,7 is the assertion that only two equal straight lines can be drawn from the circumference of a circle to an interior point F which is not the center of the circle, an assertion which is obviously equivalent to

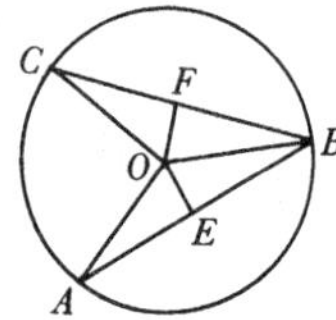

Figure 5.6

III,9 If (fig. 5.6) O is an interior point of the circle ABC, and AO, BO, CO are equal, then O is the center of the circle.

Euclid, however, proves 9 by bisecting AB, BC at E, F, connecting EO, FO, and using I,4 to argue that they are the perpendicular bisectors of AB, BC; hence by the corollary to 1, the center lies on each of EO, FO, i.e., at O. This proof shows essentially that any three noncollinear points A, B, C determine a unique circle with center at the intersection of the perpendicular bisectors of AB, BC, a fact of fundamental importance in most modern treatments of the circle. Euclid uses III,9 only once in the *Elements*, namely, in III,25, where he shows how to draw a circle of which a segment is given—a construction which itself is never used in the *Elements*. To carry out this construction Euclid could take a point C on the circumference ACB of the segment and determine the center of the circle as the intersection of the perpendicular bisectors of AC and CB. He, however, draws the perpendicular bisector of the straight line AB, intersecting the segment at C, and determines the point O on it by making angle $CBO \simeq OCB$ (fig. 5.7). Clearly OB and OC are equal, and, by I,4, so are OB and OA. Hence, by III,9, O is the center of any circle with circumference containing A, B, C. Clearly Euclid could have proved 9 and 25 right after 1. The position of 9 seems inexplicable. The position of 25 is perhaps explained by the fact that Euclid first deals with segments in 23 and 24.

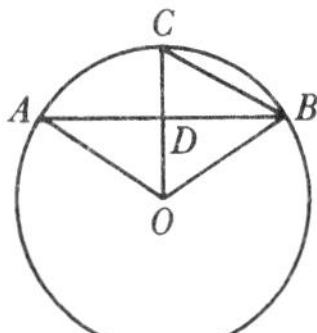

Figure 5.7

Euclid's proof of III,25 differs from the one just described in that his is divided into three separate cases depending upon the relative size of the angles DBC, DCB. Normally Euclid is content to do one case and "leave the others to the reader." Sometimes Euclid proceeds in the same way in III, but there does seem to be more treatment of separate cases in III than elsewhere. A slightly different kind of example is provided by propositions 5 and 6, in which Euclid proves that distinct circles c_1, c_2 cannot have the same center O whether they cut (5) or touch (6). The argument is the same in both cases, and establishes that circles with the same center and a common point on their circumferences coincide. For a straight line from O through a point on the circumference of c_1 but not on that of c_2 will determine unequal radii both equal to the straight line from O to the point of contact of the two circumferences.

Euclid's definitions of touching take for granted the notions of meeting and cutting, which he does not define:

III, def. 2 A straight line is said to touch a circle which, meeting the circle and being produced, does not cut the circle;

III, def. 3 Circles are said to touch one another which, meeting one another, do not cut one another.

It seems reasonably clear that Euclid intends to define touching as meeting at a point but not cutting at that point rather than as meeting at a point and not cutting at any point. In other words, he just does not consider the possibility that a straight line or circle might touch a circle at one point and cut it at another, as in fig. 5.8. For Euclid it is an obvious fact that touching at one point excludes cutting at any other. On the other hand, he attempts to prove that a straight line or circle cannot touch a circle at more than one point nor cut it at more than two.

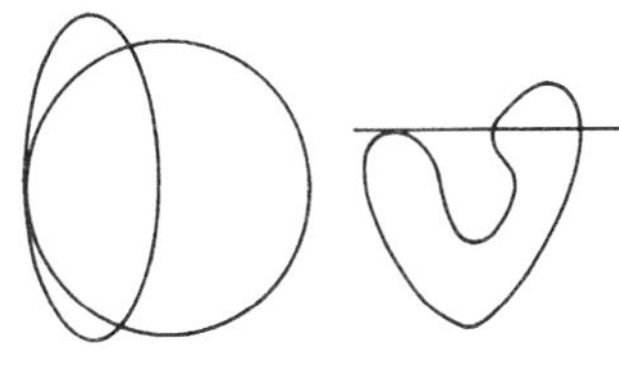
Figure 5.8

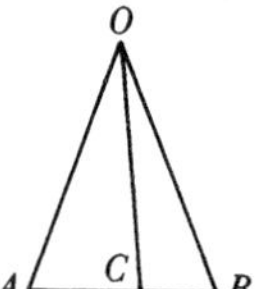

Figure 5.9

For straight lines this result is a consequence of III,2, according to which a straight line joining two distinct points on the circumference of a circle falls inside the circle. For, if (fig. 5.9) ACB is a straight line joining A, B on the circumference of the circle with center O and C is not inside the circle, then OC is greater than or equal to each of the radii AO, BO. But, by I,16, angle $OCB \succ$ angle $OAB \simeq$ angle OBA. Therefore, by I,19, $OB \succ OC$, a contradiction.

Euclid proves in III,10 that a circle does not cut a circle in more than two points. He supposes two circles to cut in three, A, B, C, and argues, using the corollary to III,1, that the point of intersection of the perpendicular bisectors of AB, BC is the center of each circle, contradicting III,5. This application of III,5 is its only use in the *Elements*. 10 is applied only in III,24, for which it is perhaps a lemma. One sees, however, that the proof of III,10 is sufficient to establish that two circles cannot have more than two common points on their circumferences—another version of the assertion that three points determine a circle.

In III,13 Euclid argues that circles do not touch at more than one point, distinguishing between external and internal touching. Since for him touching excludes cutting, we may represent his distinction by saying that two touching circles touch externally if they share no inner point, and touch internally if all internal points of one of them are internal points of the other. In proving 13 for external touching Euclid makes his only application of III,2, arguing that if there were two points of touching, the straight line connecting them would fall inside each circle, which is impossible if the circles touch externally. For internal touching Euclid makes his only application of

III,11 If two circles touch internally, a straight line through their centers will pass through the point at which they touch,

a result which is proved for externally touching circles in the probably interpolated III,12.[5] The proof of 13 for internal touching and the very formulation of 11 presuppose the distinctness of the centers of touching circles; these presuppositions are the only uses of III,6 in the *Elements*.

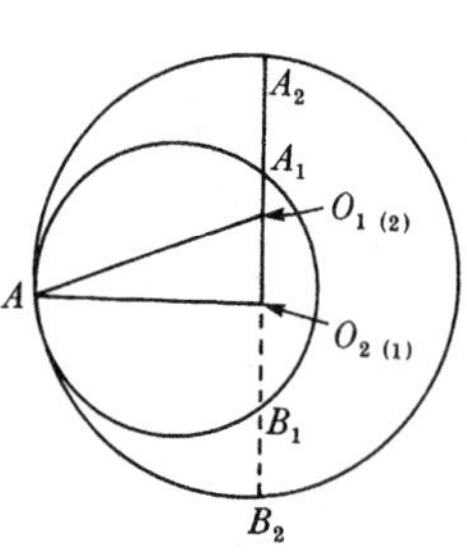

Figure 5.10

To prove 11 Euclid supposes that (fig. 5.10) the circle c_1 with center O_1 is inside the circle c_2 with center O_2 and touches it at A, and that the straight line through O_1, O_2 does not pass through A but intersects the circumferences of c_1, c_2 at A_1, A_2 on the same side of O_1, O_2. He then argues that since $AO_1 + O_1O_2 \succ O_2A \simeq O_2A_2$, $O_1A_1 \simeq AO_1 \succ O_2A_2 - O_1O_2 \simeq O_1A_2$, which is impossible. This argument presupposes that O_1 is between O_2 and A_1, but works whether or not A_1 and A_2 are distinct and whether or not O_2 is inside the circle c_1. To handle the case in which O_2 is between O_1 and A_2 it seems necessary to invoke the fact that the straight line through O_1, O_2 will meet the circumferences of c_1 and c_2 not only in one direction at possibly distinct points A_1, A_2, but also in the other at possibly distinct points B_1, B_2. One can then deal with the straight line $O_2O_1B_1B_2$ as Euclid deals with $O_2O_1A_1A_2$. The argument of III,13 against c_1 and c_2 touching internally at two points A, B is simple; for, if they did, there would be a straight line AO_1O_2B (fig. 5.11) through their centers O_1, O_2, so that $O_2A \succ O_1A \simeq O_1B \succ O_2B$, which is impossible.

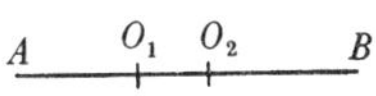

Figure 5.11

It is clear that Euclid has established that touching circles have exactly one point of contact. He, however, is not interested in this general content, but only in the fact that when circles touch they touch at a point and not along an arc. Since he never applies this result in the *Elements*, it seems that he is interested in the fact only for its own sake. This interest is perhaps connected with certain barriers which common sense encounters in connection with the circle. A common-sense understanding of the notion of a straight line makes it reasonably clear that straight lines either coincide or intersect only at a point. But no common-sense understanding of the notion of a circle will make clear what happens when a circle and a straight line or two circles appear to rest against one another. Aristotle (*Metaphysics*, B.2.998[a] 2–4) reports that Protagoras invoked common sense to "refute" the geometer's claim that circle and straight line are tangent only at a point. I am inclined to think that it is an appreciation of these limits to common sense which leads Euclid to prove many of the propositions among III,1–15 by way of kinds of argumentation for which he does not have an adequate foundation. Furthermore, the inadequate foundation forces him to deal with the various propositions in isolation from one another.

The same sort of thing can be said about III,16–19, in which Euclid shows that a straight line touches a circle at exactly one point, although these propositions are used more significantly by Euclid. A more precise statement of what Euclid establishes in 16–19 is that if (fig. 5.12) AC is a chord of a circle and BCD another straight line, any two of the following conditions imply the third:

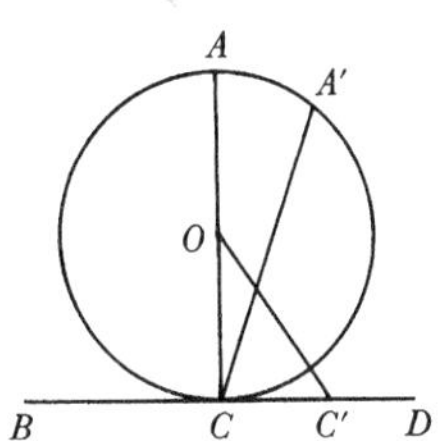

Figure 5.12

(i) AC is perpendicular to BCD;
(ii) AC is a diameter of the circle;
(iii) BCD is tangent to the circle.

Euclid shows that (i) and (ii) imply (iii) in the corollary to 16, that (ii) and (iii) imply (i) in 18, and that (i) and (iii) imply (ii) in 19. He uses the corollary to 16 in 17 to draw a straight

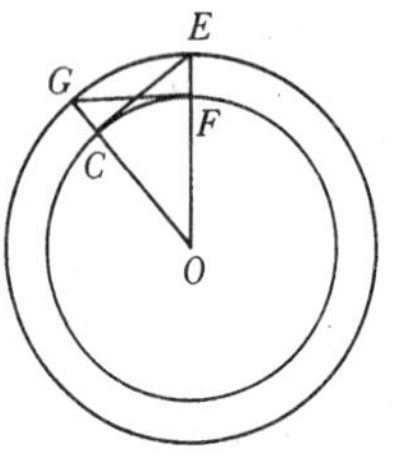

Figure 5.13

line tangent to a circle with center O and passing through a given point E outside the circle, a construction used once, in III,37. Euclid simply takes for granted the possibility of drawing a straight line tangent to a circle and passing through a point on its circumference, presumably because of the corollary to 16. To prove 17, Euclid draws (fig. 5.13) the circle with center O and radius OE intersecting the given circle at F, draws FG perpendicular to EO and intersecting the drawn circle at G, and connects GO intersecting the given circle at C. He uses I,4 to establish the congruence of triangles EOC, GOF, so that angle ECO is right and, by 16, corollary, EC is tangent to the given circle.

19 is reduced to 18, since if AC in fig. 5.12 is not the diameter through C, some other straight line $A'C$ will be, and both $A'C$, AC will be perpendicular to BCD, which is impossible. For 18 Euclid argues that if AC is not perpendicular to $BCC'D$ but that OC', which passes through the center O of the circle is, then angle $OC'C$ will be greater than OCC', so that OC will be greater than OC', which is impossible. In 16 itself Euclid establishes that if (i) and (ii) hold, then (fig. 5.12)

(a) BCD falls outside the circle,
(b) into the region (*topos*) between BCD and the circumference another straight line will not fall,
(c) the angle of a semicircle (the "angle" at C contained by AOC and arc $AA'C$) is greater and the remaining angle (contained by CD and arc $AA'C$) less than any acute rectilineal angle.

Assertion (a) is a useless form of the corollary to 16. Euclid proves it by arguing that if BCD has a second point in common with the circumference of the circle, the straight line from the center O to the point determines an isosceles triangle with two right angles, an impossibility. Since any straight line through C between CD and the circumference would not be perpendicular to AC, (b) is simply a weaker form of proposition 18 and is in fact proved in the same way. Assertion (c) is another formulation of (b) and is reduced to it; if either component of (c) did not hold, the acute angle in question would be contained by AC and a straight line incompatible with (b).

Within the *Elements* and Greek geometry generally, the only useful form of 16 is its corollary, which is proved as part (a). The comparison made by Euclid in (c) easily leads to perplexities if taken seriously, since the angle contained by CD and arc $AA'C$ of fig. 5.12 is an infinitesimal if it is smaller than any acute angle but nevertheless of some size. Proclus (121.24ff.) provides our earliest reference to these perplexities when he argues that no such angle, which he calls hornlike (*kratoeides*), will exceed a rectilineal angle. Perplexities of this kind were obviously fascinating to Greek philosophers, but there is no good evidence that they played any mathematical role. However, a passage in Aristotle (*Prior Analytics*, A.24.41[b] 13–22) makes it likely that mixed angles contained by a curved and a straight line had a place in pre-Euclidean geometry alongside of rectilineal angles.[6] In classical Greek mathematics these mixed angles occur only in 16, in the definition of the angle of a segment of a circle (III, def. 7), and at the end of III,31; in III,31 Euclid infers from the fact that (fig. 5.14) the rectilineal angle BAC in a semicircle $BDAEFC$ is right that the angle contained by AC and the arc ADB is greater than a right angle and that contained by AC and arc AEF less than one, inferences obviously based on the figure.

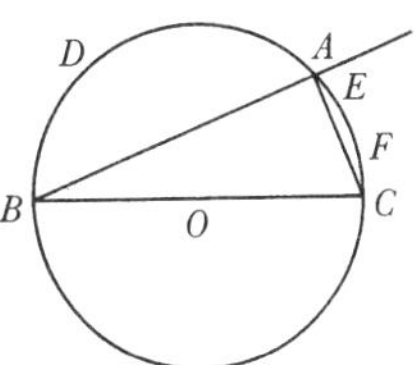

Figure 5.14

One can only speculate about the reasons why mixed angles dropped out of Greek geometry, but one obvious possibility is the difficulty of manipulating them and reasoning quantitatively about them. Even such simple operations as bisection and addition would seem to be beyond the conceptual limits of Greek mathematical reasoning. The realization that it is not necessary to use mixed angles may well have led to their elimination from geometric proofs, even though the concept of a mixed angle was preserved. Euclid's references to these angles may be due to what is often called his respect for tradition, but he could also have intended to bring out some of their counterintuitive properties. For an obvious consequence of 16 and 31 is that the angle of a segment of a circle greater than a semicircle is greater than a right angle and that of any other segment less than a right angle, the angle of no segment being exactly equal to a right angle. We cannot know whether Euclid regarded such a fact as a mere curiosity or as a reason for avoiding mixed angles.

Figure 5.2 shows that 16–19 represent a new beginning in book III which depends only on proposition 1. 20 also is a new beginning; to conclude this section I would like to treat 20–22, 31–34, all of which depend ultimately on 20. In 20 Euclid proves that if (fig. 5.15) ABC is a circle with center O, the angle BOC is twice angle BAC. He extends AO to the circumference of the circle at D and argues that since angle $DOC \simeq$ angle OAC + angle ACO, and angle $OAC \simeq$ angle ACO, angle DOC is twice angle DAC. Similarly angle DOB is twice angle DAB, and the result follows. Euclid mentions a second case in which angle BAC is again acute, but O lies outside it. However, he does not consider the case in which angle BAC is not acute because he does not acknowledge angles greater than or equal to 180°, which the relevant angle BOC would be in this case.[7] As a result,

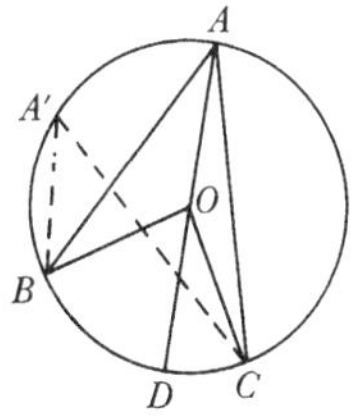

Figure 5.15

III,21 In a circle *AA′BC* (fig. 5.15), angle *BAC* ≃ angle *BA′C*,

an apparently obvious consequence of 20, is established only if arc *BA′AC* is greater than a semicircle. This restriction of 21 can be overcome by a rearrangement of Euclid's propositions. The proof of the first part of III,31, given above on p. 157, shows that if arc *BA′AC* is a semicircle, III,21 holds because angles *BAC*, *BA′C* are both right angles. But if the arc is less than a semicircle one need only form the quadrilaterals *BACD*, *BA′CD* and infer the desired equality from

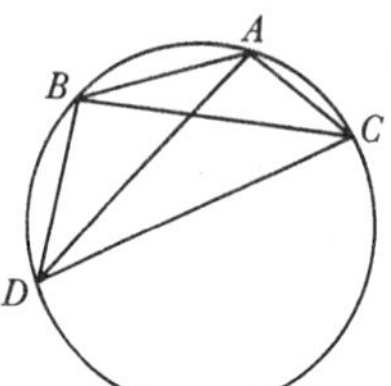

Figure 5.16

III,22 If (fig. 5.16) *BACD* is a quadrilateral inscribed in a circle, angles *BAC*, *BDC* are together equal to two right angles.

Euclid's proof of III,22 appears to apply III,21 in its general form, but it need not do so. For if *BACD* is a quadrilateral inscribed in a circle, either *BC* is a diameter, in which case both angles *BAC*, *BDC* are right, or it is not, in which case one of the arcs *BAC*, *BDC*—say *BDC*—is greater than a semicircle. One can use Euclid's proof of 22 by applying 21 to the angles *ADC*, *ABC* and to angles *BDA*, *BCA*, and inferring the consequent of 22 from the fact that the angles *ABC*, *BCA*, *BAC* are together equal to two right angles.

Euclid makes two tacit applications of III,21. In the second part of III,31, having shown that the angle *BAC* in a semicircle is right, Euclid argues that the angle in a segment greater (less) than a semicircle is acute (obtuse) by taking a point *D* on the circumference (fig. 5.17) so that *BADC* is a quadrilateral and pointing out that angle *ABC* is acute and, together with angle *ADC*, equal to two right angles. In order to make this argument general, Euclid would have to invoke III,21 to establish that any angle *AB′C* in a segment greater than a semicircle is equal to the angle *ABC* constructed by drawing the diameter *CB* and connecting *AB*.

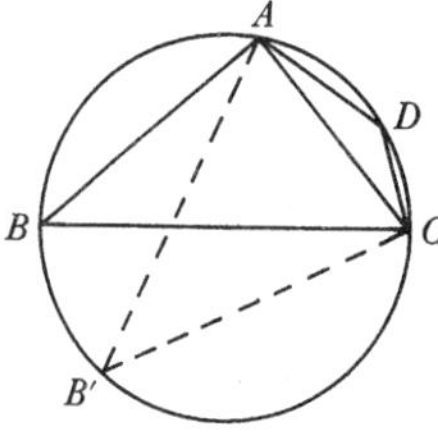

Figure 5.17

The other tacit application of 21 is in

III,32 If (fig. 5.18) the straight line *EBF* touches the circle *ABCD*, the angles *DBF*, *DBE* will be equal to the angles in the alternate (*enallax*) segments of the circle determined by *DB*.

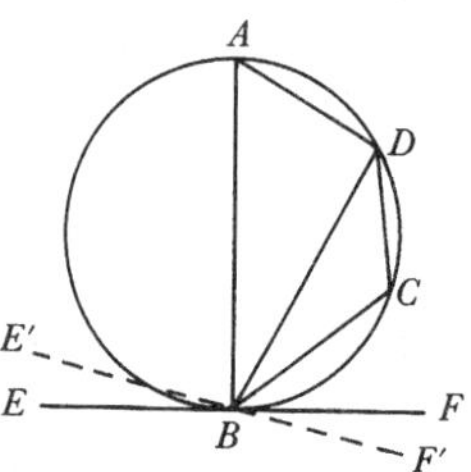

Figure 5.18

Euclid takes *AB* as diameter of the circle and argues that, since the angles *ADB*, *ABE*, *ABF* are each right and the two sums, angle *ABE* + angle *ABD* + angle *DBF* and angle *ADB* + angle *ABD* + angle *BAD*, are each equal to two right angles, angle *DBF* is equal to angle *BAD* (and therefore, by 21, to any angle in the segment *BAD*). In addition each of the sums, angle *BAD* + angle *BCD* and angle *DBF* + angle *EBD*, is equal to two right angles, so that angle *EBD* ≃ angle *BCD*.

The second part of 31 is used only in a remark at the end of the proof of IV,5. 32 is used in book III for two constructions which are not employed in the *Elements* and which I shall not discuss: the determination of a segment of a circle admitting an angle equal to a given rectilineal angle, either on a given straight line (III,33) [8] or as a segment of a given circle (III,34). 32 is also used in book IV in the inscription in a given circle of a triangle similar to a given one and of a regular pentagon. In connection with the analysis of the latter inscription it is useful to point out that the converse of 32 is indirectly reducible to it. For if $E'BF'$ (fig. 5.18) is not tangent to the circle, but angle $DBF' \simeq$ angle BAD, then, if EBF is drawn tangent to the circle, by 32, angle $DBF' \simeq$ angle $BAD \simeq$ angle DBF, which is impossible. Euclid gives an analogous reduction in deriving III,37, which he does apply in the inscription of the regular pentagon, from 36. I shall discuss these two propositions, as well as 23, 24, 26–30, and 35, in the following section.

5.2 Rectilineal Figures and the Circle

In book IV Euclid carries out the following constructions:

1.a. To inscribe in (2) or circumscribe about (3) a given circle a triangle equiangular with a given one;
1.b. to inscribe in (4) or circumscribe about (5) a given triangle a circle;
2. to inscribe in (6, 8) or circumscribe about (7, 9) a given circle (6, 7) or square (8, 9) a square or circle;
3. to inscribe in (11, 13) or circumscribe about (12, 14) a given circle (11, 12) or regular pentagon (13, 14) a regular pentagon or circle;
4. to inscribe in a given circle a regular hexagon (15) (A corollary points out that the side of the hexagon is equal to the radius of the circle and that the other cases done under 3 can be done in the same way with the hexagon);
5. to inscribe in a given circle a regular *pentekaidekagon* (fifteen-angled rectilineal figure) (16) (A remark at the end of the proof points out that the other three cases can be done).

There are two other constructions which are clearly lemmas for the main work of book IV. The first of them, placing a straight line shorter than the diameter of a given circle so that its endpoints are on the circumference of the circle (IV,1), is simple. This construction is applied in IV,16, in XII,16, in *lēmmata* after X,13 and XI,23, and in the other preliminary construction of book IV, the construction of an isosceles triangle with vertex angle of 36° (IV,10). This construction is the core of the inscription of the regular pentagon, the only very complex problem in book IV. The other inscriptions and superscriptions are straightforward and largely independent

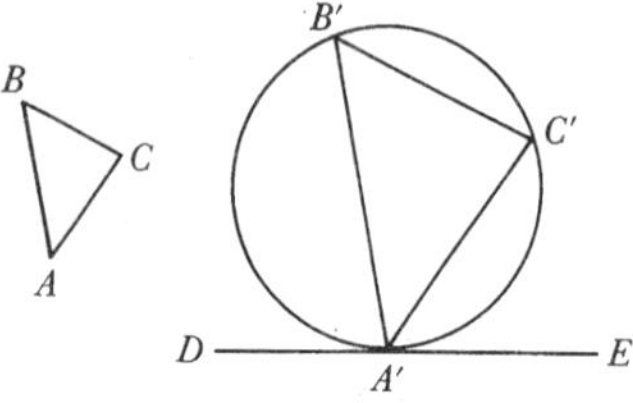

Figure 5.19

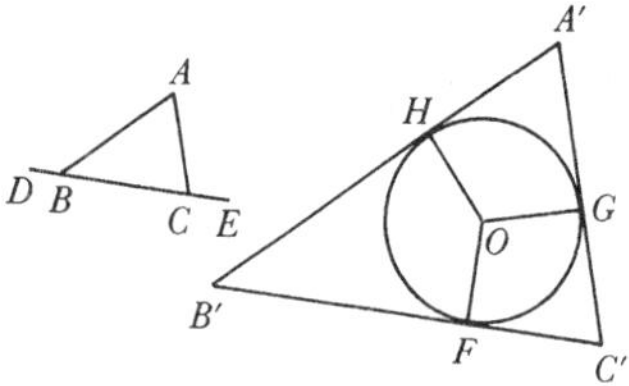

Figure 5.20

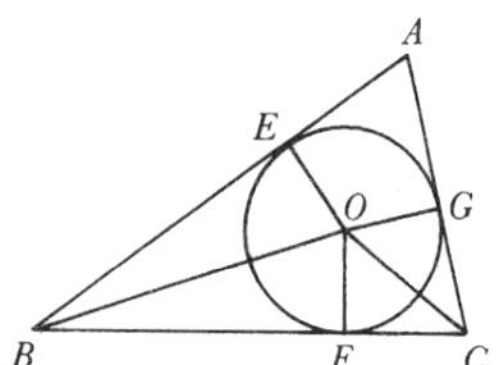

Figure 5.21

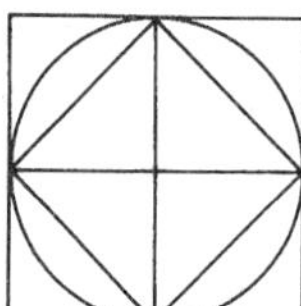
Figure 5.22

of one another. I shall describe them briefly, leaving out of account Euclid's tacit assumptions about the intersection of straight lines.

In IV,2 Euclid draws (fig. 5.19) $DA'E$ tangent to the given circle $A'B'C'$ at an arbitrary point A' and makes angles $C'A'E$, $B'A'D$ equal to the angles ABC, BCA of the given triangle ABC (I,23). The equiangularity of the triangles ABC, $A'B'C'$ is a consequence of III,32 and I,32. For IV,3 Euclid takes (fig. 5.20) the center O of the given circle FGH (III,1), makes angles FOG, FOH equal to the exterior angles ACE, ABD of the given triangle (I,23), and draws tangents $A'GC'$, $C'FB'$, $A'HB'$ to the circle. Since the interior angles of a quadrilateral are equal to four right angles (I,32, "inasmuch as the quadrilateral is divisible into two triangles") and tangents are perpendicular to radii drawn to the point of contact (III,18), by I,13, the angles $A'C'B'$, $A'B'C'$ are equal respectively to ACB, ABC; and I,32 establishes the equiangularity of the triangles ABC, $A'B'C'$.

The center of a circle which can be inscribed in the given triangle ABC of fig. 5.21 (IV,4) is the intersection O of the straight lines bisecting two of the angles, ACB, ABC (I,12). For the perpendiculars OE, OF, OG from O to the sides AB, BC, AC (I,9) are equal (I,26); and the sides will be tangent to the circle through E, F, G (III,16). It is manifest from III,1, corollary, that the intersection of the perpendicular bisectors of two sides of a given triangle is the center of the circumscribing circle (IV,5). Euclid determines this point (I,10, 11) and uses I,4 to argue that the point is the center. He does three cases depending on whether the point is inside the triangle, outside it, or on the third side, adding a remark to the effect that the three cases correspond respectively to the angle contained by the two bisected sides being acute, obtuse, or right (III,31).

Connecting the endpoints of two perpendicular diameters of a circle produces a square inscribed in it (IV,6); and tangents to those endpoints produce a circumscribed square (IV,7) (fig. 5.22). To show that the first construction works, Euclid uses I,4 for the equality of the four sides of the inscribed figure and III,31 to establish that its angles are right. For the second construction he invokes III,18 and I,28,30, and 34. Conversely (IV,8, 9), the perpendicular bisectors of the sides of a square (its diagonals) intersect at the center of an inscribed (circumscribed) circle. The arguments that these constructions work use III,16 and materials from book I.

The proofs of IV,12–14, which are referred to in the additions to 15 and 16, can be characterized in a general way as showing that (i) tangents through the vertices of an inscribed

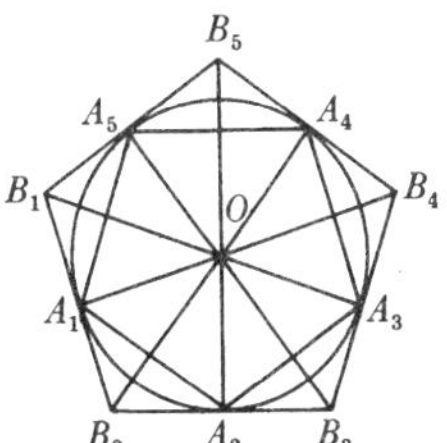

Figure 5.23

regular polygon determine a circumscribed polygon of the same kind, (ii) the intersection of the perpendicular bisectors of two consecutive sides of a regular polygon is the center of an inscribed circle, and (iii) the intersection of two bisectors of two consecutive angles of a regular polygon is the center of a circumscribed circle. The justifications for (ii) and (iii) are of the same kind that we have already seen. The justification of (i) is more elaborate. If the inscribed polygon is $A_1A_2 \ldots A_n$ (fig. 5.23) and the constructed one $B_1B_2 \ldots B_n$ with A_i between B_i and B_{i+1} and A_n between B_n and B_1, it suffices to show that $B_1B_2 \simeq B_2B_3$ and angle $B_1B_2B_3 \simeq$ angle $B_2B_3B_4$. Taking O as the center of the circle, Euclid uses the Pythagorean theorem to infer the equality of A_1B_2, A_2B_2, and then I,8 for the congruence of triangles OA_1B_2, OA_2B_2, which in turn yields that angle A_1OA_2 is twice angle A_2OB_2. Similarly, the triangles A_2OB_3, A_3OB_3 are congruent; and angle A_2OA_3 is twice angle A_2OB_3. Euclid now infers the equality of the angles A_2OB_2, A_2OB_3 from that of A_1OA_2, A_2OA_3. The latter equality could easily be derived using I,8; but Euclid invokes III,28 and 27, according to which in equal circles equal chords cut off equal corresponding arcs, and straight lines from the endpoints of equal arcs to the centers or circumferences of their respective circles contain equal angles. I shall discuss these propositions shortly. Given the equality of the angles A_2OB_2, A_2OB_3, it is a simple matter to apply I,26 to get $B_2A_2 \simeq B_3A_2$ and angle $OB_2B_3 \simeq$ angle OB_3B_2. Similarly, $B_1A_1 \simeq B_2A_1$. But it has already been shown that $B_2A_1 \simeq B_2A_2$ and that angle $OB_2B_1 \simeq$ angle OB_2B_3 and angle $OB_3B_4 \simeq$ angle OB_3B_2. Therefore, $B_1B_2 \simeq B_2B_3$ and angle $B_1B_2B_3 \simeq$ angle $B_2B_3B_4$.

Since the side of an inscribed regular hexagon is equal to a radius, Euclid is able to solve IV,15 by determining points A, B, C on the circumference of the given circle such that AB, BC are each equal to a radius and then drawing diameters AOD, BOE, COF (fig. 5.24) to determine the six vertices of the hexagon. To show that the construction works Euclid applies I,5 and 32 to infer that each of angles AOB, BOC is one-third of a right angle, I,13 to infer that angle COD is, and I,15 to infer that the remaining angles at O are as well. Euclid could now complete his justification with congruence arguments using I,4. Instead he infers that the hexagon is equilateral by invoking III,26 and 29, according to which in equal circles equal angles with vertices at the center or on the circumference stand on equal arcs, and equal arcs are subtended by equal straight lines. He then uses III,27 to infer that the hexagon is equiangular.

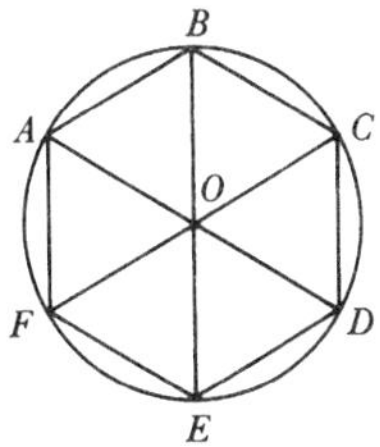

Figure 5.24

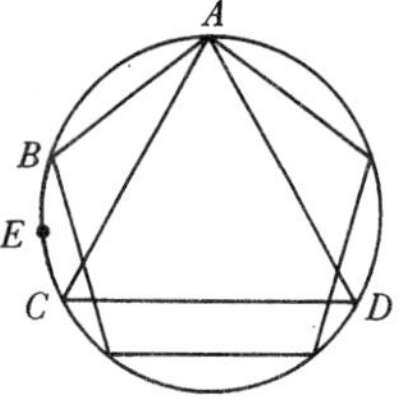

Figure 5.25

For the *pentekaidekagon* Euclid inscribes in the circle *ABCD* a regular pentagon with side *AB* and an equilateral triangle *ACD* (fig. 5.25). He reasons, tacitly applying III,28, that since the arc *ABC* is one-third of the whole circumference and arc *AB* is one-fifth, the difference *BC* is two-fifteenths, so that if the difference is bisected (III,30), the straight line subtending the half of *BC*, *BE* will be the side of a regular *pentekaidekagon* inscribable in the circle. The inscription itself involves application of IV,1 to fit fifteen straight lines equal to *BE* into the given circle.[9]

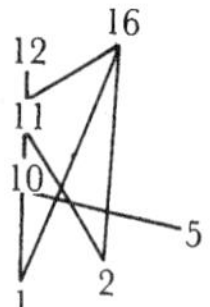

Figure 5.26

Of the propositions just discussed, 12 depends upon 11, and 16 depends upon 1, 2, and 11. The remaining dependencies in book IV all relate to the inscription of the regular pentagon. These are included in fig. 5.26, which shows how slight the internal deductive structure of book IV is. As far as dependence on earlier books is concerned, the situation is elaborate but not terribly complex. In the propositions I have discussed, Euclid employs much of the material from I,1–34, but nothing from book II. He uses I,47 in IV,12. From III he uses the taking of centers of circles (1), the elementary properties of tangents (16–19), and 26–32; III,15 can also be thought of as a *diorismos* for IV,1. IV,10 and 11 add to these dependencies II,11 and III,37, so that it is reasonable to say that much of the last part of book III is aimed at providing a basis for the constructions of book IV. As we shall now see, the inscription of the regular pentagon involves very careful preparation on Euclid's part.

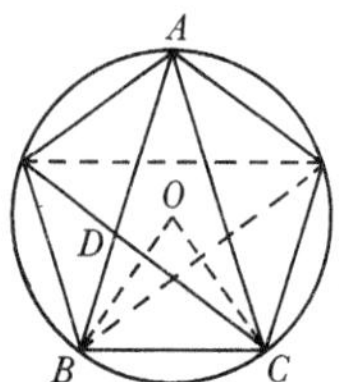

Figure 5.27

The natural way to construe the problem of inscribing a regular pentagon in a circle is as the problem of dividing the circumference of the circle into five equal parts, which is obviously equivalent to constructing an isosceles triangle *OBC* (fig. 5.27) with vertex angle at *O* equal to 72° (four-fifths of a right angle). Because of III,20 this problem is in turn equivalent to constructing an isosceles triangle *ABC* with vertex angle at *A* equal to 36°, or, in the words of IV,10, "an isosceles triangle having each angle at the base double of the remaining one." The fact that Euclid approaches the construction from this triangle makes it likely that the construction is based on the pentagram, the five-pointed star formed by joining alternate vertices of a regular pentagon.[10] In any case, suppose *ABC* is such a triangle and that the straight line *CD* from *C* to *AB* bisects angle *ACB*. Clearly, then, angle $DAC \simeq$ angle $ACD \simeq$ angle *DCB*. So triangles *ABC*, *CBD* have two angles equal to two angles and are therefore similar with $(AB, BC) = (BC, BD)$, or, since $BC \simeq CD \simeq AD$, $(AB, AD) = (AD, BD)$. Hence to construct the desired triangle with side *AB*, one need only divide *AB* in extreme and mean ratio at *D* and draw circles with center *A* and radius *AB* and with center *B* and radius equal to DA. Their intersection determines the third vertex of the triangle, as is easily proved using VI,6.[11] Since Euclid does not have the theory of proportion at his disposal, he uses II,11 for the construction, a proposition which presupposes II,6. Thus, the analysis of the problem of inscribing a regular pentagon leads to the problem of VI,30, cutting a straight line in extreme and mean ratio, which in turn gives rise to a basically geometric proof of VI,29; and the desire to avoid the theory of proportion in the inscription necessitates the proofs of II,11 and 6 which we find in the *Elements*.

Even though II,11 provides the means for constructing an isosceles triangle ABC with vertex angle at A equal to 36°, there remains the problem of showing that the constructed triangle does in fact satisfy this condition. The construction suggested leads to $\mathbf{T}(AD) \simeq \mathbf{O}(AB, BD)$, or, since $AD \simeq BC$,

(i) $\mathbf{T}(BC) \simeq \mathbf{O}(AB, BD)$.

With the theory of proportion one could use VI,17 and 6 to infer

(ii) angle $BCD \simeq$ angle BAC.

The remainder of the proof proceeds quite simply without the theory of proportion. By I,32 angle $BDC \simeq$ angle DAC + angle $ACD \simeq$ [by (ii)] angle DCB + angle $ACD \simeq$ angle ACB. But because, by construction, $AB \simeq AC$, angle $ABC \simeq$ angle $ACB \simeq$ angle BDC. Hence $DC \simeq BC \simeq AD$, and angle $DAC \simeq$ angle ACD; or, since angle $BAC \simeq$ angle DCB, angle BAC is twice angle ACB or its equal, ABC.

Thus the only difficulty remaining to be overcome is the inference of (ii) from (i) without using the theory of proportion. For this purpose Euclid uses III,32. For analysis I use its converse, which together with III,32 can be written

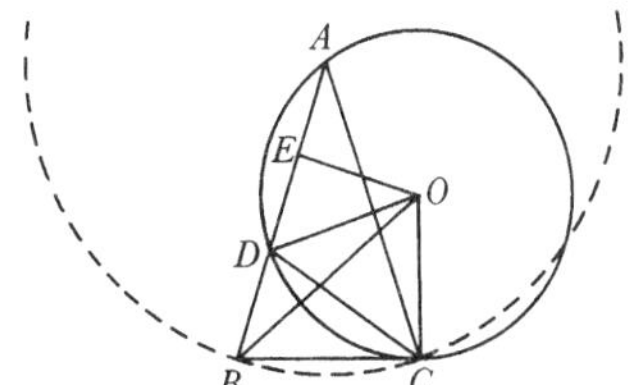

Figure 5.28

III,32 and converse If ABC (fig. 5.28) is a triangle and D is a point on AB, then if triangle ADC is inscribed in a circle, BC is tangent to the circle if and only if angle $BAC \simeq$ angle BCD.

Hence, to show that (i) implies (ii) it suffices to show that (i) implies that BC is tangent to the circle circumscribed about triangle ADC. The latter implication is straightforward if one has VI,17 and 6 and the converse of III,32. For an analysis leading to Euclid's proof, which does not depend either on the theory of proportion or on the converse of 32, one supposes that BC is tangent to the circle and draws OE from the center O of the circle perpendicular to AD at E so that (III,3) $AE \simeq ED$. Since (III,18) OC is also perpendicular to BC, applications of the Pythagorean theorem give

$$\mathbf{T}(OE) + \mathbf{T}(EB) \simeq \mathbf{T}(OB) \simeq \mathbf{T}(OC) + \mathbf{T}(BC) \simeq \mathbf{T}(OD) + \mathbf{T}(BC) \simeq \mathbf{T}(ED) + \mathbf{T}(OE) + \mathbf{T}(BC).$$

Hence $\mathbf{T}(EB) \simeq \mathbf{T}(ED) + \mathbf{T}(BC)$; and (i) is equivalent to $\mathbf{T}(EB) \simeq \mathbf{O}(AB, DB) + \mathbf{T}(ED)$, which, since E is the midpoint of AD, is simply II,6. Here again analysis of a geometric truth known to be true independently leads to recognition of a geometric-algebraic truth.

The analysis just given converts to a proof of

III,36,37 If AB cuts a circle ADC at D and BC meets it at C, then BC is tangent to the circle at C if and only if $\mathbf{T}(BC) \simeq \mathbf{O}(AB, DB)$.[12]

Euclid uses III,37, which is proved by reduction to III,36, for his construction in IV,10 of the isosceles triangle with vertex angle of 36°. He first breaks AB at D so that $\mathbf{T}(AD) \simeq \mathbf{O}(AB, BD)$, and draws a circle c with radius AB and center A. In IV,1 he has already drawn the circle with center B and radius equal to AD to solve the problem of fitting the straight line BC equal to AD into c. He therefore cites IV,1 and asserts (i). He connects AC, CD and circumscribes a circle about the triangle ACD (IV,5). III,37 gives that BC is tangent to this circle, and III,32 yields (ii).

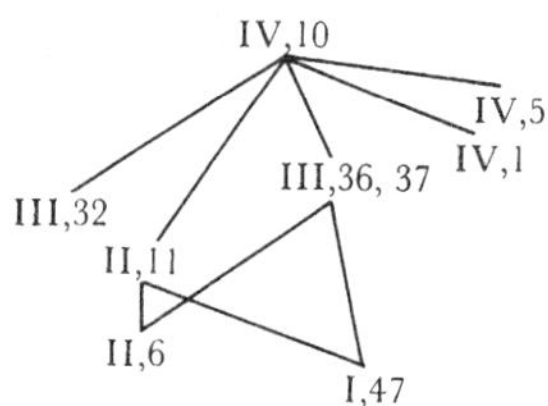

Figure 5.29

In place, then, of a relatively simple construction using the theory of proportion, Euclid substitutes an elaborate proof of IV,10 relying heavily on the Pythagorean theorem. If deductive relations internal to book III and dependencies on propositions prior to I,47 are left out of account, the important features of the structure of the proof are as represented in fig. 5.29. There seems to be a similar reworking in the case of

III,35 If (fig. 5.30) two chords AB, $A'B'$ of a circle intersect inside the circle at D, then $\mathbf{O}(AD, DB) \simeq \mathbf{O}(A'D, DB')$.

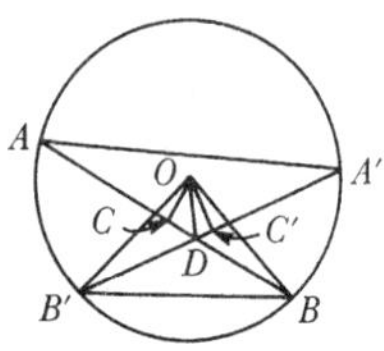

Figure 5.30

With the theory of proportion one argues simply that since, by III,21, angle $ABB' \simeq$ angle $AA'B'$ and angle $BAA' \simeq$ angle $BB'A'$, the triangles ADA', $B'DB$ are similar with $(AD, DB') = (A'D, DB)$, from which the rectangle equality follows by VI,16. Euclid's proof requires division into cases according to whether none, one, or both of AB, $A'B'$ pass through the center O of the circle. He does the first and third cases, the latter of which is trivial. For the former he drops OC, OC' perpendicular to AB, $A'B'$ with C, C' lying on AD, $A'D$ and, by III,3, bisecting AB, $A'B'$. The Pythagorean theorem yields

$$\mathbf{T}(OC') + \mathbf{T}(C'D) \simeq \mathbf{T}(OD) \simeq \mathbf{T}(OC) + \mathbf{T}(CD),$$
$$\mathbf{T}(OC') + \mathbf{T}(C'B') \simeq \mathbf{T}(OB') \simeq \mathbf{T}(OB) \simeq \mathbf{T}(OC) + \mathbf{T}(CB).$$

Subtracting equals from equals one gets

$$\mathbf{T}(C'B') - \mathbf{T}(C'D) \simeq \mathbf{T}(CB) - \mathbf{T}(CD).$$

The desired result follows from

$$\mathbf{T}(C'B') \simeq \mathbf{O}(A'D, DB') + \mathbf{T}(C'D),$$
$$\mathbf{T}(CB) \simeq \mathbf{O}(AD, DB) + \mathbf{T}(CD),$$

each of which is a direct application of II,5, for which again there is a purely geometric explanation.

There is another interesting case of the apparent avoidance of the theory of proportion in III,26–30, which we have seen to play a role in IV,12, 15, and 16 and which also plays a role in the inscription of the regular pentagon. To carry out this construction Euclid inscribes in a circle *ABCDE* (fig. 5.31) an isosceles triangle *ACD* with vertex angle at *A* one-half of either base angle (IV,10 and 2). He then bisects the two base angles with the straight lines *CE*, *DB* and argues that the angles *CAD*, *ACE*, *ECD*, *BDC*, *BDA* are all equal. He then uses III,26 to infer the equality of the lesser arcs *CD*, *AE*, *ED*, *BC*, *BA*, and applies 29 for the equality of the corresponding straight lines. The pentagon *ABCDE*, then, is equilateral. To prove that it is equiangular, Euclid adds equal arcs to get the equal arcs *BCDE*, *CDEA*, *DEAB*, *EABC*, *ABCD*, and then infers the equality of the pentagon's angles, using III,27.

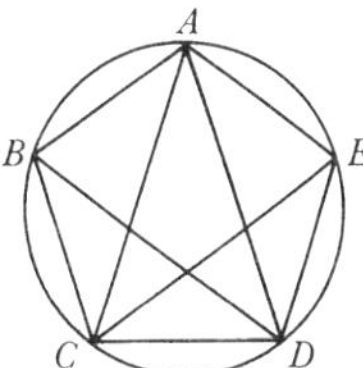

Figure 5.31

26–29 establish, for equal circles *ABCD*, *A'B'C'D'* with centers *O*, *O'*, the equivalence of the following conditions:

(a) angle *BAD* $\simeq$ angle *B'A'D'*;
(b) angle *BOD* $\simeq$ angle *B'O'D'*;
(c) arc *BCD* $\simeq$ arc *B'C'D'*;
(d) straight line *BD* $\simeq$ straight line *B'D'*.

More exactly, Euclid establishes if (a) or (b) then (c) in 26; if (c) then (a) and (b) in 27; if (d) then (c) in 28;[13] and if (c) then (d) in 29. In 30 Euclid uses 28 in a straightforward way to bisect a given arc of a circle. The major difficulties in this sequence arise in connection with 26, which depends upon III,24, which in turn depends upon

III,23 On the same straight line there cannot be constructed two similar and unequal segments of circles on the same side.

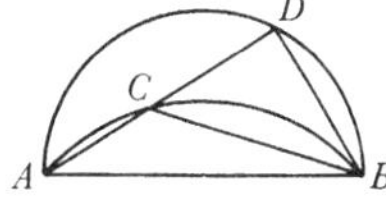

Figure 5.32

Similar segments are defined as those which "admit equal angles" (III, def. 11). Euclid argues that if *ADB*, *ACB* (fig. 5.32) are similar and unequal segments on the same side of the straight line *AB*, and *ACD* is a straight line, angle *ACB* $\simeq$ angle *ADB*, by the definition of similarity, contradicting I,16. Euclid then proves in 24 the equality of similar segments on equal straight lines by making the base *AB* of one such segment *AEB* coincide with the base *CD* of another *CFD* and arguing, "... if the straight line *AB* coincide with *CD* [of fig. 5.33], but the segment *AEB* do not coincide with *CFD*, it will either fall within it or

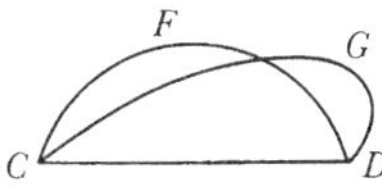

Figure 5.33

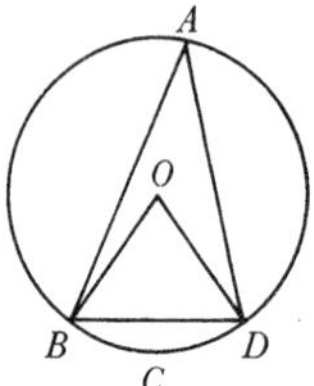

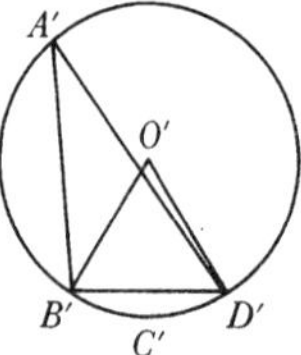

Figure 5.34

outside it; or it will fall awry, as *CGD*, and a circle cuts a circle at more points than two, which is impossible." It is apparent that 23 and 24 are essentially equivalent assertions, each given a partial proof. The proof of 23 fails to consider the possibility of one segment "falling awry" from the other (although its proof could be applied to this case), and that of 24 gives at best tacit consideration to one segment falling inside the other.[14]

The proof of 26 goes as follows: "Let [fig. 5.34] *ABCD*, *A'B'C'D'* be equal circles, and in them let there be equal angles, namely at the centers the angles *BOD*, *B'O'D'*, and at the circumferences the angles *BAD*, *B'A'D'*." Since *OB*, *OD*, *O'B'*, *O'D'* are radii of equal circles, $OB \simeq O'B'$ and $OD \simeq O'D'$, because of III,def. 1: "Equal circles are those the diameters of which are equal or the radii of which are equal." Therefore, by I,4, $BD \simeq B'D'$. But segments *BAD*, *B'A'D'* are similar by definition. Hence, by 24, the segments are equal; and, since the circles are equal, "the arc *BCD* which remains is equal to the arc *B'C'D'*."

It is clear that Euclid's definition of similar segments presupposes III,21 in its full generality and that his proof of 26 takes for granted the equivalence of (a) and (b), which he has in fact only established for the case when angles *BAD*, *B'A'D'* are acute. This assumption is reflected in the *ekthesis*, which is written as if Euclid were proving 'if (a) and (b) then (c)'. Perhaps the simplest way to rectify the argumentation is first to prove 'if (b) then (c)', assuming without loss of generality that arcs *BCD*, *B'C'D'* of fig. 5.34 are less than a semicircle and inferring the equality of angles *BAD*, *B'A'D'* via III,20, and then to prove 'if (a) then (c)', doing separately the three cases in which the arcs *BAD*, *B'A'D'* are greater than, equal to, or less than a semicircle (see fig. 5.35).[15] The first case can be done as in Euclid, using III,20 to infer the equality of angles *BOD*, *B'O'D'*; the third case can be reduced to the first, using III,22; the second case would be trivial, since Euclid would take for granted that halves of equal circles are equal. Once 26 has been correctly proved, it is a simple matter to do an indirect reduction of 27 to 26.[16] It is also easy to derive 28, 29 from 26, 27 respectively. For since $OB \simeq O'B'$ and $OD \simeq O'D'$, one

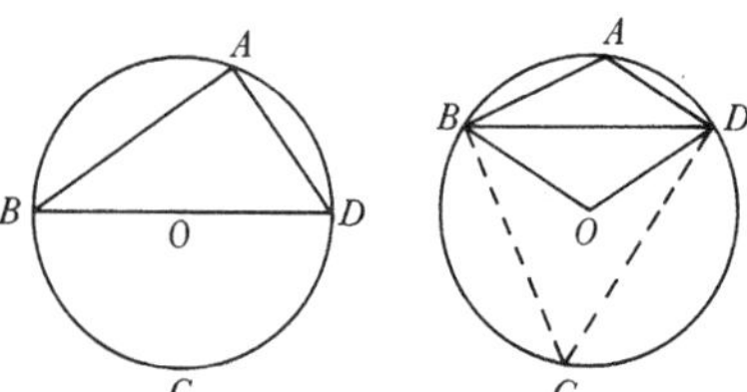

Figure 5.35

has (taking arcs BCD, $B'C'D'$ to be less than semicircles) arc $BAD \simeq$ arc $B'A'D'$ if and only if arc $BCD \simeq$ arc $B'C'D'$ if and only if (III,26, 27) angle $BOD \simeq$ angle $B'O'D'$ if and only if (I,4, 8) $BD \simeq B'D'$.

There are more perplexing difficulties involving definitions 1 and 11, both of which are invoked in the proof of 26 and are used only in the sequence of propositions now under discussion. Many commentators have claimed that 1 is not a proper definition, but an assertion to be proved or postulated.[17] Clearly this claim can be true only if 'equal' has a determinate sense when applied to circles. In fact Euclid does not invoke equal circles until book III; before that point he speaks only of equal rectilineal figures, angles, and straight lines. Since equality for rectilineal figures is equality in area in the *Elements*, it is reasonable to suppose that it means the same for circles. But then there is a slight gap in Euclid's proof of 26 (and also of 28 and 29). For in 26 Euclid infers the equality of the arcs BCD, $B'C'D'$ of fig. 5.34 directly from the equality of the two circles and of the two segments BAC, $B'A'C'$. He therefore needs a proof that circles equal in area are contained by equal circumferences.[18] The obvious way to give such a proof is to use superposition. But the same method is sufficient to prove definition 1 itself. For, given two equal circles, one can be placed on the other so that their centers coincide. Since they are equal in area, one cannot fall entirely inside the other. Hence the circumferences must coincide at one point at least, and the radii to that point must be equal. But then, by the definition of a circle, all the radii, and therefore all the diameters, of the two circles must be equal. (And clearly the circumferences will coincide entirely.) On the other hand, given two circles with equal radii or diameters, one can be placed on the other so that their centers coincide. The circumferences then must also coincide, because, if one has a point outside the other, a straight line from the centers to the point will determine a pair of unequal radii; hence, the circles are equal in area.

The provability of definition 1 makes it an awkward but not necessarily illegitimate definition. Heath, who thinks that Euclid would shun a proof by superposition of the kind just given, considers the definition to be legitimate. He writes (vol. II, p. 2),

> There is nothing technically wrong in saying "By *equal circles* I mean circles with equal radii." No flaw is thereby introduced into the system of the *Elements*; for the definition could only be objected to if it could be proved that the equality predicated of the two circles in the definition was not the same thing as the equality predicated of the other equal figures in the *Elements*, and, needless to say, this cannot be proved because it is not true.

Heath misses the point here. If 'equal circles' is a defined expression, the word 'equal' in it and the word in other contexts, e.g., in the common notions, are homonyms. One cannot indeed prove that these homonyms don't have the same sense, but the absence of a proof that they do not is not a proof that they do. In the absence of a proof that they do Euclid is not logically justified in applying the common notions to equal circles, as he does at the end of the proof of 26. He follows the same procedure in book XI when he defines (def. 10) equality for similar solid figures and then uses the common notions in connection with them. Euclid may not realize the logical difficulties in defining equality for different figures and taking for granted that the same principles can be applied to the different notions of equality. However, it seems more likely that he does not use the term 'definition' in its precise modern sense, but allows certain kinds of assumptions to count as definitions. In any case, the simplest way to rectify his procedure from a modern point of view is to assume that he does use 'equal' as a homonym and then make explicit the need to restate the principles governing equality for each sense of 'equal'.

These difficulties surrounding 26 are compounded by the role of similar segments. Euclid's definition of such segments resembles the definition of equality of circles and of proportionality in that it picks out a mathematically necessary and sufficient condition for the intuitive concept to obtain without worrying about the intuitive correctness of the definition. This procedure turns the word 'similar' into a symbol whose sole content is given by definition 11 and makes it a homonym of the same word as used in book VI. The procedure might lead to the same kind of difficulty as that incurred in connection with the definition of equal circles, if the word 'similar' were also used in a general or intuitive sense in the way that 'equal' is. However, in the *Elements*, and in Greek mathematics generally, similarity is always defined for particular kinds of objects, and only the precise sense given by the definitions is used in proofs.

Although there is no logical fallacy to be found in Euclid's use of definition 11, his use of it is peculiar. He applies it only in 23, 24, and 26, the first two of these being, as we have seen, simply lemmas for the third. But in the proof of the third Euclid invokes the similarity of the segments BAD, $B'A'D'$ of fig. 5.34 only to prove their equality. After III,26 the notion of similar segments vanishes from the *Elements*. This procedure of proving equality results in terms of similarity results stands in complete contrast to Euclid's procedure for handling rectilineal figures by using equality results from book I as the basis for proving similarity results in book VI. What makes Euclid's introduction of similar segments especially perplexing is that he could have avoided any reference to them by proving 26 directly without invoking the problematic 23 or 24. After establishing the equality of the corresponding radii BO, $B'O'$ and DO, $D'O'$ of fig. 5.34 Euclid could simply place O' on O and $B'O'$ on BO and argue, as in the proof of I,4, that the triangles OBD, $O'B'D'$ will coincide. Moreover, the circumferences of the two circles must coincide, because, if they do not, the circles will have unequal radii. Hence arcs BCD, $B'C'D'$ will coincide and be equal.

It seems to me probable that the cause of these anomalies is again Euclid's reworking of a proof based on the theory of proportion so as to avoid the theory. In order to explain in what sense this is possible and to give the interpretation some historical plausibility, it is necessary to refer to the fifth-century mathematician Hippocrates of Chios.[19] According to Proclus (66.7–8), Hippocrates was the first person known to have written a work on the elements of geometry. Proclus (213.3–11) also associates Hippocrates with the theory of proportion, ascribing to him the reduction of the problem of constructing a cube twice the size of a given cube with side s to the problem of finding two straight lines x and y such that $(s, x) = (x, y) = (y, 2 \cdot s)$. Most importantly, Hippocrates is thought to have worked on the quadrature of the circle. Simplicius gives a long description of Hippocrates' quadrature of certain lunes (figures contained by the arcs of two circles). Simplicius derives his report from Eudemus, an older contemporary of Euclid, but adds comments of his own. Although it is not always possible to distinguish Simplicius' comments with certainty, and although Eudemus is sometimes suspected of doctoring earlier proofs to conform to his own notions of rigor, there is little doubt that Hippocrates' arguments were in essence the ones given by Simplicius. For my purposes the crucial part of Simplicius' report is his discussion of Hippocrates' first principle or starting point (*archē*):

> He made his starting point and placed first among the things useful for his purposes the assertion that similar segments of circles have the same ratio to one another as their bases in square (*dunamei*). He showed this by showing that the diameters have the same ratio in square as the circles, which Euclid placed second in the twelfth book of the *Elements*; the *protasis* is formulated as follows: "Circles are to one another as the squares on their diameters." For as the circles are to one another, so are the similar segments. For similar segments are those which are the same part of the circle, e.g., semicircle is similar to semicircle and the third part of a circle to the third

part. Therefore also, similar segments admit equal angles. For the angles of all semicircles are right, and those of greater segments are less than a right and as much less as the segments are greater than semicircles; and the angles of segments less than a semicircle are greater and as much greater as the segments are less.[20]

Scholarly controversy concerning this passage has largely centered on the question whether Hippocrates could have proved XII,2. Euclid's proof of it is normally attributed to Eudoxus, who lived one or two generations after Hippocrates. Euclid first proves as a lemma, XII,1, according to which if p, p' are similar polygons inscribed in circles with diameters d, d', $(p, p') = (\mathbf{T}(d), \mathbf{T}(d'))$. He takes $ABCDE$, $A'B'C'D'E'$ (fig. 5.36) to be the similar inscribed polygons (with the obvious correspondences), BF, $B'F'$ to be the diameters of circumscribing circles. By VI,6, the triangles AEB, $A'E'B'$ are similar with angle $AEB \simeq$ angle $A'E'B'$. Euclid now infers the equality of the angles AEB, AFB and of the angles $A'E'B'$, $A'F'B'$, justifying the inference with the remark "for they stand on the same circumference"—an apparent reference to III,27; he could, of course, have used III,21. In any case, since the angles BAF, $B'A'F'$ are right, the triangles BAF, $B'A'F'$ are equiangular and, by VI,4, $(BF, B'F') = (AB, A'B')$. Euclid now invokes VI,20 and its corollary and applies VH(i) to get the desired result.

XII,2 is proved by *reductio ad absurdum*, using the assumption of the existence of a fourth proportional. One imagines two circles c, c' with diameters d, d' and that $(\mathbf{T}(d), \mathbf{T}(d')) > (c, c')$. Then, for some area s less than c', $(\mathbf{T}(d), \mathbf{T}(d')) = (c, s)$. The argument is completed by inscribing in c' a polygon p' such that $s \prec p' \prec c'$, and inscribing a similar polygon p in c. Then, by XII,1, $(p, p') = (\mathbf{T}(d), \mathbf{T}(d')) = (c, s)$, so that $(p, c) = (p', s)$, contradicting the fact that $p \prec c$ and $p' \succ s$. It is, of course, necessary to prove the possibility of inscribing p' so that $s \prec p' \prec c'$. Euclid's argument for this possibility invokes III,30 to bisect a circular arc, but can be understood

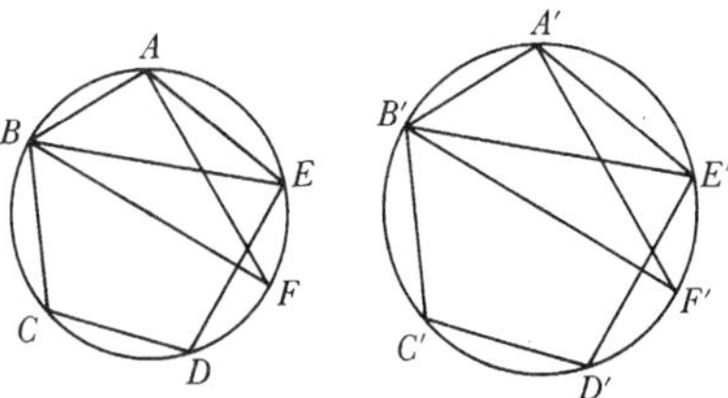

Figure 5.36

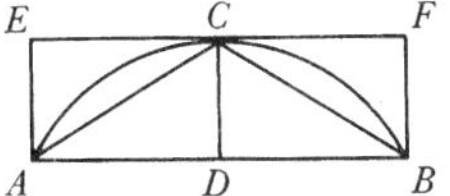

Figure 5.37

independently of this proposition. Suppose (fig. 5.37) *ACB* is a segment of a circle no bigger than a semicircle with *C* lying on the perpendicular bisector *CD* of the base *AB* of the segment. Then, if the straight line *ECF* is tangent to the segment, it will be perpendicular to *CD*, since *CD* extended passes through the center of the circle (III,1, cor. and III,18). Hence *EF* is parallel to *AB*, and if *EF* is made equal to *AB* with *C* its midpoint, *ABFE* will be a rectangle of which triangle *ACB* is half. Suppose, then, there is a regular polygon of n sides inscribed in a circle, and for each side *AB* of it the point *C* is determined as above. Then the difference between the inscribed polygon of $2 \cdot n$ sides obtained by connecting each *AC*, *BC* and the original polygon of n sides will clearly be greater than half the difference between the circle and the original polygon. Hence, by X,1, repeated applications of this procedure starting from, e.g., a square inscribed in c' will produce a polygon p' such that $c' - p' \prec c' - s$ and hence such that $s \prec p' \prec c'$.

I will discuss this proof further in section 6.3. My purpose now is only to make clear that if Hippocrates did prove XII,2 or the equivalent mentioned in Simplicius' report in anything like the way it is proved in the *Elements*, he need not have used any of III,26–30. According to the report, Hippocrates derived from a demonstration of the equivalent of XII,2,[21]

H. Similar segments are to one another as their bases in square.

Moreover, the report suggests that Hippocrates defined similar segments as those which are the same part of their circles and inferred the proposition expressed by Euclid's definition of them. Of course, the "argument" for this inference sketched by Simplicius is at best a plausibility consideration, and the definition suggested by him inadequate for Hippocrates if 'part' has the technical sense corresponding to VII, def. 3. For then any segment would be an nth part of its circle for some n. Such a restriction of the term 'segment' seems inconceivable in itself; it also fails to fit all of the segments used by Hippocrates in his quadratures. It thus seems probable that Hippocrates based his reasoning on a conception of similar segments which is captured in the definition

DH. Similar segments are those which bear the same ratio to the circles of which they are segments,

without necessarily having an adequate understanding of the term 'ratio'.[22]

It does not seem possible to determine with any certainty how Hippocrates derived H from XII,2, or III, def. 11 from DH, if indeed he did derive them in our sense. There is a short

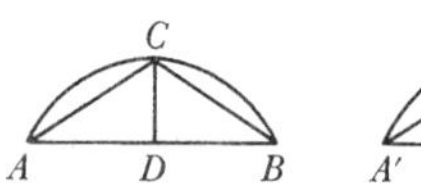

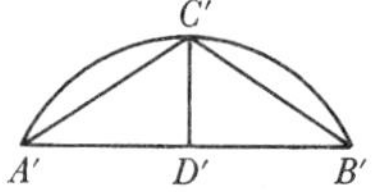

Figure 5.38

proof of H from XII,2, using XII,1, DH, and III, def. 11. Let (fig. 5.38) ACB, $A'C'B'$ be similar segments of the circles c, c' with diameters d, d', and let CD, $C'D'$ be the perpendicular bisectors of the bases ADB, $A'D'B'$ of the segments. Clearly the triangles ACD, BCD and $A'C'D'$, $B'C'D'$ are congruent, with $AC \simeq BC$ and $A'C' \simeq B'C'$, so that, by definition 11 and VI,6, triangles ACB, $A'C'B'$ are similar. By VI,19–20 and XII,1,2 then

$$(\mathbf{T}(AB), \mathbf{T}(A'B')) = (\text{triangle } ACB, \text{triangle } A'C'B') = (\mathbf{T}(d), \mathbf{T}(d')) = (c, c'),$$

but since, by DH, (segment ACB, c) = (segment $A'C'B'$, c'), (c, c') = (segment ACB, segment $A'C'B'$), and the result follows. It seems, however, that any derivation of definition 11 from DH or of DH from definition 11 requires a proof "by exhaustion" paralleling Euclid's proof of XII,2. For my purposes the existence of such derivations is less important than the fact that Hippocrates almost certainly knew both XII,2 and H, and that III, def. 1 and III,24 are trivial consequences of these two assertions. It is therefore conceivable and—in the light of the other proofs in books I–IV which seem to be revised versions of proofs using the theory of proportion—perhaps it is even likely that Euclid's postulation of definitions 1 and 11 and his use of the latter to prove III,23 and 24 represent another attempt to avoid the theory, and perhaps also the method of exhaustion.

This completes my description of the contents of books III and IV, but I would like to describe briefly some subsequent applications of this material which have not yet been discussed. In books X and XI Euclid is concerned again with rectilineal figures. In constructing a solid angle out of three plane angles (XI,23) Euclid circumscribes a circle about a triangle and also applies IV,1 and III,31. These two propositions are used in book X as well, and the latter is also used in both XII and XIII. The process of inscribing successively larger polygons in circles and applying X,1, as in XII,2, is repeated in XII,10–12. Although in all such proofs Euclid begins by inscribing a square in a circle, he draws tangents to the circle to justify the use of X,1 only in XII,2. In XII,10 and 11 he circumscribes a square about the circle to justify the claim that the inscribed square is more than half of the circle in which it is inscribed;

but in 12 he leaves the justification entirely tacit. The process of inscribing larger polygons, of course, involves the bisection of arcs. In book XIII, in which Euclid constructs the regular solids and characterizes the ratio of their edge to the diameter of a circumscribing sphere, Euclid twice inscribes an equilateral triangle in a circle, twice inscribes a regular pentagon in a circle, and twice circumscribes a circle about a regular pentagon.[23] He also makes frequent use of the fact that the side of a regular hexagon inscribed in a circle is equal to the radius of the circle. Both in book XIII and in the last two propositions of XII Euclid uses III,26–30 several times; in XII,16 he also applies IV,1 and the corollary to III,16.

For completeness I should also mention VI,33, which is a lemma for XIII,8–10 postponed to book VI because it uses the theory of proportion. In it Euclid proves that if a, a' are angles standing on the arcs c, c' of equal circles and having their vertices either at the centers or the circumferences of those circles, then $(a, a') = (c, c')$. Euclid does the case in which the vertices are at the center. He takes in the first circle n consecutive arcs $c_1, \ldots, c_n$ equal to c and with the angles $a_1, \ldots, a_n$ standing on them at the center and hence equal to a. Similarly, he takes m consecutive arcs $c'_1, \ldots, c'_m$ and angles $a'_1, \ldots, a'_m$ in the second circle, and argues in the standard way using III,26 and 27, that

$$\text{COMP}\,(c + c_1 + \ldots + c_n, c' + c'_1 + \ldots + c'_m,$$
$$a + a_1 + \ldots + a_n, a' + a'_1 + \ldots + a'_m).$$

Although it would be possible to give a constructive sense to Euclid's procedure by thinking of multiples of angles and arcs as sums which are not themselves angles or arcs,[24] it seems reasonably clear that Euclid has lost sight of the concrete geometric sense of his argument. The argument seems to be a kind of going through the motions to establish a result needed later. For this reason it is perhaps worth pointing out that the inferences in XIII which Euclid wishes to cover are from $c \simeq n \cdot c'$ to $a \simeq n \cdot a'$, and could therefore be done without the theory of proportion at all. Euclid extends 33 to angles with vertex at the circumference by invoking III,20, which again only works for angles less than a right angle. Perhaps the simplest way to make the full extension in this case is to treat an angle not less than a right angle and standing on a certain arc as the double of an acute angle standing on half the arc and apply V,15.

Although half of the propositions of book IV are used subsequently, it seems reasonable to think of book IV as the completion of one part of the *Elements*. The next part is marked by the introduction of the concept of proportionality. The postponement of the introduction of this concept may be due to the complexity of Eudoxus' treatment of it. However, it is difficult to understand why Euclid should choose to prove a proposition without using the concept when he could give a much simpler and more revealing proof by waiting until the concept had been introduced. Whatever the motivation, it seems clear that avoidance of the concept is an important factor in the structure of the first four books of the *Elements*. Indeed, if my account of III,23 and 24 is correct, avoidance of the concept is much more important for Euclid than avoidance of superposition.

Examination of books III, IV, and VI should also have made clear that the tight deductive structure of I,1–46 is by no means the general rule in the plane geometry of the *Elements*. Propositions with no future use are proved, perhaps for their own sakes, perhaps for possible uses outside the *Elements*. Needed propositions are not always supplied explicitly, so that the reader sometimes has to fill out inferences for himself with short arguments or references to previous proofs. In addition, in the first part of III and in VI, little attention is paid to logical sequence. Obviously related propositions, such as theorems and their converses, are proved together. But if these cases are left out of consideration, there is a great deal of free play in the deductive development; this free play is often associated with a failure to exploit conceptual relationships made obvious by postclassical approaches to geometry. Such failure is an important source of insight into fundamental differences between Euclidean mathematics and its modern analogues. Another feature which the *Elements* share with other Greek mathematical works is the separation of propositions from what are really lemmas for them. This separation is obvious in the case of the lemmas II,11 and III,37 for the inscription of the regular pentagon. Euclid clearly thinks of subject matter as a more important organizing principle than deductive relevance—a fact which makes it difficult to read the *Elements* straight through.

In discussing books X–XIII, it will sometimes be helpful to compare Euclid's solid-geometric methods with his plane-geometric ones. It will also be necessary to consider his uses of geometric algebra in some detail. But in general it will not be necessary to discuss specific applications of material from the first nine books. In the later books Euclid shows no concern for isolating uses of the theory of proportion, and he often uses the theory when he does not have to. In addition, Euclid's constructions and reasoning are often sketchier in the later books; thus it is not always possible to tell which proof, among alternatives compatible with what he says, he has in mind. However, it is sufficient for my purposes to have made clear in the preceding which parts of books I–IX are used in the later books.

Notes for Chapter 5

Bibliographical Note

The deductive structure of books III and IV is given a detailed representation by Neuenschwander in "Die ersten vier Bücher . . . ." He does not, however, distinguish between cases in which an object is constructed and those in which it is assumed to exist in the *ekthesis*, between applications of propositions in remarks as opposed to their use in proofs of stated propositions, and between deductive applications of a proposition and references to the proof of a proposition as something which can be copied. The materials relating to III,23, 24, 26–30 are thoroughly discussed by von Fritz in "Gleichheit . . . ."

1. See Heath, vol. II, pp. 39–40.

2. *Collectio*, pp. 648–660. The standard reconstruction of Euclid's work is by Chasles.

3. I here leave out of account the two "investigations," IX,18 and 19, which are more like theorems than problems or porisms. See pp. 98–99.

4. The positive rationals are a model for all of Euclid's assumptions about magnitudes; but since the taking of mean proportionals corresponds to the taking of square roots, they are not a model for the existence of a mean proportional between two magnitudes.

5. See Heath, vol. II, p. 28.

6. See Heath's discussion of the Aristotle passage, vol. I, pp. 152–153.

7. See Heath, vol. II, pp. 47–48, for a discussion of this point and a description of Heron's way of correctly generalizing III,20 without abandoning the Euclidean conception of an angle.

8. 33 is used in the *Data* (prop. 80) and in Euclid's *Optics* (props. 45 and 46).

9. It is perhaps worthwhile to indicate the relationship of book IV to what is now known about regular rectilineal figures and the circle. Since Euclid seems likely to have been aware of the fact that the four problems which he considers in the case of the pentagon are equivalent for any regular rectilineal figure, it suffices to consider inscriptions in circles. Gauss proved that a regular rectilineal figure of n sides is inscribable in a circle using straightedge and compass if and only if $n = 2^m \cdot p_1 \cdot \ldots \cdot p_k$, where the p_i are 0 or more distinct primes of the form $2^{2^l} + 1$. The only numbers of this form known to be prime are 3, 5, 17, 257, 65,537, representing the first five values of l. (Values of l which yield numbers known to be composite are 5–9, 11, 12, 18, 23, 36, 38, 73.) Since it seems likely that Euclid realized that a regular rectilineal figure of j sides is inscribable in a circle if and only if one with $2^m \cdot j$ is, it seems fair to say that he recognizes the inscribability

of regular rectilineal figures with 2^m, $2^m \cdot 3$, $2^m \cdot 5$, and $2^m \cdot 3 \cdot 5$ sides only.

10. See especially Dijksterhuis (vol. II, pp. 53–54), who makes clear the possible intuitive basis for the construction of the regular pentagon.

11. One determines the same triangle by drawing the second circle with center D and radius DA; but with such a construction there is no direct way to show that $DC \simeq BC$, and hence that the triangle is the desired one.

12. In a full proof of 36 following the analysis given, one must distinguish, as Euclid does, between the case in which AB passes through the center of the circle and the case in which it does not.

13. Since a chord divides a circle into two segments, 28 requires the additional condition that arcs BCD and $B'C'D'$ "correspond", i.e., that both or neither be greater than a semicircle.

14. The words 'it will either fall within it or outside it or' are missing from the Theonine manuscripts and are, in any case, a curious way of referring to 23, as Heiberg and Heath suggest they do.

15. It would probably be best to prove III,31 before 26 in order to guarantee that figs. 5.34 and 5.35 represent the only possible relative positions of the points.

16. Euclid actually does an indirect reduction for 'if (c) then (b)' and infers (a) from (b) by means of III,20. The argument is sound as it stands only when arcs BCD, $B'C'D'$ are less than semicircles.

17. For some references see Heath, vol. II, p. 2.

18. Euclid also needs a proof that similar segments on equal straight lines are contained by equal circumferences. His arguments for 23 and 24 amount to such a proof.

19. For a discussion of Hippocrates' work see Heath, *A History* . . . , vol. I, pp. 182–209.

20. Simplicius, 61.5–18.

21. Simplicius' words are *touto de edeiknuen ek tou tas diametrous deixai ton auton logon echousas dunamei tois kuklois.*

22. This view is well expressed by Dijksterhuis, vol. I, pp. 34–38.

23. In XIII,9 Euclid inscribes a regular decagon in a circle, presumably by inscribing a regular pentagon and bisecting the arcs cut off by its sides.

24. See Heath, vol. II, pp. 275–276.

6 Elementary Solid Geometry and the Method of Exhaustion

6.1 The Foundations

Like the arithmetic books, the solid-geometry books XI–XIII are conceived by Euclid as a unit. Accordingly, he sets out the list of definitions for all three books at the beginning of XI; and some of the propositions in XI are almost certainly simply lemmas for use in XII and XIII. On the other hand, the three books show much more clearly than VII–IX that they have separate provenances. XIII, in which the five regular solids are inscribed in spheres, has often been ascribed to Theaetetus, and XII, in which results concerning areas and volumes are obtained by the "method of exhaustion," to Eudoxus. The substance of XI is analogous to that of books I and VI together, although no attempt is made to separate propositions presupposing the theory of proportion from those not presupposing it. XII has no analogue in the earlier books. XIII is, of course, analogous to book IV. There is, however, no treatment of the sphere corresponding to the investigation of the circle in book III. The absence of such a treatment means that some of the argumentation in XIII lacks foundation.[1]

Book XI itself divides roughly into two parts at proposition 24, in which "parallelepipedal solids" (parallelepipeds, solids contained by three pairs of parallel parallelograms) are introduced. This proposition corresponds to I,33, in which parallelogrammic areas (parallelograms) are introduced and which makes an analogous division of book I. The extent of the analogy will become clearer as discussion proceeds, but the general organizing idea is that the first part of each book provides elementary facts about parallels, perpendiculars, angles, etc., and justifies some elementary constructions, and that the second part proves fundamental facts about areas or volumes, with parallelograms or parallelepipeds playing an essential role. One important difference between I and XI has already been mentioned: no attempt is made to avoid the theory of proportion in XI, with the result that some theorems correspond to those of book VI. A second difference is that, although parallelepipeds are essential in the second part of book XI and although they are divisible into two equal triangular prisms, there is nothing in book XI corresponding to the fundamental congruence results for triangles in book I. Moreover, the first part of XI is neither as deductively straightforward nor as foundationally satisfactory as its plane-geometric analogue. Euclid's general procedure in solid-geometric reasoning is to reduce a three-dimensional question to a two-dimensional one and then to use results already obtained. Such a reduction, however, presupposes relationships which have not been treated before in the *Elements*—relationships between planes and planes, planes and lines, planes and points. A satisfactory treatment of these relationships requires new principles analogous to Euclid's postulates for plane geometry, but he gives none. The difficulties which he incurs are perhaps made clearest by considering the simplest alternative procedures.

Plane geometry can be developed either as an independent subject or as a part of solid geometry (or, of course, as a part of something more general). Normally, when plane geometry is developed as an independent subject, it is taken for granted that all objects considered lie in a single plane which never has to be mentioned. But when it is developed as a part of solid geometry, no such assumption is made; and it is occasionally necessary to relativize propositions to particular planes. There are traces of both approaches in book I. For example, in his definition of parallel straight lines Euclid includes the condition that the lines lie in the same plane in order to exclude nonmeeting lines in different planes. But his proof of I,30 ("Straight lines parallel to the same straight line are also parallel to one another") would be invalid without the tacit assumption that the lines lie in the same plane. However, with the exception of the definitions of book I, all of Euclid's plane geometry gives the impression of being carried out within a single plane. Unfortunately, when he comes to use plane geometry in book XI, he shows no qualms about transferring results from a single plane to multiple planes. For example in XI,4 he uses I,4 to prove the congruence of two triangles in different planes and in XI,6 he uses I,3 to copy a straight line in one plane in another. In general such applications require a rewriting of plane-geometric propositions to make relevant conditions on planes explicit. Such a rewriting might not be too difficult for the *Elements*, but it would make clear the need for explicit assumptions about planes which Euclid seems to overlook. These assumptions may be stated as follows:

XIa1 if two points lie on a plane, so does the straight line through them;

XIa2 any three points not on a straight line determine a single plane;

XIa3 if two planes intersect, they intersect in a straight line;

XIa4 for any plane there is a point not on it.

The principal purpose of 4 is to generate new planes. Euclid simply takes for granted the possibility of doing this. For example, in XI,7, given a straight line *EGF*, he writes, "Let a plane be drawn through *EGF*." In the light of the analogy between planes in solid geometry and straight lines in plane geometry, one would expect Euclid to have a postulate analogous to 1 enabling him to draw such planes under appropriate conditions. The absence of such a postulate suggests that Euclid's concern for foundations in plane geometry probably had more to do with the character of the subject, perhaps its simplicity, than with any general philosophical point of view.

Euclid attempts to prove principles corresponding to XIa1–3, namely:

XI,1 A part of a straight line cannot be in the plane of reference and a part in a higher plane.[2]

XI,2 If two straight lines cut one another, they are in one plane, and every triangle is in one plane.

XI,3 The common section of two planes is a straight line.

XI,7 A straight line joining parallel straight lines lies in the same plane with them.

The proofs of these propositions are notoriously inadequate and have often been discussed.[3] The last three of these are fundamental for Euclid's arguments: 2 and 7 for generating specific planes, 3 for generating straight lines out of solid-geometric constructions. Although 1 is in a sense used constantly, its primary purpose appears to be as a lemma for 2. The only other semi-explicit uses of 1 are in XI,14 and 16 and in XIII,17. To prove 1 Euclid supposes that a straight line *AB* has a continuation *BC* in a higher plane and, apparently applying postulate 2 relativized to solid geometry, that it also has a continuation *BD* in the plane of reference. He derives a contradiction with the words "Therefore *AB* is a common segment of the two straight lines *ABC*, *ABD*, which is impossible, inasmuch as if we describe a circle with center *B* and distance *AB*, the diameters [*ABC*, *ABD*] will cut off unequal circumferences of the circle." Clearly this argument presupposes that the two straight lines *ABC*, *ABD* lie in a single plane; but Euclid would seem to have no way to justify this presupposition until he has established XI,2, which, as I have indicated, itself depends upon 1.

Proposition 1 is, of course, a consequence of the assumption XIa1; but, as the proposition is formulated, it is weaker than the assumption. However, except in XI,2, Euclid's apparent applications of 1 are in fact of XIa1. XI,7 is also a consequence of XIa1, since if *AEB*, *CFD* (fig. 6.1) are parallel, they lie in the same plane by definition; and, by XIa1, the straight line through *E*, *F* lies in the plane. Euclid, however, invokes proposition 3 to infer that if the straight line *EF* lies "in a more elevated plane as *EGF*," a plane through *EGF* will

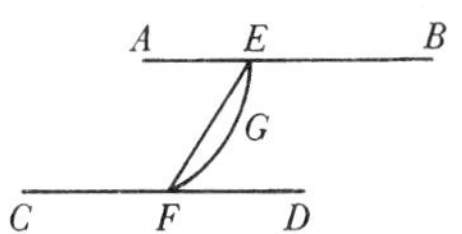

Figure 6.1

determine with the plane of *AB*, *CD* a straight line *EF* which with *EGF* will enclose a space, an impossiblity. Euclid's failure to invoke proposition 1 in proving 7 suggests that for him 1 is weaker than XIa1, so that the proof of 2 may be for him the only application of 1. His reference to the impossibility of two straight lines enclosing a space is presumably no more problematic than the reference in I,4,[4] because *EF* and *EGF* lie in the same plane. A similar reference in the proof of XI,3 is more obscure.

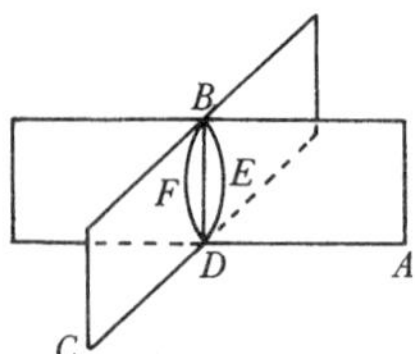

Figure 6.2

Let the two planes *AB*, *BC* [fig. 6.2] cut one another, and let the line *DB* be their common section. I say that the line *DB* is a straight line. For, if not, from *D* to *B* let the straight line *DEB* be joined in the plane *AB*, and in the plane *BC* the straight line *DFB*. Then the two straight lines *DEB*, *DFB* will have the same extremities, and will clearly enclose a space, which is absurd. Therefore *DEB*, *DFB* are not straight lines. Similarly we can prove that neither will there be any other straight line joined from *D* to *B* except *DB*, the common section of the planes *AB*, *BC*.

This proof is immediately vitiated by Euclid's assumption that the common section must be some kind of line. He then takes two points *D*, *B* of the section such that the section between them is not straight and, applying postulate 1 relativized for solid geometry, draws straight lines *DEB*, *DFB* in the two planes. Since the common section is assumed to be a non-straight line, neither straight line can coincide with it; hence they cannot coincide with each other. Euclid can conclude that the two lines enclose a space only if they are in the same plane, since, for example, semicircles with diameter *BD* could not be said to enclose a space if each was in a different plane. Therefore Euclid must be applying XI,2 to infer that *DEB*, *DFB* are in the same plane. At this point Euclid could conclude his proof, but instead he infers that *DEB* and *DFB* are not straight lines, as if he were refuting the possibility of drawing a straight line from *D* to *B* which is not the common section.

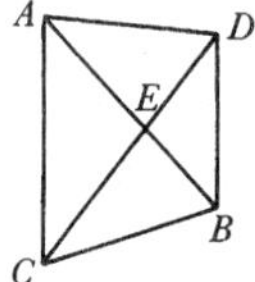

Figure 6.3

Euclid's strategy in XI,2 is to show that every triangle *ECB* (fig. 6.3) is in one plane and then to argue that if the intersecting straight lines *AEB*, *DEC* were not in one plane, neither would one of the triangles formed by connecting the ends of the lines be. His argument for the triangle *ECB* being in one plane is obscure. It is rejected by Heath (vol. III, p. 275) on the grounds that Euclid does only a particular case. However, the proof can be generalized. Suppose some part *p* of triangle *ECB* is in a different plane from the rest of it. If *p* includes part of one of the lines *EC*, *CB*, *BE* but not all of it, a straight line will have two parts in different planes, contradicting XI,1. If it does not, then a straight line can be drawn through *p* which will violate XI,1. Euclid then proves that no finite part of a triangle can fail to be in a plane with the rest of it. The difficulty is that he does not prove it in a way which is extendible to straight lines or points, which are not parts in the appropriate sense.

The generally unsatisfactory character of the proofs of these fundamental propositions of book XI makes them very difficult to interpret. Euclid clearly thinks that he both must and can prove that certain conditions determine a plane just as two points determine a straight line (postulates 1 and 2), and that the intersection of two planes is a straight line just as the intersection of two straight lines is a point (tacitly assumed in book I). However, he feels no need to postulate the possibility of drawing a plane. And rather than adopt an analogue of postulate 3 for the sphere, he gives a "constructional definition" (XI, def. 14), according to which a sphere is the figure comprehended (*perilēphthen*) by a semicircle rotated about its diameter. Euclid could have begun book I in the way in which he begins XI, or XI in the way in which he begins I. The discrepancy between the two books suggests that the need for postulates, as opposed to definitions and general principles of combinatorial reasoning, was not perceived very clearly by Euclid. As I have suggested before, the real foundation of Euclidean mathematics is clearly grasped mathematical objects characterized in definitions.

I wish now to analyze the rest of the first part of book XI. Naturally there are many applications, tacit and explicit, of propositions 2, 3, and 7 in this part, because they play the role of solid-geometric axioms. However, since there is no particular significance in their applications, I shall generally leave them out of account. Before turning to the propositions of the first part it is necessary to say a few words about the relevant definitions. After giving mathematically useless definitions of the solid and its limit, the surface (XI, defs. 1 and 2), Euclid turns to mathematically more significant matters. In definitions 3 and 4 he defines a straight line to be perpendicular to a plane when it is perpendicular to every straight line in the plane which it meets, and a first plane to be perpendicular to a second when straight lines in the first plane which are perpendicular to the common section of the planes are perpendicular to the second plane. In this latter definition Euclid takes for granted in a harmless way the result proved as XI,3. Applications of this definition are considerably simplified by Euclid's proof in XI,4 that a straight line *FE* is perpendicular to a plane if it is perpendicular to two straight lines *AEB*, *DEC* in the plane (fig. 6.4). The proof is a good example of the reduction of a solid proposition to plane ones. Euclid assumes that *AEB*, *DEC* are equal with *E* as their midpoint, and that *GEH* is another

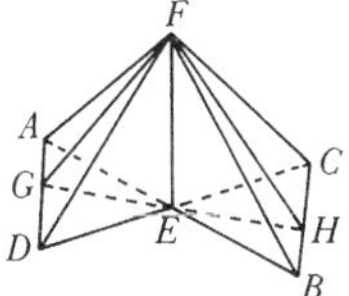

Figure 6.4

straight line in their plane intersecting *AD*, *BC* at *G*, *H* respectively. Successive plane arguments yield the congruence of the triangles *AED* and *BEC*, *AGE* and *BHE*, *AEF* and *BEF*, *DEF* and *CEF*, *ADF* and *BCF*, *FAG* and *FBH*, and *GEF* and *HEF*. The last of these congruences implies the equality of angles *FEG* and *FEH*, and hence the perpendicularity of *FE* to *GH*.

In definition 11 Euclid gives two accounts of the solid angle: first as "the inclination by more than two lines which meet one another and are not in the same surface towards all the lines," and second as "the [angle] which is contained by more than two plane angles which are not in the same plane and are constructed to one point." We have seen that in book I Euclid defines an angle as an inclination between two lines, including curved lines. The definition is mathematically useless, and the ultimate foundation for size comparisons of angles *ABC*, *A′B′C′* is whether *AB* falls inside, coincides with, or falls outside *A′B′* when *BC* is made to coincide in an appropriate way with *B′C′*. Euclid deals only with solid angles enclosed by planes, and his tacit notion of equality for them is equality of the corresponding rectilineal angles. In XI,26, for example, he infers that (fig. 6.5) the solid angle at *A* contained by the rectilineal angles *DAC*, *BAC*, *BAD* is equal to the one at *A′* contained by the angles *D′A′C′*, *B′A′C′*, *B′A′D′* from the equality of *DAC* to *D′A′C′*, of *BAC* to *B′A′C′*, and of *BAD* to *B′A′D′*. The second definition of a solid angle, which appears to presuppose that the reader already knows what sort of thing a solid angle is, furnishes no help here. The first definition may, however, be applicable because it seems to identify the solid angle with the totality of the plane angles formed by the lines making up the solid angle. Unfortunately, Euclid provides no clue to his intended justification of the inference in XI,26.

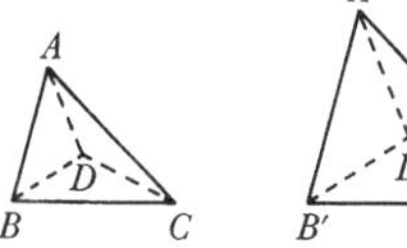

Figure 6.5

The deductive structure of the first part of book XI is depicted in fig. 6.6, in which dependence on definitions and on XI,1–3 and 7 is left out of account.[5] Here a broken line above a number indicates that the corresponding proposition is first used after book XI. What is perhaps most striking about this diagram is the number of terminal points. 17 and 18 appear to have been included with an eye to their later use in XII,4 and 17 respectively. Although 16 is subsequently used in XI,24, it is primarily a lemma for 17. Similarly, 21 is primarily a lemma for 23, although it is used once again at the end of the *Elements* in an argument to show that all possible regular solids have been constructed in book XIII. 15, 19, and 23 and the propositions used only in their proofs, namely 13, 14, 20, and 22, have no further role in the *Elements*; and 15 and 19 appear to have no special intrinsic interest. The first part of book XI is, then, unlike its analogue book I in that its contents are not

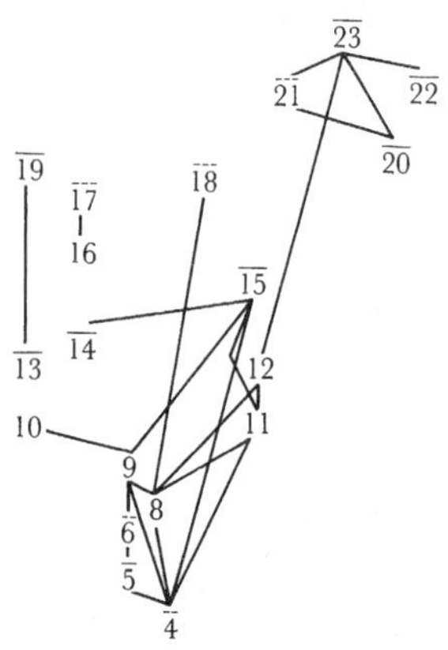

Figure 6.6

more or less completely determined by an immediate goal and subsequent applicability. Only the first twelve propositions of book XI have much subsequent use. I shall content myself with describing those among these first twelve which have not already been discussed, namely:

XI,5 If *AB* is perpendicular to *BC*, *BD*, and *BE*, then *BC*, *BD*, and *BE* are in one plane.

XI,6,8 If *AB* is perpendicular to a plane, then *CD* is perpendicular to the plane if (8) and only if (6) *AB* and *CD* are parallel.

XI,9 Two straight lines parallel to a third not in the same plane with them are parallel to each other.

XI,10 If *AB*, *BC* in one plane are parallel to *A'B'*, *B'C'* in another, angle *ABC* $\simeq$ angle *A'B'C'*.

XI,11 To drop a perpendicular to a plane from a given point *A* outside the plane.

XI,12 To draw a perpendicular to a plane from a given point *B* on the plane.

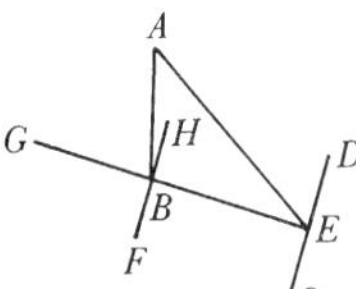

Figure 6.7

Euclid reduces 12 to 11 by "conceiving" (*noein*) a point *C* not on the plane, dropping a perpendicular *CD* to the plane from *C*, and drawing *AB* through *B* and parallel to *CD*. That *AB* solves the problem follows from 8. For 11 Euclid draws a straight line *CED* in the plane and drops *AE* perpendicular to *CD* (fig. 6.7). If *AE* is perpendicular to the plane, the problem is solved. If it is not, Euclid draws *EBG* in the plane perpendicular to *CED*, *AB* perpendicular to *EBG*, and *FBH* parallel to *CED*. The construction is justified by showing that *AB* is perpendicular to *FBH*. But, by XI,4, *CED* is perpendicular to the plane through *GE*, *EA*. Hence, by XI,8, so is *FBH*. But *AB* is also in this plane (XI,2), so that, by definition 3, *AB* and *FBH* are perpendicular.

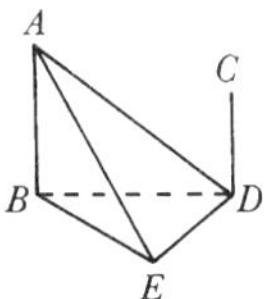

Figure 6.8

For Euclid's proofs of 6 and 8 one assumes that (fig. 6.8) *AB* is perpendicular to the plane of reference at *B* and that *CD* intersects it at *D*.[6] Then, if *DE* in the plane of reference is perpendicular to *BD* and equal to *AB*, successive plane arguments give the congruence of triangles *ABD*, *EDB* and of *ABE*, *EDA*. Hence angle *EDA* is right. But, by XI,2, *AB*, *BD*, *AD* are in one plane. Suppose then that *CD* is perpendicular to the plane of reference; by definition it is perpendicular to *DE*, and, by XI,5, *CD*, *AD*, *DB* are in one plane; thus *AB*, *BD*, *CD* are in one plane, and *AB*, *CD*, being perpendicular to *BD*, are parallel. On the other hand, if *CD* is parallel to *AB*, then *AB*, *CD*, *AD*, *BD* are in one plane (XI,7) and *CD* is perpendicular to *BD*; but since *DE* is perpendicular to *AD*, *BD*, it is perpendicular to *CD*, so that *CD* is perpendicular to both *BD* and *DE* and hence to the plane of reference.

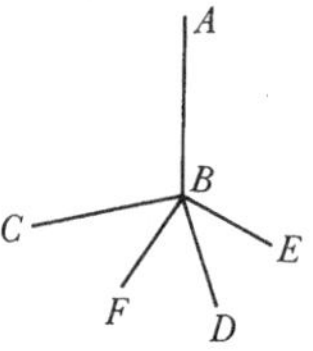

Figure 6.9

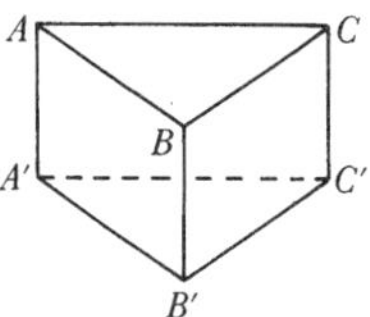

Figure 6.10

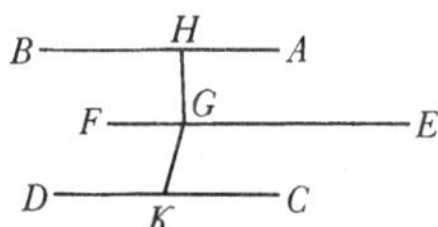

Figure 6.11

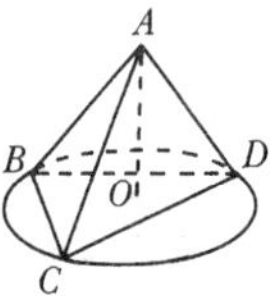

Figure 6.12

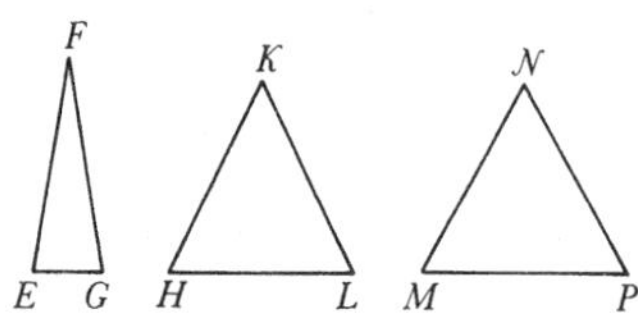

Figure 6.13

The proof of 5 is a simple *reductio*. Euclid supposes that (fig. 6.9) BC does not lie in the plane of reference in which BD, BE lie and that BF is the common section of the plane of reference and the plane of AB, BC. It is now easy to argue, using XI,4 and definition 3, that AB is perpendicular to both BC and BF, which is impossible, since all three are in one plane.

Euclid reduces 10 to 9. He makes (fig. 6.10) AB, BC, $A'B'$, $B'C'$ all equal. I,33 gives that AA' and CC' are each equal and parallel to BB', and hence equal and, by 9, parallel to each other. Therefore, by I,33 again, AC and $A'C'$ are equal; and, by I,8, angles ABC, $A'B'C'$ are equal. For 9 Euclid takes (fig. 6.11) straight lines AHB, DKC to be parallel to FGE in a distinct plane from them and drops GH, GK perpendicular to AHB, DKC. By XI,4, EF is perpendicular to the plane through GH, GK; and, by XI,8, AB and CD are perpendicular to this plane and, hence (XI,6), parallel.

Although the construction of a solid angle out of three plane angles (XI,23) is never employed in the *Elements*, it merits some discussion because of its intrinsic interest. The central idea of the construction may be understood by imagining (fig. 6.12) angles BAC, CAD, BAD to enclose a solid angle at A with $AB \simeq AC \simeq AD$. If AO is drawn perpendicular to the plane of triangle BCD at O, BO, CO, and DO will all be equal because the square on any one of them added to the square on AO is equal to the square on any one of AB, AC, AD. Hence, to construct a solid angle out of angles equal to EFG, HKL, MNP (fig. 6.13), it suffices to set the sides of the angles all equal, to construct a triangle BCD with $BC \simeq EG$, $CD \simeq HL$, and $BD \simeq MP$, to inscribe it in a circle with center O, to erect a perpendicular AO to the plane of the circle with $\mathbf{T}(AO) + \mathbf{T}(BO) \simeq \mathbf{T}(EF)$, and to connect BA, CA, DA. For this construction to be completable two conditions must be satisfied:

(i) It must be possible to form a triangle out of EG, HL, and MP, i.e., any two of them must be greater than the third;
(ii) EF must be greater than OB.

Euclid reduces these two conditions to two others which, it is intuitively clear, any three plane angles making up a solid angle must satisfy:

(i′) any two of the angles must be greater than the third;
(ii′) the three together must be less than four right angles.

The reduction of (i) to (i′) can be formulated as

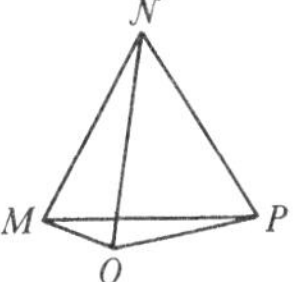

Figure 6.14

XI,22 If any two of the angles EFG, HKL, MNP are together greater than the third and the sides of the angles are all equal, then any two of EG, HL, MP are together greater than the third.

The manuscripts of the *Elements* give two proofs of this proposition. The second, although logically preferable to the first, is put in an appendix by Heiberg. I give the second, changing only its division into cases.[7] Let (figs. 6.13 and 6.14) angle MNP be no smaller than either of EFG, HKL. Clearly, by I,4 and 24, MP is no smaller than either of EG, HL, so that either added to MP is greater than the other. Hence one need only show that $EG + HL \succ MP$, and, because of I,4, it suffices to do the case in which both angles EFG, HKL are less than MNP. If angle MNQ is made equal to EFG, as in fig. 6.14, and NQ is made equal to MN, one has $MQ \simeq EG$. Moreover, by I,20, $MQ + QP \succ MP$. But since the angles MNQ, HKL are together greater than angle MNP, angle $HKL \succ$ angle QNP; or, by I,24, $HL \succ QP$; so that $EG + HL \succ MP$.

It is to be noticed that 22 is a theorem of plane geometry. Its position in the solid-geometry books is undoubtedly to be explained by its being a lemma for 23, but clearly its position is not congruous with Euclid's standard practice of grouping propositions by subject. There are other examples of such incongruity in books X and XIII. One may see in this situation evidence of unreflective following of independent sources, but it is difficult to be certain. Euclid may make some intuitive distinction between preliminary results of some general interest and those of interest only for a particular application, just as he distinguishes between those preliminary results to be proved as separate propositions and those which are to be labeled *lēmmata*, and also between those trivial consequences which are formulated as propositions and those which are called corollaries.

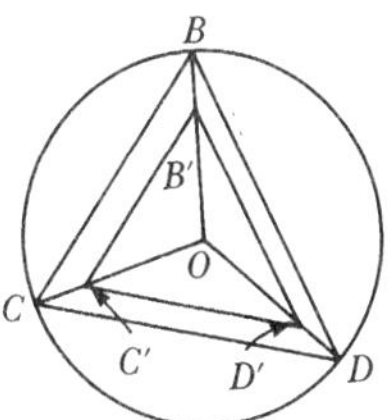

Figure 6.15

Euclid reduces (ii) to (ii′) in the course of proving 23. There are three cases to consider, depending on whether O falls inside triangle BCD, on its perimeter, or outside it. I do the first by *reductio*.[8] If (figs. 6.12, 6.13, and 6.15) $EF \simeq OB$, then, by I,8, angle $EFG \simeq$ angle BOC, angle $HKL \simeq$ angle COD, and angle $MNP \simeq$ angle DOB; but if $EF \prec OB$ and B', C', D' are found on OB, OC, OD so that $EF \simeq OB' \simeq OC' \simeq OD'$, then since $*B'C' \prec BC$, $C'D' \prec CD$, and $D'B' \prec DB$, angle $EFG \succ$ angle BOC, angle $HKL \succ$ angle COD, angle $MNP \succ$ angle DOB, by I,25. In either case the angles EFG, HKL, MNP are together not less than angles BOC, COD, and DOB, i.e., not less than four right angles. To establish the intuitively obvious inequalities marked with an asterisk Euclid makes his first unnecessary application of the theory of proportion. He argues that since $OB \simeq OC$ and $OB' \simeq OC'$, $BB' \simeq CC'$ and BC is parallel to $B'C'$, by VI,2; hence, by I,29, triangles OBC, $O'B'C'$ are equiangular and, by VI,4, $(OB, BC) = (OB', B'C')$ or $(OB, OB') = (BC, B'C')$, and since $OB \succ OB'$, $BC \succ B'C'$.

The construction then is completed for angles satisfying (i′) and (ii′). It remains to show that these conditions are satisfied by any plane angles BAC, CAD, BAD containing a solid angle. Euclid does this for (i′) in 20 and for (ii′) in 21. For 20 Euclid dismisses the trivial case in which the three angles are equal and says, "[If they are not equal] let BAC be the greater." The more precise condition is that angle BAC be greater than angle BAD and no smaller than angle CAD. Euclid makes (fig. 6.16) angle BAE equal to angle BAD and $AE \simeq AD$, and completes the tetrahedron $ABCD$ with E lying on BC. Clearly triangles BAD, BAE are congruent. But also $BE + DC \simeq BD + DC \succ BC$, so that $DC \succ EC$, and, by I,25, angle $CAD \succ$ angle EAC. Hence, angle CAD + angle $BAD \simeq$ angle CAD + angle $BAE \succ$ angle EAC + angle $BAE \simeq$ angle BAC.

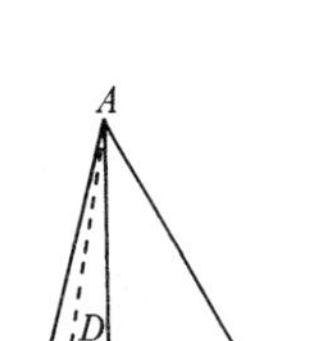

Figure 6.16

For 21 Euclid uses the fact that a tetrahedron contains four solid angles. Hence, by XI,20 (see fig. 6.16), the angle sum $(ABD + ABC) + (ADC + ADB) + (DCA + BCA)$, i.e., $(ABC + BCA) + (ABD + ADB) + (DCA + ADC)$, is greater than the sum of the angles CBD, BDC, DCB, which is equal to two right angles. But then the angles BAC, BAD, CAD together must be less than four right angles because, by I,32, the result of adding them to $(ABC + BCA) + (ABD + ADB) + (DCA) + ADC)$ is equal to six right angles. Although Euclid proves 21 only for solid angles contained by three plane angles, he states it for any number of plane angles and uses this more general form at the end of XIII,18. Heath (vol. III, pp. 310–311) indicates how this more general result is established.

6.2 Solids and Their Volumes

For the discussion of the second part of book XI and, to a lesser extent, of book XII it is useful to recall Euclid's treatment of the areas of triangles, parallelograms, polygons, and circles. The fundamental materials having to do directly with these figures are:

A1 the basic equality theorems for triangles (I,4,8,26),
A2 the definition of equal circles (III, def. 1),
A3 the definition of similar segments of circles (III, def. 11),
A4 the definition of similar polygons (VI, def. 1),
B1 the fact that a diagonal of a parallelogram divides it into two congruent and hence equal triangles, (I,34),

B2 the fact that any polygon is divisible into triangles and "similar polygons are divided into similar triangles and into triangles equal in multitude and in the same ratio as the wholes."

To give a general representation of the propositions Euclid derives from these fundamentals, I let f_i be a triangle or parallelogram with base b_i, and height h_i, c_i be a circle with diameter d_i, and p_i, p_i' be similar rectilineal figures with corresponding sides s_i, s_i'. Euclid could prove all of the following propositions by means of easy transformations of propositions he does prove:

C if $b_1 \simeq b_2$ and $h_1 \simeq h_2$, then $f_1 \simeq f_2$;
D if $h_1 \simeq h_2$, then $(f_1, f_2) = (b_1, b_2)$;
E if $b_1 \simeq b_2$, then $(f_1, f_2) = (h_1, h_2)$;
F $f_1 \simeq f_2$ if and only if $(b_1, b_2) = (h_2, h_1)$;
G p_1 is to p_1' in the duplicate of the ratio of s_1 to s_1';
H $(p_1, p_1') = (p_2, p_2')$ if and only if $(s_1, s_1') = (s_2, s_2')$;
J the ratio of f_1 to f_2 is compounded of the ratios of b_1 to b_2 and of h_1 to h_2;
K if p_1, p_1' are inscribed in c_1, c_1', then $(p_1, p_1') = (\mathbf{T}(d_1), \mathbf{T}(d_1'))$;
L $(c_1, c_2) = (\mathbf{T}(d_1), \mathbf{T}(d_2))$.

In fact, however, Euclid proves only **G** (VI,19–20), **H** (VI,22), **K** (XII,1), and **L** (XII,2) in the form stated here. **C** is restricted to triangles and parallelograms whose heights are contained by the same parallel lines, and proved for identical as well as equal bases (I,35–38). **D** is restricted to identical heights (VI,1), **E** not proved at all. **F** (VI,14–17) and **J** (VI,23) are proved for equiangular parallelograms or triangles with a pair of equal angles, b_i and h_i being sides containing an equal angle. If these divergences are borne in mind, the broad deductive structure of these propositions in the *Elements* can be seen to be as represented in fig. 6.17.

In XI,24–40 and XII,3–18 Euclid is concerned with parallelepipeds, pyramids, (right) cones, (right) cylinders, and spheres. In the course of his argumentation he also establishes

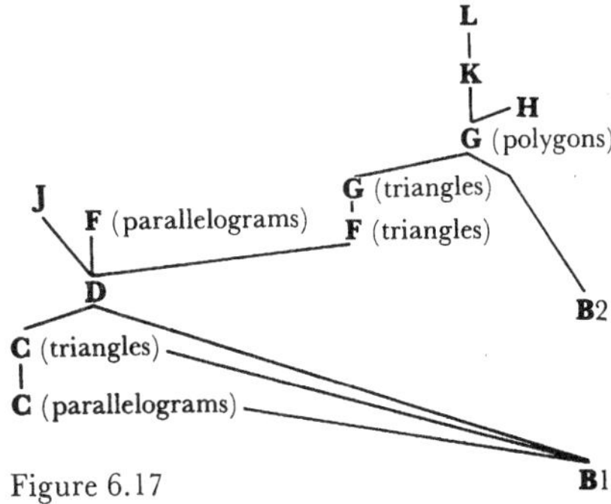

Figure 6.17

some facts about prisms. One of the goals of this section is to compare Euclid's treatment of plane and solid geometry. One fundamental difference is the dependence of his treatment of equal solid figures upon the notion of similar solid figures. He defines

XI, def. 9 Similar solid figures are those contained by similar planes equal in multitude.

XI, def. 10 Equal and similar solid figures are those contained by similar planes equal in multitude and in magnitude.

In these definitions, which correspond to **A** materials in his treatment of plane figures, Euclid reiterates two conceptual shortcomings found in his definition of similar polygons: he tacitly assumes a restriction to convex polyhedra; and he fails to consider the ordering of the plane faces.[9] If allowance is made for these shortcomings, 9 can be interpreted formally as a definition, but 10 cannot be. 10 states a condition under which similar solid figures are equal in the ordinary sense, i.e., equal in volume. Hence Euclid is able to infer that, since a plane through parallel diagonals of opposite faces of a parallelepiped divides the parallelepiped into two equal and similar solids, the plane bisects the solid (XI,28), or that, if the same solid is added to two equal and similar solids, the result is two equal solids (XI,29). Since in the second of these cases the resulting equal solids are not similar, Euclid clearly does not think of definition 10 as specifying the sense of the word 'equal' for solids and therefore does not think of it as a definition in the modern sense. Definition 10 is rather like the basic equality theorems for triangles (**A**1), which are used to establish further equalities for both triangles and parallelograms as well as proportionality relations for certain figures. On the other hand, it is an at least initial source of puzzlement that Euclid makes definition 10 depend upon the theory of proportion when he need not. For he could have characterized equal and similar solid figures as those contained by congruent plane figures equal in multitude, where congruent plane figures are those which "have their angles severally equal and the sides about the equal angles equal." (Compare VI, def. 1.)

Parallelepipeds are the solid analogues of parallelograms, as is made clear in the most significant application of definition 10: the proof of the theorem giving the fundamental relationship between parallelepipeds and triangular prisms, the (partial) analogues of triangles:

XI,28 A plane through (parallel) diagonals of opposite faces of a parallelepiped bisects the parallelepiped.

Together with

XI,24 The opposite faces of parallelepipeds are equal (and similar) parallelograms,

from which it is derived, XI,28 is an analogue of I,34. In the proof of 24 Euclid uses the obvious fact that the sections made by a plane cutting two parallel planes are parallel straight lines (XI,16; if the sections met, the planes would also meet); clearly, then, the opposite faces of a parallelepiped are parallelograms. Euclid proceeds by taking two opposite faces $ABCD$, $A'B'C'D'$, arranged as in fig. 6.18, and connecting BD, $B'D'$. By I,34 one has $AB \simeq A'B'$ and $AD \simeq A'D'$. XI,10 yields the equality of the angles BAD, $B'A'D'$. Euclid could now argue that the two parallelograms $ABCD$ and $A'B'C'D'$ are similar, since they are equiangular and the sides about the equal angles are proportional, and then infer that the two parallelograms are equal, by VI,14. Instead he uses I,4 to infer the equality of the two triangles ABC, $A'B'D'$ and then the equality of their doubles, the parallelograms, without bothering to prove their similarity.

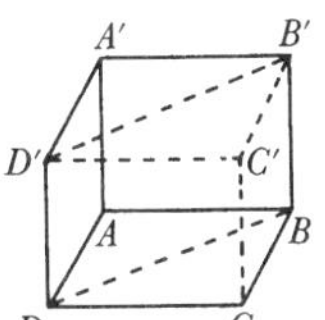

Figure 6.18

There is some significance in the fact that Euclid states and proves 24 without reference to the similarity of the opposite faces. For Euclid's applications of 24 are in conjunction with definition 10 or, on one occasion, definition 9; and therefore presuppose the similarity as well as the equality of the opposite faces. Yet in three of the six applications of XI,24 (25, 28, 29) he does not bother to refer to similarity, whereas in the other three (27, 31, 33) he explicitly includes it. For example, in the proof of 28, having inferred by XI,24 and I,34 that the prisms determined by the plane through DB and $D'B'$ (fig. 6.18) are contained by pairwise equal faces, Euclid infers that the prisms are equal with the words, "for they are contained by planes equal both in multitude and in magnitude." Since it does not seem possible to attribute the unsatisfactoriness of Euclid's procedure here to carelessness or ignorance of the distinction between equality and equality plus similarity, it is natural to ask what alternative procedure he might have adopted. One alternative not open to him in the case of XI,28 is superposition, because the two prisms are "symmetric": they can be maneuvered so that one is a mirror image of the other, but there is no way to make one coincide with the other. However, if a distinction is made between symmetric and congruent solids, it is possible to prove the equality of congruent prisms by superposition and to reduce the equality of symmetric prisms to that of congruent ones, without invoking definition 10 in any form. (See Heath, vol. III, pp. 331–333.)

It seems reasonable to suppose that Euclid had some sense of the distinction between symmetry and congruence.[10] The question is why he did not attempt to formulate it. The obvious way to do so is in terms of direction or perspective. To make a similar distinction with respect to solid angles Heath (vol. III, p. 312) suggests that "we may suppose ourselves to be placed at the vertices and to take the faces in clockwise direction, or the reverse, for *both* angles." However, it has already been made clear that Greek mathematicians tend to leave positional and directional considerations to intuition. But, although the distinction between congruence and symmetry is intuitively clear, it would not seem feasible to leave the distinction at the intuitive level if it were going to serve as the foundation of a proof. Perhaps, then, Euclid chose to obviate the need for making the distinction by "defining" similar and equal solid figures to include both congruent and symmetric ones. One cannot know whether the choice is based on a belief that the distinction would be impossible to make or on a belief that it would be overly complicated. Nor is it possible to explain the inconsistent way in which Euclid sometimes invokes and sometimes ignores the requirement in definition 10 that the corresponding faces of equal and similar solids be similar.

In the cases other than proposition 28, in which Euclid proves the equality of two solids by using definition 10 (XI,25 and 29 and XII,3), the solids in question are congruent so that Euclid could use superposition. 28 itself is not used until the proof of 39. There and in book XII it is used in much the same way as **B**1. The simplest way to state **B**1 and **B**2 for parallelepipeds and prisms with triangular bases is to take for granted that a parallelepiped is determined by any two contiguous faces:

B1p ('p' for 'parallelepiped') Given a prism with triangular base ABC, the parallelepiped determined by the parallelogram determined by AB, BC and the parallelogrammic face of the prism with side AB is twice the original prism and has a base twice as big.

B2p Given two similar triangular prisms with bases ABC, $A'B'C'$ and AB corresponding to $A'B'$, the parallelpipeds determined as in **B**1p are similar.

In book XI Euclid proves analogues of **C**–**J** only for parallelepipeds; but of course, **B**1p and **B**2p make it easy to extend the results to prisms with triangular bases. To understand the analogues one need only read **C**–**J** taking f_i to be a parallelepiped with base b_i and height h_i; p_i, p_i' to be similar parallelepipeds with corresponding sides s_i, s_i'; and substituting 'triplicate' for 'duplicate'. The resulting assertions correspond to propositions in the *Elements* as follows: **C**p, XI,29–31; **D**p, XI,25, 32; **F**p, XI,34–36; **G**p, XI,33; **H**p, XI,37. **E** and **J** are not represented in XI. **E** is not included in the plane books either, and is in fact proved only for cylinders and cones. **J** is the powerful result using the compounding of ratios. Euclid makes no use of its power in either plane geometry or arithmetic, and does not even use the notion of compounding in the solid books. His failure to do so is perhaps the strongest evidence that he does not construe compounding ratios as multiplication.

The only propositions from the second part of book XI which have not been accounted for are 26, 27, and 38–40.[11] The last three are clearly lemmas for XII,17, XIII,17, and XII,4. 26 is the construction, on a straight line and at a point on it, of a solid angle equal to a given one. Euclid does the case of a given solid angle contained by three plane angles *BAC*, *BAD*, *CAD* (fig. 6.19). He makes *AB* equal to the given line *A′B′* and drops *CE* perpendicular to the plane of *AD*, *DB* at *E*. On *A′B′* and in the plane of reference he makes angle *B′A′D′* equal to angle *BAD* with *A′D′* equal to *AD*, angle *B′A′E′* equal to angle *BAE*, and *A′E′* equal to *AE*. At *E′* he erects *C′E′* perpendicular to the plane of reference and equal to *CE*. Congruence arguments show that the angles *B′A′C′*, *C′A′D′* are equal to the angles *BAC*, *CAD* respectively, so that the problem is solved. This construction, which seems more appropriate to the first part of book XI, is probably postponed to this point because it is first applied in 27, the construction on a given straight line of a parallelepiped similar and similarly situated to a given one. The latter is a simple construction and, like its analogue, VI,18, is used in the proof of **H**.

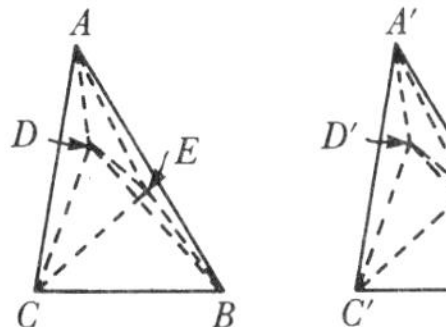

Figure 6.19

Before discussing Euclid's handling of parallelepipeds I would like to indicate the analogues of **C**–**J** which are proved by Euclid for pyramids and for cones and cylinders. For the first, one takes f_i to be a pyramid with base b_i and height h_i; p_i, p_i' to be similar pyramids with corresponding sides s_i, s_i'; and reads 'duplicate' as 'triplicate'. Then Euclid proves **D**py (XII,5, 6), **F**py (XII,9), and **G**py (XII,8), but the latter two only for pyramids with triangular bases. For cones and cylinders, one takes f_i to be a (right) cone or cylinder with circular base b_i and height h_i, and p_i, p_i' to be similar cones or cylinders whose bases have diameters s_i, s_i'. (For the notion of similarity involved, see XI, def. 24.) Again 'duplicate' is to be read as 'triplicate'. Euclid proves **D**c (XII,11), **E**c (XII,13, 14), **F**c (XII,15), and **G**c (XII,12). Of the other propositions in book XII, 3, 4, 7, 10 are lemmas for these results; 18 is the analogue of **L** for spheres, for which 16 and 17 are preliminary constructions. It appears that for Euclid the crucial propositions with regard to figures contained by straight lines or planes are **F** and **G**, **F** being the strongest formulation of the condition of equality for figures of the same kind, **G** being the expression of the ratio between two similar figures in terms of the ratio between corresponding straight lines in each figure. **F** and **G**, and therefore **C**–**E** and **H**, are all proportion-theoretic consequences of **J**, so that Euclid's failure to prove **J** for solids makes his argumentation much more complicated than it needs to be.

In order to investigate the deductive structure of the second part of book XI, it is necessary to separate out the components of **C**p, **D**p, and **F**p. In XI,29 and 30, the analogues of I,35, b_1 and b_2 are taken to be identical. In 29 Euclid considers the case in which the vertices of f_1 and f_2 not on the base lie on two straight lines. The proof is exactly like the proof of I,35, although Euclid does a different case. To prove (fig. 6.20) the equality of f_1, with base $ABCD$ and opposite face $A_1B_1C_1D_1$, to f_2, with the same base and opposite face $A_2B_2C_2D_2$, Euclid proves the equality of the prism p_1, with parallel triangular faces BB_1B_2, CC_1C_2, to prism p_2, with parallel triangular faces AA_1A_2, DD_1D_2; he uses XI,24 and plane arguments to be able to apply def. 10, but again omits to mention the pairwise similarity of the faces. Since f_2 results from f_1 by adding p_1 and subtracting p_2, the desired equality follows. In XI,30, to deal with the case in which the vertices A_1, A_2, B_1, B_2 and D_1, D_2, C_1, C_2 do not lie on straight lines, Euclid constructs (fig. 6.21) a third parallelepiped f_3 on the same base and with opposite face $A_3B_3C_3D_3$, which satisfies the conditions of 29 relative to each of f_1 and f_2.

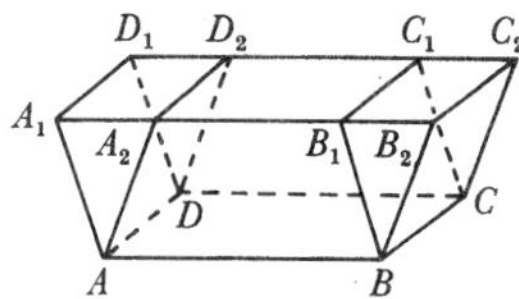

Figure 6.20

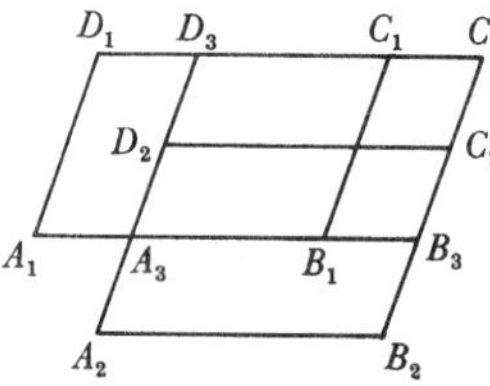

Figure 6.21

Although 29 and 30, like 28, do not require any concepts from the theory of proportion for their formulation, they do depend on this theory in the *Elements* because they depend on the definition of equal and similar figures. Clearly Euclid could have avoided this dependence by proving equality theorems for triangular prisms analogous to **A**1. That he does not do so is another indication that the concern to avoid the theory of proportion, which is so prominent in books I–IV, is dropped in the treatment of solid geometry. One reason for this change may be that XI,31, the analogue of I,36, is not provable by a direct extension of the method of proof of I,36. Euclid reduces I,36 to I,35 by constructing for two parallelograms p_1, p_2, with equal bases and in the same parallels, a third one p satisfying the conditions of I,35 with respect to each of p_1, p_2 (fig. 6.22). XI,31 cannot be reduced to 29 and 30 in the same way because equal bases of parallelepipeds need not be congruent. In the course of his proof of 31 Euclid establishes what amount to two corollaries of 29 and 30; these enable him to transform a parallelepiped with base $ABCD$ and height h into an equal parallelepiped with the same height, with four faces perpendi-

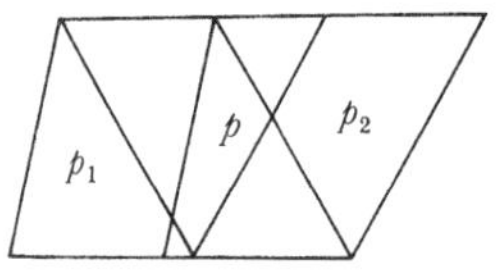

Figure 6.22

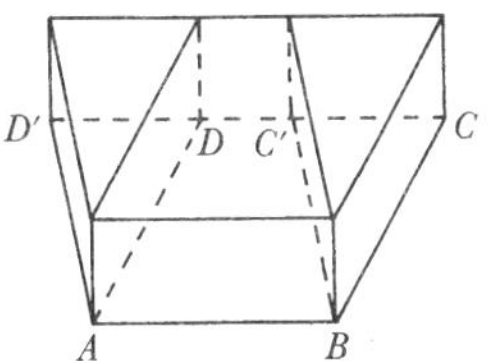

Figure 6.23

cular to its base, and with base $ABC'D'$ equal to $ABCD$ and equiangular with a given parallelogram (see fig. 6.23).

Because of the possibility of this transformation it suffices to prove 31 for parallelepipeds with equal height, equal and equiangular bases, and with four faces perpendicular to their bases. The strategy of the proof is now almost exactly like that of the proof of VI,14. Let the parallelepipeds f_1, f_2 satisfy these conditions, and let them be positioned as in fig. 6.24 on the bases $A_1B_1C_1D$, $A_2B_2C_2D$ and sharing the edge DE equal to their height.[12] One completes the parallelogram $A_1A_3A_2D$ determined by A_1D and A_2D and then the parallelepiped f with this parallelogram as base and DE as an edge. Euclid now invokes

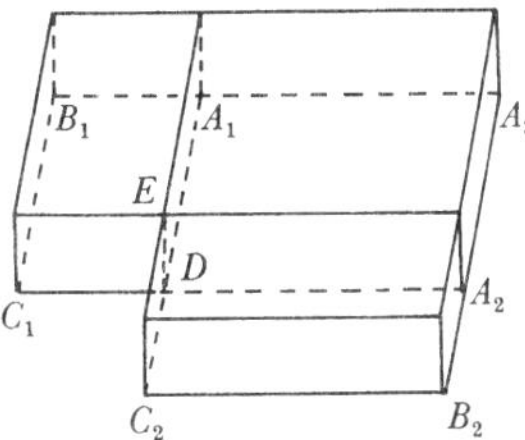

Figure 6.24

XI,25 If a parallelepiped with base $F_1F_2G_2G_1$ and parallel faces $F_1G_1H_1K_1$, $F_2G_2H_2K_2$ (fig. 6.25) is cut by a plane $FGHK$ parallel to the two faces and dividing it into two parallelepipeds p_1, p_2 with bases FF_1G_1G, FF_2G_2G, then $(p_1, p_2) = (FF_1G_1G, FF_2G_2G)$.

In particular then $(f_1, f) = (A_1B_1C_1D, A_1A_3A_2D)$ and $(f_2, f) = (A_2B_2C_2D, A_1A_3A_2D)$; and, since $A_1B_1C_1D \simeq A_2B_2C_2D$, $f_1 \simeq f_2$.

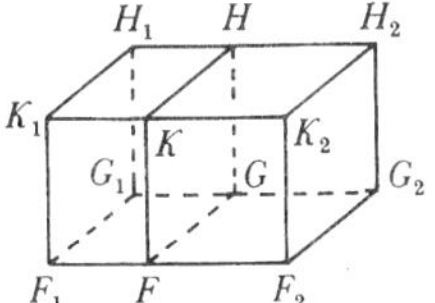

Figure 6.25

The proof of 25 is quite like the proof of VI,1. The two would be even more alike if Euclid had proved VI,1 directly for parallelograms rather than proving it for triangles and extending it to parallelograms by means of I,34 or 41. To prove XI,24 Euclid constructs parallelepipeds $p_1', p_1'', p_1''', \ldots, p_2', p_2'', p_2''', \ldots$, with bases $b_1', b_1'', b_1''', \ldots, b_2', b_2'', b_2''', \ldots$, as in fig. 6.26, in such a way that he can prove the equality of the bases b_1 and b_2 of p_1 and p_2 to each of $b_1', b_1'', b_1''', \ldots$ and $b_2', b_2'', b_2''', \ldots$, respectively; thus the bases $b_1^* (\simeq b_1' + b_1'' + b_1''' + \ldots)$ and $b_2^* (\simeq b_2' + b_2'' + b_2''' + \ldots)$ of the parallelepipeds $p_1^* (\simeq p_1' + p_1'' + p_1''' + \ldots)$ and $p_2^* (\simeq p_2' + p_2'' + p_2''' + \ldots)$ are the same multiples of b_1 and b_2 that p_1^* and p_2^* are of p_1 and p_2. Euclid completes the proof by invoking the definition of proportionality and assuming

(i) if $b_1^* \simeq b_2^*$, then $p_1^* \simeq p_2^*$;
(ii) if $b_1^* \prec b_2^*$, then $p_1^* \prec p_2^*$.

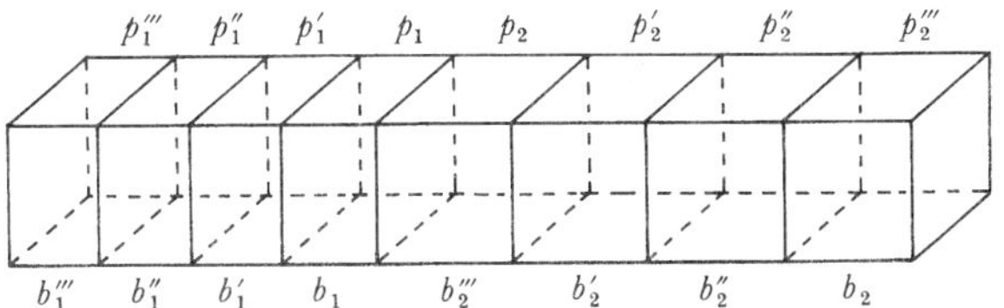

Figure 6.26

(ii) corresponds to VIA, (i) to I,36; VIA and I,36 are both used in the proof of VI,1.[13] (ii) is easily reduced to (i). Euclid could prove (i) by showing that when $b_1^* \simeq b_2^*$, then p_1^* and p_2^* are contained by equally many equal and similar parallelograms. His failure to prove this obscures a fundamental similarity between his treatment of parallelepipeds and parallelograms, even more than does his proving of 25 well before it is needed in 31. Using 31 and 25, Euclid is able to establish, in 32, **D**p in its general form. The proof consists in applying 31 to transform f_1 into an equal parallelepiped f_3 with base equal to b_1, but such that 25 can be applied to f_2 and f_3.

However, both 31 and 32 are trivial consequences of **J**p, which can be proved without them. Suppose one has two equiangular parallelepipeds f_1 and f_2 with sides s_1, s_1', s_1'' and s_2, s_2', s_2'' positioned as in fig. 6.27. It is easy enough to construct two parallelepipeds f' and f'' with sides s_1, s_2', s_1'' and s_2, s_2', s_1''. Application of 25 and VI,1 yields that $(f_1, f') = (s_1', s_2')$, $(f'', f_2) = (s_1'', s_2'')$, and $(f', f'') = (s_1, s_2)$. Given the obvious definition of a ratio being compounded of three ratios, one can conclude

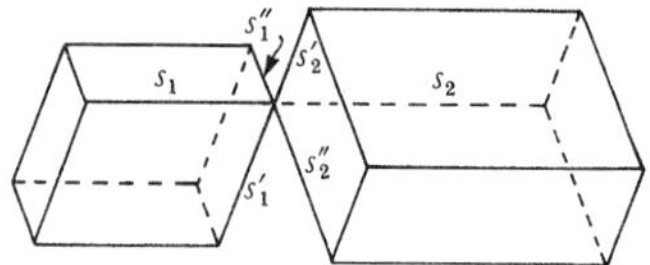

Figure 6.27

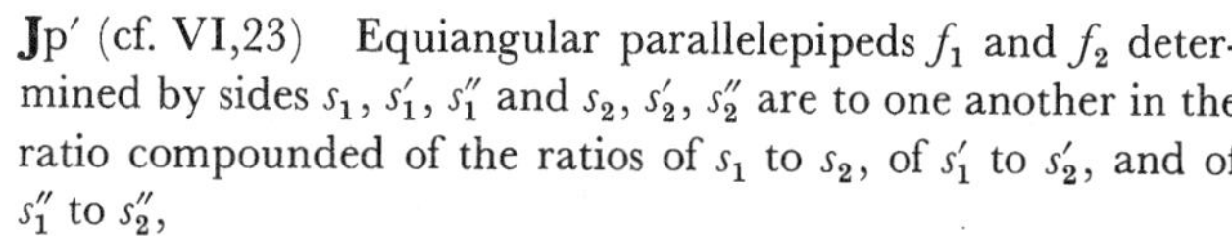
Jp′ (cf. VI,23) Equiangular parallelepipeds f_1 and f_2 determined by sides s_1, s_1', s_1'' and s_2, s_2', s_2'' are to one another in the ratio compounded of the ratios of s_1 to s_2, of s_1' to s_2', and of s_1'' to s_2'',

and also

XI,33 (**G**p) Similar parallelepipeds are to one another in the triplicate ratio of their corresponding sides,

of which 37 (**H**p) is a direct consequence, using VH. Euclid's proof of 33 follows the reasoning given here, although he does not, of course, prove the stronger **J**p′.

Suppose, now, that one calls the parallelograms contained by s_1, s_1'' and by s_2, s_2'' the bases b_1 and b_2 of f_1 and f_2; and s_1' and s_2' the third edges of f_1 and f_2. From the proportionalities established in the proof of **J**p′, one has that f' is to f_2 in the ratio compounded of the ratios of s_1 to s_2 and of s_1'' to s_2''. But, by VI,23, b_1 is to b_2 in the ratio compounded in the same way. Hence,

Jp″ (cf. VI,23) Equiangular parallelepipeds are to one another in the ratio compounded of the ratios of their bases and of their third edges.

Given any two parallelepipeds f_1, f_2 with bases b_1, b_2 and heights h_1, h_2, the techniques of Euclid's proof of 31 can be used to transform f_1 into an equal parallelepiped f with base b equal to b_1 and equiangular with b_2, and then f, f_2 into equal

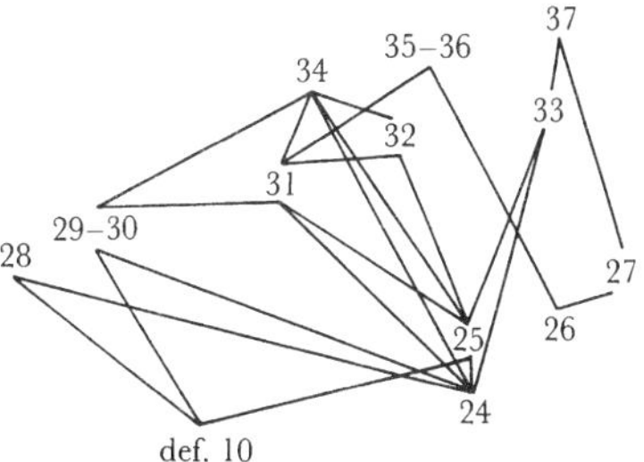

Figure 6.28

parallelepipeds f', f'_2 with the same bases and heights, but each with four faces perpendicular to its base. Since f' and f'_2 are equiangular, and their third edges are their heights, one has a proof of **Jp**.

Among the direct consequences of **Jp** are XI,31 (**C**p), 32 (**D**p), and 34 (**F**p). **Jp**″ simplifies the proof of the unused

XI,36 (cf. VI,14) If s_1, s_2, s_3 are three straight lines and $(s_1, s_2) = (s_2, s_3)$, then if f_1, determined by three sides equal to s_1, s_2, s_3, and f_2, determined by three sides equal to s_2, are equiangular parallelepipeds, $f_1 \simeq f_2$.

For, if f_1 and f_2 are equiangular, then a base b_1 of f_1 contained by sides equal to s_1 and s_3 is equiangular with a base b_2 of f_2 contained by sides equal to s_2. Since $(s_1, s_2) = (s_2, s_3)$, $b_1 \simeq b_2$, by VI,14. But, by **Jp**″, f_1 is to f_2 in the ratio compounded of the ratios of b_1 to b_2 and of s_2 to s_2, i.e., $f_1 \simeq f_2$. Since Euclid has only XI,31 or 34 at his disposal for the proof of 36, he is forced to prove an elaborate but elementary lemma, XI,35 and corollary, establishing that f_1 and f_2 have the same height. He also has to give an independent argument for **F**p (XI,34). He uses the same kind of construction as in 31 to reduce the general case to the one in which the heights of f_1 and f_2 are edges. He divides the proof into two subcases, depending on whether or not b_1 and b_2 are equal. If they are equal, then, by XI,32, $h_1 \simeq h_2$ if and only if $f_1 \simeq f_2$; hence, since $b_1 \simeq b_2$, $f_1 \simeq f_2$ if and only if $(b_1, b_2) = (h_2, h_1)$. On the other hand, if $b_1 \succ b_2$ and f is a parallelepiped with base b_2 and height h_1, then $(f_2, f) = (h_2, h_1)$, and, by XI,32, $(f_1, f) = (b_1, b_2)$; hence, $f_1 \simeq f_2$ if and only if $(b_1, b_2) = (h_2, h_1)$.[14]

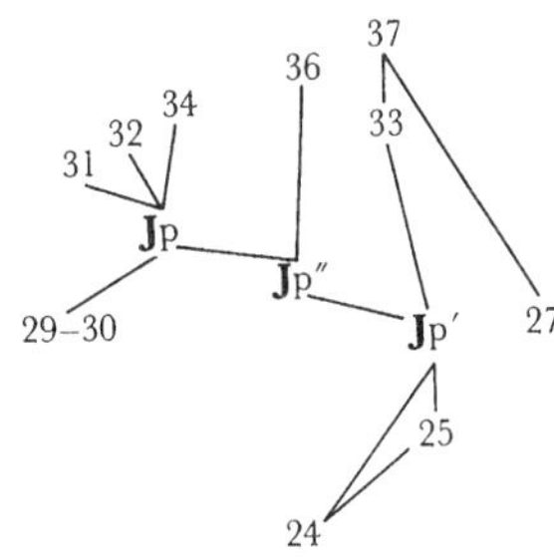

Figure 6.29

Figure 6.28 indicates the deductive structure of the last half of book XI. Figure 6.29 shows the simpler deductive structure made possible for 31–37 if the power of the concept of compounding ratios is exploited. In addition, the proofs of 31–36 corresponding to fig. 6.29 are straightforward applications of the theory of proportion as opposed to Euclid's quite complicated geometric arguments. Neither structure eliminates the use of the theory of proportion in the proof of 31 (**C**p), although it would be possible to eliminate it by complicated but essentially Euclidean arguments.[15] Similarly, the use of the theory involved in the applications of definition 10 can be eliminated in favor of complicated arguments within Euclid's reach. The general situation might perhaps be summarized by saying that Euclid will use the theory of proportion in an elementary way to avoid a complex solid-geometrical argument, but he will not use it in an abstract computational way to substitute for geometrical argument.

28 is not used in the material I have been describing, but, as already indicated, it and 24 amount to **B**1p and **B**2p and can be used to derive XI,29–34, **J**p, **J**p′, and **J**p″ for triangular prisms, results which I shall refer to by adding the letters 'pr'. Some of these results are used in book XII. For example, XII,8 and 9, the fundamental propositions **G** and **F** stated for pyramids with triangular bases (tpy), follow from 33pr, 34pr, and

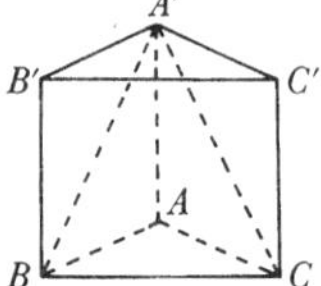

Figure 6.30

XII,7 (**B**1tpy) If (fig. 6.30) *ABCA′* is a pyramid with a triangular base *ABC*, and at the vertex *A′* in the plane parallel to that of *ABC* one constructs a triangle *A′B′C′* equal and similar to *ABC* by drawing *A′B′*, *B′C′*, *A′C′* parallel and equal to *AB*, *BC*, *AC* and connects *BB′*, *CC′*, the result is a triangular prism of which the pyramid is one-third.

B2tpy (proved in the course of the proof of XII,8) If one carries out the construction of **B**1tpy for two similar triangular pyramids, the result is two similar triangular prisms.

To prove XII,8 and 9 Euclid takes triangular pyramids p_1, p_2 with bases b_1, b_2 and heights h_1, h_2, and carries out the construction to determine two triangular prisms f_1, f_2 with bases b_1, b_2 and heights h_1, h_2. By **B**1tpy the pyramids are equal if and only if the prisms are; but, by 34pr, the prisms are equal if and only if their bases and heights are reciprocally proportional. In the same way, by **B**1tpy, $(p_1, p_2) = (f_1, f_2)$. But if p_1 and p_2 are similar with s_1 and s_2 corresponding sides of their bases, by **B**2tpy and 33pr, p_1 is to p_2 in the triplicate of the ratio of s_1 to s_2. Obviously all of 29–34, **J**p, **J**p′, **J**p″ can be extended in this way to pyramids with triangular bases.

B1tpy is itself a consequence of **C**tpy. Imagine the construction of **B**1tpy carried out and *B′C* connected (fig. 6.31). The prism is then divided into three pyramids *ABCA′*, *A′B′C′C*, and *BB′CA′*. **C**tpy enables one to infer the equality of these three pyramids. $ABCA' \simeq A'B'C'C$ because their bases *ABC*, *A′B′C′* are equal and their height is the distance between the two bases; and $A'B'C'C \simeq BB'CA'$ because their bases *B′C′C*, *BB′C* are equal by I,34; and their heights are equal since their

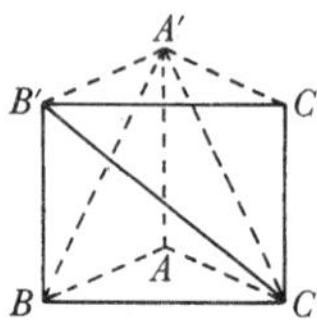

Figure 6.31

bases are in the same plane and they have the common vertex A'. Since, then, the three pyramids are equal, each is one-third of the constructed prism.

If should be clear that the three pyramids of the preceding argument are neither congruent nor symmetric and that there is no way to correlate their faces as pairwise similar planes equal in magnitude. In fact, neither **B**1tpy nor **C**tpy is provable by elementary geometric arguments.[16] Euclid invokes **C**tpy as an obvious proportion-theoretic consequence of **D**tpy in the course of proving **B**1tpy. **D**tpy itself he proves as XII,5, using the method of exhaustion, which I will be discussing in the following section. Hence the overall structure of Euclid's treatment of pyramids with triangular bases is as indicated in fig. 6.32, in which an asterisk indicates a proof by the method of exhaustion. As will become clear in the following section, the proof of XII,5 could easily be transformed into a proof of **J**tpy, of which **D**tpy, **F**tpy, and **G**tpy are all direct consequences. XII,3 and 4 are lemmas for XII,5; and XII,6 is an extension of 5 to pyramids with polygonal bases. To prove 6 Euclid does the case in which both pyramids have pentagonal bases, although nothing in the argument turns on the bases having the same number of sides. He lets (fig. 6.33) *ABCDE*, *FGHKL* be bases of two pyramids of the same height with vertices *M*, *N*, and connects *AC*, *AD*, *FH*, *FK*. He then argues as follows. By XII,5 one has

(i) $(ABCM, ACDM) = (ABC, ADC)$,
(ii) $(ACDM, ADEM) = (ADC, ADE)$,
(iii) $(ADEM, FGHN) = (ADE, FGH)$.

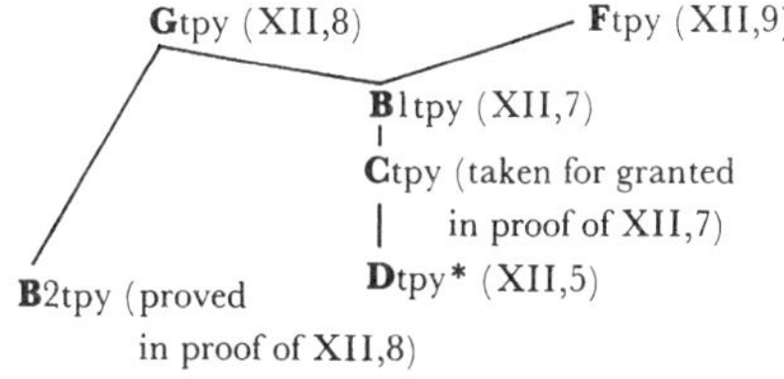

Figure 6.32

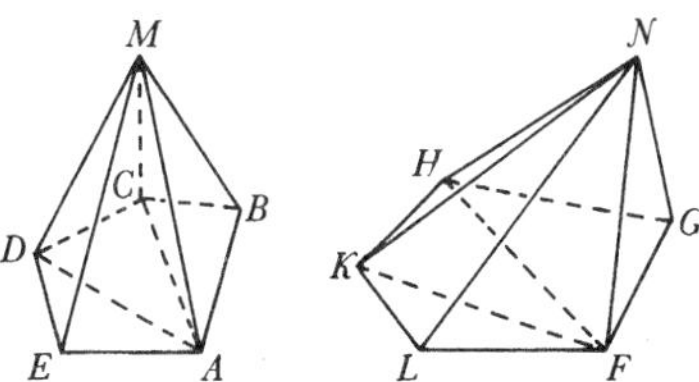

Figure 6.33

By V,18 it follows from (i) that $(ABCM + ACDM, ACDM) = (ABC + ADC, ADC)$, so that ((ii) and V,22) $(ABCM + ACDM, ADEM) = (ABC + ADC, ADE)$ and (V,18 again) $(ABCM + ACDM + ADEM, ADEM) = (ABC + ADC + ADE, ADE)$, i.e.,

(iv) $(ABCDEM, ADEM) = (ABCDE, ADE)$.

The same reasoning shows that $(FGHKLN, FGHN) = (FGHKL, FGH)$, or

(v) $(FGHN, FGHKLN) = (FGH, FGHKL)$.

(iii)–(v) and V,22 yield the desired result:

$(ABCDEM, FGHKLN) = (ABCDE, FGHKL)$.

XII,7–9 can also be extended to pyramids with polygonal bases; but Euclid appears to make the extension only in the case of 7, with a corollary which runs, "From this it is manifest that every pyramid is a third part of the prism having the same base as it and an equal height." To describe a proof of this corollary, I shall speak of the division of a pyramid p with polygonal base b $(A_1A_2 \ldots A_n)$ into partial pyramids $p_1, p_2, \ldots, p_{n-2}$ at A_1 to refer to the division of p into triangular pyramids $p_1, p_2, \ldots, p_{n-2}$ with bases $b_1, b_2, \ldots, b_{n-2}$ $(A_1A_2A_3, A_1A_3A_4, \ldots, A_1A_{n-1}A_n)$, which results from connecting A_1 with all the vertices of b. The simplest proof of the corollary to XII,7 involves constructing for a given pyramid a prism on the same base and with the face opposite the base having a vertex of the pyramid as one of its vertices, and dividing the pyramid into partial pyramids at the point in the base corresponding to the vertex of the pyramid in the opposite face of the prism.

In the margin of P and in most other manuscripts there is a corollary extending XII,8 to arbitrary pyramids. The proof of this corollary takes for granted that if similar pyramids are divided into partial pyramids at corresponding vertices, the corresponding partial pyramids are similar. Although the corollary is quoted in a problematic remark at the end of XII,17, its position in the margin of P perhaps justifies treating it as a later insertion. The total absence of an extension of XII,9 to arbitrary polygons is perhaps to be explained by the fact that Euclid has no use for such a result; but it is also of some interest to note that there is no direct extension of XII,9 on the basis of what is proved in the *Elements*. For there is no necessary relation between the equality of two pyramids and the equality of any of their partial pyramids. Perhaps the simplest way to prove XII,9 for pyramids with polygonal bases is as a corollary to **J**py. In the following section I show how Euclid could have proved **J**tpy; here I indicate how **J**py is derived from **J**tpy, using the commutativity of compounding. Suppose p, p' are pyramids with heights h, h' and polygonal bases b, b'; and let them be divided into partial pyramids including p_1, p'_1 with triangular bases b_1, b'_1. One has, as in the proof of XII,6, (i) $(p, p_1) = (b, b_1)$ and (ii) $(p', p'_1) = (b', b'_1)$. If, then, $(b_1, b'_1) = (x, y)$ and $(h, h') = (y, z)$, one has, by **J**tpy, $(p_1, p'_1) = (x, z)$; hence, if $(p, p_1) = (u, x)$, then, by (i) and V,22, $(p, p'_1) = (u, z)$, and (p, p'_1) is equal to the ratio compounded of (b, b'_1) and (h, h'), i.e., to the ratio compounded of (h, h') and (b, b'_1); clearly then, if (iii) $(h, h') = (u, w)$, $(b, b'_1) = (w, z)$. Therefore, if $(p'_1, p') = (z, v)$, then, by (ii) and V,22, $(p, p') = (u, v)$, and (p, p') is equal to the ratio compounded of (h, h') and (b, b'), i.e. of (b, b') and (h, h'). Euclid's failure to prove this extension of XII,9 is perhaps some further confirmation of the view that the connections among compounding, multiplying, and volumes were not so immediately clear to him as they are to us.

Euclid makes no special assumptions concerning equality for (right) cones and cylinders, but he does define their similarity by

XI, def. 24 If c, c' are cones or cylinders having heights h, h' and bases b, b' with diameters d, d', then c and c' are similar if and only if $(h, h') = (d, d')$.

The fundamental theorem enabling Euclid to treat cones and cylinders simultaneously is **B**1c (XII,10), according to which a cone is one-third of the cylinder with the same base and height. This proposition enables Euclid to demonstrate an assertion for one solid and extend it immediately to the other. Its proof uses the method of exhaustion and will be described in the following section. Clearly one can prove **C**c by superposition, and hence establish by the standard proof using the Eudoxean definition of proportionality that

XII,13 If a cylinder is cut by a plane which is parallel to its opposite planes, then as the cylinder is to the cylinder so will the axis be to the axis,

or, more generally, **E**c (XII,14). Euclid does not prove **C**c at all, but uses **D**c (XII,11, also requiring the method of exhaustion) to do the job of **C**c in the demonstration of XII,13. **D**c is used again in the proof of **E**c to construct a cylinder c_3 equal to c_1—or, rather, to justify conceiving (*noein*) c_3 to be "about" a constructed axis—instead of placing c_1 so that its base coincides with the base of c_2. Why Euclid proves **E** for cones and cylinders when he does not prove it for other figures[17] is unclear, for **E**c plays no further role in the *Elements*.

Euclid's proof of **F**c (XII,15) may be represented as follows. Given cylinders c_1, c_2 with bases b_1, b_2 and heights h_1, h_2, he constructs a cylinder c with height h_1 and base b_2, so that $(c_1, c) = (b_1, b_2)$ by XII,11 and $(c_2, c) = (h_2, h_1)$ by XII,13. Hence, $c_1 \simeq c_2$ if and only if $(b_1, b_2) = (h_2, h_1)$. (Euclid, of course, proves two conditionals rather than a biconditional; and in arguing for one of them he does the case $h_1 \simeq h_2$ separately.) The only other proposition which Euclid proves for cones and cylinders is **G**c (XII,12), for which he gives a lengthy demonstration by exhaustion. Since **G**c is a direct consequence of **J**c, which, as will be clear in the following section, he could readily have established, the failure to take advantage of the power of the notion of compounding is especially uneconomical in this case.

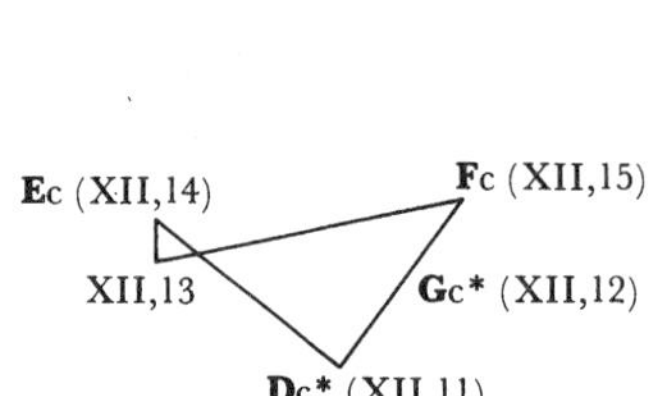

Figure 6.34

Figure 6.34 shows the deductive structure of Euclid's treatment of cones and cylinders, where each of **D**c–**G**c is proved for cylinders and extended to cones by means of **B**1c (XII,10). Euclid could, of course, have proved **J**c for cylinders, extended it to cones by means of **B**1c, and then taken each of **D**c–**G**c as a proportion-theoretic corollary.

It should be clear from this section that Euclid's treatment of prisms, parallelepipeds, pyramids, cones, and cylinders is not as standardized as one would expect in a modern treatise. He does not necessarily repeat a pattern of argumentation when it is available to him, and his choices of theorems to prove are not always clearly motivated. In general he does not seem to have a clear sense of the proportion-theoretic relations between various propositions, so that cumbersome geometric argument is sometimes used when a simple proportionality argument would suffice. I have chosen not to go into the details of the geometric argument at such junctures in the belief that its complexity hinders more than it helps in understanding Euclid's solid geometry. It should be clear, however, that the argument is generally not up to the standard of Euclid's plane geometry. Finally, it should perhaps be mentioned that Heiberg's text of book XII may deviate more considerably from Euclid's original than the text of any other book.[18] I shall, however, adhere to the policy of following the manuscript P wherever possible.

6.3 The Method of Exhaustion

The distinctive feature of book XII is its application of the so-called method of exhaustion in the proofs of propositions 2, 5, 10–12, and 18. Eudoxus is usually credited with the discovery of this method and with its use in proving these propositions, primarily on the basis of remarks by Archimedes.[19] Much of Archimedes' work involves ingenious applications of the method, which unfortunately must be left out of account in this discussion.[20] I shall content myself with a general characterization of the method as it is used in book XII and with a description of Euclid's proofs in terms of the characterization. This description involves some adaptation of the proofs because Euclid does not follow a prescribed uniform procedure but adjusts his techniques to the problem at hand. In general, however, the changes required to represent his proofs in a reasonably uniform way are not of any great significance and are indicated in the notes.

The propositions proved by Euclid using the method of exhaustion can all be represented in the form

(i) $(x', x) = (y', y)$.

In XII,10 x' is a cone, x a cylinder with the same base and height, y' and y the positive integers 1 and 3; Euclid expresses the proposition as the assertion that x' is a third part of x. The remaining cases are described in the following list.

XII,2 x', x: circles,
y', y: squares on the diameters of x', x;
XII,5 x', x: triangular prisms of the same height,
y', y: the bases of x', x;
XII,11 x', x: cones or cylinders of the same height,
y', y: the bases of x', x;
XII,12 x', x: similar cones or cylinders,
(y', y): the ratio triplicate of the ratio of the diameters of the bases of x', x;
XII,18 x', x: spheres,
(y', y): the ratio triplicate of the ratio of the diameters of x', x.

Clearly XII,10 differs from these other propositions because in its case x' and x are not objects of the same kind, and y' and y have no direct geometric connection with x' and x. There are corresponding differences in its proof which require differences in exposition.

In all six cases Euclid proceeds indirectly by assuming that (i) is false. In 2, 5, 11, 12, and 18 he then infers the existence of an x^* satisfying

(ii) $(x', x^*) = (y', y)$ and either (a) $x^* \prec x$ or (b) $x^* \succ x$.

This x^* is an arbitrary plane figure in 2; an arbitrary solid in 5, 11, and 12; and a sphere in 18. Each inference then involves an application of Vc: the assumption of the existence of a fourth proportional, of which an especially strong form is used in 18. In the proof of 10 Euclid uses what amounts to the assumption of trichotomy for ratios, but in the form (ii′a) $x \prec 3 \cdot x'$ or (ii′b) $x \succ 3 \cdot x'$.

This application of trichotomy is in fact constructive; and, as I shall show after describing the method of exhaustion in more detail, all the uses of Vc can be replaced with constructive applications of trichotomy. Now I would like to point out that Euclid's use of Vc is a good indication that a constructive point of view plays no real role in book XII.

A second form of disregard for the constructive point of view can be illustrated by reference to XII,2. Because of the definition of proportionality the assumption (ii) makes constructive sense only if the notion of multiples of x', x^*, y', and y does. For squares this notion does make constructive sense, since it is an elementary matter to reproduce a square n times and represent the result as a rectilineal figure or even a square. There is, of course, no general method of reproducing an arbitrary figure. And even for circles, which can obviously be reproduced n times, there is no way of representing the result as an "ordinary" figure until XII,2 has been proved. One might attempt to preserve the constructive point of view by construing multiplication of a figure as n-fold reproduction of it; but it seems much more likely that multiplication should be taken abstractly, as it is in book V. There is no good reason to suppose that the abandonment of the constructive point of view represents a conscious philosophical decision, since the nonconstructive one is as appropriate for the material in book XII as the constructive one is for more elementary geometry and arithmetic.

The bulk of the proofs is devoted to eliminating the two alternatives, (ii) or (ii′). In 10 Euclid treats each alternative separately, but in the other five cases he reduces (b) to (a). The reduction is argued for most explicitly in XII,2. Euclid applies inversion to (iib) to get $(x^*, x') = (y, y')$ and invokes Vc to infer the existence of an x^{**} satisfying $(x^*, x') = (x, x^{**})$, so that, since $x^* \succ x$, $x' \succ x^{**}$;[21] therefore $(x, x^{**}) = (y, y')$ with $x^{**} \prec x'$; and the refutation of (iia) carries over to (iib). It remains, then, to consider the refutation of (iia). This involves the construction of objects z', z such that

(iii) $(z', z) = (y', y)$,
(iva) $z' \prec x'$
(ivb) $x^* \prec z$.

From (ii) and (iii) one infers $(x', x^*) = (z', z)$, which is incompatible with (iva) and (ivb). More specifically Euclid constructs a z inside x and copies the construction to get a z' inside x'.[22] Because z' is inside x', (iva) is satisfied; known results and the copying procedure are used to establish that (iii) is fulfilled.

For (ivb), z must not only lie inside x but be greater than x^*. Obviously no single z will do for every x^*. Hence one must describe a general method which can be applied for any particular x^* to produce a z satisfying (ivb). What Euclid does is to describe a method which produces an intial figure v_1 inside x and not less than half of x, and which, given a figure v_n inside x, produces an increment u_{n+1} to v_n inside x, satisfying the condition $u_{n+1} \succeq \frac{1}{2}(x - v_n)$; he takes v_{n+1} to be $v_n + u_{n+1}$. That some v_n will be a z satisfying (ivb)—or, as Euclid puts it, that $x - v_n$ will be less than $x - x^*$—follows from X,1.

It should be clear from this account of the method of exhaustion that an ordinary application of it is sufficiently characterized by describing a construction process in terms of a step 1 producing v_1 and a step $n + 1$ in which, given v_n, u_{n+1} is produced, and then showing

(A1) $v_1 \succeq \frac{1}{2}(x)$;
(A2) $u_{n+1} \succeq \frac{1}{2}(x - v_n)$, for any n;
(B) $(v'_n, v_n) = (y', y)$, for any n.

I shall do this for XII,2, 5, and 10–12, substituting more explicitly inductive considerations for at least some of Euclid's uses of the word 'similarly'. 18 requires a somewhat different treatment in which (iii), (iva), and (ivb) are established more directly. However, before turning to this topic, I would like to speak more generally about the method of exhaustion.

I remarked above that the applications of Vc in XII can be eliminated in favor of constructive applications of trichotomy. One need only substitute for the refutations of (iia) and (iib) a refutation of

(iia′) $(y', y) > (x', x)$ or (iib′) $(x', x) > (y', y)$.

If (iia′) holds, there are, by VL, m and n such that $m \cdot y' \succ n \cdot y$ and $m \cdot x' \prec n \cdot x$. Find s so that $s \prec n \cdot x - m \cdot x'$; and let s^* be an nth part of s. The Euclidean construction process can now be used to produce z and z' satisfying (iii), (iva) (so that also $m \cdot x' \succ m \cdot z'$) and

(ivb′) $x - s^* \prec z$ (or $x - z \prec s^*$).

But then $n \cdot x - n \cdot z \simeq n \cdot (x - z) \prec n \cdot s^* \simeq s \prec n \cdot x - m \cdot x'$, so that $n \cdot z \succ m \cdot x' \succ m \cdot z'$. Hence, since (iii) is satisfied, $n \cdot y \succ m \cdot y'$, which contradicts $m \cdot y' \succ n \cdot y$. The same argument with x and x' as well as y and y' interchanged refutes (iib′).[23]

The argument just given depends essentially on definitions 5 and 7 of book V, but nothing in book XII itself does so. Book XII does, of course, presuppose the laws of proportion; but if these laws are taken for granted, the only substantive foundational connection between V and XII is a reliance on

Vc and some form of Vd, the Archimedean condition. As Archimedes himself realized, the latter is required for the method of exhaustion to work. The attempt to make a historical connection between Vd or X,1 and the exclusion of infinitesimals is especially intriguing in the case of the method of exhaustion because of the close relation of the method to modern techniques developed in the nineteenth century for eliminating the use of infinitesimals in the calculus. Whether or not the Greeks ever attempted to use infinitesimals in serious mathematical argument, it seems reasonably clear that they could not have been successful because they saw no way out of the Zenonian argument that an infinite sum of things is either infinite (if the things summed have size) or null (if the things summed do not have size). We now know that this argument can be refuted through the use of transfinite sums. As a result, the question whether reasoning with infinitesimals is legitimate or not reduces to a basically philosophical question about the legitimacy of reasoning about the infinite. The Greeks may, of course, have raised the philosophical question without possessing rigorous means of dealing with infinite sums. The important point is, however, that this lack of means by itself provides sufficient motivation for adopting the method of exhaustion independently of any philosophical scruples about the infinite. Here again, then, the notion of rigorous proof suffices to account for what might seem to be a philosophically motivated feature of the *Elements*.

It is sometimes suggested that at least some of the results proved by the method of exhaustion must first have been discovered by the use of infinitesimals or infinite sums of some kind. Heath[24] takes XII,2 to be an example, citing Simplicius' report according to which the fifth-century sophist Antiphon claimed that a polygon of sufficiently many sides inscribed in a circle would exhaust (*dapanan*) the circle. Even if some such reasoning were the basis of the discovery of XII,2, there is no mention in Simplicius' report of infinite-sided polygons. It seems much more natural to suppose that Antiphon argued on the basis of the Protagorean suggestion (see p. 185) that straight line and curve can coincide at more than a point. Moreover, the realization that this suggestion is false does not necessitate thinking in terms of infinite-sided polygons in order to see that XII,2 is very probably true. The recognition that polygons of any finite number of sides inscribed in circles satisfy XII,2 is sufficient for this purpose without any "passage to the limit."

The relationship between the method of exhaustion and the integral calculus can be brought out by reformulating (A) and (B) in more modern and general terminology. (A) establishes for a particular kind of series v_i that it converges to x, i.e.,

$$\lim_{i\to\infty} (v_i) = x.$$

Since Euclid's argumentation also establishes that the series v_i' converges to x', and since

$$\lim\left(\frac{v_i'}{v_i}\right) = \frac{\lim(v_i')}{\lim(v_i)},$$

this result can be expressed as

$$\lim\left(\frac{v_i'}{v_i}\right) = \frac{x'}{x}.$$

(B) simply gives the value of $\frac{v_i'}{v_i}$, for each i as the constant $\frac{y'}{y}$. Euclid's uses of the method of exhaustion can be said to involve the inference to the conclusion that

$$\lim\left(\frac{v_i'}{v_i}\right) = \frac{y'}{y},$$

i.e., as an application of the law that the limit of a constant series is the constant,

$$\forall i(w_i = c) \rightarrow \lim(w_i) = c.$$

There can be no question that Euclid's arguments, conceived as instances of integration, are extremely simple. However, it does not seem totally anachronistic to construe the arguments as integrations, despite the absence of a terminology of limits and sequences and despite the fact that Euclid's reasoning is geometric rather than arithmetical. For there is no important difference between showing for an increasing series v_i that $\lim(v_i) = x$, and showing for any $x^* \prec x$ that there is a v_n such that $x^* \prec v_n \prec x$.

Nevertheless, it does seem to me that the difference between the Euclidean method of exhaustion and the integral calculus is conceptual, and not merely a matter of terminology and relative simplicity. This difference lies at the theoretical rather than the practical level. Euclid applies one technique of integration, but for him there are no laws of integration. Hence, for different problems he has to go through the same steps in geometrically different forms. In many cases the difference between applying a mathematical technique and knowing a mathematical law justifying the technique is not great, because the law and its proof are simply generalizations of the technique itself. However, the laws justifying even simple forms of integration depend for their proofs on a theory of limits in which one proves the existence of limits for series satisfying certain conditions, e.g., constant series. Ultimately, these proofs of existence depend upon the assumption of continuity, which we have seen to be left entirely at the intuitive level in the *Elements*. Moreover, Euclid does not need the assumption of continuity to prove the existence of limits, because in the cases he deals with the limit is given in advance as the geometric object x. One might say that in applications of the method of exhaustion the limit is given and the problem is to construct a certain kind of sequence converging to it, whereas in the integral calculus one is usually given a sequence and the problem is to determine whether it has a limit and, if it does, what the limit is. Since in the *Elements* the limit always has a simple description, the construction of a sequence converging to it can be done within the bounds of elementary geometry; and the question of constructing a sequence for any given arbitrary limit never arises. In the calculus one is concerned with arbitrary sequences, and their treatment leads to the kind of abstract systematic considerations which are foreign to the *Elements*.

I turn now to the description of Euclid's applications of the method of exhaustion. Since the proof of XII,2 has already been given in section 5.2, it suffices to indicate its more formal representation briefly. The construction process may be described as follows:

Step 1 Inscribe a square v_1 in the circle x.

Step $n + 1$ Given a rectilineal figure $v_n (A_1 A_2 \ldots A_{2^{n+1}})$ inscribed in x, bisect each of the smaller arcs $A_1 A_{2^{n+1}}$ and $A_i A_{i+1}$ at $B_{2^{n+1}}$ and B_i. Construct the triangles $t_{2^{n+1}}$ $(A_1 B_{2^{n+1}} A_{2^{n+1}})$ and t_i $(A_i B_i A_{i+1})$, and take u_{n+1} as their sum.

(A), the condition of convergence, was established on pp. 200–201; (B), the condition that $(v_n', v_n) = (y', y)$, is simply XII,1.

XII,5

Let p be a triangular pyramid with base ABC and vertex D (fig. 6.35). By the Euclidean division of p, I mean the bisection of its edges AB, BC, CA, AD, DB, DC at E, F, G, H, K, L, and the connection of HE, EG, GH, HK, KL, LH, GF, FK, dividing p into two triangular pyramids and two triangular prisms. I set

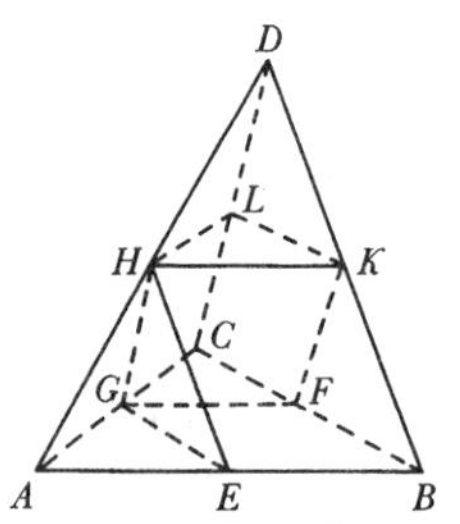

Figure 6.35

$\sigma(p)$ = the pyramid with vertex D, base HKL,
$\tau(p)$ = the pyramid with vertex H, base AGE,
$\pi(p)$ = the prism contained by the triangles HKL, GFC,
$\rho(p)$ = the prism contained by the triangles HGE, KFB.[25]

What I have called the Euclidean division is described by Euclid in XII,3, where he also shows that $\sigma(p)$ and $\tau(p)$ are equal to each other and similar to p and that $\pi(p) + \rho(p)$ is greater than half of p. The construction process may be described as follows.

Step 1 Given a triangular pyramid x, perform the Euclidean division, and set

$$v_1 = \pi(x) + \rho(x),$$
$$s_1 = \sigma(x).^{26}$$

Step $n + 1$ Given a triangular pyramid x divided into $2^{n+1} - 2$ prisms together called v_n, and 2^n triangular pyramids $x_1, x_2, \ldots x_{2^n}$, one of them also called s_n, with bases parallel to the base of x, perform the Euclidean division on each x_i, and set

$$u_{n+1} = \pi(x_1) + \rho(x_1) + \ldots + \pi(x_{2^n}) + \rho(x_{2^n}),$$
$$s_{n+1} = \sigma(s_n).$$

The construction process, then, consists of applying the Euclidean division to a triangular pyramid producing two triangular pyramids and two prisms, then applying it to the two pyramids producing four pyramids and a total of six prisms, then applying it to the four pyramids producing eight pyramids and a total of fourteen prisms, and so on. As a preliminary to establishing (A) I show

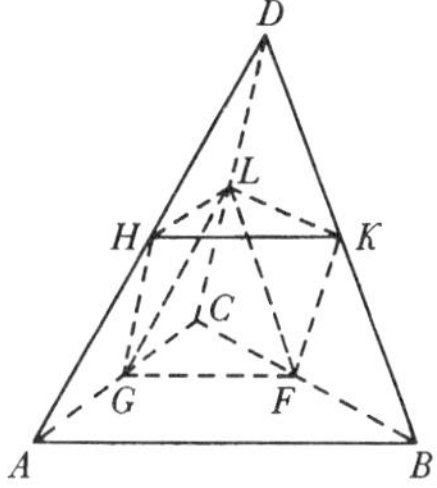

XII,3(a) $\sigma(p)$ and $\tau(p)$ are similar to p and equal to each other.

It is easy enough to establish the similarity of the faces of each of the small pyramids to their correspondents in p, and therefore the similarity of the three pyramids by XI, def. 9. In order to establish the equality of the two smaller pyramids using def. 10, it suffices, because of I,26, to establish the equality of one corresponding side in each pair of corresponding faces. Starting from the equality of AH and HD, one easily establishes in succession the congruence of the triangles AHE and HDK, HAG and DHL, AGE and HLK, and EGH and LKD.[27]

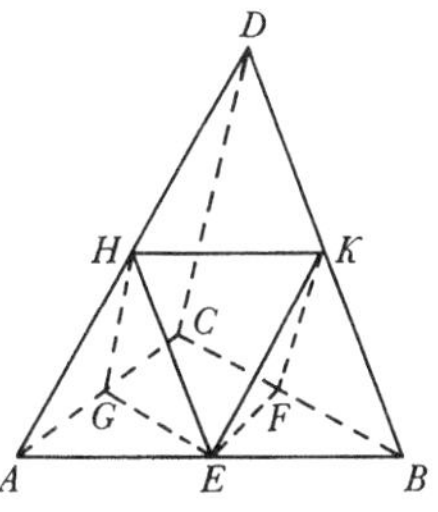

Figure 6.36

If one connects GL, LF and EK, EF, one determines (fig. 6.36) two pyramids, $GFLC$ contained in $\pi(p)$ and $BEFK$ contained in $\rho(p)$, each similar and equal to the two pyramids $\sigma(p)$ and $\tau(p)$. Clearly, then, the two prisms are greater than half of p. Hence

(A1) (XII,3(b)) $\pi(x) + \rho(x) \succ \frac{1}{2}(x)$.

The same argument generalized yields (A2). For $x - v_n$ is $x_1 + x_2 + \ldots + x_{2^n}$ and u_{n+1} is $\pi(x_1) + \rho(x_1) + \ldots + \pi(x_{2^n}) + \rho(x_{2^n})$. But the argument just given shows that $\pi(x_i) + \rho(x_i) \succ \frac{1}{2}(x_i)$. Hence $u_{n+1} \succ \frac{1}{2}(x - v_n)$.

It remains to establish (B). Naturally, Euclid does this for pyramids x' and x of the same height, taking (y', y) to be the ratio of their bases. I revise his argument slightly to prove **J**tpy, i.e., XII,5 for the case in which x' and x are arbitrary triangular pyramids and (y', y) is the ratio compounded of the ratios of their heights and bases. Some lemmas are required to establish (B) for this case.

XII,3(c) $\pi(p) \simeq \rho(p)$.

$\pi(p)$ is half of the parallelepiped determined by the parallelograms *CGEF* and *HGCL* (cf. fig. 6.36). Also, $\rho(p)$ is half of the parallelepiped determined by the parallelograms *GEBF* and *FGHK*. But the bases *CGEF* and *GEBF* are equal, since they are each twice the triangle *GEF*. Also, the heights of the two parallelepipeds are equal, since the parallelepipeds are contained between the same parallel planes *HLK* and *ABC*. The result follows by XI,31.[28]

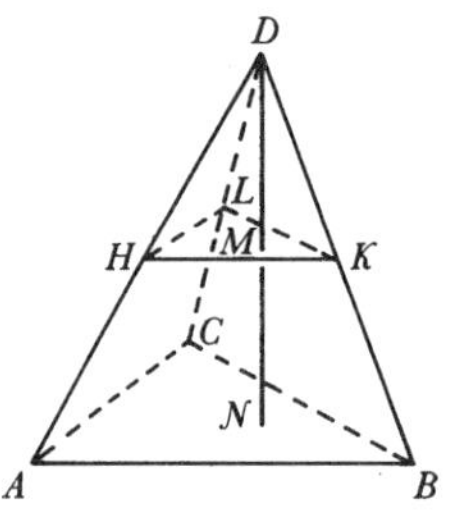

Figure 6.37

Now let $\eta(p)$ be a perpendicular *DMN* dropped from *D* to the plane of *ABC* at *N* and meeting the plane of *HLK* at *M* (fig. 6.37). Clearly, $\eta(p)$ is the height of p; but also, by XI,17, one has

$$(DM, MN) = (DH, HA),$$

or, since *H* is the midpoint of *DA*,

$$DM \simeq MN.$$

Hence, if we set $\theta(p) = DM$, $\theta(p)$ is the height of $\sigma(p)$, $\tau(p)$, and $\pi(p)$, and, in addition, $\theta(p)$ is half of $\eta(p)$ and is identical with $\eta(\sigma(p))$.

Suppose step 1 of the method is applied to two triangular pyramids x' $(A'B'C'D')$ and x $(ABCD)$. Clearly

$$(\eta(x'), \theta(x')) = (\eta(x), \theta(x)),$$

or

$$(\eta(x'), \eta(x)) = (\theta(x'), \theta(x)).$$[29]

The similar triangles $A'B'C'$, $H'K'L'$ (ABC, HKL) are to one another in the duplicate of the ratio of $B'C'$ to $L'K'$ (BC to LK). But the latter ratio is the duplicate of the ratio of $B'C'$ to $F'C'$ (BC to FC), which is (2, 1). Hence the triangles $A'B'C'$, $H'K'L'$ and ABC, HKL are in the same ratio, or (triangle $A'B'C'$, triangle ABC) = (triangle $H'K'L'$, triangle HKL). Since $\pi(p) \simeq \rho(p)$,

$$(\pi(x') + \rho(x'), \pi(x) + \rho(x)) = (\pi(x'), \pi(x)).$$

But, by **J**pr,

$$\begin{aligned}(\pi(x'), \pi(x)) &= \text{the ratio compounded of } (\theta(x'), \theta(x)) \text{ and (triangle } H'K'L', \text{ triangle } HKL)\\ &= \text{the ratio compounded of } (\eta(x'), \eta(x)) \text{ and (triangle } A'B'C', \text{ triangle } ABC).\end{aligned}$$[30]

This argument, which parallels Euclid's proof of XII,4, establishes that (B) holds for $n = 1$, i.e.,

(B_1) $(\pi(x') + \rho(x'), \pi(x) + \rho(x)) = (y', y)$.

After establishing this result Euclid asserts that (B) holds generally with the words

And similarly, if the pyramids $H'K'L'D'$, $HKLD$ be divided into two prisms and two pyramids, as the base $H'K'L'$ is to the base HKL so will the two prisms in the pyramid $H'K'L'D'$ be to the two prisms in the pyramid $HKLD$. But as the base $H'K'L'$ is to the base HKL so is the base $A'B'C'$ to the base ABC, for the triangles $H'K'L'$, HKL are equal to the triangles $G'C'F'$, GCF respectively. Therefore also as the base $A'B'C'$ is to the base ABC so are the four prisms to the four prisms.

And similarly also, if we divide the remaining pyramids into two pyramids and into two prisms, then as the base $A'B'C'$ is to the base ABC so will all the prisms in the pyramid $A'B'C'D'$ be to all the prisms, being equal in multitude, in the pyramid $ABCD$.

The remainder of my discussion of XII,5 is simply a more formal representation of these words adapted to the case I am considering. I first note that

If x is divided by n steps of the construction process into v_n and pyramids $x_1, \ldots, x_{2^n}$, x_i and x_j are similar and equal.

This result has already been proved for $n = 1$ in XII,3. So suppose it is true for n, and let x be divided by $n + 1$ steps of the construction process into v_{n+1} and pyramids $\sigma(x_1), \ldots, \sigma(x_{2^n}), \tau(x_1), \ldots, \tau(x_{2^n})$, where, by inductive assumption, $x_1, \ldots, x_{2^n}$ are similar and equal. We know that $\sigma(x_i)$ and $\tau(x_i)$ are similar and equal; so it suffices to show that $\sigma(x_i)$ and $\sigma(x_j)$ are. One can think of x_i, x_j, $\sigma(x_i)$, $\sigma(x_j)$ as the pyramids $D_iA_iB_iC_i$, $D_jA_jB_jC_j$, $D_iH_iK_iL_i$, $D_jH_jK_jL_j$, with corresponding vertices D_i, D_j, D_i, D_j and A_i, A_j, H_i, H_j and B_i, B_j, K_i, K_j and C_i, C_j, L_i, L_j. Each face of $\sigma(x_i)$ and $\sigma(x_j)$ is similar to its correspondent in x_i and x_j, and, by VI,19–20, has to its correspondent the ratio duplicate of the ratio of the corresponding sides, i.e., it has to its correspondent the ratio of $(1, 4)$. Since, then, x_i and x_j are similar and equal by inductive hypothesis, so are $\sigma(x_i)$ and $\sigma(x_j)$.

Since x_i and x_j are similar and equal, $\eta(x_i) \simeq \eta(x_j)$, so that also $\theta(x_i) \simeq \theta(x_j)$. The result proved as (B_1) may be expressed

$(v_1', v_1) = (y', y) =$ the ratio compounded of (triangle $A'B'C'$, triangle ABC) and $(\eta(x'), \eta(x))$
$=$ the ratio compounded of (base of s_1', base of s_1) and $(\eta(s_1'), \eta(s_1))$.

To establish (B) in general, suppose

$(v'_n, v_n) = (y', y) =$ the ratio compounded of (triangle $A'B'C'$, triangle ABC) and $(\eta(x'), \eta(x))$
$=$ the ratio compounded of (base of s'_n, base of s_n) and $(\eta(s'_n), \eta(s_n))$.

At step $n + 1$ of the construction procedure, x' and x are divided into the prisms v'_n, $\pi(x'_1), \ldots, \pi(x'_{2^n})$, $\rho(x'_1), \ldots, \rho(x'_{2^n})$, and v_n, $\pi(x_1), \ldots, \pi(x_{2^n})$, $\rho(x_1), \ldots, \rho(x_{2^n})$, and pyramids $\sigma(x'_1), \ldots, \sigma(x'_{2^n})$, $\tau(x'_1), \ldots, \tau(x'_{2^n})$ and $\sigma(x_1), \ldots, \sigma(x_{2^n})$, $\tau(x_1), \ldots, \tau(x_{2^n})$. By **Jpr**,

$(\pi(x'_i), \pi(x_i)) =$ the ratio compounded of (base of $\pi(x'_i)$, base of $\pi(x_i)$) and $(\theta(x'_i), \theta(x_i))$.

But since for any i and j the bases of $\pi(x'_i)$ and $\pi(x_i)$ are equal to the bases of $\sigma(x'_j)$ and $\sigma(x_j)$ and

$$\eta(\sigma(x'_i)) \simeq \theta(x'_i) \simeq \theta(x'_j) \simeq \eta(\sigma(x'_j)),$$
$$\eta(\sigma(x_i)) \simeq \theta(x_i) \simeq \theta(x_j) \simeq \eta(\sigma(x_j)),$$

we may conclude that

$(\pi(x'_i), \pi(x_i)) =$ the ratio compounded of (base of s'_{n+1}, base of s_{n+1}) and $(\eta(s'_{n+1}), \eta(s_{n+1}))$.

Because $\pi(x'_i) \simeq \rho(x'_i)$ and $\pi(x_i) \simeq \rho(x_i)$, $(\rho(x'_i), \rho(x_i)) = (\pi(x'_i), \pi(x_i))$, and, by V,12,

$(u'_{n+1}, u_{n+1}) =$ the ratio compounded of (base of s'_{n+1}, base of s_{n+1}) and $(\eta(s'_{n+1}), \eta(s_{n+1}))$.

Since s'_{n+1}, s'_n and s_{n+1}, s_n are pairs of similar pyramids, we have that

(base of s'_{n+1}, base of s_{n+1}) $=$ (base of s'_n, base of s_n);

and since $\eta(s'_{n+1})$, $\eta(s_{n+1})$ are equal to $\theta(s'_n)$, $\theta(s_n)$, which in turn are half of $\eta(s'_n)$, $\eta(s_n)$, therefore

$$(\eta(s'_n), \eta(s_n)) = (\eta(s'_{n+1}), \eta(s_{n+1})),$$

and

$(u'_{n+1}, u_{n+1}) =$ the ratio compounded of (base of s'_n, base of s_n) and $(\eta(s'_n), \eta(s_n)) = (y', y) = (v'_n, v_n)$.

Hence, by V,12,

$$(v'_{n+1}, v_{n+1}) = (v'_n + u'_{n+1}, v_n + u_{n+1}) = (y', y).$$ [31]

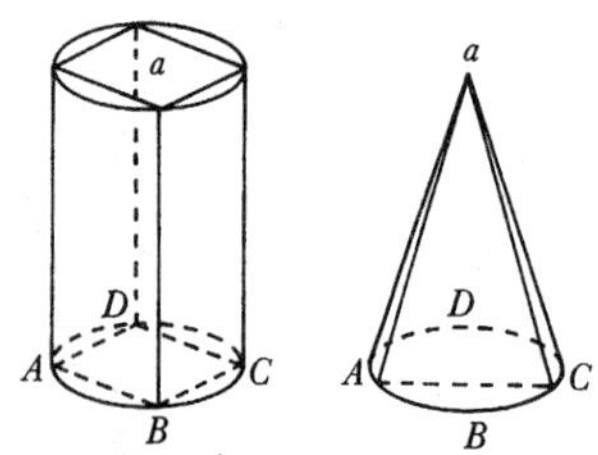

Figure 6.38

XII, 10–12

For these propositions consider (fig. 6.38) a cylinder or cone c with base $ABCD$ and opposite face or vertex a in the plane p

parallel to the plane of $ABCD$. Given a rectilineal figure f in the plane of $ABCD$, or a segment ABC of $ABCD$, I set

$\pi(f)$ = the prism with base f, opposite face in the plane p, and remaining faces perpendicular to f,

$\sigma(ABC)$ = the segment of c containing B and determined by passing a plane through the straight line AC perpendicular to $ABCD$,

for c a cylinder, and

$\pi(f)$ = the pyramid with base f and vertex a,

$\sigma(ABC)$ = the segment of c containing B and determined by passing a plane through the straight line AC and the vertex a,

for c a cone.

It is now possible to describe the construction process for XII,10–12 with x a cone or cylinder of the kind described. In the case of 10 the process for a cylinder (cone) will be applied to x and then the process for the cone (cylinder) applied equally often to x'.

Step 1 Inscribe a square s in the base $ABCD$. Set

$v_1 = \pi(s)$.

Step $n + 1$ Given a prism or pyramid v_n inscribed in x and having a polygonal base $A_1A_2 \ldots A_{2^{n+1}}$, bisect the smaller arcs $A_1A_{2^{n+1}}$ and A_iA_{i+1} of the circle $ABCD$ at $B_{2^{n+1}}$ and B_i. Construct the triangles $A_1B_{2^{n+1}}A_{2^{n+1}}$ and $A_iB_iA_{i+1}$, and set

$u_{n+1} = \pi(A_1B_1A_2) + \pi(A_2B_2A_3) + \ldots + \pi(A_{2^{n+1}}B_{2^{n+1}}A_1)$.

(A) is a consequence of the following lemma which Euclid in effect establishes in the course of proving XII,10:

If c is a cone or cylinder of the kind described and the segment ABC of the base is not greater than a semicircle and B is the midpoint of the arc ABC, then

$\pi(\text{triangle } ABC) \succ \frac{1}{2}(\sigma(\text{segment } ABC))$.

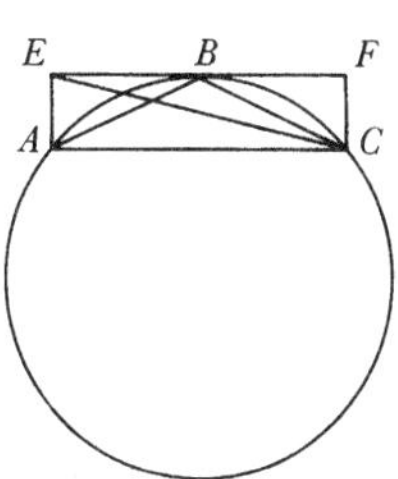

Figure 6.39

Euclid completes the rectangle $EACF$ (fig. 6.39) with EF tangent to the base at B, and connects EC. Since the triangles EAC, ABC are equal, so are $\pi(EAC)$, $\pi(ABC)$, by XI,31pr and XII,7, corollary. By XI,28 and XII,7, corollary, $\pi(EAC) \simeq \frac{1}{2}\pi(EACF)$. The result follows because $\sigma(\text{segment } ABC) \prec \pi(EACF)$.

(B) for XII,10 is an immediate consequence of the corollary to XII,7. Given 10, Euclid can do 11 and 12 only for cones. (B) for 11 is an immediate consequence of XII,6; **J**py, which I derived from **J**tpy in the previous section, yields (B) for **J**c, from which 11 and 12 follow directly. Euclid establishes (B)

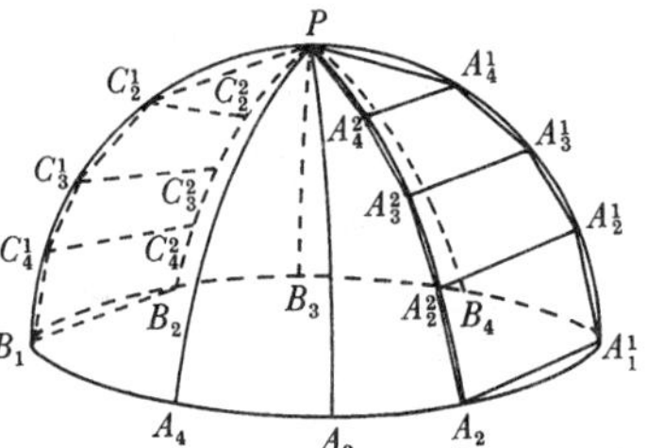

Figure 6.40

for 12 by connecting all the vertices of the bases of v'_n, v_n to the centers of the circles o', o with radii r', r, in which they are inscribed, producing triangles $t'_1, \ldots, t'_{2^n+1}, t_1, \ldots, t_{2^n+1}$; he then argues for the similarity of $\pi(t'_i)$ to $\pi(t_i)$. By XII,8 $\pi(t'_i)$ is to $\pi(t_i)$ in the triplicate of the ratio of r' to r, i.e., of d' to d. (B) follows because

$$v'_n = \pi(t'_1) + \ldots + \pi(t'_{2^n+1})$$

and

$$v_n = \pi(t_1) + \ldots + \pi(t_{2^n+1}).$$ [32]

XII,18

Let (fig. 6.40) s be a sphere with center O, t a greatest circle of s with poles P, Q, and let $A_1A_2 \ldots A_{2^n-1}B_1B_2 \ldots B_{2^n-1}$ be a regular polygon of 2^n sides inscribed in t. The method for XII,18 involves the construction of a polyhedron $\pi(A_1A_2 \ldots A_{2^n-1}B_1B_2 \ldots B_{2^n-1})$ in s. In describing the construction I follow Euclid in presupposing a number of geometric facts about the sphere. Let $r_1, \ldots, r_{2^n-1}$ be greatest circles with r_i passing through A_i and P, hence through B_i and Q, and hence also perpendicular to t. Inscribe in each r_i a regular polygon $A^i_1A^i_2 \ldots A^i_{2^n-1}C^i_1C^i_2 \ldots C^i_{2^n-1}B^i_1B^i_2 \ldots B^i_{2^n-1}D^i_1D^i_2 \ldots D^i_{2^n-1}$, with $A^i_1 = A_i$, $B^i_1 = B_i$, $C^i_1 = P$, $D^i_1 = Q$.

The greatest circles r_i divide s into congruent "slices" and t divides each slice into congruent parts. Thus, to show that the vertices of these polygons determine a polyhedron inscribed in s and contained by triangular faces with vertex P or Q and quadrilateral faces otherwise, it suffices to show that the quadrilateral $A^i_jA^i_{j+1}A^{i+1}_{j+1}A^{i+1}_j$ lies in one plane.[33] To simplify notation I call this quadrilateral $ABCD$ (fig. 6.41). Let BEG, CFH be perpendiculars to the plane of t interesting OA_i, OA_{i+1} in G, H and OA, OD in E, F.[34] Since all greatest circles in a sphere are equal and arc $BAA_i \simeq$ arc CDA_{i+1}, angle $BOG \simeq$ angle COH. But angles BGO, CHO are right, and OB,

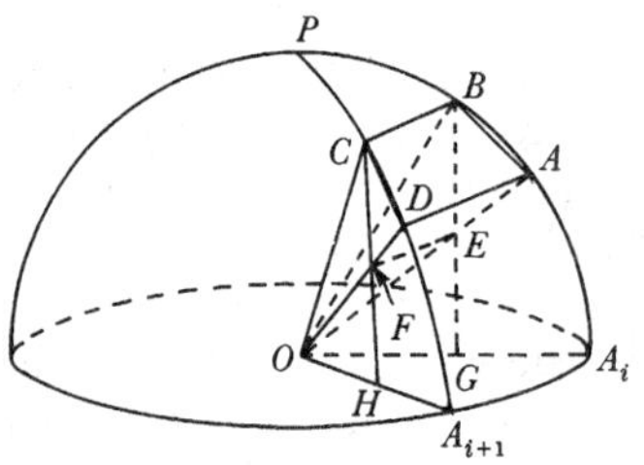

Figure 6.41

OC are equal; therefore, by I,26, $BG \simeq CH$ and $OG \simeq OH$. Again, by III,27, angle $EOG \simeq$ angle FOH; so, by I,26, $EG \simeq FH$; and, subtracting equals from equals, BE, OE are equal to CF, OF. But BE is parallel to CF, by XI,6. Therefore, by I,33, BC is parallel to EF. VI,2 applied to triangle AOD and straight line EF yields that AD is also parallel to EF, so that AD and BC are parallel and $ABCD$ is in one plane. I note as a corollary of this argment that $BC \simeq EF \prec AD$. More generally, $A_j^iA_j^{i+1} \succ A_{j+1}^iA_{j+1}^{i+1}$; and, since $A_j^iA_j^{i+1} \simeq A_j^kA_j^{k+1}$,

(a) If $j \prec l$, then $A_j^iA_j^{i+1} \succ A_l^kA_l^{k+1}$.

It is now possible to describe the construction process for a sphere x with greatest circle t.

Step 1 Perform step 1 of the method of XII,2 with respect to t, producing a square p_1 inscribed in t; set $v_1 = \pi(p_1)$.

Step $n + 1$ Given a regular polygon p_n of 2^{n+1} sides inscribed in t, perform step $n + 1$ of the method of XII,2, producing a regular polygon p_{n+1} of 2^{n+2} sides inscribed in t; set $v_{n+1} = \pi(p_{n+1})$.

Obviously in this description v_{n+1} is not represented as the result of an increment u_{n+1} to v_n, presumably because of the great complexity such a representation would involve. Euclid avoids the need for one by taking x^* to be a sphere concentric with x and showing in XII,17 that the construction process applied to x will yield a v_n which is a z completely containing x^*. Clearly, then, (ivb) on p. 232 is satisfied; and (iva) will be taken care of if a similar polyhedron z' is inscribed in x'. (iii) is a consequence of

XII,17, addition Similar polyhedrons inscribed in spheres are in the triplicate of the ratios of the diameters of the spheres.

Euclid does not really prove this "corollary."[35] He simply remarks that if the vertices of the polyhedrons are connected to the centers of the spheres in which they are inscribed, the polyhedrons will be divided into "pyramids similar in multitude and in arrangement," and then quotes the extension of XII,8 to arbitrary pyramids. If this result is accepted, the proof of XII,18 is reduced to showing

(b) If x and x^* are spheres with the same center O and $x \succ x^*$, then some v_n will completely contain x^*, i.e., the perpendicular OX from O to any face of v_n will be greater than the radius of x^*.[36]

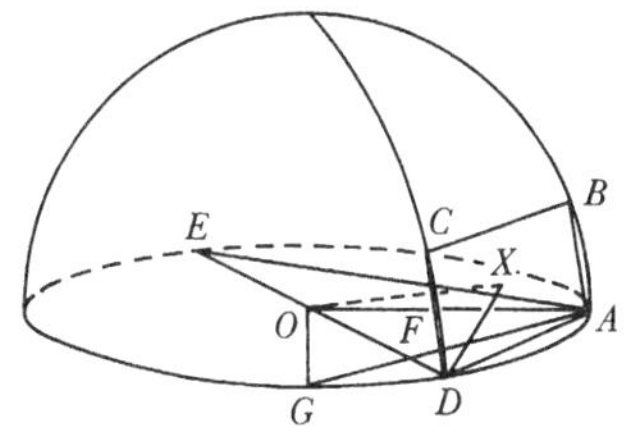

Figure 6.42

To explain Euclid's proof of this assertion, I first consider (fig. 6.42) the face $A_1^iA_2^iA_2^{i+1}A_1^{i+1}$ ($ABCD$) of an arbitrary v_n. If EOD is a diameter of x intersecting $A_1^iA_1^{i+2}$ (AG) at F, a congruence argument, using I,4, yields that $AF \simeq FG$ so that (III,3) ED is also perpendicular to AG. If then OX is perpendicular to $ABCD$ at X, $\mathbf{T}(OF) + \mathbf{T}(FA) \simeq \mathbf{T}(OA) \simeq \mathbf{T}(OD) \simeq \mathbf{T}(OX) + \mathbf{T}(XD)$. Obviously then $OX \succ OF$ if and only if $\mathbf{T}(XD) \prec \mathbf{T}(FA)$. Euclid establishes the latter inequality by arguing that $2 \cdot \mathbf{T}(FA) \succeq \mathbf{T}(AD) \succ 2 \cdot \mathbf{T}(XD)$. He first connects AE, producing, by III,31, a right triangle EAD. By VI,8 and 17 one has $\mathbf{T}(FA) \simeq \mathbf{O}(EF, FD)$ and $\mathbf{T}(AD) \simeq \mathbf{O}(ED, FD)$, so that $(\mathbf{T}(FA), \mathbf{T}(AD)) = (EF, ED)$. But the latter ratio is (1, 2) for v_1, since p_1 is a square, and more than (1, 2) for any other v_n.[37] To show that $\mathbf{T}(AD) \succ 2 \cdot \mathbf{T}(XD)$, Euclid connects the straight lines as in fig. 6.43. Since the four radii of the sphere are all equal, and OX is perpendicular to all the straight lines in the plane of $ABCD$ which it meets, the Pythagorean theorem yields that AX, BX, CX, DX are all equal. By the construction of v_n, we have that AB, AD, CD are all equal; but, by (a), $AD \succ BC$. Hence, by I,8 and 25, the angles AXB, DXA, CXD are equal and each greater than angle CXB. But all four are together equal to four right angles; hence angle DXA is obtuse; and, since $\mathbf{T}(XD) \simeq \mathbf{T}(AX)$, the desired result follows from

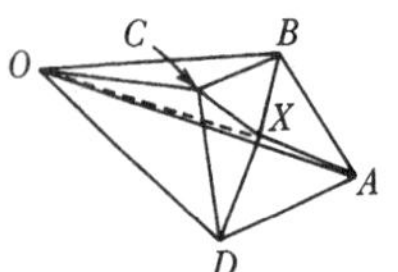

Figure 6.43

II,12 If angle AXD is obtuse,
$\mathbf{T}(AD) \succ \mathbf{T}(AX) + \mathbf{T}(DX)$....

Euclid does not explicitly cite II,12, but says, somewhat mysteriously, "Since $ABCD$ is a quadrilateral in a circle, and AD, AB, CD are equal and BC less, and DX is the radius of the circle, therefore the square on AD is greater than double the square on DX." The argument I have given is supplied in scholia 85 and 86 on XII,17. If it correctly fills in Euclid's reasoning, the argument would represent Euclid's only application of II,12, which is actually a stronger result than just indicated. Together with its companion, the unused II,13, it may be stated

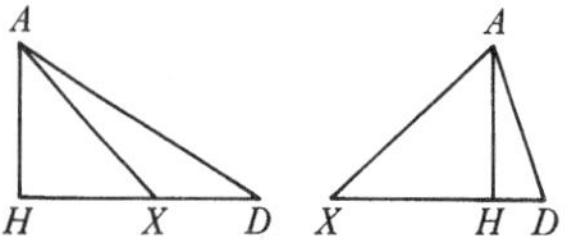

Figure 6.44

II,12(13) If (fig. 6.44) AXD is an obtuse (acute) angle and if AH is perpendicular to DX extended (DX) at H, then

$$\mathbf{T}(AD) \simeq \mathbf{T}(AX) + \mathbf{T}(XD) + (-)2 \cdot \mathbf{O}(XH, XD).^{38}$$

These propositions are direct consequences of the Pythagorean theorem, according to which $\mathbf{T}(AD) \simeq \mathbf{T}(AH) + \mathbf{T}(DH)$ and $\mathbf{T}(AH) \simeq \mathbf{T}(AX) - \mathbf{T}(XH)$, and II,4(7), according to which $\mathbf{T}(DH) \simeq \mathbf{T}(XH) + \mathbf{T}(XD) + (-)2 \cdot \mathbf{O}(XH, XD)$. Although these proofs can be represented algebraically and numerical values assigned to the straight lines involved, there seems to be no reason to treat them as anything but geometric extensions of the Pythagorean theorem which are made obvious by II,4 and 7.

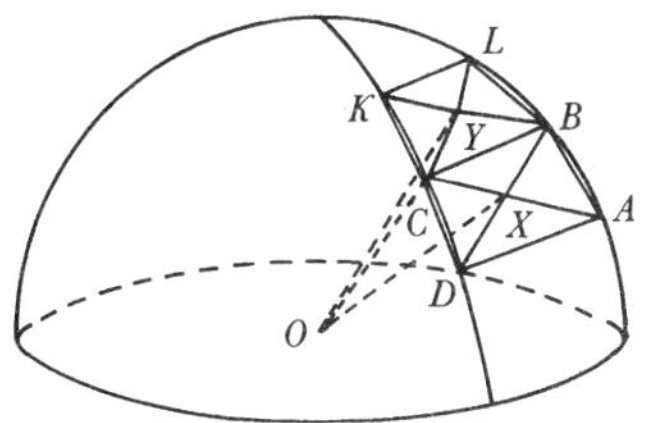

Figure 6.45

The result proved in the previous paragraph establishes that if OF is greater than the radius of x^*, none of the faces $A_1^iA_2^iA_2^{i+1}A_1^{i+1}$ will meet x^*. This can be extended to an arbitrary face $A_j^iA_{j+1}^iA_{j+1}^{i+1}A_j^{i+1}$ by showing (fig. 6.45)

(c) If OX is perpendicular to the face $A_j^iA_{j+1}^iA_{j+1}^{i+1}A_{j+1}^i$ ($ABCD$) and OY to the face $A_{j+1}^iA_{j+2}^iA_{j+2}^{i+1}A_{j+1}^{i+1}$ ($BLKC$), then $YC \preceq XC$.[39]

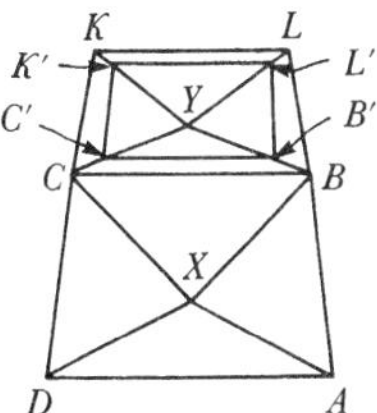

Figure 6.46

For then, since $\mathbf{T}(OY) + \mathbf{T}(YC) \simeq \mathbf{T}(OC) \simeq \mathbf{T}(OX) + \mathbf{T}(XC)$, if $YC \preceq XC$, $OY \succeq OX \succ OF$. To prove (c) one notes that $DC \simeq CK \simeq BL \simeq AB$, by the construction of v_n, and $AX \simeq BX \simeq CX \simeq DX$ and $BY \simeq CY \simeq KY \simeq LY$, as in the proof of the preceding paragraph, and then argues indirectly. One supposes that (fig. 6.46) $YC \succ XC$ and sets $YC' \simeq YK' \simeq YL' \simeq YB' \simeq XC$. By I,8 the angles AXB, CXD as well as the angles BYL, CYK are equal, and, since $B'L' \prec BL \simeq AB$ and $C'K' \prec CK \simeq CD$, angle $BYL \prec$ angle AXB and angle $KYC \prec$ angle CXD, by I,25. Similarly, since, by (a), $AD \succ BC \succ B'C'$, angle $BYC \prec$ angle $BXC \prec$ angle AXD. However, the four angles at Y together and the four angles at X together are each equal to four right angles. Therefore angle $KYL \succ$ angle CXB, so that, by I,24, $CB \prec K'L' \prec KL$, contradicting (a).

Therefore, if t^* and t are taken to be great circles of x^* and x in the same plane, the solution of XII,18 is reduced to the demonstration that

XII,16 Some p_n is such that neither it nor the straight line connecting two alternate vertices of it meets t^*.[40]

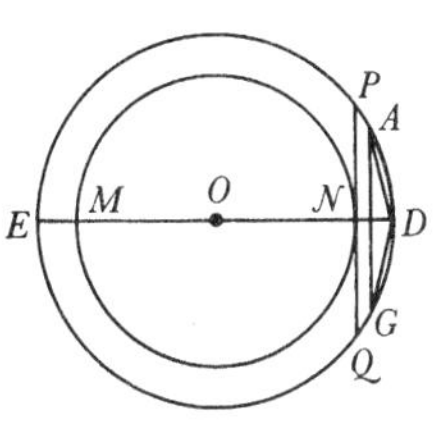

Figure 6.47

To see that this is so, one lets (fig. 6.47) $EMOND$ be a diameter of t intersecting t^* at M, N, and lets PNQ be the chord of t tangent to t^* at N. It is easy to see that D bisects arc PDQ. The construction of the p_n can be said to involve the fixing of vertices $R_1, R_2, \ldots, R_n$ on arc EPD such that R_1 bisects arc EPD and R_{i+1} bisects arc $R_iR_{i+1}D$. By X,1 some R_iD will be an arc AD smaller than arc PD. Let p_n have the consecutive sides AD, DG. Clearly G lies between D and Q on arc PDQ, and none of AD, DG, AG meet t^*. By III,14 no other side of p_n and no other straight line connecting alternate vertices of p_n touches t^*.

Thus Euclid's procedure for proving XII,18 may be summarized as follows. Given x^*, one places it so that its center coincides with the center of x, passes a plane through the center to determine great circles t^* and t, and inscribes in t a polygon p_n satisfying XII,16. Construction of $\pi(p_n)$ in x and a similar polyhedron in x' yields the desired result by the argument indicated previously. One notes that this procedure enables Euclid to avoid having to apply X,1 to show either that the volumes of the $\pi(p_i)$ converge toward that of the sphere x or that the distances of the faces of the $\pi(p_i)$ from the center of x converge to the radius of x; instead he shows that the lengths of the longest edges of $\pi(p_i)$ converge to 0. Despite the apparent complexity of this procedure, it actually simplifies the reasoning from the standpoint of Euclidean geometry. On the other hand, the reasoning completely eliminates any obvious and direct connection with the calculation of the volumes of spheres.

The elimination of the aspect of calculation is a striking feature of books XI and XII. One can, of course, derive proofs of **J** for parallelepipeds, prisms, pyramids, cylinders, and cones from Euclid's proofs of weaker propositions and use the interpretation of compounding as multiplication to represent **J** as a law of volume calculation; but for understanding Euclid the more significant fact is that he does not attempt to prove **J** in these cases. As for circles and spheres, one could adopt Euclid's treatment of them in book XII to derive techniques for computing their areas and volumes. Archimedes uses a variant of the method of XII,2 in his *Measurement of a Circle* to establish that a circle is equal to the right triangle with legs equal to the circumference and radius of the circle and to give an approximation for the ratio of a circle to the square on its diameter. However, from what we know of Archimedes' work and how he conceived it, there can be no doubt that for him and his contemporaries these results were substantive new contributions to mathematics rather than relatively direct extensions of known theorems. Archimedes also proves (*On the Sphere and Cylinder* I,34) that a sphere is equal to the cone with base equal to a great circle of the sphere and height equal to its radius, but in this case the process of construction is quite different from the Euclidean one for XII,18.

Book XII can be seen as a culminating point of the *Elements* in two respects. First of all, it is a deductive terminus in the sense that its results are not applied in XIII. In addition, book XII seems to represent a kind of outer limit to the Euclidean style in mathematics, at least in one direction. Subsequent development of the method of exhaustion involved the treatment of conceptually more complex figures and the introduction of calculational as opposed to geometric techniques. Archimedes is a chronologically isolated figure in this further development, which picks up again in the seventeenth century. By the nineteenth century there is a recognition of the difference between the logical beauty of the method of exhaustion and the unnecessary cumbersomeness of the geometric mode of its Greek geometric formulation. It should be clear that this distinction is not really applicable to the *Elements*.

Notes for Chapter 6

Bibliographical Note

The deductive structure of the three books on solid geometry in the *Elements* is presented in a convenient form by Neuenschwander in "Die stereometrischen Bücher . . . ," pp. 93–97; the looseness of Euclid's reasoning in these books makes it unlikely that any account of their deductive structure can be completely satisfactory. Heller's "Ueber Euklids Definitionen . . ." is a useful treatment of the problems involved in Euclid's handling of equality and similarity for polyhedrons. One can find discussions of the method of exhaustion in almost any work dealing with Greek mathematics. I have found the following useful: Baron, ch. 1; Boyer, ch. 2; Dijksterhuis, vol. II, pp. 228–237; and Zeuthen, *Histoire* . . . , ch. 20.

1. There survives from antiquity a *Spherica* in three books by Theodosius. Although Theodosius lived in the first century B.C., this work may be based on a pre-Euclidean original. (See Heath, *A History* . . . , vol. I, pp. 349–350.) However, although Theodosius starts from fundamentals, he uses a different definition of the sphere than Euclid does, and he clearly intends his work for astronomical rather than geometric purposes.

2. Euclid uses the phrase 'plane of reference' (*hupokeimenon epipedon*) for a plane understood to be fixed for a given argument. One plane is higher (*meteōroteros*) than another when the two intersect in an unspecified way appropriate to the problem at hand. In the present case, the plane of reference and the higher plane cannot have the original straight line as common section.

3. See, for example, Heath's commentary on them and the references he gives.

4. See above, pp. 31–32.

5. Neuenschwander ("Die stereometrischen Bücher . . . ," p. 94) adds a tacit dependence of 23 on 13, presumably because a proof of the uniqueness of the construction in 23 would invoke 13.

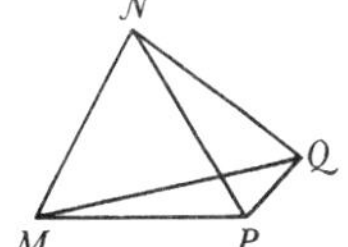

6. Euclid assumes in 8 that CD will intersect any plane which its parallel AB cuts. He should probably give an argument to the effect that if CD did not intersect such a plane, the common section of the plane of reference and the plane through BCD would be both parallel to and intersect AB. In 6 Euclid assumes that AB and CD could not meet at the same point on the plane to which they are perpendicular. He argues against this possibility in XI,13. Obviously it would have been preferable to have proved 13 first.

Figure 6.48

7. In this second proof, division is made into (1) the case in which all three angles are equal and (2) the one in which one of the angles is greater than each of the other two. In the first proof the first of these cases is handled and then a second case in which relative size is left out of account. Euclid draws angle PNQ equal to angle EFG as in fig. 6.48 (cf. fig. 6.13), makes $NQ \simeq NP$, and argues that $MP + EG \simeq MP + PQ \succ$ (I,20) $MQ \succ$ (I,24) HL. The lower part of fig. 6.48 shows that this argument does not work when the angles MNP, EFG are together greater than 180° and the sum of all three angles is greater

than 360°. This possibility is, of course, ruled out by condition (ii′), but not by the *protasis* of 22.

8. The other cases are done in what Heiberg takes to be an interpolation. See Heath, vol. III, pp. 319–321.

9. Compare the discussion of VI, def. 1 above, pp. 157–158. See also Heath, vol. III, pp. 265–267 and Heller, "Ueber Euklids Definitionen . . . ."

10. An example of symmetric noncongruent tetrahedrons is given in scholium 43 to XI,27. The tetrahedrons are said to be equal and similar, but reciprocally rather than similarly situated.

11. Heiberg rejects 38 as an interpolation for reasons given by Heath, vol. III, p. 360.

12. In VI,14 Euclid takes for granted the possibility of positioning equiangular parallelograms in the appropriate way; in XI,31 he carries out a construction to justify the positioning.

13. More accurately, I,36 would be used in a direct proof of VI,1 for parallelograms. Since Euclid proves it directly for triangles, he uses I,38.

14. In the case in which $b_1 \simeq b_2$, Euclid needs a slight argument based on XI,32 to get that $h_1 \simeq h_2$ iff $f_1 \simeq f_2$; see Heath, vol. III, pp. 349–350. To get in the second case that $(f_2, f) = (h_2, h_1)$, Euclid considers f_2 and f as parallelepipeds with bases b_2', b' each determined by the same side of b_2 and h_2, h_1. By XI,25 and VI,1, $(f_2, f) = (b_2', b') = (h_2, h_1)$. Heath (*loc. cit.*) points out that in his actual argument Euclid asserts without justification that if $b_1 \succ b_2$ and $f_1 \simeq f_2$, $h_1 \prec h_2$; the assertion, however, plays no role in the proof.

15. See Boltyanskii, pp. 57–58.

16. See Boltyanskii, p. 37ff.

17. Heath's claim (vol. III, p. 419) that for parallelepipeds "XI,25 really contains the property corresponding to that in" XII,14 is very misleading, because XI,25 applies directly only to parallelepipeds with congruent bases.

18. Heiberg prints in an appendix a version of book XII and the end of book XI based on what he calls manuscript b. Thaer ("Die Euklid-Ueberlieferung . . .") argues that this manscript, which corresponds more closely to Arabic versions of the *Elements*, may reflect Euclid's original more faithfully than does P or the other Greek manuscripts. The fundamental difference between b and P in XII is that b leaves to the reader geometric details and elementary inferences which are spelled out in P.

19. For a brief discussion and references see Heath, vol. III, pp. 365–368.

20. Dijksterhuis gives a concise account of Archimedes' uses of the method on pp. 130–133 of his book *Archimedes*, and more details in his subsequent discussion of particular Archimedean results.

21. The argument beginning with the invocation of Vc is carried out

by Euclid in a separate *lēmma* at the end of XII,2. The inference to $x' > x^{**}$ is a direct application of V,14; but here and in similar cases throughout book XII Euclid uses alternation and then asserts the inequality.

22. Euclid does not actually copy the construction; he simply gives the instruction that a figure similar to z be constructed inside x'. Although this construction can be carried out in more than one way, copying of the construction of z seems to make formal argumentation easiest.

23. This proof is derived from one given by Hasse and Scholz, p. 27. Becker, "Das Prinzip . . . ," pp. 376–378, gives a more complicated argument which does not use any form of trichotomy. To carry out the argument given in the text for XII,18, one must make an assumption corresponding to Euclid's assumption that the fourth proportional exists as a sphere, namely, that $x - s^*$ can be represented as a sphere.

24. Heath, "Greek geometry . . . ," pp. 252–253. The other examples given by Heath seem to me no more persuasive. Simplicius' account of Antiphon's reasoning is found on pp. 54.20–55.8 of his commentary on Aristotle's *Physics*.

25. That $\pi(p)$ and $\rho(p)$ actually are prisms is easily proved using VI,2. Euclid foregoes a proof.

26. The formal device s_n is introduced to make it possible to refer to a unique pyramid generated at a given stage of the construction. Although the device plays no role in the description of the construction process, it is useful for proofs given subsequently.

27. Euclid's proof is somewhat more elaborate than the one sketched here, but nothing important turns on the difference.

28. This proof is taken from XI,40, which Euclid simply cites.

29. Euclid proves the analogue of this, namely, 'if $\eta(x') \simeq \eta(x)$, then $\theta(x') \simeq \theta(x)$' in proving the *lēmma* at the end of XII,4.

30. This assertion corresponds to Euclid's *lēmma* at the end of XII,4.

31. It is perhaps of some interest to indicate a more modern approach to the volume of the triangular pyramid. One considers the prism *pr* which results from p by drawing a triangular face DOP parallel to ABC. (Compare XII,7, p. 226.) It is easy to establish that $\pi(p)$ is similar to *pr*, so that, by **G**pr, $\pi(p)$ is to *pr* in the triplicate of the ratio of corresponding sides, i.e., in the ratio of 1 to 8. Since $\pi(p) \simeq \rho(p)$, one has $v_1 = \frac{1}{4}(pr)$, and similar arguments show that $u_{n+1} = \frac{1}{4}(v_n)$, so that $v_{n+1} = v_n + \frac{1}{4}(v_n)$, and p is the infinite sum

$$\frac{1}{4}(pr) + \frac{1}{4^2}(pr) + \frac{1}{4^3}(pr) + \dots,$$

i.e., $p = \frac{1}{3}(pr)$. With this information all of Euclid's results for triangular prisms can be transferred directly to triangular pyramids. In his *Quadrature of the Parabola* Archimedes does the equivalent of summing an infinite series of the form

$$\frac{1}{4}(x) + \frac{1}{4^2}(x) + \frac{1}{4^3}(x) + \dots.$$

32. This argument amounts to a proof of the extension of XII,8 to pyramids with polygonal bases of a special kind. If Euclid had the full extension of XII,8, he could have established (B) for XII,12 by showing that if x' and x are similar cones, v'_n and v_n are similar pyramids.

33. Euclid establishes this result for $j = 1$ in the course of proving XII,17, and remarks that the same proof works for all faces. Although, in discussing this last application of the method of exhaustion in the *Elements*, I include some details omitted by Euclid, I have made no attempt at completeness; in particular, I do not describe changes needed to adjust certain proofs for quadrilateral faces to triangular ones.

34. To show that BG, CH intersect OA_i, OA_{i+1}, Euclid makes his only application of XI,38. For this proposition see Heath, vol. III, p. 360.

35. Heiberg labels the addition a corollary, although the Greek word *porismos* occurs only in the margin of P and is not found at all in the Theonine manuscripts. Dijksterhuis (vol. II, p. 247) points out that the addition is completely independent of the construction of XII,17.

36. Euclid proves this result in XII,17 as the solution of the problem of inscribing in the greater of two concentric spheres a polyhedron which does not come in contact with the lesser sphere. It would not, of course, matter if some of the faces of the polyhedron were tangent to the inner sphere. But the method of construction would produce such faces only in exceptional circumstances, so that it is simpler to leave the possibility out of consideration.

37. Euclid does not consider the case of v_1, and so asserts that twice the square on FA is greater than the square on AD.

38. Euclid actually states II,13 for triangles with three acute angles. See Heath, vol. I, pp. 406–407.

39. In fact $YC \prec XC$, but the stronger result is not needed. Euclid proves no analogue of (c), apparently taking a proof for faces with edges on the circumference of t to be sufficient.

40. Euclid formulates 16 as a problem, but does not do so very accurately: "Given two circles about the same center, to inscribe in the greater circle an equilateral polygon with an even number of sides which does not touch the lesser circle." The regular polygon with side AG constructed in the text to prove XII,16 would satisfy this condition, but it would not serve Euclid's purposes, as the polygon actually constructed by Euclid does.

7 The Investigation of the Platonic Solids

7.1 Pyramid, Cube, Octahedron, Icosahedron

In propositions 13 to 17 of book XIII Euclid deals in succession with each of the five so-called Platonic or regular solids: the pyramid, contained by four equilateral triangles; the octahedron, contained by eight; the cube, contained by six squares; the icosahedron, contained by twenty equilateral triangles; and the dodecahedron, contained by twelve regular pentagons. In each proposition Euclid does three things: (1) he constructs (*sunistasthai*) the solid; (2) he comprehends (*perilambanein*) it in a given sphere; and (3) he characterizes, sometimes quantitatively, sometimes qualitatively, what I shall call the edge value of the figure—the mathematical relation between the length of an edge of the figure and that of the diameter of the comprehending sphere. In proposition 18, the final proposition of the *Elements*, Euclid "sets out the edges of the five figures and compares them," and concludes with an argument that no other regular solid can be constructed. The remainder of the book, propositions 1 to 12, consists—with the exception of 2, the unused converse of 1—of lemmas for the principal propositions 13 to 17.

There seems to be general agreement among scholars that the mathematical essentials of book XIII are due to Plato's younger contemporary Theaetetus, who is also responsible for at least some part of book X.[1] The close connection between these two books is apparent because book X is essential to the characterization of the edge value of the icosahedron and dodecahedron. For my purposes the question of historical source is not as important as the question of logical relationship to other books. It is sometimes suggested that book XIII is an independent treatise incorporated in the *Elements* with very little revision.[2] Such claims are very hard to evaluate. In general, the logical dependence of XIII on earlier books is as strong as that of other geometric books in the *Elements*. However, the first five propositions are a striking example of failure to exploit deductive connections with earlier results.

These propositions concern the division of a straight line AB into extreme and mean proportion at C. Although Euclid formulates them in terms of proportionality, in their proofs he moves immediately from 'AB is divided into extreme and mean proportion at C with AC as greater segment' to '$\mathbf{T}(AC) \simeq \mathbf{O}(AB, BC)$'; hence it is simpler to use the latter expression directly in representing them.

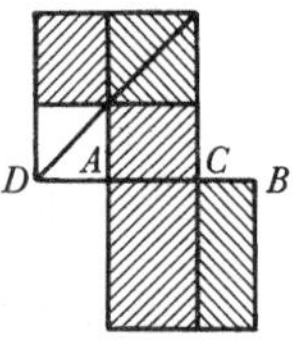

Figure 7.1

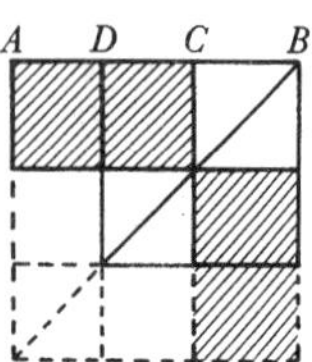

Figure 7.2

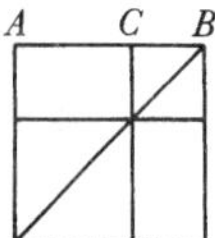

Figure 7.3

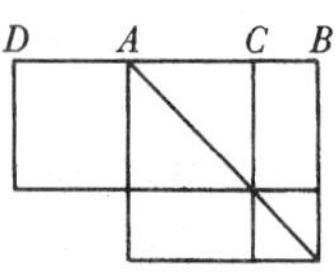

Figure 7.4

XIII,1,2 If $DACB$ (fig. 7.1) is a straight line with $DA \simeq \frac{1}{2}(AB)$, then

$$\mathbf{T}(AC) \simeq \mathbf{O}(AB, BC) \leftrightarrow \mathbf{T}(CD) \simeq 5 \cdot \mathbf{T}(AD).$$

XIII,3 and converse If $ADCB$ (fig. 7.2) is a straight line and AC is bisected at D, then

$$\mathbf{T}(AC) \simeq \mathbf{O}(AB, BC) \leftrightarrow \mathbf{T}(BD) \simeq 5 \cdot \mathbf{T}(AD).$$

XIII,4 and converse If ACB (fig. 7.3) is a straight line, then

$$\mathbf{T}(AC) \simeq \mathbf{O}(AB, BC) \leftrightarrow \mathbf{T}(AB) + \mathbf{T}(BC) \simeq 3 \cdot \mathbf{T}(AC).$$

XIII,5 and converse If $DACB$ (fig. 7.4) is a straight line and $DA \simeq AC$, then

$$\mathbf{T}(AC) \simeq \mathbf{O}(AB, BC) \leftrightarrow \mathbf{T}(AB) \simeq \mathbf{O}(BD, AD).$$

The manuscripts include two treatments of these propositions. In the first and presumably Euclidean treatment, they are proved by the methods of book II from primitive geometric constructions and area comparisons, which are perhaps made sufficiently clear by the figures. The second treatment gives analyses and proofs like those given by Heron for II,1–10 and discussed in section 1.3; the analyses, of course, constitute proofs of the converses of the propositions. I shall follow Heiberg in attributing this second treatment to Heron.[3] To describe it I give algebraic representations of the propositions, taking AB as x, AC as y, so that BC is $x - y$. The propositions and converses then assert the equivalence of

$$y^2 = x(x - y) \quad \text{and} \quad \left(y + \frac{x}{2}\right)^2 = 5\left(\frac{x}{2}\right)^2 \qquad \text{(XIII,1 and 2)},$$

$$\left(x - \frac{y}{2}\right)^2 = 5\left(\frac{y}{2}\right)^2 \qquad \text{(XIII,3)},$$

$$x^2 + (x - y)^2 = 3y^2 \qquad \text{(XIII,4)},$$

$$x^2 = (x + y)y \qquad \text{(XIII,5)}.$$

The quickest of Heron's proofs is the reduction of XIII,4 to II,7, which may be represented as the assertion that

$$x^2 + (x - y)^2 = y^2 + 2x(x - y).$$

Heron's proof of XIII,5 uses the theory of proportion. Invoking inversion and V,17 and 18, he argues that

$$(x, y) = (y, x - y) \leftrightarrow (y, x) = (x - y, y) \leftrightarrow (x + y, x) = (x, y).$$

From an algebraic point of view XIII,5 is very obvious, since

$$y^2 = x(x - y) = x^2 - xy \leftrightarrow y^2 + xy = y(y + x) = x^2.$$

Here, then, is another case in which neither Heron nor Euclid proceeds in the way one would expect of an algebraist. Moreover, neither recognizes the fundamental importance of XIII,5, which, with its converse, establishes that a straight line is divided into extreme and mean ratio if and only if its longer segment is so divided at a point corresponding to the length of the lesser segment.

XIII,3 is an immediate consequence of XIII,1, 2, and 5, as can be seen by setting $y_1 = x - y$, $x_1 = y$, so that $x = x_1 + y_1$. Then

$$\begin{aligned} y^2 = x(x-y) \leftrightarrow x_1^2 = (x_1 + y_1)y_1 \leftrightarrow \\ \text{(XIII,5 and converse) } y_1^2 = x_1(x_1 - y_1) \leftrightarrow \\ \text{(XIII,1 and 2) } \left(y_1 + \frac{x_1}{2}\right)^2 = 5\left(\frac{x_1}{2}\right)^2 \leftrightarrow \\ \left(x - \frac{y}{2}\right)^2 = \left(x - y + \frac{y}{2}\right)^2 = 5\left(\frac{y}{2}\right)^2. \end{aligned}$$

The solid lines in the figure for XIII,3 amount to the figure for II,6, the proof of which Euclid essentially repeats in arguing for XIII,3. Heron invokes II,6 in the form

$$x(x-y) + \left(\frac{y}{2}\right)^2 = \left(x - \frac{y}{2}\right)^2.$$

Finally, Heron's proof of XIII,1,2 uses II,4 and 2, according to which

$$\left(y + \frac{x}{2}\right)^2 = y^2 + \left(\frac{x}{2}\right)^2 + xy$$

and

$$4\left(\frac{x}{2}\right)^2 = x^2 = yx + (x-y)x,$$

so that

$$\begin{aligned} y^2 = x(x-y) \leftrightarrow \left(y + \frac{x}{2}\right)^2 &= x(x-y) + \left(\frac{x}{2}\right)^2 + xy \\ &= \left(\frac{x}{2}\right)^2 + 4\left(\frac{x}{2}\right)^2 \\ &= 5\left(\frac{x}{2}\right)^2. \end{aligned}$$

Euclid's failure to make use of any of these alternative simpler proofs is clear evidence of his obliviousness to certain logical connections, but this is hardly sufficient evidence for concluding that Euclid is more or less blindly reproducing a source in which, for some reason or other, the author was unable to presuppose the relevant contents of book II. Indeed, Euclid's procedure in XIII,1–5 reduplicates the procedure of II,1–10; he proves propositions from scratch when they are in fact consequences of others that he has already proved. The logical relationships are, of course, particularly obvious on the algebraic interpretation of the propositions involved. Yet this fact would seem to count at least as much against the hypothesis that Euclid reasons algebraically as it does in favor of the hypothesis that he is simply reproducing someone else's work. There remains the question why 1–5 are placed at the beginning of book XIII when they could have been proven in book VI or even—in the formulation I have given them—in book II. I can make no fully satisfactory answer to this question; it is possible, however, that Euclid does not prove 1–5 in II because he conceives them to involve the theory of proportion, and does not prove them in VI because the reasoning involved in their proofs is not appropriate there. In any case, as this chapter will make clear, 1–5 are lemmas for the principal results of book XIII and could have been reached by geometric analysis aimed at attaining those results.

Before turning to the discussion of these results, I wish to say a word about Euclid's procedure of constructing a figure and then comprehending it in a sphere. This procedure involves constructing the figure outside the sphere, relative to a straight line equal to the diameter of the given sphere, and then arguing that the semicircle with the straight line as diameter will pass through all the vertices of the figure as it revolves around the diameter.[4] There is no real mathematical difference between this procedure and inscribing the figure in the sphere. Apparently Euclid adopts the procedure and, hence, the generative definition of the sphere (XI, def. 14) as a means of avoiding a treatment of the sphere analogous to the treatment of the circle in book III. From a foundational point of view the advantage gained is only apparent, since the procedure depends upon tacit assumptions about the properties of a semicircle revolving about its diameter. Euclid's willingness to take for granted these properties, but not those of a straight line revolving about an endpoint, is another example of the curious asymmetry between Euclid's presentation of plane and solid geometry.

Euclid's treatment of the regular solids in terms of a construction and comprehension within the sphere generated by a revolving semicircle obscures the intuitive basis of what he is doing. Similar obscurity results from his piling up of the series of lemmas at the beginning of book XIII. Since it seems quite likely that the ultimate source of the principal results of the book was an analysis based on an intuitive picture of the solids already inscribed in a sphere, I shall approach the solids from this point of view, analyzing the conditions which make

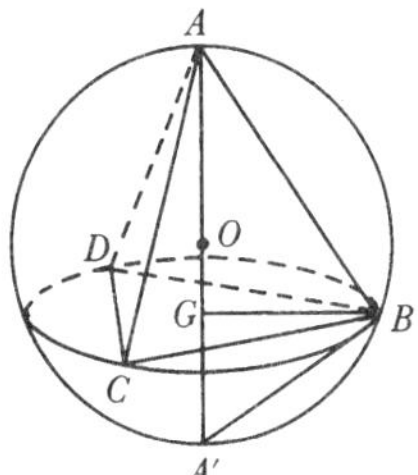

Figure 7.5

a constructive inscription possible.[5] These analyses lead to inscriptions which correspond quite closely to Euclid's constructions and comprehensions. In general I shall not pay much attention to the plane-geometric results used by Euclid; I shall also take for granted obvious facts about the sphere.

The Pyramid (XIII,13)

If a regular pyramid with vertex A and base BCD (fig. 7.5) is inscribed in a sphere with center O, the diameter AOA' will be perpendicular to the plane of the base at a point G, which is also the center of the circle determined by the plane and the sphere, and circumscribing triangle BCD. The problem of inscribing a regular pyramid then reduces to the problem of determining the point G on the diameter AA'. For, given a point A on the surface of the sphere, one need only draw the diameter AA', determine G on it, pass a plane through G perpendicular to the diameter, and inscribe an equilateral triangle in the resulting circular section of the sphere. The pyramid is completed by connecting the vertices of the triangle with A.

The following analysis of the completed construction shows how G is determined. Arc ABA' is a semicircle, so that ABA' is a right triangle. Since BG is perpendicular to AGA', VI,9 gives that $(AG, BG) = (BG, GA')$, or $(\mathbf{T}(AG), \mathbf{T}(BG)) = (AG, GA')$, Hence, to determine G, it suffices to determine $(\mathbf{T}(AG), \mathbf{T}(BG))$, where it is known that

(i) $\mathbf{T}(AG) + \mathbf{T}(BG) \simeq \mathbf{T}(AB) \simeq \mathbf{T}(BD)$.

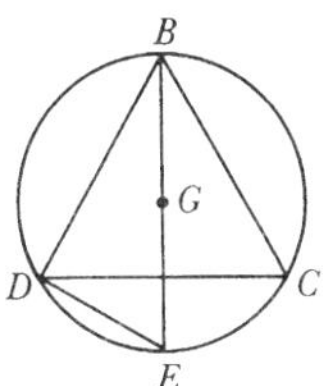

Figure 7.6

In the circle BCD imagine the diameter BGE drawn and DE connected (fig. 7.6). Then BDE is a right triangle; and DE, being equal to the side of an inscribed regular hexagon, equals BG. Hence, $\mathbf{T}(BD) + \mathbf{T}(BG) \simeq \mathbf{T}(BD) + \mathbf{T}(DE) \simeq \mathbf{T}(BE) \simeq 4 \cdot \mathbf{T}(BG)$, and

(ii) $\mathbf{T}(BD) \simeq 3 \cdot \mathbf{T}(BG)$.[6]

Steps (i) and (ii) together give $\mathbf{T}(AG) \simeq 2 \cdot \mathbf{T}(BG)$, so that $AG \simeq 2 \cdot GA'$, and G is determined. Clearly the edge value of the pyramid is also determined, because $(AA', AB) = (AB, AG)$ and $(\mathbf{T}(AA'), \mathbf{T}(AB)) = (AA', AG) = (AG + GA', AG) = (2 + 1, 2) + (3, 2)$. In modern terms the length of the edge of an inscribed regular pyramid is $(\frac{2}{3})^{\frac{1}{2}}\, d$, where d is the length of the diameter of the circumscribing sphere.

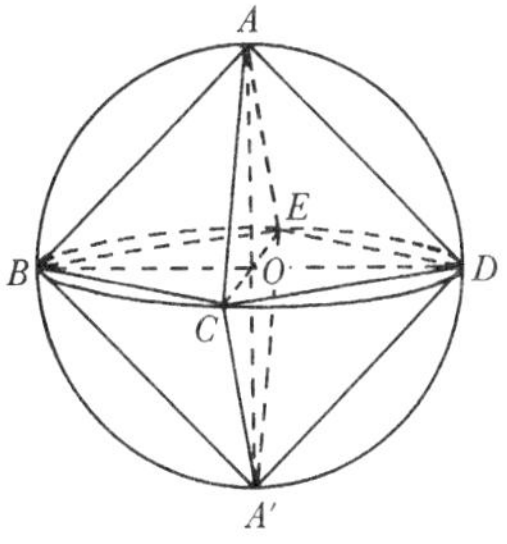

Figure 7.7

The Octahedron (XIII,14)

An inscribed octahedron can be thought of as the result of joining the vertices of a square $BCDE$ (fig. 7.7) inscribed in a great circle to the ends of the diameter AA' perpendicular to the plane of $BCDE$ and intersecting the plane at the center O of the sphere. The equality of the edges of such a figure is clear because each edge is a base of an isosceles right triangle having two radii for sides. Also the edge value of the octahedron is determined by the fact that the square on the diameter is equal to the sum of the squares on two edges; in modern terms the length of the edge of the inscribed regular octahedron is $2^{-\frac{1}{2}}d$.

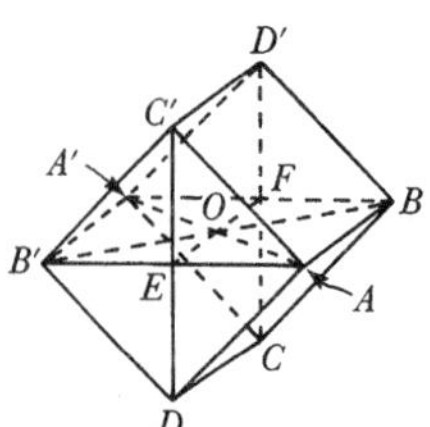

Figure 7.8

The Cube (XIII,15)

It is intuitively clear that if (fig. 7.8) a cube contained by parallel squares $ABCD$, $A'B'C'D'$ and with edges AC', BD', CA', DB' is inscribed in a sphere, the center of the sphere will be the intersection of the diagonals AA', BB', CC', DD' of the cube. To prove this fact in a Euclidean way, it is perhaps simplest to take O as the midpoint of the straight line FE which is the common section of the parallelograms $ABA'B'$, $CDC'D'$. Congruence arguments establish the equality of AE, $A'F$, and, by I,29, angle $OEA \simeq$ angle OFA'; hence, because $FO \simeq OE$, angle $FOA' \simeq$ angle EOA (I,4). It is an easy inference from I,15 that AA' is a straight line; and obviously it is bisected at O. Similarly, BB', CC', DD' are straight lines bisected at O, so that O is the center of the sphere and AA' its diameter. In addition,

$$\mathbf{T}(AA') \simeq \mathbf{T}(AB) + \mathbf{T}(A'B) \simeq \mathbf{T}(AB) + \mathbf{T}(BC) + \mathbf{T}(A'C) \simeq 3\cdot\mathbf{T}(AB),$$

and the edge value of the cube is determined; in modern terms it is $3^{-\frac{1}{3}}d$.

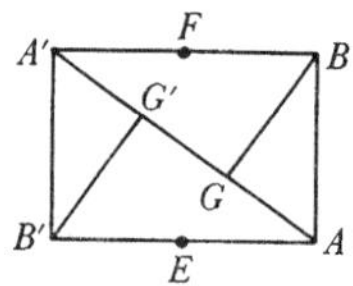

Figure 7.9

Given a diameter AA' of a sphere, the simplest way to inscribe a cube in the sphere is perhaps to determine first the points G and G' on AA' to which the perpendiculars to AA' from B and B' fall (fig. 7.9). These points are easily determined because, by similar triangles, $(AA', AB) = (AB, AG)$ and $(\mathbf{T}(AA'), \mathbf{T}(AB)) = (AA', AG)$. Hence $AA' \simeq 3\cdot AG \simeq$ (by the same argument) $3\cdot A'G'$. Thus, to inscribe a cube in the sphere one need only determine G and G' using VI,9, and, in the plane of reference through AA', draw perpendiculars to AA' at G and G' in opposite directions which intersect the sphere at B and B'. If AB' and $A'B$ are joined and bisected at E and F, perpendiculars to the plane of reference at E and F will intersect the sphere at C' and D' on one side of the plane, and at D and C on the other. Elaborate but elementary arguments establish that the points A, B, C, D, A', B', C', D', determine a cube.[7]

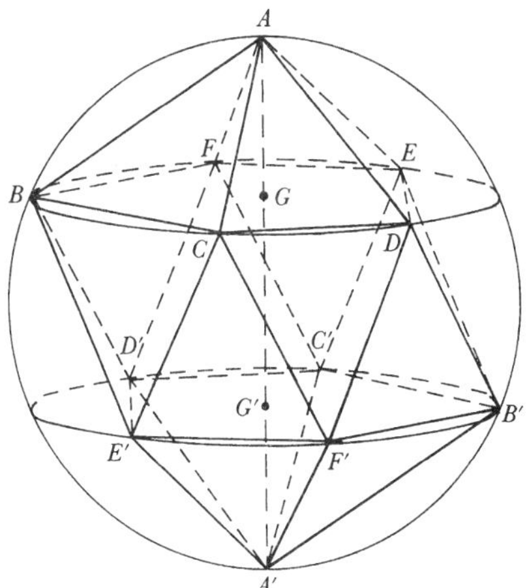

Figure 7.10

The Icosahedron (XIII,16)

Figure 7.10 shows an icosahedron inscribed in a sphere with center O. AA' is a diameter of the sphere; and $BCDEF$, $B'C'D'E'F'$ are two parallel regular pentagons with vertices on the surface of the sphere. The twenty faces of the icosahedron are equilateral triangles formed by joining A to B, C, D, E, F and A' to B', C', D', E', F', and then forming the equilateral triangles $BE'C$, $E'CF'$, $CF'D$, $F'DB'$, etc. The crux of the analysis of the inscription of the icosahedron is to determine the symmetrical points G, G' at which AA' intersects the planes of $BCDEF$, $B'C'D'E'F'$. One first notes that these two planes are perpendicular to AA', so that a perpendicular to the plane of $B'C'D'E'F'$ through E' will be perpendicular to the plane of $BCDEF$ at a point R on the lesser arc BC, and $E'R$ (of fig. 7.11) will be equal to GG' (of fig. 7.10). Clearly, also, the triangles BRE', CRE' are congruent as right triangles with two equal sides; hence R will lie at the midpoint of the arc BC, and BR will be the side d_s of a regular decagon inscribed in the circle $BCDEF$. Since BE' is the side p_s of a regular pentagon inscribed in the same circle and the angle BRE' is right, one has

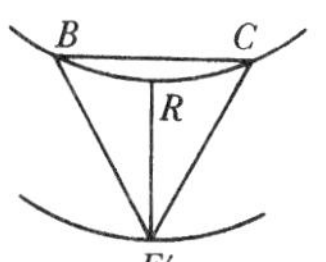

Figure 7.11

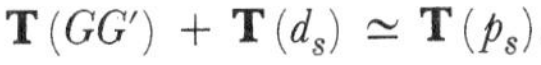

$$\mathbf{T}(GG') + \mathbf{T}(d_s) \simeq \mathbf{T}(p_s).$$

Again, the straight line AB (figs. 7.12 and 7.10) is the side p_s of such a pentagon, and GB is the radius of the circle in which it is inscribed—or, equivalently (IV,15, cor.), the side h_s of a hexagon inscribed in the circle. And since the angle AGB is right,

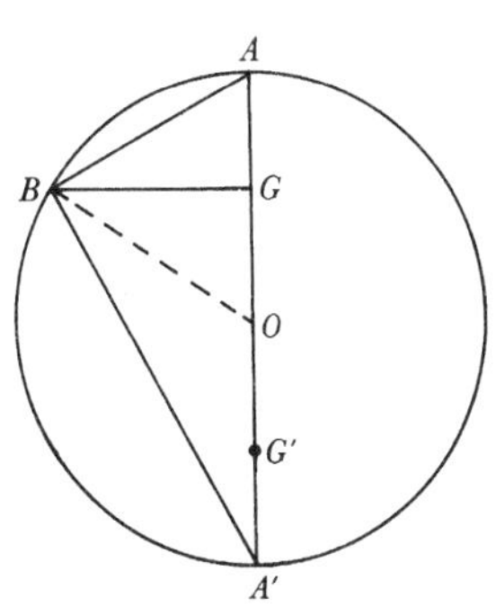

Figure 7.12

$$\mathbf{T}(h_s) + \mathbf{T}(AG) \simeq \mathbf{T}(p_s).$$

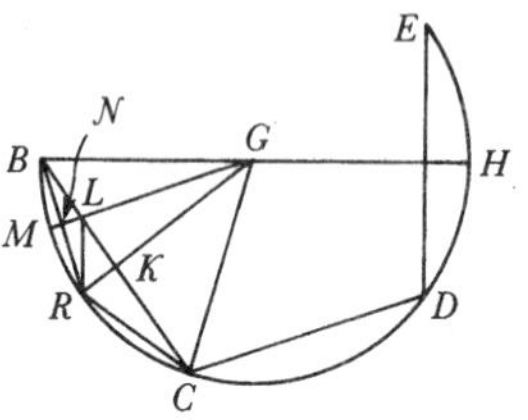

Figure 7.13

In what I can only call a *tour de force*[8] Euclid establishes that $GG' \simeq BG \simeq h_s$ and $AG \simeq d_s$ by proving that for any circle $BCDEF$

XIII,10 $\mathbf{T}(h_s) + \mathbf{T}(d_s) \simeq \mathbf{T}(p_s)$.

It is clear that (fig. 7.13) the radius GR, intersecting BC at K, is the perpendicular bisector of BC. Let the radius GM be the perpendicular bisector of BR, intersecting it at N and BC at L. Clearly the angles RCB, RBC are equal; but since the triangles LBN, LRN are equal in all respects, the angles RBC, LRN are equal and the triangles BRL, BCR are similar, with

$$(BC, BR) = (BR, BL), \quad \text{i.e.,} \quad \mathbf{T}(d_s) \simeq \mathbf{T}(BR) \simeq \mathbf{O}(BC, BL).$$

But since

$$\mathbf{T}(p_s) \simeq \mathbf{T}(BC) \simeq (\text{II},2) \quad \mathbf{O}(BC, BL) + \mathbf{O}(BC, CL),$$

it remains only to show that

$$\mathbf{T}(h_s) \simeq \mathbf{T}(CG) \simeq \mathbf{O}(BC, CL).$$

If the diameter BGH is drawn, it is clear that the angle CBH stands on $\frac{3}{10}$ of the circumference of the circle while CGM stands on $\frac{3}{20}$. Hence (III,20 and VI,33), the two angles are equal and the triangles CGL, CBG are similar, with

$$(BC, CG) = (CG, CL), \quad \text{i.e.,} \quad \mathbf{T}(CG) \simeq \mathbf{O}(BC, CL).$$

The knowledge that AG and $A'G'$ are the sides of a regular decagon inscribed in a circle with radius GG' enables one to determine G and G' on AA' constructively, because it is possible to inscribe a regular decagon in a circle and get a geometric representation of the relevant ratio; indeed, the ratio is given in fig. 7.13. However, it is also clear that the triangle GBR is an isosceles triangle with vertex angle of 36°. One knows from the inscription of the regular pentagon that, with such a triangle, the base BR is equal to the greater segment when a leg BG is divided in extreme and mean ratio, i.e., $\mathbf{T}(d_s) \simeq \mathbf{O}(h_s, h_s - d_s)$, or, by XIII,5,

XIII,9 $\mathbf{T}(h_s) \simeq \mathbf{O}(h_s + d_s, d_s)$.

Thus AG' is divided in extreme and mean ratio at G; and, by XIII,3, G is determined by the fact that $\mathbf{T}(AO) \simeq 5 \cdot \mathbf{T}(GO)$.

Euclid's reasoning differs from that just described in one significant respect. He does not use the fact that GBR is an isosceles triangle with vertex angle of 36°, and he shows no sense of the relationship of XIII,9 to material proved in connection with the inscription of the regular pentagon. In fact, he establishes

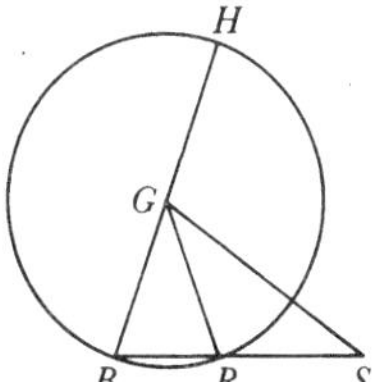

Figure 7.14

XIII,9 and converse If (fig. 7.14) BRS is a straight line, BR is equal to the side of a regular decagon inscribed in the circle with radius RS if and only if $\mathbf{T}(RS) \simeq \mathbf{O}(BS, BR)$.

Euclid takes GB, GR, each equal to RS, to be radii of a circle of which BGH is a diameter. Then BR is the side of a regular decagon inscribed in this circle if and only if (III,28,29, VI,33) angle $RGH \simeq 4 \cdot$ angle BGR. But, since $GB \simeq GR \simeq RS$, angle $RGH \simeq 2 \cdot$ angle GRB and angle $GRB \simeq 2 \cdot$ angle GSR. Hence, angle $RGH \simeq 4 \cdot$ angle GSR; and BR is the side of an inscribed decagon if and only if angle $BGR \simeq$ angle BSG if and only if triangles BGR and BSG are similar if and only if $(BS, GB) = (GB, BR)$ if and only if $(BS, RS) = (RS, BR)$ if and only if $\mathbf{T}(RS) \simeq \mathbf{O}(BS, BR)$.

In this case Euclid's failure to exploit a connection with work already done in book IV may be due to a failure to recognize the direct connection between the regular decagon and the isosceles triangle with vertex angle of 36°.[9] However, in the following section we will see a failure to exploit an even more obvious connection involving the same material. This failure may, of course, be taken as further evidence for Euclid's slavish following of independent sources. However, it is also possible to suppose that Euclid chose to prove the minimum number of results needed for the inscription of the regular pentagon and accordingly did not put that material in a form directly usable in book XIII. In any case, there is always a danger in historical interpretation of mistaking the obvious to us for the eternally obvious.

The analysis I have given leads directly to the inscription of the icosahedron. One determines (fig. 7.10 again) the points G, G' on the diameter AA' of a sphere with center O by the condition $\mathbf{T}(AO) \simeq 5 \cdot \mathbf{T}(GO) \simeq 5 \cdot \mathbf{T}(G'O)$[10] and passes planes through G, G' perpendicular to AA', which determine equal circles $BCDEF$, $B'C'D'E'F'$. Then a regular pentagon $BCDEF$ is inscribed in the first circle, and the smaller arc BC is bisected at R (fig. 7.11). The perpendicular to the plane of $BCDEF$ at R intersects the circle $B'C'D'E'F'$ at a point E' which is taken as the vertex for a regular pentagon $B'C'D'E'F'$ inscribed in that circle. Connection of the appropriate points among A, B, C, D, E, F, A', B', C', D', E', F' determines an icosahedron. Each triangular face of this icosahedron has as one side a side of a regular pentagon inscribed in a circle with radius equal to BG. To show that the triangular faces are equilateral it suffices to show that AB and BE' are sides of regular pentagons of the same kind. However, one knows that

$$\mathbf{T}(AB) \simeq \mathbf{T}(BG) + \mathbf{T}(AG)$$

and that

$$\mathbf{T}(BE') \simeq \mathbf{T}(RE') + \mathbf{T}(BR) \simeq \mathbf{T}(GG') + \mathbf{T}(BR),$$

i.e.,

$$\mathbf{T}(AB) \simeq \mathbf{T}(h_s) + \mathbf{T}(AG)$$

and

$$\mathbf{T}(BE') \simeq \mathbf{T}(GG') + \mathbf{T}(d_s);$$

hence, by XIII,10, it suffices to show that AG is the side of a regular decagon inscribed in a circle with radius GG' and that $GG' \simeq BG$. For the first, one argues that since $\mathbf{T}(AO) \simeq 5\cdot\mathbf{T}(GO)$ and GG' is bisected at O, then $\mathbf{T}(GG') \simeq \mathbf{O}(AG', AG)$, by the converse of 3, so that, by the converse of 9, AG is the side of such a decagon. For the second, one need only point out that, since (see fig. 7.12) $\mathbf{T}(GO) + \mathbf{T}(BG) \simeq \mathbf{T}(OB) \simeq \mathbf{T}(OA) \simeq 5\cdot\mathbf{T}(GO)$, then $\mathbf{T}(BG) \simeq 4\cdot\mathbf{T}(GO)$, $BG \simeq 2\cdot GO$, and $BG \simeq GG'$.

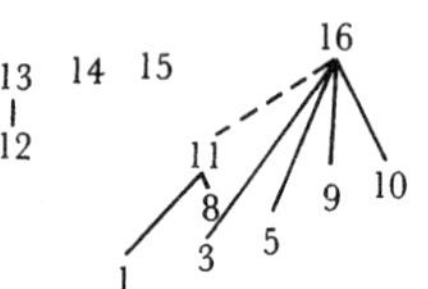

Figure 7.15

Figure 7.15 indicates the internal deductive structure of book XIII insofar as the direct treatment of the regular pyramid, cube, octahedron, and icosahedron are concerned. The broken line indicates that 11 is relevant only to the determination of the edge value of the icosahedron, a problem which will be discussed in the following section.

7.2 The Edge Value of the Icosahedron and Book X

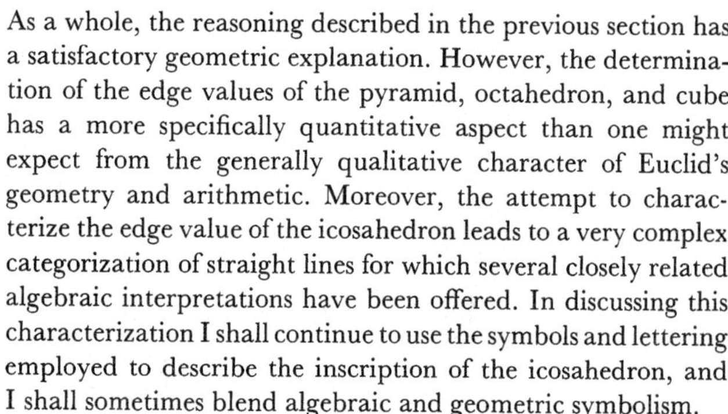

As a whole, the reasoning described in the previous section has a satisfactory geometric explanation. However, the determination of the edge values of the pyramid, octahedron, and cube has a more specifically quantitative aspect than one might expect from the generally qualitative character of Euclid's geometry and arithmetic. Moreover, the attempt to characterize the edge value of the icosahedron leads to a very complex categorization of straight lines for which several closely related algebraic interpretations have been offered. In discussing this characterization I shall continue to use the symbols and lettering employed to describe the inscription of the icosahedron, and I shall sometimes blend algebraic and geometric symbolism.

Since (figs. 7.12 and 7.10)$\mathbf{T}(AO) \simeq 5\cdot\mathbf{T}(GO)$, $AA' \simeq 2\cdot AO$ and $GB \simeq GG' \simeq 2\cdot GO$, then $(\mathbf{T}(AA'), \mathbf{T}(GB)) = (5, 1)$; hence, to determine the edge value of the icosahedron, it is only necessary to determine $(\mathbf{T}(FB), \mathbf{T}(GB))$, i.e., (p_s, h_s), the ratio of the side of an inscribed regular pentagon to the radius of the circle in which it is inscribed. It is convenient to begin by considering a direct modern numerical algebraic way of dealing with this problem. One has

$$h_s^2 = (h_s + d_s)\cdot d_s = h_s\cdot d_s + d_s^2,$$

by XIII,9. Completing the square gives

$$(d_s + \tfrac{1}{2}h_s)^2 = d_s^2 + h_s \cdot d_s + \tfrac{1}{4}h_s^2 = h_s^2 + \tfrac{1}{4}h_s^2 = 5 \cdot \tfrac{1}{4}h_s^2.$$

Hence

$$d_s + \tfrac{1}{2}h_s = \sqrt{5} \cdot \tfrac{1}{2}h_s,$$

and

$$d_s = \tfrac{1}{2}h_s \cdot (\sqrt{5} - 1).$$

Substituting this result in $p_s^2 = h_s^2 + d_s^2$ (XIII,10) yields that

$$p_s^2 = h_s^2 + (\tfrac{1}{2}h_s \cdot (\sqrt{5} - 1))^2 = h_s^2 + \tfrac{1}{4}h_s^2 \cdot (6 - 2\sqrt{5}) = \tfrac{1}{4}h_s^2 \cdot (10 - 2\sqrt{5});$$

and, hence,

$$p_s = \tfrac{1}{2}h_s \cdot \sqrt{10 - 2\sqrt{5}}.$$

Moreover, since the square on the diameter d of the sphere is five times the square on h_s, the edge value in modern terms is

$$\tfrac{1}{10}d \cdot \sqrt{10 \cdot (5 - \sqrt{5})}.$$

In quasi-geometric terms the relations between p_s and h_s and between p_s and d are expressed by

$$\text{(i)} \quad \mathbf{T}(p_s) \simeq \mathbf{T}\left(\frac{h_s}{2}\right) \cdot (10 - 2 \cdot \sqrt{5}),$$

$$\text{(ii)} \quad \mathbf{T}(p_s) \simeq \mathbf{T}(d) \cdot \left(\frac{1}{2} - \frac{1}{2} \cdot \frac{1}{\sqrt{5}}\right);$$

there is consequently no arithmetic expression for either relation. To indicate one way in which Euclid could have overcome this fact, I rewrite (i) and (ii) as

$$\text{(i}'\text{)} \quad \mathbf{T}(p_s) \simeq \mathbf{O}\left(\frac{h_s}{2}, 10 \cdot \frac{h_s}{2} - 2 \cdot \sqrt{5} \cdot \frac{h_s}{2}\right),$$

$$\text{(ii}'\text{)} \quad \mathbf{T}(p_s) \simeq \mathbf{O}\left(d, \frac{d}{2} - \frac{1}{2} \cdot \frac{1}{\sqrt{5}} \cdot d\right).$$

In order to obtain a geometric representation of these equalities, Euclid could "solve"

$$\mathbf{T}(d_s + \tfrac{1}{2}h_s) \simeq 5 \cdot \mathbf{T}(\tfrac{1}{2}h_s)$$

by constructing a straight line x such that

$$\mathbf{T}(x) \simeq 5 \cdot \mathbf{T}(\tfrac{1}{2}h_s)\,(\simeq \mathbf{T}(\tfrac{1}{2}d));$$

x, then, is a representation of $\sqrt{5} \cdot (\tfrac{1}{2}h_s)$. Moreover, since $d_s \simeq x - \tfrac{1}{2}h_s$, one has, by XIII,10 and II,7,

$$\mathbf{T}(p_s) \simeq \mathbf{T}(h_s) + \mathbf{T}(x) + \mathbf{T}(\tfrac{1}{2}h_s) - 2 \cdot \mathbf{O}(x, \tfrac{1}{2}h_s) \simeq 10 \cdot \mathbf{T}(\tfrac{1}{2}h_s) - \mathbf{O}(2 \cdot x, \tfrac{1}{2}h_s) \simeq \mathbf{O}(\tfrac{1}{2}h_s, 10 \cdot (\tfrac{1}{2}h_s) - 2 \cdot x),$$

a fully geometric representation of (i′). Similarly, since $x \simeq \frac{1}{2}d$,

$$\mathbf{T}(p_s) \simeq 2\cdot\mathbf{T}(x) - 2\cdot\mathbf{O}(x, \tfrac{1}{2}h_s) \simeq 2\cdot\mathbf{T}(\tfrac{1}{2}d) - 2\cdot\mathbf{O}(\tfrac{1}{2}d, \tfrac{1}{2}h_s) \simeq 2\cdot(\mathbf{O}(\tfrac{1}{2}d, \tfrac{1}{2}d - \tfrac{1}{2}h_s)) \simeq \mathbf{O}(d, \tfrac{1}{2}d - \tfrac{1}{2}h_s),$$

where h_s is a geometric representation of $5^{-\frac{1}{2}}d$.

Euclid's determination of the edge value of the icosahedron does not really correspond to this geometric version of our algebraic calculation. In the first place he does not use XIII,9 or 10, but what is essentially a corollary of the technique for inscribing a regular pentagon in a circle:

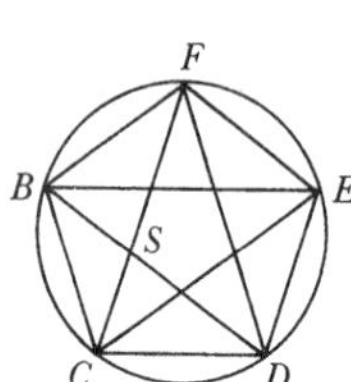

Figure 7.16

XIII,8 If (fig. 7.16) $BCDEF$ is a regular pentagon inscribed in a circle and BD, CF intersect at S, then $\mathbf{T}(SD) \simeq \mathbf{O}(BD, BS)$ and $SD \simeq CD$.

Again Euclid shows no sense of a connection between this assertion and his earlier inscription of the pentagon. He argues that BCD, CBF are congruent triangles with $FB \simeq BC \simeq CD$; hence angles SBC, SCB are equal, and angle CSD is twice angle FCB. But angle FCD is twice the same angle, since it stands on twice the circumference. Therefore, angle $CSD \simeq$ angle SCD, and $SD \simeq CD$. In addition, because $BC \simeq CD$ and $BS \simeq SC$, triangles BCD, BSC are similar with $(BD, CD) = (BC, BS)$ or $(BD, SD) = (SD, BS)$.

The crucial proposition for the characterization of the edge value of the icosahedron is XIII,11, the exposition of which will lead into a discussion of book X. In 11 Euclid characterizes the relation between p_s and the diameter of the circle of which h_s is the radius, i.e., he works with $(p_s, 2\cdot h_s)$ rather than with (p_s, h_s). In fig. 7.17 the angle BFH and the angles at T are right, so that $(BH, FB) = (FB, BT)$ and $\mathbf{T}(FB) \simeq \mathbf{O}(BH, BT)$. We may express this equality as $p_s^2 = BT\cdot(2\cdot h_s)$; hence the algebraic analysis at the beginning of this section can be said to have shown that $BT = \frac{1}{8}h_s\cdot(10 - 2\sqrt{5})$. Euclid's characterization of the relation of BT to BH can be given an algebraic interpretation, but his formulation appears to be geometrically motivated. He uses the fact that $BT \simeq \frac{1}{2}BH - GT$, where $FG \simeq \frac{1}{2}BH$ and, since the triangles CFU, GFT are similar, $(FG, GT) = (FC, CU)$. But $CU \simeq \frac{1}{2}CD$; and, by XIII,8,

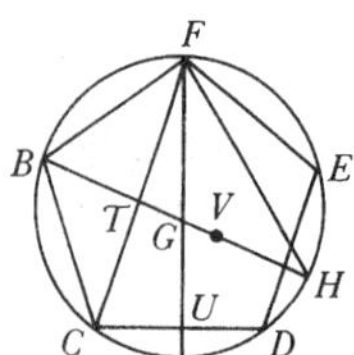

Figure 7.17

$$\mathbf{T}(CD) \simeq \mathbf{O}(FC, FC\text{–}CD).$$

Therefore, by XIII,1,

$$\mathbf{T}(CD + CT) \simeq 5\cdot\mathbf{T}(CT).$$

Since CT is half of FC, one has

$$(CT, CD) = (\tfrac{1}{2}FC, 2\cdot CU) = (\tfrac{1}{2}FG, 2\cdot GT) = (\tfrac{1}{4}FG, GT) = (\tfrac{1}{8}BH, GT)$$

and

$$\mathbf{T}(\tfrac{5}{8}BH - BT) \simeq \mathbf{T}(\tfrac{1}{2}BH - BT + \tfrac{1}{8}BH) \simeq \mathbf{T}(GT + \tfrac{1}{8}BH) \simeq 5 \cdot \mathbf{T}(\tfrac{1}{8}BH).$$

Euclid gives a geometric representation of this equality by finding V on GH with $GV \simeq \frac{1}{8}BH$, so that $\mathbf{T}(TV) \simeq 5 \cdot \mathbf{T}(GV)$; or, since $BV \simeq 5 \cdot GV$ and $\mathbf{T}(BV) \simeq 25 \cdot \mathbf{T}(GV)$, then $\mathbf{T}(BV) \simeq 5 \cdot \mathbf{T}(TV)$, or

$$\mathbf{T}(\tfrac{5}{8}BH) \simeq 5 \cdot \mathbf{T}(\tfrac{5}{8}BH - BT).$$

Again, it is possible to extract from this equality the numerical value of p_s. For it can be expressed

$$(\tfrac{5}{8}BH)^2 = 5 \cdot (\tfrac{5}{8}BH - BT)^2,$$

so that

$$\frac{5}{8}BH = \sqrt{5} \cdot \left(\frac{5}{8}BH - BT\right), \qquad \frac{\sqrt{5}}{8}BH = \frac{5}{8}BH - BT,$$

$$BT = \frac{5}{8}BH - \frac{\sqrt{5}}{8}BH = \frac{1}{8}BH(5 - \sqrt{5}) = \frac{1}{16}BH(10 - 2\sqrt{5}) = \frac{1}{8}h_s \cdot (10 - 2\sqrt{5}).$$

However, although the last geometric equality is crucial for Euclid's account of the relation of p_s to h_s, he does not further pursue anything resembling computation. In order to understand what he does do, it is necessary to look at book X, which begins with a definition of commensurability and incommensurability, terms which I have used previously.

X, def. 1 Magnitudes measured by the same measure are commensurable, but those for which no common measure can come-to-be (*ōn mēden endechetai koinon metron genesthai*) are called incommensurable.

Becker[11] interprets the definition of incommensurability in a constructive manner to mean "Magnitudes are incommensurable when there is a law showing that they can have no common measure." He points out that on this interpretation an inference from noncommensurability to incommensurability is constructive, whereas one from nonincommensurability to commensurability is not. Only in X,16 does Euclid attempt to make the nonconstructive inference; but one would hardly expect many instances of such an inference, since indirect proofs of commensurability are unnatural in elementary cases in which a measure can be described constructively. In 16 Euclid proves

X,16 (i) If x and y are incommensurable so are x and $x + y$;

(ii) If x and $x + y$ are incommensurable, so are x and y.

Euclid could reduce these two assertions to X,15, which asserts that x and y are commensurable if and only if x and $x + y$ are, but he chooses to argue from the definition. For (ii) he argues constructively that a measure for x and y would be one for x and $x + y$. He could proceed in the same way for (i), but instead he says that "if x, $x + y$ are not incommensurable, some magnitude will measure them; let it measure them, if possible, and let it be z." To preserve his thesis that Euclid is a conscious constructivist, Becker lays stress on the word 'possible' and asserts that Euclid is inferring from the non-incommensurability of x and $x + y$ only the possibility, not the actuality, of a common measure. Since Euclid only deals with this possibility to deny rather than affirm commensurability, Becker's suggestion is irrefutable. However, because Euclid could have avoided even the appearance of nonconstructive reasoning by proceeding as he does in case (ii), it seems more natural to assume that he makes no distinction between non-incommensurability and commensurability. Hence I shall render definition 1 as

X, def. 1 $\text{COM}(x, y) \leftrightarrow \exists z(z \text{ measures } x \;\&\; z \text{ measures } y)$,
$\text{INCOM}(x, y) \leftrightarrow \neg\, \text{COM}(x, y)$.

This definition, of course, applies to magnitudes in general. Euclid extends the notion of commensurability to potential commensurability or commensurability in square (*dunamei summetros*) for straight lines in a definition which may be rendered

X, def. 2 $\text{TCOM}(x, y) \leftrightarrow \text{COM}(\mathbf{T}(x), \mathbf{T}(y))$,
$\text{TINCOM}(x, y) \leftrightarrow \neg\, \text{TCOM}(x, y)$,

where it is taken for granted that these notions apply only to straight lines. It should be clear that a separate symbol is not needed for either incommensurability or incommensurability in square, although Euclid proves a number of results by the substitution of 'not commensurable (in square)' for 'incommensurable (in square)' and simple logical manipulation. Nor, from the point of view of economy, is a separate symbol needed for commensurability in square, but the concept involved is of fundamental importance for book X and for the characterization of the edge value of the icosahedron. For the problem of book X is to develop a way of dealing with those straight lines which, like p_s and h_s, are neither commensurable nor commensurable in square. It is also convenient to have a symbol for the frequently employed Euclidean expression 'commensurable in square only'. I shall use TCOM* (x, y) to abbreviate TCOM (x, y) & $\neg$ COM (x, y).

Since some of the basic material in book X has already been discussed and some of the rest of this material has quite simple proofs, here I shall only summarize what is established. After proving X,1, which was discussed above on p. 140ff., Euclid shows in proposition 2 that alternate subtraction (see p. 78) applied to magnitudes produces a measure if one exists; he argues that a common measure measures any result of alternate subtraction and, by X,1, alternate subtraction will eventually produce a result smaller than a given magnitude. Euclid then applies alternate subtraction to pairs (X,3) and triples (X,4) of commensurable magnitudes to find their greatest common measure (cf. VII,2 and 3). Euclid never uses any of these results, although he might have taken greatest common measures in place of arbitrary ones in X,5–8, where he shows that magnitudes are commensurable if and only if they have the ratio of a number to a number. The weakness from a foundational point of view of these propositions and their proofs has been discussed in terms of X,5 on pp. 136–138.

X,9 and 10 have been discussed on pp. 112–113. Proposition 9 is of fundamental importance, since it establishes as a criterion for commensurability of straight lines which are commensurable in square that the squares on the straight lines have the ratio of a square number to a square number. Of course, in any application of X,9 to show that a given x and y are commensurable in square only, one would have to show that $\mathbf{T}(x)$ and $\mathbf{T}(y)$ do not have the ratio of a square number to a square number. In such cases Euclid is content to assert truly but without proof that a given numerical ratio is not that of a square number to a square number. As was indicated in the earlier discussion of 9, the tacit justification for such an assertion is that the two numbers expressing the ratio are not similar plane numbers. In propositions 11–13, 15, and 16 Euclid proves some frequently used elementary facts about commensurability:

X,11 $(x, y) = (z, w) \rightarrow$ (COM $(x, y) \leftrightarrow$ COM (z, w));

X,12,13 Two magnitudes which are commensurable (in square) are each commensurable (in square) with the same magnitudes;

X,15,16 Two magnitudes are commensurable if and only if their sum (difference) is.

The words in parentheses in X,12,13 and 15,16 are my trivial additions to what Euclid explicitly proves. In general I shall use these laws without citation. A final elementary law which

Euclid proves for application in X,31, 32, 66, 103, 112, and 113 is

X,14 $(x,y) = (x',y')\ \&\ \mathbf{T}(x) - \mathbf{T}(y) \simeq \mathbf{T}(z)\ \&$
$\mathbf{T}(x') - \mathbf{T}(y') \simeq \mathbf{T}(z') \rightarrow (\text{COM}(x, z) \leftrightarrow \text{COM}(x', z'))$.

Euclid argues that, if the antecedent holds, then

$$(\mathbf{T}(x), \mathbf{T}(y)) = (\mathbf{T}(x'), \mathbf{T}(y')),$$
$$(\mathbf{T}(x), \mathbf{T}(z)) = (\mathbf{T}(x'), \mathbf{T}(z')),$$
$$(x, z) = (x', z');$$

and the result follows by X,11.

The results described in the previous paragraph suffice for the next step in Euclid's determination of the edge value of the icosahedron. Since $(\mathbf{T}(AA'), \mathbf{T}(BH)) = (5, 4)$, $(BH, BV) = (8, 5)$, and $(\mathbf{T}(BV), \mathbf{T}(TV)) = (5, 1)$, it is clear that any two of AA', BH, BV, TV, except BH, BV, are commensurable in square only. One can also show that BT is incommensurable in square with BV (and hence with BH and AA'). For since $\text{TCOM}(BV, TV)$, then $\text{COM}(\mathbf{T}(BV), \mathbf{T}(BV) + \mathbf{T}(TV))$. However, by II,7, $\mathbf{T}(BV) + \mathbf{T}(TV) \simeq 2 \cdot \mathbf{O}(TV, BV) + \mathbf{T}(BT)$, so that $\text{COM}(\mathbf{T}(BV), 2 \cdot \mathbf{O}(BV, TV) + \mathbf{T}(BT))$. But, since $(\mathbf{T}(BV), \mathbf{O}(BV, TV)) = (BV, TV)$ and $\neg\,\text{COM}(BV, TV)$, then $\neg\,\text{COM}(\mathbf{T}(BV), 2 \cdot \mathbf{O}(BV, TV))$; and therefore

$$\neg\,\text{COM}(\mathbf{T}(BV), \mathbf{T}(BT)).$$

A general form of this proof establishes X,73; but to explain this proposition one needs more terminology from the beginning of book X.

In definitions 3 and 4 of book X Euclid speaks of an assigned (*protetheisa*) straight line r_1. A straight line x or area equal to $\mathbf{T}(x)$ is said to be rational (*rhētos*) if and only if x and r_1 are commensurable in square. Irrationality is defined in the obvious way. The difference between the Euclidean notion of rationality and ours should be clear. First of all, the Euclidean notion is relative to a straight line r_1, whereas ours is the absolute numerical notion of being a fraction. More importantly, given a choice of r_1, for Euclid a straight line the length of which is expressible in our terms in either the form $\frac{m}{n} \cdot r_1$ or in the form $\sqrt{\frac{m}{n}} \cdot r_1$ is rational; we, however, would presumably consider only a length expressible in the former way rational. In other words, the Euclidean notion of rationality cuts right across the distinction we draw between rationality and irrationality.

Although rationality is relative to a choice of r_1, because of X,12 and 13 any straight line rational relative to r_1 will determine the same class of rational straight lines that r_1 does

For many purposes it is appropriate to think of r_1 simply as an arbitrarily chosen member of a class R of straight lines which is closed and connected with respect to commensurability in square and whose members are called rational; in the remainder of this section I will use the symbol r_1 in conformity with this conception to designate an arbitrarily chosen rational straight line. However, some of the distinctions made by Euclid do involve the difference between commensurability and commensurability in square with a rational straight line; and obviously how a line will be classified with respect to such a distinction will depend upon the choice of a member of R.

This point will become clearer when I describe the categorization of straight lines developed in X for the purpose of characterizing the edge value of the icosahedron. Before turning to that description, I want to make a general point about the nature of the categorization in terms of X,73, which can be stated, using an obvious abbreviation, as

X,73 $\text{RAT}(u)\ \&\ \text{RAT}(v)\ \&\ \text{TCOM}^*(u, v) \rightarrow \neg\,\text{RAT}(u - v)$.

It is clear that BV and TV are u and v satisfying the antecedent of this proposition if BH is taken as rational. Euclid calls the difference between such lines an apotome. It should be clear that the important numerical content of $\mathbf{T}(BV) \simeq 5 \cdot \mathbf{T}(TV)$ is lost in the description of BT as an apotome, because it would be one so long as the ratio of BV to TV were not expressible as the ratio of a square number to a square number. Of course, the numerical content is in the proof that BT is an apotome, but it would presumably have to be in any natural proof that $(\mathbf{T}(BV), \mathbf{T}(TV))$ is equal to a numerical ratio of the appropriate kind. The important thing to observe is that the problem Euclid solves is one of qualitative description and classification, not numerical evaluation, even though his solution includes steps corresponding to numerical evaluation.

I shall explain the categorization of straight lines developed in book X from the general point of view adopted there, but I shall also try to make clear the way in which this point of view relates to the edge value of the icosahedron. In place of BV, TV, BT, one considers straight lines a, b, c satisfying the condition $c \simeq a - b$; and, in place of BH, an arbitrary straight line r. Just as the problem of XIII,11 is to characterize the side BF of the regular pentagon in terms of the diameter BH of the circumscribing circle, where $\mathbf{T}(BF) \simeq \mathbf{O}(BV{-}TV, BH)$, so in the general case one wants to characterize the straight line x satisfying $\mathbf{T}(x) \simeq \mathbf{O}(c, r)$. What Euclid does is to place c in one of six coordinated categories of apotome and x in one of six other coordinated categories, while bringing out the correlation between the two sets of categories. His order of presentation is not completely perspicuous, and, in any case, the correlation is quite elaborate.

I begin by explaining the conditions which make the correlation possible. The first two have already been mentioned.

C1 $c \simeq \mathrm{a} - b$;
C2 $\mathbf{T}(x) \simeq \mathbf{O}(c, r)$.

Euclid treats x as the difference between two straight lines y and z, i.e.,

C3 $x \simeq y - z$.

There is no clear intuitive motivation for this choice, which is fundamental to the classificatory scheme. One can say, however, that simple variations of it introduce substantial complexity. By II,7 one has that $\mathbf{T}(x) \simeq \mathbf{T}(y - z) \simeq \mathbf{T}(y) + \mathbf{T}(z) - 2 \cdot \mathbf{O}(y, z) \simeq \mathbf{O}(c, r) \simeq \mathbf{O}(a, r) - \mathbf{O}(b, r)$. The next conditions arise naturally from these equalities. Euclid sets

C4 $\mathbf{T}(y) + \mathbf{T}(z) \simeq \mathbf{O}(a, r)$,

or equivalently,

C5 $2 \cdot \mathbf{O}(y, z) \simeq \mathbf{O}(b, r)$, i.e., $\mathbf{O}(y, z) \simeq \mathbf{O}(\frac{1}{2}b, r)$.

In the treatment of the side of the regular pentagon Euclid takes BH (r) to be rational, and shows that it is commensurable with BV (a) and commensurable in square only with TV (b). In the general case in which r is rational and a and b are rational straight lines commensurable in square only, there are three possibilities:

(αi) COM (a, r) & TCOM* (b, r);
(αii) TCOM* (a, r) & COM (b, r);
(αiii) TCOM* (a, r) & TCOM* (b, r).

It should be noted that, even with a fixed class of rational straight lines, which of these possibilities is realized by rational a and b depends upon which rational straight line is taken as r. In particular, BV and TV satisfy (αi) when BH is taken as r, but (αiii) when AA' is.

C4 and 5 determine equalities between certain areas and the rectangles $\mathbf{O}(a, r)$ and $\mathbf{O}(b, r)$. Euclid proves as a general result

X,19–21 RAT (r) & $\mathbf{T}(w) \simeq \mathbf{O}(u, r) \rightarrow$ (RAT $(w) \leftrightarrow$ COM (u, r))

on the grounds that, if the antecedent is true,

RAT $(w) \leftrightarrow$ TCOM $(w, r) \leftrightarrow$ COM $(\mathbf{O}(u, r), \mathbf{T}(r)) \leftrightarrow$ COM (u, r).

In the particular case in which the antecedent holds, and u is rational and commensurable in square only with r, Euclid calls w and an area equal to the square on it medial (*mesos*), presumably because w is the mean proportional between u and r. Because of C4 and 5 one has corresponding to (αi), (αii), and (αiii),

(βi) $\text{RAT}(\mathbf{T}(y) + \mathbf{T}(z))$ & $\text{MED}(2 \cdot \mathbf{O}(y, z))$;
(βii) $\text{MED}(\mathbf{T}(y) + \mathbf{T}(z))$ & $\text{RAT}(2 \cdot \mathbf{O}(y, z))$;
(βiii) $\text{MED}(\mathbf{T}(y) + \mathbf{T}(z))$ & $\text{MED}(2 \cdot \mathbf{O}(y, z))$.

The α conditions enable one to place BT in a threefold classification of straight lines; and the β conditions do the same for BF. Euclid further refines the classificatory schemes by introducing straight lines a' and b' satisfying

C6 $\mathbf{T}(y) \simeq \mathbf{O}(a', r)$ & $\mathbf{T}(z) \simeq \mathbf{O}(b', r)$.

C4 and 6 together imply

C7 $a \simeq a' + b'$.

The following argument, which is frequently repeated in book X, establishes that, if C6 holds, C5 is equivalent to

C8 $\mathbf{O}(a', b') \simeq \mathbf{T}(\tfrac{1}{2}b)$.

Since $(\mathbf{T}(y), \mathbf{O}(y, z)) = (\mathbf{O}(y, z), \mathbf{T}(z))$, then

$$\begin{aligned}&\mathbf{O}(y, z) \simeq \mathbf{O}(\tfrac{1}{2}b, r) \leftrightarrow\\ &\quad (\mathbf{T}(y), \mathbf{O}(\tfrac{1}{2}b, r)) = (\mathbf{O}(\tfrac{1}{2}b, r), \mathbf{T}(z)) \leftrightarrow\\ &\quad (\mathbf{O}(a', r), \mathbf{O}(\tfrac{1}{2}b, r)) = (\mathbf{O}(\tfrac{1}{2}b, r), \mathbf{O}(b', r)) \leftrightarrow\\ &\quad (a', \tfrac{1}{2}b) = (\tfrac{1}{2}b, b') \leftrightarrow \mathbf{O}(a', b') \simeq \mathbf{T}(\tfrac{1}{2}b).\end{aligned}$$

The algebraic analogue of this argument is very simple; it amounts to the assertion that if $y^2 = a'r$ and $z^2 = b'r$, then

$$yz = \tfrac{1}{2}br \leftrightarrow (\tfrac{1}{2}br)^2 = y^2z^2 = a'rb'r \leftrightarrow a'b' = (\tfrac{1}{2}b)^2.$$

Nevertheless, there does not seem to be any particular reason to think that Euclid's argument is based in some way on algebraic considerations. However, his argument does reduce the determination of a' and b' to the geometric-algebraic VI,28, because C7 and 8 say in geometric terms that a' and b' are quantities with given sum and product. (Since C1 implies that $b \prec a$, it implies that the precondition of 28 holds.) Because of C6, the question of the character of y and z becomes a question about the character of a' and b'. To bring out this relationship Euclid uses II,5, whose close relation in algebraic content to VI,28 has been shown. According to II,5, if C7 and 8 are satisfied, then

$$\mathbf{T}(\tfrac{1}{2}a) \simeq \mathbf{O}(a', b') + \mathbf{T}(a' - \tfrac{1}{2}a),$$

or, since $a' - \frac{1}{2}a \simeq a - b' - \frac{1}{2}a \simeq \frac{1}{2}a - b' \simeq \frac{1}{2}(a - 2 \cdot b')$,

$$\mathbf{T}(a) \simeq 4 \cdot \mathbf{T}(\tfrac{1}{2}a) \simeq 4 \cdot \mathbf{O}(a', b') + 4 \cdot \mathbf{T}(a' - \tfrac{1}{2}a) \simeq 4 \cdot \mathbf{T}(\tfrac{1}{2}b) + 4 \cdot \mathbf{T}(\tfrac{1}{2}(a - 2 \cdot b')) \simeq \mathbf{T}(b) + \mathbf{T}(a - 2 \cdot b'),$$

i.e.,

$$\mathbf{T}(a) - \mathbf{T}(b) \simeq \mathbf{T}(a - 2 \cdot b').$$

Since

$$\text{COM}(a', b') \leftrightarrow \text{COM}(a' + b', b') \leftrightarrow \text{COM}(a, b') \leftrightarrow \text{COM}(a, a - 2 \cdot b'),$$

if one sets

C9 $\mathbf{T}(a) - \mathbf{T}(b) \simeq \mathbf{T}(e)$;

one has

X,17,18 If C7, 8, and 9 are satisfied, then
$\text{COM}(a', b') \leftrightarrow \text{COM}(a, e)$.[12]

Heath (vol. III, pp. 43–44) and others have pointed out the algebraic significance of this proposition. For, since $\mathbf{T}(\frac{1}{2}b)$ can be thought of as an arbitrary quantity b, and

$$\text{COM}(a', b') \leftrightarrow \text{COM}(a', a) \leftrightarrow \text{COM}(b', a) \leftrightarrow \exists k \exists l \exists m \exists n \left(a' = \frac{k}{l}a \;\&\; b' = \frac{m}{n}a \right),$$

X,17, 18 can be interpreted as specifying a condition under which the roots a', b' of the equation $ax - x^2 = b$ are rational in our sense relative to a. However, this fact seems unlikely to be relevant to the interpretation of X,17, 18, since the conceptual apparatus of book X obliterates the distinction between rationality and irrationality in our sense. On the other hand, it must be admitted that the reasoning involved in the proof of this proposition goes smoothly when interpreted algebraically, and that the reasoning almost certainly presupposes a knowledge of VI,28 or II,5. But although the reasoning has an algebraic explanation, it does not seem to have an algebraic motivation. In other words one can give a simple representation of Euclid's reasoning and conclusions in algebraic terms, but the representation does not enable one to account for Euclid's desire to establish those conclusions. I believe that a satisfactory account is provided by attributing to Euclid the desire to fit the edge of the inscribed regular icosahedron into a classificational scheme; indeed, the qualitative scheme evolved in book X is itself a sufficient explanation of the reasoning gone through to reach it. One would, of course, prefer an explanation that invoked a clear mathematical goal intelligible to us in terms of our own notions of mathematics and which, under analysis, would lead univocally to the reasoning in book X. Unfortunately, book X has never been explicated successfully in this way nor does it appear amenable to explication of this sort. Rather, book X appears to be an expedient for dealing with a particular problem and at the same time a mathematical blind alley.

Proposition C9 enables Euclid to double the threefold classification of apotomes given by (αi)–(αiii). For if a and b are rational and commensurable in square only, and C9 holds, TCOM (a, e); but each of (αi)–(αiii) is compatible with COM (a, e) and with $\neg$ COM (a, e). Euclid uses this fact in the classification of what he calls first, second, third, fourth, fifth, and sixth apotomes, but what I shall call r-apotomes because the classification is relative to the choice of a rational straight line r.

X, defs. III If C1 and 9 hold, TCOM* (a, b), and RAT (r), then if COM (a, e), c is

a first r-apotome if (αi) holds,
a second r-apotome if (αii) holds,
a third r-apotome if (αiii) holds;

but if TCOM* (a, e), c is

a fourth r-apotome if (αi) holds,
a fifth r-apotome if (αii) holds,
a sixth r-apotome if (αiii) holds.

It should be clear that this classification could be refined even further, because when a is incommensurable with e, then b may be either commensurable or incommensurable with e. Euclid does not introduce any further refinements. BT is a fourth BH-apotome and a sixth AA'-apotome since COM (BV, BH), TCOM* (TV, BH), TCOM* (BV, AA'), TCOM* (TV, AA'); and, because $(\mathbf{T}(BV), \mathbf{T}(BV) - \mathbf{T}(TV)) = (5, 5 - 1)$,

TCOM* $(\mathbf{T}(BV), \mathbf{T}(BV) - \mathbf{T}(TV))$.

In order to see what this means for the classification of BF, it is convenient to bring out some consequences of X,17, 18 which lead to Euclid's complications of (βi)–(βiii). For this purpose, I take for granted that conditions C1–9 hold.

Ca TCOM $(y, z) \leftrightarrow$ COM (a, e).
Cb COM $(a, e) \leftrightarrow$ COM $(a, a') \leftrightarrow$ COM (a, b').
Cc COM $(a, e) \rightarrow$ (COM $(a, r) \leftrightarrow$ COM $(a', r) \leftrightarrow$ COM (b', r)) & (TCOM* $(a, r) \leftrightarrow$ TCOM* $(a', r) \leftrightarrow$ TCOM* (b', r)).
Cd COM $(a, e) \rightarrow$ (COM $(a, r) \leftrightarrow$ TCOM $(y, r) \leftrightarrow$ TCOM (z, r)).

Ca simply makes explicit the point of X,17, 18, whose algebraic significance is expressed by Cb. Cc is a trivial consequence of Cb, and in turn implies Cd since

$\text{TCOM}(y, r) \leftrightarrow \text{COM}(\mathbf{O}(a', r), \mathbf{T}(r)) \leftrightarrow \text{COM}(a', r)$

and

$\text{TCOM}(z, r) \leftrightarrow \text{COM}(\mathbf{O}(b', r), \mathbf{T}(r)) \leftrightarrow \text{COM}(b', r)$.

Finally I mention

Ce $\text{TCOM}^*(a, b) \rightarrow \neg\,\text{COM}(y, z)$.

Since this assertion is trivially true if $\neg\,\text{TCOM}(y, z)$, it need only be proved for the case in which $\text{TCOM}(y, z)$. But then $\text{COM}(a, e)$, by Ca; and $\text{COM}(a', a)$, by Cb. Since

$(z, y) = (\mathbf{O}(y, z), \mathbf{T}(y)) = (\mathbf{O}(\tfrac{1}{2}b, r), \mathbf{O}(a', r) = (\tfrac{1}{2}b, a')$,

if $\text{COM}(y, z)$, then $\text{COM}(b, a')$; but then $\text{COM}(a, b)$, contradicting the fact that $\text{TCOM}^*(a, b)$.

Ca and e suggest doubling the classification of (βi)–(βiii) by introducing the consideration of whether $\text{TCOM}^*(y, z)$ or $\neg\,\text{TCOM}(y, z)$. This is essentially what Euclid does for the three cases in which $\neg\,\text{TCOM}(y, z)$, i.e., the three cases corresponding to fourth, fifth, and sixth r-apotomes. But he is able to make certain simplifications for the cases corresponding to first, second, and third r-apotomes. Indeed, no new category needs to be introduced for the case in which c is a first r-apotome. For then, by Cd and e, y and z are rational straight lines commensurable in square only, so that x is an apotome.

For the case of c a second or third r-apotome, Euclid uses the fact that an area commensurable with a rational area is rational and

X,23, cor. An area commensurable with a medial area is medial.

To describe Euclid's proof of this corollary I first describe his proof of

X,22 $\text{MED}(w) \;\&\; \text{RAT}(v) \;\&\; \mathbf{T}(w) \simeq \mathbf{O}(u, v) \rightarrow$
$\text{RAT}(u) \;\&\; \text{TCOM}^*(u, v)$.

If w is medial, $\mathbf{T}(w) \simeq \mathbf{O}(u', v')$ for some u', v' rational and commensurable in square only. Therefore, if the antecedent holds,

$\mathbf{O}(u, v) \simeq \mathbf{O}(u', v')$,
$(u, u') = (v', v)$,
$(\mathbf{T}(u), \mathbf{T}(u')) = (\mathbf{T}(v'), \mathbf{T}(v))$.

But, since u', v', and v are all rational, $\text{RAT}(u)$ and $\text{TCOM}(u, v)$; on the other hand, since $\neg\,\text{RAT}(w)$, $\neg\,\text{COM}(u, v)$. To prove the corollary to 23 one need only suppose that (i) $\mathbf{T}(w) \simeq \mathbf{O}(u, r_1)$ so that, by 22, u is rational and commensurable with r_1 in square only, (ii) COM$(\mathbf{T}(w), \mathbf{T}(w'))$, and (iii) $\mathbf{T}(w') \simeq \mathbf{O}(u', r_1)$. Then, since

$$(\mathbf{T}(w), \mathbf{T}(w')) = (u, u'),$$

COM(u, u') and TCOM*(u', r_1), so that $\mathbf{T}(w')$ is medial. This argument also establishes that a straight line commensurable (or commensurable in square) with a medial straight line is medial (X,23).

Since an area commensurable with a rational or medial area is itself rational or medial, it obviously does not matter whether or not the 2 is included in any of (βi)–(βiii). Similarly, by Ca, when c is a second or third r-apotome, $\mathbf{T}(y)$ and $\mathbf{T}(z)$ are commensurable; and therefore, by X,15 and 23, cor., the first conjunct of (βii) and (βiii) implies that each of $\mathbf{T}(y)$ and $\mathbf{T}(z)$ is medial, i.e., that each of y and z is. Clearly too, if y and z are medial and commensurable in square, $\mathbf{T}(y) + \mathbf{T}(z)$ is also medial. Hence Euclid can define (in X,74 and 75)

If C3 holds, then

x is a first (second) apotome of a medial (*mesēs apotome*) $\leftrightarrow$ MED(y) & MED(z) & TCOM*(y, z) & RAT$(\mathbf{O}(y, z))$ (MED$(\mathbf{O}(y, z))$).

The other three categories are dealt with in X,76–78:

If C3 holds and $\neg$ TCOM(y, z), then

x is a minor (*elassōn*) $\leftrightarrow$ RAT$(\mathbf{T}(y) + \mathbf{T}(z))$ & MED$(\mathbf{O}(y, z))$;

x is that which produces with a rational area a medial whole (*hē meta rhētou meson to holon poiousa*) $\leftrightarrow$ MED$(\mathbf{T}(y) + \mathbf{T}(z))$ & RAT$(2 \cdot \mathbf{O}(y, z))$;

x is that which produces with a medial area a medial whole (*hē meta mesou meson to holon poiousa*) $\leftrightarrow$ MED$(\mathbf{T}(y) + \mathbf{T}(z))$ & MED$(2 \cdot \mathbf{O}(y, z))$ & $\neg$ COM$(\mathbf{T}(y) + \mathbf{T}(z), 2 \cdot \mathbf{O}(y, z))$.

There is no mathematical reason for Euclid to use $\mathbf{O}(y, z)$ in three of these definitions and $2 \cdot \mathbf{O}(y, z)$ in the last two. The final conjunct of the last definition is needed for the proof of 102, which I discuss below (pp. 275–276). It is redundant when C1, 4, and 5 hold, no matter what kind of apotome c is. Then

$$(\mathbf{T}(y) + \mathbf{T}(z), 2 \cdot \mathbf{O}(y, z)) = (\mathbf{O}(a, r), \mathbf{O}(b, r)) = (a, b),$$

and

$$\neg \text{COM}(\mathbf{T}(y) + \mathbf{T}(z), 2 \cdot \mathbf{O}(y, z)).$$

Euclid's vocabulary in these definitions is not entirely transparent. The terminology of the last two definitions is determined by the fact that

$$\mathbf{T}(x) + 2 \cdot \mathbf{O}(y, z) \simeq \mathbf{T}(y) + \mathbf{T}(z).$$

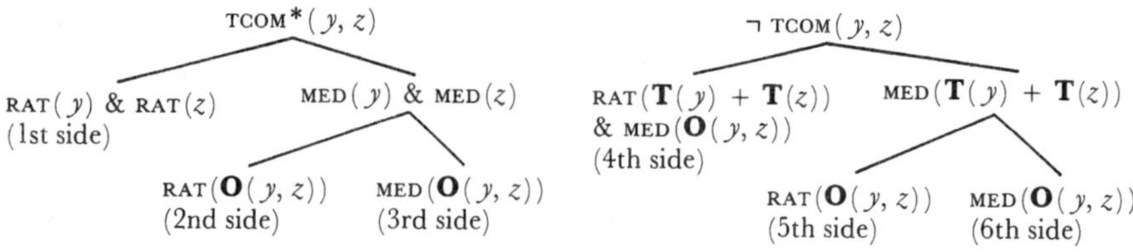

Figure 7.18

The phrase 'apotome of a medial' presumably refers to the fact that the apotomes of a medial are determined by "cutting off" (*apotemnein*) one medial straight line from another; however, it is important to see that apotomes of a medial are not apotomes in the technical sense of the term, since medial straight lines are irrational. I know of no satisfactory explanation for the term 'minor'.[13] In any case, it simplifies the representation of the results of book X to introduce a more uniform terminology. I shall call an apotome a first side, a first apotome of a medial a second side, a second apotome of a medial a third side, a minor a fourth side, a producer with a rational area of a medial whole a fifth side, and a producer with a medial area of a medial whole a sixth side. The word 'side' is used because it is Heath's translation of the Greek *dunamenē*; when $\mathbf{T}(w) \simeq \mathbf{O}(u, v)$ Euclid calls *w hē dunamenē* of $\mathbf{O}(u, v)$.

The essentials of Euclid's classification of sides $y - z$ are depicted in fig. 7.18. It is clear that this categorization makes no sense as an independent classification of straight lines $y - z$; it could easily be expanded and it would be natural to do so. The basis of the classification is obviously the fact that

X,91–96 $\text{RAT}(r) \,\&\, \mathbf{T}(x) \simeq \mathbf{O}(c, r) \rightarrow$ (c is an nth r-apotome $\rightarrow x$ is an nth side),

so that in particular, since BT is a fourth BH-apotome, BF is a fourth side, a minor (XIII,11). And although BT is also a sixth AA'-apotome, BF is not necessarily a sixth side, because $\neg(\mathbf{T}(BF) \simeq \mathbf{O}(BT, AA'))$. In fact, a consequence of X,100 and the fact that no straight line can be two kinds of r-apotome is that BF is a fourth side only, relative to the class of rational lines including AA' and BH. For now the point to be made is that the classification of sides, unlike the classification of r-apotomes, is relative only to the class of rational straight lines R and not to the choice of a particular straight line from the class.

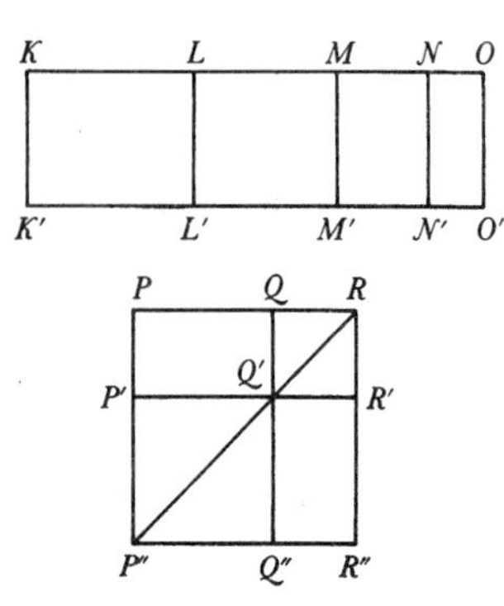

Figure 7.19

To establish X,91–96 it is only necessary to show that if C1 and 2 are satisfied, C3–8 can also be satisfied.[14] To describe Euclid's demonstration of this fact, it is convenient to adopt his geometric style but to dispense with the relatively obvious details of his geometric argumentation. In fig. 7.19, KK' corresponds to r, KO to a, LO to b, so that KL corresponds to c. Euclid bisects LO at M so that LM, MO each represent $\frac{1}{2}b$. He uses VI,28 to construct, as a' and b' satisfying C7 and 8, KN and NO. Now he constructs as $\mathbf{T}(y)$ and $\mathbf{T}(z)$ satisfying C6 the squares PR'' and QR' arranged as shown. VI,26 guarantees that they are "about the same diagonal," so that the figure can be completed. Euclid now cites the third part of a *lēmma* proved before X,54 to the effect that (square PR'', rectangle $P'R$) = (rectangle $P'R$, square QR') in order to use the argument given on p. 269 for the equivalence of C8 and 5 to establish C5 in the form rectangle $QR'' \simeq$ rectangle MO', i.e., rectangle PR' + rectangle $Q'R''$ + square $QR' \simeq$ rectangle PR' + rectangle $QR'' \simeq$ rectangle LO'. C4 (square PR'' + square $QR' \simeq$ rectangle KO') is, of course, true by construction, so that square $P'Q'' \simeq$ rectangle KL', i.e., C3 is satisfied.

Euclid also proves

X,97–102 $\text{RAT}(r)\ \&\ \mathbf{T}(x) \simeq \mathbf{O}(c, r) \rightarrow$
$(x \text{ is an } n\text{th side} \rightarrow c \text{ is an } n\text{th } r\text{-apotome})$.

In order to reproduce his proofs I first point out a consequence of C4 and 5 and X,20 and 22.

Cf $\text{RAT}(r) \rightarrow (\text{RAT}(\mathbf{T}(y) + \mathbf{T}(z)) \rightarrow \text{COM}(a, r))\ \&$
$(\text{MED}(\mathbf{T}(y) + \mathbf{T}(z)) \rightarrow \text{TCOM}^*(a, r))\ \&$
$(\text{RAT}(\mathbf{O}(y, z)) \rightarrow \text{COM}(b, r))\ \&$
$(\text{MED}(\mathbf{O}(y, z)) \rightarrow \text{TCOM}^*(b, r))$.

Suppose now that r is rational and that C2 and C3 hold. One can construct a' and b' satisfying C6, a and b satisfying C4 and 5, and e satisfying C9. Since

$$\mathbf{T}(x) \simeq \mathbf{T}(y) + \mathbf{T}(z) - 2\cdot\mathbf{O}(y, z),$$

C1 is satisfied, and since C,4, 5, and 6 imply C7 and 8, they are satisfied as well. It is now fairly simple to confirm X,97–102. I outline the steps.

(i) Ca implies that $\text{COM}(a, e)$ for x a first, second, or third side and that $\text{TCOM}^*(a, e)$ in the other cases.

(ii) Cf implies that $\text{COM}(a, r)$ for x a fourth side and that $\text{TCOM}^*(a, r)$ for x a fifth or sixth side. If x is a first (second or third) side, y and z are rational (medial) and commensurable in square; hence $\mathbf{T}(y)$ and $\mathbf{T}(z)$ are rational (medial) and commensurable, so that $\mathbf{T}(y) + \mathbf{T}(z)$ is rational (medial), and, by Cf again, $\text{COM}(a, r)$ ($\text{TCOM}^*(a, r)$).

(iii) Cf implies that $\text{COM}(b, r)$ for x a second or fifth side and that $\text{TCOM}^*(b, r)$ for x a third, fourth, or sixth side. If x is a first side, then y and z are rational and commensurable in square only, so that $\mathbf{O}(y, z)$ is medial and, again by Cf, $\text{TCOM}^*(b, r)$.

These results suffice to show that a and b are rational in every case and commensurable in square only when x is a first, second, fourth, or fifth side. For the other two cases Euclid invokes the fact that

$$(a, b) = (\mathbf{O}(a, r), \mathbf{O}(b, r)) = (\mathbf{T}(y) + \mathbf{T}(z), 2 \cdot \mathbf{O}(y, z)).$$

When x is a third side, one has that $\text{COM}(\mathbf{T}(y) + \mathbf{T}(z), \mathbf{T}(y))$, but $\neg\,\text{COM}(\mathbf{T}(y), \mathbf{O}(y, z))$, since $\neg\,\text{COM}(y, z)$; hence,

$$\neg\text{COM}(\mathbf{T}(y) + \mathbf{T}(z), 2 \cdot \mathbf{O}(y, z)).$$

When x is a sixth side Euclid is able to point out that this last assertion is true by the definition of a sixth side—perhaps the most direct indication that, whatever its symmetry, Euclid's categorization of r-apotomes and sides is not exhaustive even in terms of its own way of making distinctions. In any case, one has that $\neg\,\text{COM}(a, b)$ and hence $\text{TCOM}^*(a, b)$ for x a third or sixth side; and the proof of X,97–102 is completed.[15]

X,91–102 are the essential propositions for understanding the categorization of straight lines developed by Euclid for dealing with the edge value of the icosahedron. However, Euclid proves other propositions which clarify the nature of the categorization, but he does so in an order which increases the surface mystery of what he is doing. The structure of his whole treatment may be represented as follows:

X,10,27–35 construction of each side (without naming it);

X,73–78 proof that each side is irrational and introduction of its name;

X,79–84 if x, y_1, z_1 and x, y_2, z_2 satisfy the definition of nth side, then $y_1 \simeq y_2$ and $z_1 \simeq z_2$;

X, defs. III definition of the six r-apotomes;

X,85–90 construction of the six r-apotomes;

X,91–102 $\text{RAT}(r)\ \&\ \mathbf{T}(x) \simeq \mathbf{O}(c, r) \rightarrow$ (c is an nth r-apotome $\leftrightarrow x$ is an nth side);

X,103–107 w is an nth r-apotome (side) & $\text{COM}(w, w') \rightarrow w'$ is an nth r-apotome (side);

X,108–110 $\mathbf{T}(x) \simeq u - v \rightarrow$

$(\text{RAT}(u)\ \&\ \text{MED}(v) \rightarrow x$ is a first or fourth side) &
$(\text{MED}(u)\ \&\ \text{RAT}(v) \rightarrow x$ is a second or fifth side) &
$(\text{MED}(u)\ \&\ \text{MED}(v)\ \&\ \neg\,\text{COM}(u, v) \rightarrow x$ is a third or sixth side).

Before discussing the proofs of these propositions I would like to describe the remaining propositions of book X. In X,36–72 Euclid develops in complete parallel with X,73–110 a classification of straight lines which might be called r-binomial (*ek duo honomatōn*) and additive sides. Roughly speaking, $a +$

is an nth r-binomial if and only if a and b satisfy the conditions which would make $a - b$ an nth r-apotome; and $y + z$ is an nth additive side if and only if y and z satisfy the conditions which would make $y - z$ an nth side.[16] The qualification 'roughly speaking' is needed only because addition, unlike subtraction, is commutative. To avoid the complexity which this fact introduces, I shall employ the convention that, in a description of a binomial or additive side as a sum of two straight lines, the greater line is referred to with the alphabetically earlier letter. If one substitutes 'binomial' for 'apotome' and 'additive side' for 'side' in my formulations of X,73–107, 110, one gets formulations of X,36–70, 72. Because of the commutativity of addition, the analogue of X,108, 109 can be stated as a single assertion (X,71) that the side of a square equal to the sum of a rational and a medial area is either a first, second, fourth, or fifth additive side.

After X,72 Euclid points out that, relative to a given r_1, the medials and the six additive sides constitute mutually exclusive classes, as do the six r-binomials. In 111 he proves that no straight line is both a binomial and an apotome and argues that the medials, the six additive sides, and the six sides are all mutually exclusive classes, as are the six r-binomials and six r-apotomes. The last four propositions of book X, considered spurious by Heiberg, are an addition to the main work of the book and will be treated at the end of this section.[17] The only propositions in book X not yet mentioned are 24–26. Proposition 26 is a lemma for 79, 80, 82, 83 and their analogues, 42, 43, 45, 46, and will be introduced in connection with their proofs. Propositions 24 and 25, although not actually used in the *Elements*, are closely connected with the constructions of the sides and additive sides; they will be discussed in connection with those constructions.

The perfect parallelism between X,36–72 and X,73–110 makes it possible to describe only the proofs of the latter and then to indicate the adjustments needed to turn them into proofs of the former. Moreover, although Euclid gives more or less separate proofs for each proposition, there are many repetitions which permit consolidation. An example is provided by the proofs of the irrationality of the sides. The proof for the first side has already been indicated. The proofs for the second, fourth, and fifth sides turn on the fact (II,7) that

$$\mathbf{T}(y - z) + 2\cdot\mathbf{O}(y, z) \simeq \mathbf{T}(y) + \mathbf{T}(z).$$

If $y - z$ is a second or fifth (fourth) side, then $\mathbf{T}(y) + \mathbf{T}(z)$ is medial (rational), and $2\cdot\mathbf{O}(y, z)$ is rational (medial). Therefore

$$\neg\text{COM}(\mathbf{T}(y - z) + 2\cdot\mathbf{O}(y, z), 2\cdot\mathbf{O}(y, z)),$$

and hence

$$\neg\text{COM}(\mathbf{T}(y - z), 2\cdot\mathbf{O}(y, z))$$

$$(\neg\text{COM}(\mathbf{T}(y - z) + 2\cdot\mathbf{O}(y, z)\mathbf{T}(y - z)),$$

and hence

$$\neg\text{COM}(\mathbf{T}(y) + \mathbf{T}(z), \mathbf{T}(y - z))).$$

Therefore, $\mathbf{T}(y - z)$ is irrational, and so is $y - z$. For the third and sixth sides Euclid carries out much of the construction of X,97–102 and argues that

$$\mathbf{T}(y - z) \simeq \mathbf{T}(x) \simeq \mathbf{O}(c, r)$$

for an apotome c and rational straight line r, so that, by X,20, $y - z$ is irrational.

What I have called the construction of the sides is at the same time the construction of the additive sides. These constructions, which involve applications of arithmetic discussed in section 2.3, are in some respects problematic. In a corollary to X,6 Euclid remarks that, given numbers m and n and a straight line u, it is possible to construct v so that $(u, v) = (m, n)$; and, if w is taken as a mean proportional between u and v (VI,13), one has $(\mathbf{T}(u), \mathbf{T}(w)) = (m, n)$. This corollary is very loosely connected with X,6, which concerns magnitudes in general, but the corollary plays a very important role in book X. It is first used in X,10, in which Euclid shows how to find u, v such that $\text{TCOM}^*(u, r_1)$ and $\neg\,\text{TCOM}(v, r_1)$. He takes nonsimilar plane numbers m, n and uses the corollary to find u such that $(\mathbf{T}(r_1), \mathbf{T}(u)) = (m, n)$. By X,9, $\text{TCOM}^*(r_1, u)$. He then takes v as a mean proportional between r_1 and u, so that $(\mathbf{T}(r_1), \mathbf{T}(v)) = (r_1, u)$ and $\neg\,\text{TCOM}(r_1, v)$. Although Heiberg does not bracket X,10, he expresses great doubt about its authenticity. The principal reason for considering 10 spurious is that the last inference in it depends upon the proposition that follows 10 in our texts. This is obviously a very strong reason for doubt about 10, but the construction of u in 10 is the only explicit construction of rational straight lines commensurable in square only. Such a construction is, of course, needed for the apotome and binomial; and, as we shall see, such lines are used in Euclid's constructions of the other sides and additive sides. I should perhaps add that in X,28 and 32 Euclid takes for granted the possibility of finding three rational straight lines which are pairwise commensurable in square only.

It is convenient to look next at the constructions of the fourth, fifth, and sixth sides and additive sides, which are

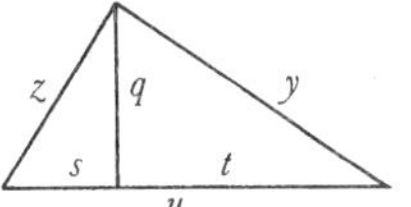

Figure 7.20

carried out in 33–35 but are really the concern of 29–35. The best way to explain the constructions is to explain the analysis which gives rise to them. However, it is useful to mention first some simple consequences of VI,8, 16, and 17 which Euclid has used previously but which he now draws together in a simple *lēmma* before X,33.

If (fig. 7.20) y and z are the legs of a right triangle with $y \succ z$, and q is a perpendicular from the right angle to the hypotenuse u, and q divides u into t and s with $t \succ s$, then $\mathbf{O}(u, s) \simeq \mathbf{T}(z)$, $\mathbf{O}(u, t) \simeq \mathbf{T}(y)$, $\mathbf{O}(s, t) \simeq \mathbf{T}(q)$, and $\mathbf{O}(u, q) \simeq \mathbf{O}(z, y)$.

Now suppose that $y - z\ (y + z)$ is a

fourth (33) fifth (34) sixth (35)

side (additive side). The condition that $\mathbf{T}(y) + \mathbf{T}(z)$ be rational or medial suggests the construction of a right triangle like that of fig. 7.20, with

(a) $\mathrm{RAT}(u)$ $\mathrm{MED}(u)$ $\mathrm{MED}(u)$.

Since $\mathbf{O}(y, z) \simeq \mathbf{O}(u, q)$, one also wants

(b) $\mathrm{MED}(\mathbf{O}(u, q))$ $\mathrm{RAT}(\mathbf{O}(u, q))$ $\mathrm{MED}(\mathbf{O}(u, q))$;

and, since $(\mathbf{T}(y) + \mathbf{T}(z), \mathbf{O}(y, z)) = (\mathbf{T}(u), \mathbf{O}(u, q)) = (u, q)$, to satisfy the additional condition in the definition of a sixth side (additive side) one wants

(c) $\neg\,\mathrm{COM}(u, q)$.

It is not difficult to construct u and q satisfying these three conditions; but it is also necessary that y and z be incommensurable in square, i.e., since $\mathbf{T}(y) \simeq \mathbf{O}(u, t)$ and $\mathbf{T}(z) \simeq \mathbf{O}(u, s)$,

(d) $\neg\,\mathrm{COM}(s, t)$ $\neg\,\mathrm{COM}(s, t)$ $\neg\,\mathrm{COM}(s, t)$.

However, since $\mathbf{T}(q) \simeq \mathbf{O}(s, t)$ and $s + t \simeq u$, if one sets $v \simeq 2 \cdot q$ and $\mathbf{T}(u) - \mathbf{T}(v) \simeq \mathbf{T}(w)$, one has by X,17,18 that $\mathrm{COM}(s, t) \leftrightarrow \mathrm{COM}(u, w)$.[18] Moreover, because such a v is commensurable with q, the effect of (b) and (c) remains the same with v substituted for q. Therefore, the problem of satisfying (a)–(d) is reduced to finding straight lines u and v satisfying

(e) if $\mathbf{T}(u) - \mathbf{T}(v) \simeq \mathbf{T}(w)$, then $\neg\,\mathrm{COM}(u, w)$,

(f) $\mathrm{RAT}(u)$ & $\mathrm{MED}(\mathbf{O}(u, v))$ $\mathrm{MED}(u)$ & $\mathrm{RAT}(\mathbf{O}(u, v))$ $\mathrm{MED}(u)$ & $\mathrm{MED}(\mathbf{O}(u, v))$ & $\neg\,\mathrm{COM}(u, v)$.

Because of X,21 and 22, 33 reduces to

X,30 (29) To find straight lines u and v such that

(i) TCOM* (u, v),
(ii) if $\mathbf{T}(u) - \mathbf{T}(v) \simeq \mathbf{T}(w)$, then $\neg$ COM(u, w) (COM(u, w)),
(iii) RAT(u) & RAT(v).

The proof of 30 (29) depends upon the arithmetic *lēmmata* discussed in section 2.3. Using these *lēmmata*, Euclid finds numbers m^2 and n^2 such that $m^2 + n^2$ $(m^2 - n^2)$ is not a square number. He takes a rational straight line u and constructs v to satisfy

$$(\mathbf{T}(u), \mathbf{T}(v)) = (m^2 + n^2, m^2)$$
$$((\mathbf{T}(u), \mathbf{T}(v)) = (m^2, m^2 - n^2)),$$

so that v is rational and commensurable with u in square only. (ii) is also satisfied because

$$(\mathbf{T}(u) - \mathbf{T}(v), \mathbf{T}(u)) = (n^2, m^2 + n^2)$$
$$((\mathbf{T}(u) - \mathbf{T}(v), \mathbf{T}(u)) = (n^2, m^2)).$$[19]

34 (35) can also be reduced further. For suppose (f) holds and

(f1a) $\mathbf{T}(u) \simeq \mathbf{O}(u_1, r_1)$,
(f1b) TCOM* (u_1, r_1),
(f2) $\mathbf{O}(u, v) \simeq \mathbf{O}(u_2, r_1)$.

Then RAT(u_2) and COM(u_2, r_1) (TCOM* (u_2, r_1)), and so TCOM* (u_1, u_2);[20] but since

(f3) $(u, v) = (\mathbf{T}(u), \mathbf{O}(u, v)) = (\mathbf{O}(u_1, r_1), \mathbf{O}(u_2, r_1)) = (u_1, u_2)$,

u and v satisfy (e) if and only if u_1 and u_2 do, by X,14. Therefore, 34 (35) is reduced to finding rational straight lines r_1, u_1, u_2 such that

TCOM* (u_1, u_2), COM (u_2, r_1) (TCOM* (u_2, r_1) & TCOM* (u_1, r_1)),

and

if $\mathbf{T}(u_1) - \mathbf{T}(u_2) \simeq \mathbf{T}(w)$, then $\neg$ COM (u_1, w);

but this problem is already solved in X,30.

Euclid's procedure is somewhat different from the one suggested by this analysis. First of all he shows how

X,31 (32) To find straight lines u and v such that

(i) TCOM* (u, v),
(ii) if $\mathbf{T}(u) - \mathbf{T}(v) \simeq \mathbf{T}(w)$, then $\neg$ COM (u, w),
(iii) MED (u) & MED (v) & RAT $(\mathbf{O}(u, v))$ (MED $(\mathbf{O}(u, v))$).[21]

Such u and v obviously satisfy (e) and (f), and therefore enable one to solve 34 (35). It is to be noticed that the construction of u and v satisfying (i) and (iii) of 31 (32) is the construction of a second (third) side or additive side. Euclid himself does not appear to notice this fact. For he uses virtually the same construction as in 31 (32) to show how

X,27 (28) To find straight lines u and v satisfying (i) and (iii) for X,31 (32).

X,24 and 25, which are never used in the *Elements*, appear to be analyses leading to these constructions. In 24 Euclid shows that if u and v are medial and commensurable, $\mathbf{O}(u, v)$ is medial on the grounds that $(\mathbf{O}(u, v), \mathbf{T}(u)) = (u, v)$ and an area commensurable with a medial area is medial. The more important analysis is given in 25 and closely parallels the analysis I gave above for 34 and 35. Euclid supposes that $\text{MED}(u)$ & $\text{MED}(v)$ & $\text{TCOM}^*(u, v)$. Then there are u_1, u_2 and v_1 satisfying (f1), (f2), and

(f1a′) $\mathbf{T}(v) \simeq \mathbf{O}(v_1, r_1)$,
(f1b′) $\text{TCOM}^*(v_1, r_1)$.

Since $\text{TCOM}(u, v)$ and $(\mathbf{T}(u), \mathbf{T}(v)) = (\mathbf{O}(u_1, r_1), \mathbf{O}(v_1, r_1)) = (u_1, v_1)$,

(f4) $\text{COM}(u_1, v_1)$ & $\text{RAT}(\mathbf{O}(u_1, v_1))$.

Finally, since $(\mathbf{T}(u), \mathbf{O}(u, v)) = (\mathbf{O}(u, v), \mathbf{T}(v))$, (f1), (f1′), and (f2) give that

$$(\mathbf{O}(u_1, r_1), \mathbf{O}(u_2, r_1)) = (\mathbf{O}(u_2, r_1), \mathbf{O}(v_1, r_1)),$$

or

(f5) $(u_1, u_2) = (u_2, v_1)$, i.e., $\mathbf{T}(u_2) \simeq \mathbf{O}(u_1, v_1)$.

Steps (f4) and (f5) show that u_2 is rational and hence either commensurable with r_1, which is the case if and only if $\text{RAT}(\mathbf{O}(u, v))$, or commensurable in square only with r_1, which is the case if and only if $\text{MED}(\mathbf{O}(u, v))$. Euclid does not show that both of these alternatives can be realized, but simply asserts

X,25 $\text{MED}(u)$ & $\text{MED}(v)$ & $\text{TCOM}^*(u, v) \rightarrow$
$\text{RAT}(\mathbf{O}(u, v)) \vee \text{MED}(\mathbf{O}(u, v))$.

It is clear from (f2) that $\text{RAT}(\mathbf{O}(u, v)) \leftrightarrow \text{COM}(u_2, r_1)$; hence, the constructions for 27 and 28 are the same as those for 31 and 32 except that one need not add the condition 'if $\mathbf{T}(u_1) - \mathbf{T}(u_2) \simeq \mathbf{T}(w)$, then $\neg\,\text{COM}(u_1, w)$'. Euclid's actual constructions differ in a number of relatively superficial ways. I shall describe one construction, based on Euclid's proof of 32, and then indicate some of the variations he introduces in the other three cases. Parenthetical phrases mention slight alterations which are needed for 31. One begins by using X,10 to take rational straight lines r_1, u_1, u_2 satisfying (f1b), $\text{TCOM}^*(u_1, u_2)$, and $\text{TCOM}^*(u_2, r_1)$ $(\text{COM}(u_2, r_1))$. For 31 and 32, but not for

27 and 28, one also uses X,30 to make $\mathbf{T}(u_1)$ greater than $\mathbf{T}(u_2)$ by a square on a straight line incommensurable with u_1. If one now constructs u to satisfy (f1a), u is medial. If then v is constructed to satisfy (f2), one will satisfy (f3) as well, so that TCOM* (u, v), MED (v), and if $\mathbf{T}(u) - \mathbf{T}(v) \simeq \mathbf{T}(w)$, we have $\neg$COM (u, w). Finally, since $\mathbf{O}(u_2, r_1)$ is medial (rational), (f2) implies that $\mathbf{O}(u, v)$ is as well.

In X,27 and 28 Euclid uses the theory of proportion more heavily, e.g., by taking u to satisfy $(u_1, u) = (u, r_1)$ rather than (f1a). A more substantial modification comes when, instead of constructing v to satisfy (f2), he constructs it to satisfy $(u_1, u_2) = (u, v)$, so that TCOM* (u, v) and MED (v). He then argues that $(u_2, v) = (u_1, u) = (u, r_1)$, satisfying (f2); and, since $\mathbf{O}(u_2, r_1)$ is medial (rational), so is $\mathbf{O}(u, v)$. I should also mention one other simplification which Euclid introduces in 27 and 31. Because r_1 and u_2 are commensurable in the solution of those problems which I have given, it is not necessary to distinguish the two. Euclid's solutions are represented by replacing u_2 with r_1 and $\mathbf{O}(r_1, r_1)$ with $\mathbf{T}(r_1)$ in my description, and eliminating the resulting redundancies.

Although I have referred to the construction of the sides and the r-apotomes, Euclid speaks of finding them; moreover, in the manuscripts his proofs end with either 'Q.E.D.' or with 'That which was to be found' or with nothing. The lack of uniformity in the concluding tags makes it seem unlikely that this vocabulary is invested with any philosophical significance; indeed, the fact that some manuscripts lack a tag and that the others vary in their tag suggests that Euclid may not have written any tag at all.

The construction of the six r-apotomes is very straightforward. I summarize the arguments for the

first	second	fourth	fifth.

Given r, Euclid first constructs a and b such that

COM (a, r)	COM (b, r)	COM (a, r)	COM (b, r)

and then numbers m and n such that

m, n are square; $m - n$ is not square.	$(m + n, m)$ and $(m + n, n)$ are not the ratios of a square number to a square number.

He finds the other of a and b, using X,6, cor., by setting $(\mathbf{T}(a), \mathbf{T}(b))$ equal to

$(m, m - n)$	$(m + n, n)$,

so that $(\mathbf{T}(a), \mathbf{T}(a) - \mathbf{T}(b))$ is equal to

(m, n)	$(m + n, m)$,

and all the appropriate defining conditions are fulfilled. For the third (sixth) r-apotome, Euclid takes k, m, n such that (m, n) (m, k), (n, k) are not the ratios of a square number to a square number and $(n, n - k)$ is (is not either). He sets $(\mathbf{T}(r), \mathbf{T}(a)) = (m, n)$, $(\mathbf{T}(a), \mathbf{T}(b)) = (n, k)$, so that $(\mathbf{T}(r), \mathbf{T}(b)) = (m, k)$ and $(\mathbf{T}(a), \mathbf{T}(a) - \mathbf{T}(b)) = (n, n - k)$, and the defining conditions are fulfilled.

Propositions 103–110 are very simple; they seem to serve no purpose except to add to the aesthetic elegance of the classification of r-apotomes and sides. For 103 (104–107) Euclid imagines that w, u, v satisfy the definition of nth r-apotome (side) and constructs v' satisfying $(w, w') = (v, v')$ and u' satisfying $u' \simeq v' + w'$. Since $\text{COM}(w, w')$, $\text{COM}(v, v')$ and $\text{COM}(u, u')$; and it is easy to argue that w', u', v' must satisfy the definition of nth r-apotome (side). Although 103–107 are never applied in the *Elements*, Euclid takes for granted a stronger result than 104–107 in XIII,18, namely,

X,104′–107′ w is an nth side & $\text{TCOM}(w, w') \rightarrow w'$ is an nth side.

The analogous assertion for r-apotomes is false. The simplest way to prove 104′–107′ is to set $\mathbf{T}(w) \simeq \mathbf{O}(c, r)$ and $\mathbf{T}(w') \simeq \mathbf{O}(c', r)$ with r rational, so that, by X,97–102, c is an nth r-apotome. In addition, since

$$(\mathbf{T}(w), \mathbf{T}(w')) = (\mathbf{O}(c, r), \mathbf{O}(c', r) = (c, c'),$$

$\text{COM}(c, c')$, so that, by X,103, c' is an nth r-apotome and, by X,91–96, w' is an nth side.

For

108	109	110

Euclid assumes the antecedent and sets $u \simeq \mathbf{O}(a, r_1)$ and $v \simeq \mathbf{O}(b, r_1)$, so that

$\text{COM}(a, r_1)$	$\text{TCOM}^*(a, r_1)$	$\text{TCOM}^*(a, r_1)$,
$\text{TCOM}^*(b, r_1)$	$\text{COM}(b, r_1)$	$\text{TCOM}^*(b, r_1)$,
$\text{TCOM}^*(a, b)$	$\text{TCOM}^*(a, b)$	$\text{TCOM}^*(a, b)$.

If $\mathbf{T}(a) - \mathbf{T}(b) \simeq \mathbf{T}(e)$, then, if $\text{COM}(a, e)$, $a - b$ is a

first	second	third

r_1-apotome, and, since $\mathbf{T}(x) \simeq \mathbf{O}(a - b, r_1)$, x is a corresponding side. Similarly, if $\neg\,\text{COM}(a, e)$, then $a - b$ is a

fourth	fifth	sixth

r_1-apotome, and x is a corresponding side.

The same construction, or a simple application of X,15, 16 and X,23, cor., can be used to establish

$$\text{X,110}' \quad \mathbf{T}(x) \simeq u - v \rightarrow (\text{MED}(u) \ \& \ \text{MED}(v) \ \& \ \text{COM}(u, v) \rightarrow \text{MED}(x)).$$

Euclid carries out a somewhat more complicated form of the argument for 108–110 to prove as a lemma for X,79, 80, 82, and 83 the following result, which is obviously weaker than 110,110′:

X,26 The difference between two medial areas is irrational.

To establish the propositions Euclid points out that if x, y_1, z_1 and x, y_2, z_2 satisfy the definition of first, second, fourth, or fifth side, then $\mathbf{T}(y_1) + \mathbf{T}(z_1) - 2 \cdot \mathbf{O}(y_1, z_1) \simeq \mathbf{T}(y_2) + \mathbf{T}(z_2) - 2 \cdot \mathbf{O}(y_2, z_2)$. Hence, if $y_1 \succ y_2$, so that also $z_1 \succ z_2$, $(\mathbf{T}(y_1) + \mathbf{T}(z_1)) - (\mathbf{T}(y_2) + \mathbf{T}(z_2)) \simeq 2 \cdot \mathbf{O}(y_1, z_1) - 2 \cdot \mathbf{O}(y_2, z_2)$. However, this last equality is impossible, because one side of it is the difference between two rational areas and is therefore rational, whereas the other side is the difference between two medial areas and is therefore irrational. Euclid reduces X,81 and 84 to 79. He carries out much of the construction of X,97–102, setting

$$\mathbf{O}(a_1, r) \simeq \mathbf{T}(y_1) + \mathbf{T}(z_1),$$
$$\mathbf{O}(b_1, r) \simeq 2 \cdot \mathbf{O}(y_1, z_1),$$
$$\mathbf{O}(a_2, r) \simeq \mathbf{T}(y_2) + \mathbf{T}(z_2),$$

and making $b_2 \simeq a_2 - (a_1 - b_1)$, so that $a_1 - b_1 \simeq a_2 - b_2$ and

$$\mathbf{O}(b_2, r) \simeq 2 \cdot \mathbf{O}(y_2, z_2).$$

He then argues that $a_1 - b_1, a_1, b_1$ and $a_1 - b_1, a_2, b_2$ both satisfy the definition of an apotome or first side and that if $y_1 \succ y_2$, $z_1 \succ z_2$, so that $a_1 \succ a_2$—which is incompatible with X,79.

This completes my discussion of the classification of r-apotomes and sides. In general Euclid's proof of a proposition among X,73–110 is transformed into a proof of its analogue for r-binomials and additive sides in X,36–72 by making appropriate substitutions of + for − and using II,4 in place of II,7. Most of the deviations from this correlation are basically matters of style rather than substance. It is, however, perhaps worthwhile to mention three variations which depend largely on Euclid's concrete conceptions of addition and subtraction. The first comes in the proofs of X,54–59, the analogues of X,91–96. To indicate this variation I shall simply go through the steps of the argument for X,91–96 given on pp. 274–275, making the changes required to turn it into Euclid's argument for X,54–59. First it is necessary to remark that among C1–9 only C1—which becomes $c \simeq a + b$ (C1′)—and C3—which becomes $x \simeq y + z$ (C3′)—require alteration for r-binomials and additive sides.

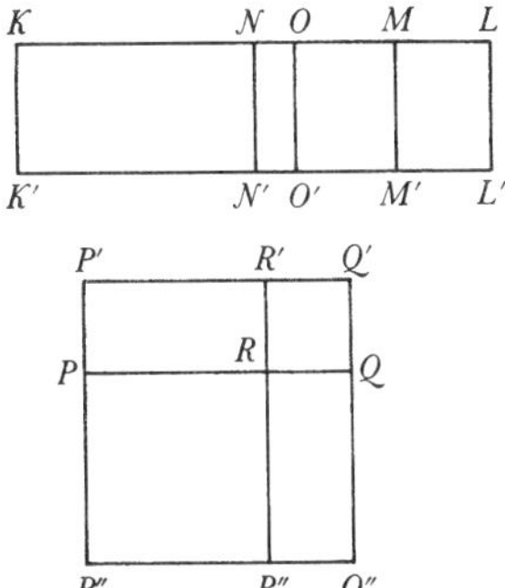

Figure 7.21

To establish X,54–59 it is only necessary to show that if C1′ and 2 are satisfied, conditions C3′–8 can also be satisfied. In fig. 7.21, *KK′* corresponds to *r*, *KO* to *a*, *LO* to *b*, so that *KL* corresponds to *c*. Euclid bisects *LO* at *M* so that *LM*, *MO* each represent $\frac{1}{2}b$. He uses VI,28 to construct, as a' and b' satisfying C7 and 8, *KN* and *NO*. Now he constructs as $\mathbf{T}(y)$ and $\mathbf{T}(z)$ satisfying C6 the squares *PR″* and *QR′* arranged as shown. Euclid now cites the first two parts of a *lēmma* proved before X,54, to the effect that the parallelogram *P′Q″* determined by the squares *PR″* and *QR′* is a square and that (square *PR″*, rectangle *P′R*) = (rectangle *P′R*, square *QR′*). He then uses the argument given on p. 269 for the equivalence of C8 and 5 to establish C5 in the form rectangle *PR′* $\simeq$ rectangle *QR″* $\simeq$ rectangle *MO′* $\simeq$ rectangle *ML′*. C4 (square *PR″* + square *QR′* $\simeq$ rectangle *KO′*) is, of course, true by construction, so that square *P′Q″* $\simeq$ rectangle *KL′*, i.e., C3′ is satisfied.

The other two variations occur in X,44 and 47, the analogues of X,81 and 84, and in X,60–65, the analogues of X,97–102. These two variations are to a certain extent related to one another. For 44 and 47 it is possible to paraphrase with minimal variation the argument for 81 and 84 given on p. 284. Euclid carries out much of the construction of X,60–65, setting

$$\mathbf{O}(a_1, r) \simeq \mathbf{T}(y_1) + \mathbf{T}(z_1),$$
$$\mathbf{O}(b_1, r) \simeq 2\cdot\mathbf{O}(y_1, z_1),$$
$$\mathbf{O}(a_2, r) \simeq \mathbf{T}(y_2) + \mathbf{T}(z_2),$$

and making $b_2 \simeq (a_1 + b_1) - a_2$, so that $a_1 + b_1 \simeq a_2 + b_2$ and

$$\mathbf{O}(b_2, r) \simeq 2\cdot\mathbf{O}(y_2, z_2).$$

He then argues that $a_1 + b_1$, a_1, b_1 and $a_1 + b_1$, a_2, b_2 both satisfy the definition of a binomial. However, since Euclid knows only that $y_1 + z_1 \simeq y_2 + z_2$ and (by the notational convention for terms of additive sides) $y_1 \succ z_1$ and $y_2 \succ z_2$, he can only infer from $y_1 \succ y_2$ that $y_1 \succ y_2 \succ z_2 \succ z_1$. In order to be able to show that $a_1 \succ a_2$, Euclid establishes as a *lēmma* before X,42

$$y_1 \succ y_2 \succ z_2 \succ z_1 \;\&\; y_1 + z_1 \simeq y_2 + z_2 \rightarrow \mathbf{T}(y_1) + \mathbf{T}(z_1) \succ \mathbf{T}(y_2) + \mathbf{T}(z_2).$$ [22]

The proof of this *lēmma* is in the style of Heron's proofs of II,1–10 and uses II,4 and 5. According to the latter

$$\mathbf{O}(y_1, z_1) + \mathbf{T}(\tfrac{1}{2}(y_1 + z_1) - z_1) \simeq \mathbf{T}(\tfrac{1}{2}(y_1 + z_1)) \simeq \mathbf{O}(y_2, z_2) + \mathbf{T}(\tfrac{1}{2}(y_1 + z_1) - z_2).$$

However, if $z_2 \succ z_1$, then

$$\mathbf{T}(\tfrac{1}{2}(y_1 + z_1) - z_1) \succ \mathbf{T}(\tfrac{1}{2}(y_1 + z_1) - z_2),$$

so that $\mathbf{O}(y_1, z_1) \prec \mathbf{O}(y_2, z_2)$; and the result follows from II,4, according to which

$$\mathbf{T}(y_1) + \mathbf{T}(z_1) + 2\cdot\mathbf{O}(y_1, z_1) \simeq \mathbf{T}(y_1 + z_1) \simeq \mathbf{T}(y_2) + \mathbf{T}(z_2) + 2\cdot\mathbf{O}(y_2, z_2).$$

The *lēmma* then establishes that $a_1 \succ a_2$; this, however, does not complete the proof of X,44 and 47, because the possibility that $a_1 \simeq b_2$ and $a_2 \simeq b_1$ has not yet been excluded. This possibility would be excluded if it were legitimate to apply our notational convention according to which $a_1 \succ b_1$ and $a_2 \succ b_2$; but since these straight lines have been constructed in a particular way, it is necessary to prove that the inequalities hold. Euclid in fact asserts that $a_2 \succ b_2$ on the grounds that $\mathbf{T}(y_2) + \mathbf{T}(z_2) \succ 2\cdot\mathbf{O}(y_2, z_2)$, apparently taking for granted the general truth

$$y \succ z \rightarrow \mathbf{T}(y) + \mathbf{T}(z) \succ 2\cdot\mathbf{O}(y, z).$$

This proposition is a direct consequence of II,7, according to which, if $y \succ z$, $\mathbf{T}(y) + \mathbf{T}(z) \simeq \mathbf{T}(y - z) + 2\cdot\mathbf{O}(y, z)$—a relationship which Euclid acknowledges fairly explicitly in his proofs of X,79–84.

However, although Euclid takes the proposition for granted in 44 and 47, he proves it as a *lēmma* for X,60–65. In the analogues of X,60–65, X,97–102, Euclid constructs a and b to satisfy C4 and 5 by constructing $\mathbf{O}(c, r)$ equal to $\mathbf{T}(y - z)$ and $\mathbf{O}(a, r) \simeq \mathbf{T}(y) + \mathbf{T}(z)$; he then takes b as $a - c$, so that C5 is satisfied, by II,7, and clearly $b \prec a$. The corresponding construction for X,60–65 leaves b as $c - a$, and hence not necessarily less than a. The *lēmma*, together with II,4, suffices to establish that $b \prec a$. However, Euclid does not use II,7 to prove the *lēmma*, but instead makes his only application of

$$\text{II,9} \quad y \succ z \rightarrow \mathbf{T}(y) + \mathbf{T}(z) \simeq 2\cdot(\mathbf{T}(\tfrac{1}{2}(y + z)) + \mathbf{T}(\tfrac{1}{2}(y - z))).$$

For, by II,5, if $y \succ z$, then $\mathbf{O}(y, z) \prec \mathbf{T}(\frac{1}{2}(y + z))$, and the result follows. Heiberg brackets the *lēmma* as an interpolation, partly because of its unnecessarily complicated proof and, more importantly, because the result is used without justification in X,44. The adequacy of these reasons has been doubted,[23] and in any case the *lēmma* provides the only internal explanation for the inclusion of II,9.

Heron gives a simple derivation of II,9 from II,4 and 7, according to which

$$\mathbf{T}(\tfrac{1}{2}y + \tfrac{1}{2}z) \simeq \mathbf{T}(\tfrac{1}{2}y) + \mathbf{T}(\tfrac{1}{2}z) + 2\cdot\mathbf{O}(\tfrac{1}{2}y, \tfrac{1}{2}z)$$

and

$$\mathbf{T}(\tfrac{1}{2}y - \tfrac{1}{2}z) \simeq \mathbf{T}(\tfrac{1}{2}y) + \mathbf{T}(\tfrac{1}{2}z) - 2\cdot\mathbf{O}(\tfrac{1}{2}y, \tfrac{1}{2}z),$$

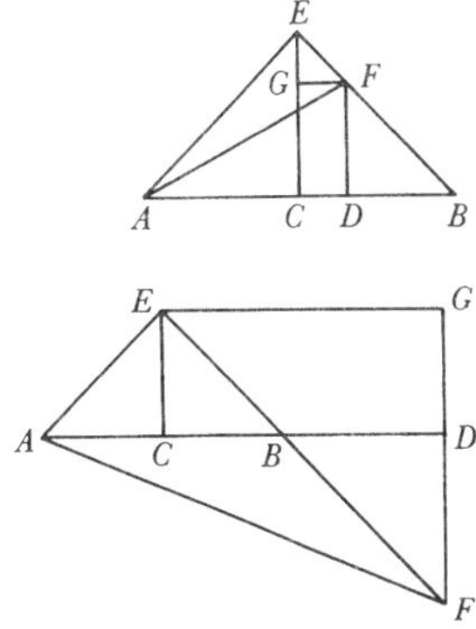

Figure 7.22

so that

$$\mathbf{T}(\tfrac{1}{2}(y+z)) + \mathbf{T}(\tfrac{1}{2}(y-z)) \simeq 2\cdot\mathbf{T}(\tfrac{1}{2}y) + 2\cdot\mathbf{T}(\tfrac{1}{2}z) \simeq \tfrac{1}{2}\mathbf{T}(y) + \tfrac{1}{2}\mathbf{T}(z).$$

Heath (vol. I, pp. 394–395) gives a rather cumbersome geometric-algebraic argument along the lines of some of the earlier proofs in book II. Euclid's procedure is much more geometric. It is convenient to describe it in connection with its algebraic equivalent, the unused II,10. Euclid's formulation of these two propositions is represented by

II,9 (10) If (fig. 7.22) $ACDB$ ($ACBD$) is a straight line and AB is bisected at C, then

$$\mathbf{T}(AD) + \mathbf{T}(BD) \simeq 2\cdot(\mathbf{T}(AC) + \mathbf{T}(CD)).$$

Euclid draws EC perpendicular to AB and equal to AC, DF parallel to EC and intersecting EB (EB extended) at F, and draws GF (EG) parallel to CD and intersecting EC (FD extended) at G. It is not difficult to show that BDF, AEB, ACE, EGF are isosceles right triangles with right angles at D, E, C, and G and that $EG \simeq CD$, so that, by the Pythagorean theorem,

$$\mathbf{T}(AD) + \mathbf{T}(DB) \simeq \mathbf{T}(AD) + \mathbf{T}(DF) \simeq \mathbf{T}(AE) + \mathbf{T}(EF) \simeq 2\cdot\mathbf{T}(AC) + 2\cdot\mathbf{T}(CD).$$

II,9 and 10 have played a significant role in algebraic interpretations of Greek geometric algebra. In his commentary on Plato's *Republic*,[24] Proclus points out that the conclusion of II,10 can be stated as

$$\mathbf{T}(2\cdot CB + BD) + \mathbf{T}(BD) \simeq 2\cdot\mathbf{T}(CB) + 2\cdot\mathbf{T}(CB + BD);$$

but if $BD \simeq EB$, then

$$\mathbf{T}(BD) \simeq 2\cdot\mathbf{T}(CB),$$

so that

$$\mathbf{T}(2\cdot CB + BD) \simeq 2\cdot\mathbf{T}(CB + BD).$$

Proclus is interested in this result because it establishes that if t (BD) is the diagonal of a square with side s (CB), $2\cdot s + t$ is the diagonal of a square with side $s + t$. In the same context Proclus describes a procedure for finding "rational" diagonals of squares. One sets

$$s_1 = 1, \quad t_1 = 1, \quad s_{n+1} = s_n + t_n, \quad t_{n+1} = 2\cdot s_n + t_n.$$

Proclus points out by means of examples that

$$t_n^2 = 2\cdot s_n^2 \pm 1;$$

so that, in our terms, the sequence of fractions $\frac{t_n}{s_n}$ provides ever-improving approximations to the square root of 2. The algebraic interpretation of II,10 provides an explicit connection between it and the arithmetic procedure described by Proclus. For, if one takes *CB* as s and *BD* as t, the equality established in II,10 may be written as the algebraic formula

$$(2s + t)^2 + t^2 = 2s^2 + 2(s + t)^2;$$

or, assuming $t^2 < 2s^2$, as

$$(2s + t)^2 - 2(s + t)^2 = 2s^2 - t^2,$$

a formula which provides a justification for the arithmetic procedure described by Proclus and a proof of his claim about its outcome. Unfortunately, Proclus does not use II,9 and 10 in this way. For him they are geometric results which can be proved exactly and to which the arithmetic procedure produces approximations. He makes no attempt to show that II,10 or the corresponding result about sides and diagonals of squares leads to an arithmetic truth when lines are interpreted as numbers. In other words, he does not take II,10 as an algebraic law.

The role of r-apotomes and sides in the determination of the edge value of the icosahedron makes it seem likely that their categorization is fundamental; while that of r-binomials and additive sides, which plays no role in the *Elements* or elsewhere, is derived, despite the order of Euclid's presentation. Further confirmation of this suggestion is perhaps provided by the fact that in each of X,91–102 Euclid carries out a full construction and argument, but in 54–65 he does so only for 54, 55, and 60; in the other cases he speaks of carrying out the same construction and proving things "in manner similar to the foregoing." The treatment of r-binomials and additive sides would seem, then, to represent categorization for its own sake even more than the treatment of r-apotomes and sides from which it is easily derived and with which it shares the same defect of arbitrariness.

The definitions make clear that no straight line can be two kinds of r-apotome or two kinds of r-binomial. Hence X,60–65 and 97–102 establish that no straight line can be two kinds of side or two kinds of additive side. In addition, because of X,22, the medial straight line cannot be a side or additive side. Euclid proves that sides and additive sides are distinct by proving in X,111 the distinctness of apotomes and binomials. He supposes that x is both an apotome and a binomial, and sets $\mathbf{T}(x) \simeq \mathbf{O}(c, r)$ with r rational. Then c is both a first r-apotome $a_1 - b_1$ and a first r-binomial $a_2 + b_2$. Since, then, a_1 and a_2 are commensurable with r, so is $a_1 - a_2$; and since TCOM* (b_1, r), TCOM* $(a_1 - a_2, b_1)$. Therefore, $(a_1 - a_2) - b_1$, i.e., $c - a_2$, i.e., b_2, is an apotome, contradicting the fact that b_2 is rational.

Thus, as Euclid states at the end of proposition 111, he has succeeded in defining thirteen distinct genera of irrational straight lines. He has also shown that none of the genera are empty by showing how to construct members of each; and he has subdivided two of the genera into six distinct species, each of which he has proved to be nonempty. In addition he has demonstrated that the species and genera are closed with respect to commensurability and established an interesting correlation between the species and the genera. From a purely formalistic point of view one could hardly ask for more in a classification.

There are, however, four more propositions in book X. In 115 Euclid shows how to generate an infinite series of disjoint nonempty classes of irrational straight lines. His reasoning may be represented as follows: Let class 0 contain the sides and additive sides, class 1 the medials, and class $n+1$ (for $n \geq 1$) the straight lines x satisfying $\mathbf{T}(x) \simeq \mathbf{O}(y, r_1)$, for some y in class n. Clearly these classes contain only irrational straight lines, by X,20; and classes 0 and 1 are disjoint. So suppose classes 0 to n are disjoint, but that x is a member of class $n+1$ and an earlier class; then $\mathbf{T}(x)$ is equal to $\mathbf{O}(y_1, r_1)$ and to $\mathbf{O}(y_2, r_1)$ for some y_1 in class n and y_2 in an earlier class; but then $y_1 \simeq y_2$, and the classes 0 to n are not disjoint. Proposition 115 gives a clear demonstration of one of the limitations of the classificatory scheme developed in book X by showing that there are infinitely many kinds of straight lines which it fails to categorize.

In 112–114 Euclid considers the conditions under which $\mathbf{O}(c_1, c_2)$ contained by a binomial c_1 and apotome c_2 is rational. Clearly, if c_1 and c_2 are determined by the same terms a and b, $\mathbf{O}(c_1, c_2)$ is rational, because then $\mathbf{O}(c_1, c_2) \simeq \mathbf{O}(a-b, a+b) \simeq \mathbf{T}(a) - \mathbf{T}(b)$ with a and b rational. It is not difficult to extend this result to

X,114 If $\mathbf{T}(r) \simeq \mathbf{O}(c_1, c_2)$ and c_1, a_1, b_1 and c_2, a_2, b_2 satisfy the definitions of apotome and binomial respectively, and $(a_1, a_2) = (b_1, b_2)$ and COM(a_1, a_2), then RAT(r).

For, if the antecedent holds, one has

(i) $(a_1, a_2) = (a_1 - b_1, a_2 - b_2) =$
$$(\mathbf{O}(a_1 + b_1, a_1 - b_1), \mathbf{O}(a_1 + b_1, a_2 - b_2));$$

and therefore COM$(\mathbf{O}(a_1 + b_1, a_1 - b_1), \mathbf{O}(c_1, c_2))$. But since, as the argument just given shows, RAT$(\mathbf{O}(a_1 + b_1, a_1 - b_1))$, RAT$(\mathbf{O}(c_1, c_2))$ and RAT(r).

It should be clear that, when the antecedent of 114 holds, c_1 and c_2 are "of the same order," i.e., for any rational straight line r, c_1 is an nth r-binomial if and only if c_2 is an nth r-apotome.

Euclid brings out this fact in the converse of X,114; but I shall leave it out of account, and formulate the converse as

X,112 (113) If $\text{RAT}(r)$ & $\mathbf{T}(r) \simeq \mathbf{O}(c_1, c_2)$ and c_1, a_1, b_1 (c_2, a_2, b_2) satisfy the definition of binomial (apotome), then there exist a_2, b_2 (a_1, b_1) such that $(a_1, a_2) = (b_1, b_2)$ and $\text{COM}(a_1, a_2)$ and c_2, a_2, b_2 (c_1, a_1, b_1) satisfy the definition of apotome (binomial).

The preceding considerations suggest the following proof of these propositions. Given the antecedent, one constructs a_2 (a_1) to satisfy $(a_2, c_2) = (a_1, a_1 - b_1)$ ($(a_1, c_1) = (a_2, a_2 + b_2)$) and sets $b_2 \simeq c_2 + a_2$ ($b_1 \simeq c_1 - a_1$), so that $c_2 \simeq a_2 - b_2$ ($c_1 \simeq a_1 + b_1$) and $(a_1, a_2) = (b_1, b_2)$. In addition, since (i) holds and both $\mathbf{O}(a_1 + b_1, a_1 - b_1)$ and $\mathbf{O}(a_1 + b_1, a_2 - b_2)$ are rational, $\text{COM}(a_1, a_2)$.

Euclid's proofs of 112–114 are much more complicated than the ones just given. Since they do not seem to have an intuitive motivation,[25] I shall simply describe what Euclid does. He reduces 114 to 112 by setting $\mathbf{T}(r_1) \simeq \mathbf{O}(c_2, c_3)$, so that there are a_3, b_3 such that c_3, a_3, b_3 satisfy the definition of an apotome; $\text{COM}(a_2, a_3)$, so that $\text{COM}(a_1, a_3)$; and also $(a_1, b_1) = (a_2, b_2) = (a_3, b_3)$, so that $(a_1, a_3) = (a_1 - b_1, a_3 - b_3)$ and $\text{COM}(a_1 - b_1, a_3 - b_3)$. The rationality of r follows from the fact that $(\mathbf{T}(r), \mathbf{T}(r_1)) = (\mathbf{O}(c_1, c_2), \mathbf{O}(c_2, c_3)) = (c_1, c_3) = (a_1 - b_1, a_3 - b_3)$.

In order to describe Euclid's proofs of 112 and 113 it is sufficient to describe his construction of a_2 and b_2 or a_1 and b_1 to satisfy both $(a_1, a_2) = (b_1, b_2)$ or, equivalently, $(a_1, b_1) = (a_2, b_2)$, and also $c_2 \simeq a_2 - b_2$ or $c_1 \simeq a_1 + b_1$, and then his demonstration that a_1 and a_2 or, equivalently, b_1 and b_2 are commensurable. Although there are many parallels between Euclid's proofs of 112 and 113, it is simplest to handle each proof separately.

Construction for 112. Euclid sets $\mathbf{T}(r) \simeq \mathbf{O}(b_1, x)$, so that

$$(c_1, b_1) = (x, c_2), \text{ i.e., } (a_1 + b_1, b_1) = (x, c_2),$$

and

$$(a_1, b_1) = (x - c_2, c_2).$$

Since what is wanted is that

$$(a_1, b_1) = (a_2, b_2)$$

and

$$a_2 - b_2 \simeq c_2,$$

it suffices to take b_2 to satisfy

$$(x - c_2, c_2) = (c_2 + b_2, b_2)$$

and to set

$$a_2 \simeq b_2 + c_2.$$

Euclid does not explain how such a b_2 is to be found; but the procedure is clear, because

$$(x - c_2, c_2) = (c_2 + b_2, b_2) \leftrightarrow \mathbf{T}(c_2) + \mathbf{O}(c_2, b_2) \simeq \mathbf{O}(b_2, x - c_2) \simeq \mathbf{O}(b_2, x) - \mathbf{O}(b_2, c_2) \leftrightarrow \mathbf{T}(c_2) \simeq \mathbf{O}(b_2, x) - 2\cdot\mathbf{O}(b_2, c_2) \simeq \mathbf{O}(b_2, x - 2\cdot c_2).$$

Proof of commensurability for 112. Since $\mathbf{O}(b_1, x)$ is rational, b_1 and x are commensurable; and to establish the commensurability of b_1 and b_2 it suffices to show that $\text{COM}(x, b_2)$ or, equivalently, $\text{COM}(x + b_2, b_2)$, i.e., $\text{COM}(x + a_2 - c_2, b_2)$. But since

$$(x - c_2, c_2) = (c_2 + b_2, b_2),$$

then

$$(x - c_2, a_2) = (x - c_2, c_2 + b_2) = (c_2, b_2),$$

and therefore

$$(x - c_2 + a_2, a_2) = (x - c_2 + a_2, c_2 + b_2) = (c_2 + b_2, b_2) = (a_2, b_2) = (a_1, b_1).$$

Hence, $\text{TCOM}(x - c_2 + a_2, c_2 + b_2)$ and

$$(\mathbf{T}(x - c_2 + a_2), \mathbf{T}(c_2 + b_2)) = (x - c_2 + a_2, b_2),$$

so that $\text{COM}(x - c_2 + a_2, b_2)$.

Construction for 113. Euclid sets $\mathbf{T}(r) \simeq \mathbf{O}(a_2, x)$, so that

$$(a_2, c_2) = (c_1, x)$$

and

$$(a_2, b_2) = (a_2, a_2 - c_2) = (c_1, c_1 - x).$$

Since what is wanted is that

$(a_1, b_1) = (a_2, b_2)$ and $a_1 + b_1 \simeq c_1$,

it suffices to take b_1 satisfying

$$(c_1, c_1 - x) = (c_1 - b_1, b_1),$$

and set

$$a_1 \simeq c_1 - b_1.$$

To simplify the argument for the commensurability of a_1 and a_2, Euclid takes b_1 satisfying $(c_1, c_1 - x) = (b_1, (c_1 - x) - b_1)$ and argues that

$$(c_1, c_1 - x) = (c_1 - b_1, (c_1 - x) - ((c_1 - x) - b_1)) = (c_1 - b_1, b_1).$$

Again Euclid does not indicate how such a b_1 is to be found, apparently expecting the reader to realize that

$$(c_1, c_1 - x) = (b_1, (c_1 - x) - b_1) \leftrightarrow$$
$$\mathbf{O}(c_1, (c_1 - x) - b_1) \simeq \mathbf{O}(b_1, c_1 - x) \leftrightarrow$$
$$\mathbf{T}(c_1) - \mathbf{O}(c_1, x) - \mathbf{O}(c_1, b_1) \simeq \mathbf{O}(b_1, c_1) - \mathbf{O}(b_1, x) \leftrightarrow$$
$$\mathbf{O}(c_1, c_1 - x) \simeq 2\cdot\mathbf{O}(b_1, c_1) - \mathbf{O}(b_1, x) \simeq \mathbf{O}(b_1, 2\cdot c_1 - x).$$

Proof of commensurability for 113. Since $\mathbf{O}(a_2, x)$ is rational, a_2 and x are commensurable; and to establish the commensurability of a_1 and a_2, it suffices to show that $\text{COM}(a_1, x)$ or, equivalently, $\text{COM}(a_1, a_1 - x)$, i.e., $\text{COM}(a_1, (c_1 - b_1) - x)$. But $(a_2, b_2) = (a_1, b_1) = (c_1 - b_1, b_1) = (c_1, c_1 - x) = (b_1, (c_1 - x) - b_1)$. Hence, $\text{TCOM}(a_1, b_1)$ and $(\mathbf{T}(a_1), \mathbf{T}(b_1)) = (a_1, (c_1 - x) - b_1)$, so that $\text{COM}(a_1, (c_1 - b_1) - x)$.

This completes my discussion of book X. I have left out of account only three trivial *lēmmata* before X,14, 17, and 22, in addition to several curious explanatory remarks, most of which are put in appendices by Heiberg. Book X is too elaborate to be given a legible diagrammatic representation, and, in any case, the broad outlines of its deductive structure and its relation to other materials in the *Elements* are of more interest than many of the details. However, even in discussing these things it is useful to make certain simplifications. The important foundations of book X, other than definitions, may be characterized as follows:

(i) Standard proportion theory (the laws of book V) and its geometric applications in VI,1, 8, 16, 17 (14 in X,22), 19–20, 22.

(ii) The arithmetic results which may be characterized as the assertion that all and only similar plane numbers have the ratio of a square number to a square number (VIII,26 and converse) and the construction of square numbers $m_1, \ldots, m_n$ any two of which satisfy specified conditions regarding the question whether or not their sum or difference is square (*lēmmata* before X,29).

(iii) The correlation involving arithmetic, geometry, and proportion theory argued for in X,5–9. The important results are that commensurability is having the ratio of a number to a number, and that two straight lines are commensurable in square only if the squares on them have the ratio of a nonsquare number to a nonsquare number. It is also shown how to construct straight lines or squares standing in given numerical ratios. These materials of course depend upon materials from (i) and (ii).

(iv) General laws of commensurability (X,11–16), which, in some cases, depend on materials from (i)–(iii).

(v) Geometric algebra, notably the construction of a rectangle on a given straight line and equal to a given area (I,44–45, but sometimes VI,12 is used), the Pythagorean theorem, the construction of a square equal to a given area (II,14, but sometimes VI,13), II,4 and 7, VI,28, and X,17,18, which depends upon previous materials and on II,5.

(vi) X,19–23, which establish the foundations for dealing with the notions of being rational or medial.

In summarizing the deductive structure of book X I shall make no mention of the use of materials from (i), (iv), and (vi), which are used throughout the book.

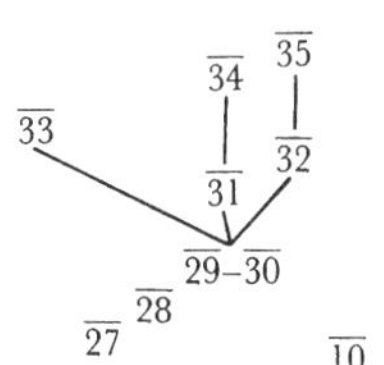

Figure 7.23

X,10,27–35, the construction of the sides and additive sides, depend upon all the foundational materials. Their internal deductive structure is represented in fig. 7.23. None of the propositions is used subsequently.

X,73–78 (36–41), the proofs of the irrationality of the sides (additive sides), depend upon II,7 (II,4). The proofs for the third and sixth sides (additive sides) are reductions to the proposition for the first and make use of I,44,45. The only uses of these propositions in the rest of book X are due to the fact that they are also definitions introducing the names of the sides.

X,79–84 (42–47), in which it is shown that equal *n*th sides (additive sides) are determined by equal terms only, have the same deductive structure as 73–78 (36–41); but in addition these propositions use X,26 as a lemma for the first, second, fourth, and fifth sides (additive sides). (42–47 also use the *lēmma* before 42, which depends on II,4 and 5.) These propositions are not applied subsequently.

X,85–90 (48–53), the construction of the six *r*-apotomes (*r*-binomials), depend on materials from (ii) and (iii) and are not used subsequently.

X,91–96 (54–59), the proofs that the side of a square equal to a rectangle contained by a rational straight line *r* and an *n*th *r*-apotome (*r*-binomial) is an *n*th side (additive side), use II,14, VI,28, X,17,18, and the elementary *lēmma* before X,54. These propositions are subsequently used in X,108–110 (71–72).

X,97–102 (60–65), the converses of the preceding six propositions, use I,44,45, II,7 (II,4), and X,17,18. (The *lēmma* before 60, the proof of which invokes II,5 and 9, is also needed for 60–65.) These propositions are used to show the mutual exclusiveness of the various categories of irrational straight line in X,111 and the remark following it (the remark after X,72).

X,103–107 (66–70), in which it is proved that a straight line commensurable with an *n*th *r*-apotome or side (*r*-binomial or additive side) is one as well, depend only on elementary materials and definitions; they are never used in book X.

X,108–110 (71–72), in which Euclid shows that the side of a square equal to the difference (sum) of a rational and a medial area is a first, second, fourth, or fifth side (additive side) and that the side of a square equal to the difference (sum) of two medial and incommensurable areas is a third or sixth side (additive side), depend upon 91–96 (54–59) and I,44,45. They too are never used.

X,111 and the remark following it, in which it is brought out that no straight line can be more than one of medial and first, second, third, fourth, fifth, or sixth r-apotome, side, r-binomial, or additive side, depend on X,60, 97–102, and the remark after 72, which in turn depends upon 60–65. This fact about the categorization of irrational straight lines is tacitly used in XIII,18.

There is no reason to recapitulate here the deductive information concerning X,1–4, 24, 25, 112–115, since these propositions, except for the previously discussed X,1, play no further role in the *Elements*. It remains to describe the applications of propositions from book X in book XIII. The most significant of these applications is the use of X,94 in XIII,11 to show that the side of a regular pentagon inscribed in a circle with rational diameter r is a fourth side. To be able to apply 94 Euclid has to show first that another straight line is a fourth r-apotome, for which purpose he uses elementary materials from (iii) and (iv) and definitions. The same situation recurs in XIII,6. There Euclid argues that, when a rational straight line r is divided in extreme and mean ratio, the greater segment is an apotome and then that the lesser segment is a first r-apotome. The first argument is again elementary, the second an application of X,97. However, only the first result is applied subsequently; and in the *protasis* of XIII,6 Euclid records only the fact that both segments are apotomes. Euclid uses XIII,6 in XIII,17, and XIII,11 in XIII,16. In each case he uses the fact that if $\exists m \exists n((\mathbf{T}(x), \mathbf{T}(y)) = (m, n))$ and $\text{RAT}(x)$, then $\text{RAT}(y)$—a definitional consequence of X,6. Euclid uses a similar principle in XIII,18 where he introduces without explanation the notion of being in a rational ratio (*logos rhētos*), by which he apparently means being commensurable in square. He tacitly assumes three facts: $\exists m \exists n((\mathbf{T}(x), \mathbf{T}(y)) = (m, n)) \rightarrow \text{TCOM}(x, y)$; $\text{RAT}(x) \ \& \ \neg \text{RAT}(y) \rightarrow \neg \text{TCOM}(x, y)$, which is a definitional consequence of X,13; and the proposition which I called X,104′–107′on p. 283 and which I derived from X,91–103.

Geometric algebra obviously plays a very significant role in book X; and although there are many places in the book where algebraic argument in a more modern sense would simplify the proceedings, it is also clear that in book X the actual geometric positions of straight lines are ultimately of less interest than their lengths and the areas of squares and rectangles which they generate. However, in book X there is no concern with specific lengths or areas; all that matters is various forms of relative commensurability and incommensurability. The interest in these relations is basically qualitative and classificatory; it arises out of the problem of characterizing the edge value of the regular icosahedron inscribed in a sphere, as does the way in which the relations are treated. I do not know why this problem of characterization was thought to be mathematically significant, but I see no good evidence that the problem was construed as a quantitative or computational one.

7.3 The Dodecahedron

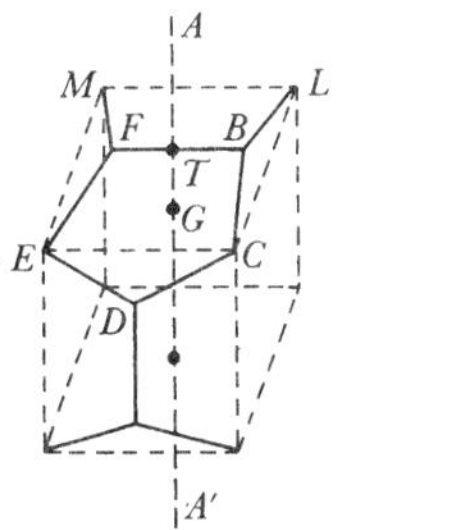

Figure 7.24

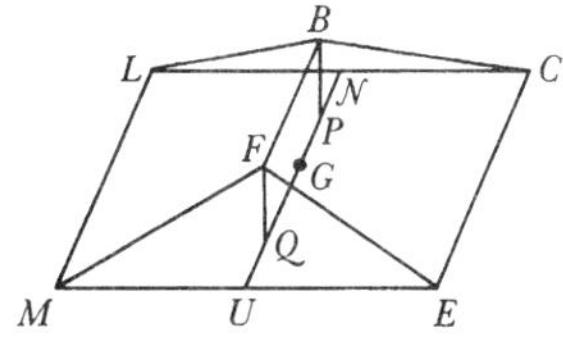

Figure 7.25

The regular dodecahedron is contained by twelve regular pentagons. The crux of Euclid's treatment of it is the recognition of the relation between the dodecahedron and a cube inscribed in the same sphere. Figure 7.24 is an attempt to convey this relation. In it AA' is a diameter of the sphere, passing through the midpoint T of BF, one edge of the dodecahedron, and G, the intersection of the diagonals of the face $ECLM$ of the cube inscribed in the same sphere. Each face of the cube then supports a figure like the roof of a house (fig. 7.25), containing four edges of the cube and five of the dodecahedron. The only problem for the construction is the determination of B and F; but, given the inscription of the cube in the sphere, this problem reduces to the determination of (CE, BF), which obviously yields the length of BF. Hence, one need only draw the straight line NU through G and parallel to CE, and mark off GP, GQ both equal to $\frac{1}{2}BF$. Perpendiculars to the plane of $ECLM$ through P and Q will intersect the surface of the sphere in B and F. Moreover, the ratio (CE, BF) is easy to determine because BF is the side of the regular pentagon $BCDEF$; and, by XIII,8,

$$(CE, BF) = (BF, CE - BF),$$

or

$$(\tfrac{1}{2}CE, \tfrac{1}{2}BF) = (\tfrac{1}{2}BF, \tfrac{1}{2}CE - \tfrac{1}{2}BF).$$

Hence, to inscribe a dodecahedron in a sphere, one need only inscribe a cube in the sphere, draw straight lines connecting the midpoints of the opposite sides of each face of the cube, and divide appropriate segments of these straight lines between their intersections and the sides into extreme and mean ratio, the longer division being toward the intersections. Perpendiculars to the faces of the cube through the points so determined

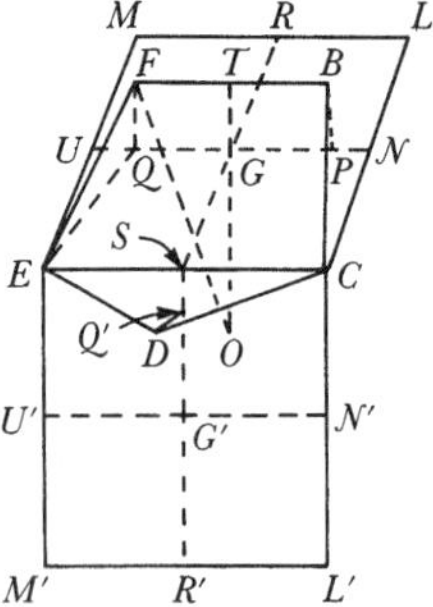

Figure 7.26

produce the remaining vertices of the dodecahedron, the others being the vertices of the cube. Euclid does not fully describe the appropriate segments, and doing so would be rather elaborate; but the character of the construction should be clear.

It remains to show that the construction works. For Euclid this amounts to showing that *BCDEF* lies in one plane and has equal sides and angles. I prove these facts in the order in which Euclid proves them. In fig. 7.26, *ECLM*, *ECL'M'* are faces of a cube inscribed in a sphere with center *O*; *S*, *N*, *R*, *U*, *N'*, *R'*, *U'* are midpoints of the edges *EC*, *CL*, *LM*, *ME*, *CL'*, *L'M'*, *M'E*; *G* and *G'* are the points of intersection of *RS*, *NU* and of *R'S*, *N'U'*; *P*, *Q*, *Q'* divide *NG*, *UG*, *SG'* into extreme and mean ratio with $GP \succ PN$, $GQ \succ QU$, $G'Q' \succ Q'S$; *B*, *F*, *D* are points on the surface of the sphere with *BP*, *FQ* perpendicular to the plane of *ECLM* and *DQ'* perpendicular to the plane of *ECL'M'*. The first step is to show that if *OG* is connected and extended, it is perpendicular to *PQ* and *BF*, and hence bisects *BF* at the point of intersection *T*. This step is not really carried through satisfactorily by Euclid;[26] but rather than going through the details required, I shall simply take for granted the following facts: $PG \simeq QG \simeq FT \simeq BT \simeq G'Q'$; *OGT* is perpendicular to *PQ* and *BF*; $OG \simeq GU$.

Euclid proves that *BCDEF* is equilateral by showing that $FB \simeq FE$ and asserting that the equality of all the sides can be established similarly. His argument for the equality of *FB* and *FE* may be represented as follows. Since $FB \simeq 2 \cdot QG$,

(i) $\mathbf{T}(FB) \simeq 4 \cdot \mathbf{T}(QG)$.

But also, $\mathbf{T}(FE) \simeq \mathbf{T}(EQ) + \mathbf{T}(QF)$ and $\mathbf{T}(EQ) \simeq \mathbf{T}(EU) + \mathbf{T}(UQ) \simeq \mathbf{T}(UG) + \mathbf{T}(UQ) \simeq$ [by XIII,4] $3 \cdot \mathbf{T}(QG)$, so that

(ii) $\mathbf{T}(FE) \simeq 3 \cdot \mathbf{T}(QG) + \mathbf{T}(QF)$,

and it is only necessary to prove that $QF \simeq QG$ or, equivalently, $TG \simeq GP$ or, again equivalently, $TO \simeq UP$. But

(iii) $\mathbf{T}(TO) + \mathbf{T}(FT) \simeq \mathbf{T}(FO)$;

and, since, by XIII,5, *UP* is cut in extreme and mean ratio at *G*, $\mathbf{T}(UP) + \mathbf{T}(FT) \simeq \mathbf{T}(UP) + \mathbf{T}(GP) \simeq$ [by XIII,4] $3 \cdot \mathbf{T}(UG) \simeq 3 \cdot \mathbf{T}(EU)$; and it is only necessary to show

(iv) $\mathbf{T}(FO) \simeq 3 \cdot \mathbf{T}(EU)$.

This equality is, however, equivalent to the characterization of the edge value of the cube. (See above, p. 256.)

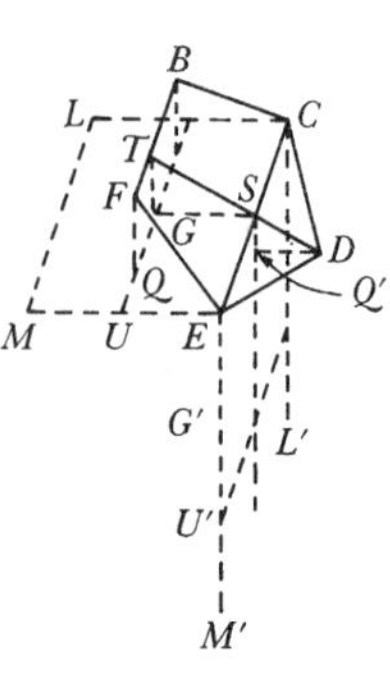

Figure 7.27

To show that *BCDEF* lies in one plane Euclid connects *TS* and *SD* (fig. 7.27). Since *FBCE* and *EDC* are each in one plane and *TS*, *SD* are in the planes of *FBCE*, *EDC* respectively, it suffices to show that TS, SD lie in a straight line. For this purpose Euclid uses VI,32, which is obviously included in the *Elements* only because of this application. According to it, TS, SD are in a straight line if TG is parallel to SQ', GS to $Q'D$, and $(GS, TG) = (Q'D, SQ')$. That the parallelisms hold is a consequence of XI,6, because TG and SQ' are perpendicular to the plane of $ECLM$; and GS and $Q'D$ are perpendicular to the plane of $ECL'M'$. Moreover, since $TG \simeq Q'D \simeq Q'G'$ and $GS \simeq G'S$, $(GS, TG) = (Q'D, SQ')$ if and only if $(G'S, Q'G') = (Q'G', SQ')$, which holds by construction.

To prove that $BCDEF$ is equiangular, Euclid establishes the equality of the angles at B, F, D and cites

XIII,7 If $BCDEF$ is an equilateral pentagon with the angles at B, F, D (or B, F, E) equal, it is equiangular.

The arguments for the equality of the angles at F and B to the angle at D are the same; therefore I do only the argument for the angle at F. Since $BCDEF$ is equilateral, it suffices to show that $BE \simeq EC$; and, since $EC \simeq 2 \cdot UE$, it suffices to show (fig. 7.28) that $BE \simeq 2 \cdot UE$, i.e., that $\mathbf{T}(BE) \simeq 4 \cdot \mathbf{T}(UE)$. However, as in the argument for the equilaterality of $BCDEF$, one has $PG \simeq BP$ and $\mathbf{T}(UP) + \mathbf{T}(BP) \simeq 3 \cdot \mathbf{T}(UG)$, so that $\mathbf{T}(BE) \simeq \mathbf{T}(BP) + \mathbf{T}(PE) \simeq \mathbf{T}(BP) + \mathbf{T}(UP) + \mathbf{T}(UE) \simeq 3 \cdot \mathbf{T}(UG) + \mathbf{T}(UE) \simeq 4 \cdot \mathbf{T}(UE)$.

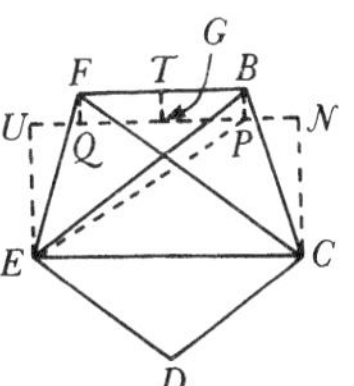

Figure 7.28

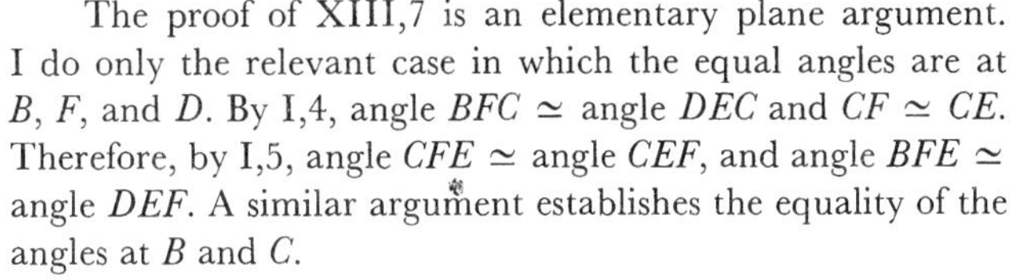

The proof of XIII,7 is an elementary plane argument. I do only the relevant case in which the equal angles are at B, F, and D. By I,4, angle $BFC \simeq$ angle DEC and $CF \simeq CE$. Therefore, by I,5, angle $CFE \simeq$ angle CEF, and angle $BFE \simeq$ angle DEF. A similar argument establishes the equality of the angles at B and C.

It remains to characterize the edge value of the dodecahedron. We know by XIII,15 that $(\mathbf{T}(AA'), \mathbf{T}(EC)) = (3,1)$, and, by construction, that $(EC, FB) = (FB, EC - FB)$. Hence if the diameter AA' of the sphere is taken to be rational, so is EC; and the edge FB of the regular dodecahedron inscribed in the sphere is the greater segment when EC is divided in extreme and mean ratio. Euclid proves

XIII,6 If EVC is a rational straight line and $\mathbf{T}(EV) \simeq \mathbf{O}(EC, VC)$, then EV and VC are apotomes.[27]

For this purpose Euclid uses XIII,1, according to which, if the antecedent holds,

$$\mathbf{T}(\tfrac{1}{2}EC + EV) \simeq 5 \cdot \mathbf{T}(\tfrac{1}{2}EC),$$

so that, since $\frac{1}{2}EC$ is rational, so are $5 \cdot \mathbf{T}(\frac{1}{2}EC)$ and $\frac{1}{2}EC + EV$; clearly also $\frac{1}{2}EC + EV$ and $\frac{1}{2}EC$ are commensurable in square only, and their difference EV is an apotome; by X,97 then

VC is a first *EC*-apotome. Euclid does not go into any more detail on *EV*, i.e., *FB*, and is content to assert that the edge *FB* of a regular dodecahedron inscribed in a sphere with rational diameter *AA′* is an apotome. I shall show that *FB* is a fifth *EC*-apotome and a sixth *AA′*-apotome. If one sets $a \simeq \frac{1}{2}EC + FB$ and $b \simeq \frac{1}{2}EC$, one has, since $(\mathbf{T}(a), \mathbf{T}(b)) = (5, 1)$,

$$(\mathbf{T}(a), \mathbf{T}(a) - \mathbf{T}(b)) = (5, 4),$$
$$(\mathbf{T}(a), \mathbf{T}(EC)) = (5 \cdot \mathbf{T}(\tfrac{1}{2}EC), 4 \cdot \mathbf{T}(\tfrac{1}{2}EC)) = (5, 4),$$
$$(\mathbf{T}(b), \mathbf{T}(EC)) = (1, 4),$$

so that $a - b$, i.e., *FB*, is a fifth *EC*-apotome. Moreover, since $(\mathbf{T}(EC), \mathbf{T}(AA')) = (1, 3)$,

$$(\mathbf{T}(a), \mathbf{T}(AA')) = (5, 12),$$
$$(\mathbf{T}(b), \mathbf{T}(AA')) = (1, 12);$$

and *FB* is a sixth *AA′*-apotome. Euclid's failure to prove either of these more precise results clearly indicates that the subdivision of apotomes into species is merely a device for the determination of the edge value of the icosahedron.

Euclid completes his discussion of the five regular solids in XIII,18 by "setting out" (*ekthesthai*) the edges e_p, e_o, e_c, e_i, e_d of the regular pyramid, octahedron, cube, icosahedron, dodecahedron, and "comparing" (*sugkrinai pros allēlas*) them. The setting out involves constructing them as sides or, in the case of the dodecahedron, part of a side of right triangles with hypotenuse *AA′*, the diameter of the sphere in which the solids were imagined to be comprehended. The only relatively nontrivial new information gained by the comparison of the edges is that $e_i \succ e_d$. For the three rational edges e_p, e_o, e_c the basis of the construction is the fact that (fig. 7.29) if *AA′P* is a right triangle and *PB* is perpendicular to its hypotenuse at *B*, then

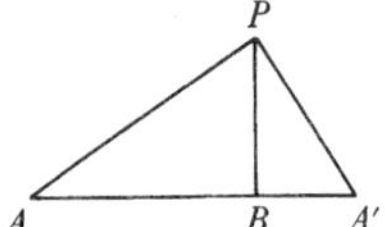

Figure 7.29

$$(\mathbf{T}(AA'), \mathbf{T}(AP)) = (AA', AB).$$

Since $\mathbf{T}(AA') \simeq \frac{3}{2} \cdot \mathbf{T}(e_p)$, $\mathbf{T}(AA') \simeq 2 \cdot \mathbf{T}(e_o)$, and $\mathbf{T}(AA') \simeq 3 \cdot \mathbf{T}(e_c)$, the three edges can be constructed by drawing (fig 7.30) a semicircle on *AA′*, cutting *AA′* at *C* and *D* so tha $AC \simeq CA' \simeq \frac{1}{2}AA'$ and $AD \simeq \frac{2}{3}AA'$ ($DA' \simeq \frac{1}{3}AA'$), and erecting perpendiculars to *AA′* from *C*, *D* intersecting the circle a *E*, *F*. It is then easy to see that $e_p \simeq AF$, $e_o \simeq A'E$, and $e_c \simeq A'F$ Euclid also expresses the ratios of the squares on these edges t one another in numerical terms and asserts that the edges ar to one another "in rational ratios" (*en logois rhētois*), i.e., tha they are commensurable in square. For e_d Euclid, citing a corollary to XIII,17, simply divides *A′F* into extreme and mea ratio at *N*, and takes the greater segment to represent th edge.

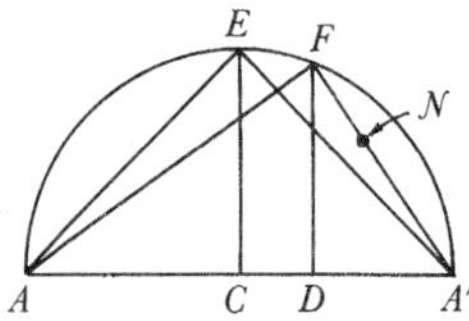

Figure 7.30

For the icosahedron Euclid uses three facts made clear in its inscription, two of which he brings out in a corollary to XIII,16. The corollary is formulated in terms of "the circle from which the icosahedron has been described," the circle which I called $BCDEF$ in my description of the inscription in section 7.1. In order to express these facts I shall use h_s to denote the radius of this circle or (IV,15, cor.) its equal, the side of a regular hexagon inscribed in the circle, p_s to denote the side of a regular pentagon inscribed in it, and d_s to denote the side of a regular decagon so inscribed. The facts expressed in the corollary are that $\mathbf{T}(AA') \simeq 5 \cdot \mathbf{T}(h_s)$ and that $AA' \simeq h_s + 2 \cdot d_s$; the other fact is that $e_i \simeq p_s$. Euclid first constructs h_s by drawing (fig. 7.31) GA perpendicular and equal to AA', connecting GC intersecting the semicircle at H, and dropping HK perpendicular to AA' at K. Since $\mathbf{T}(GA) \simeq 4 \cdot \mathbf{T}(AC)$, $\mathbf{T}(HK) \simeq 4 \cdot \mathbf{T}(KC)$. But $\mathbf{T}(HK) + \mathbf{T}(KC) \simeq \mathbf{T}(HC)$, i.e., $\mathbf{T}(HC) \simeq 5 \cdot \mathbf{T}(KC)$ or $\mathbf{T}(AA') \simeq 5 \cdot \mathbf{T}(2 \cdot KC)$. Hence, if CL is made equal to KC, $KL \simeq h_s$ and $LA' \simeq d_s$. Moreover, since $\mathbf{T}(KL) \simeq 4 \cdot \mathbf{T}(KC)$, $KL \simeq HK$; and, if ML is drawn perpendicular to AA' and intersects the semicircle at M, $ML \simeq HK$, by III,14, so that $ML \simeq KL \simeq h_s$, and, by XIII,10, $MA' \simeq p_s \simeq e_i$.

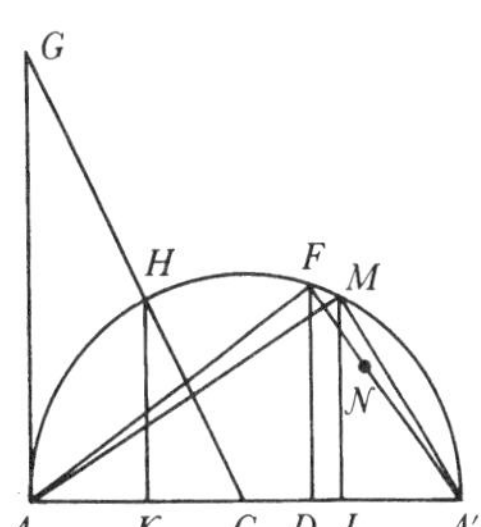

Figure 7.31

Euclid goes on to assert that e_i and e_d are not "in rational ratios" either to any of e_p, e_o, e_c, because e_i and e_d are irrational but e_p, e_o, e_c are each rational relative to AA', or to one another, since e_i is a minor and e_d is an apotome. The first of these inferences makes use of the fact that the class of rational straight lines is closed with respect to commensurability in square; the second presupposes the exclusiveness of the classes of sides and the proposition I called X,104′–107′ on p. 283. Euclid does not proceed any further in terms of explicit calculation of the relative lengths of e_i and e_d, but he does argue that $e_i \succ e_d$, i.e., that $MA' \succ NA'$. He points out that $MA' \succ ML$ and argues as follows that $ML \succ NA'$. Since $\mathbf{T}(A'F) \simeq 3 \cdot \mathbf{T}(A'D)$ and $\mathbf{T}(AD) \simeq 4 \cdot \mathbf{T}(A'D)$, $AD \succ A'F$. Euclid concludes that $ML \succ NA'$ on the grounds that NA' is equal to the greater segment when $A'F$ is divided in extreme and mean ratio; and, by XIII,9, ML is equal to the greater segment when AL is divided in the same way. Obviously he is taking for granted that

$$\text{(i)}\quad \mathbf{T}(u) \simeq \mathbf{O}(v, v-u) \;\&\; \mathbf{T}(u') \simeq \mathbf{O}(v', v'-u') \;\&\; v \succ v' \rightarrow u \succ u'.$$

The principal manuscripts of the *Elements*, including P, contain an alternative proof that $MA' \succ NA'$ which avoids this assumption. The proof may be summarized as follows.

Since it has been established that $3 \cdot \mathbf{T}(A'F) \simeq \mathbf{T}(AA') \simeq 5 \cdot \mathbf{T}(KL) \simeq 5 \cdot \mathbf{T}(ML)$ and $MA' \succ ML$, it is sufficient to show that $3 \cdot \mathbf{T}(A'F) \succ 5 \cdot \mathbf{T}(NA')$. Since by construction, $NA' \succ NF$, then

$$\mathbf{O}(NA', A'F) \succ \mathbf{O}(NF, A'F)$$

and

$$\mathbf{T}(A'F) \simeq \mathbf{O}(NA', A'F) + \mathbf{O}(NF, A'F) \succ 2 \cdot \mathbf{O}(NF, A'F) \simeq 2 \cdot \mathbf{T}(NA');$$

therefore

$$3 \cdot \mathbf{T}(A'F) \succ 6 \cdot \mathbf{T}(NA') \succ 5 \cdot \mathbf{T}(NA').$$

Heiberg puts this proof into an appendix; perhaps it was inserted by someone who found the assumption of (i) objectionable. Assertion (i) is a consequence of a more general proposition formulated and proved almost identically both by Hypsicles in what used to be called book XIV of the *Elements* and by Pappus.[28] It may be rendered

(ii) $\mathbf{T}(u) \simeq \mathbf{O}(v, v - u)$ & $\mathbf{T}(u') \simeq \mathbf{O}(v', v' - u') \rightarrow (v, u) = (v', u')$.

A simple proof of this assertion uses XIII,4, according to which, if

$$\mathbf{T}(u) \simeq \mathbf{O}(v, v - u),$$

then

$$\mathbf{T}(v) + \mathbf{T}(v - u) \simeq 3 \cdot \mathbf{T}(u).$$

Hence, by II,4,

$$\mathbf{T}(v + (v - u)) \simeq \mathbf{T}(v) + \mathbf{T}(v - u) + 2 \cdot \mathbf{O}(v, v - u) \simeq 5 \cdot \mathbf{T}(u).$$

Thus if the antecedent of (ii) holds, then

$$\mathbf{T}(2 \cdot v - u), \mathbf{T}(u)) = (\mathbf{T}(2 \cdot v' - u'), \mathbf{T}(u')),$$
$$(2 \cdot v - u, u) = (2 \cdot v' - u', u'),$$
$$(2 \cdot v, u) = (2 \cdot v', u');$$

and the consequent follows. The chief difference between this argument and the proof given by Hypsicles and Pappus is that they substitute for XIII,4 the more general

II,8 If (fig. 7.32) $ACBD$ is a straight line with $CB \simeq BD$ then $4 \cdot \mathbf{O}(AB, BC) + \mathbf{T}(AC) \simeq \mathbf{T}(AD)$,

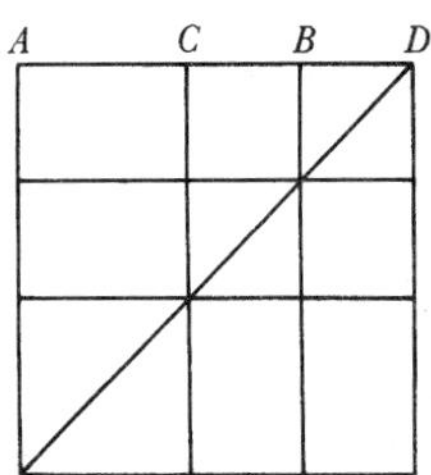

Figure 7.32

which, when AB is divided at C so that $\mathbf{T}(AC) \simeq \mathbf{O}(AB, BC)$, says simply that

$$5 \cdot \mathbf{T}(AC) \simeq \mathbf{T}(AB + BC) \simeq \mathbf{T}(AB + (AB - AC)).$$

Euclid's proof of this proposition is clear from fig. 7.32. Heron reduces it to II,4 and 7, according to which

$$\mathbf{T}(AD) \simeq \mathbf{T}(AB) + \mathbf{T}(BC) + 2 \cdot \mathbf{O}(AB, BC)$$

and

$$\mathbf{T}(AB) + \mathbf{T}(BC) \simeq 2 \cdot \mathbf{O}(AB, BC) + \mathbf{T}(AC).$$

II,8, then, although not explicitly cited in the *Elements*, is used in what appears to be the only classical proof of an assumption made by Euclid in XIII,18. Admittedly, this fact is hardly conclusive evidence that II,8 is included in the *Elements* because it justifies the assumption; but the fact would seem to be as good an explanation of its presence in the *Elements* as, for example, the fact that it is used by Euclid in the *Data* or that it could be used to prove important properties of the parabola.[29] In this respect II,8 is a good example of the problematic character of book II as a whole. II,9, of which the unused II,10 is a counterpart, is cited only in the problematic *lēmma* before X,60; and II,12, of which the unused II,13 is a counterpart, provides a possible explanation (given by a scholiast) for a problematic inference in XII,17. However, the presence and position of propositions 12 and 13 may be sufficiently explained by the fact that they are completions of the Pythagorean theorem but are proved by methods appropriate to book II. The position of II,14 is probably to be explained in much the same way, because, although it is not used until book X, the representation of an arbitrary rectilineal area as a square is of sufficient intrinsic interest to merit inclusion in book II as a culmination of the handling of rectilineal areas. It may also have been important to Euclid to show that this important construction can be carried out without the theory of proportion. For it seems almost certain that the position of II,11 is to be explained in this manner.

If one accepts II,11–14 as a goal of book II, one has an explanation for the presence of II,4–7, which are used in their proofs. 4–7 are, of course, fundamental examples of geometric algebra and are used frequently in the *Elements*. As far as I am able to determine, there is nothing in the *Elements* themselves which makes the algebraic interpretation of these propositions more natural than the straightforward geometric one. On the other hand, the minimal use of II,1–3,[30] 8–10, together with the generally loose connection between book II and books X and XIII, makes it difficult to feel confident about book II.

In general, however, it seems to me that the algebraic interpretation of book II raises as many problem as it purports to solve, and that at least the main propositions of the book, 4–7 and 11–14, are satisfactorily understood either as propositions proved for their own sake (12–14) or as geometric preliminaries to other results (4–7, 11). What unites all of book II is the methods employed: the addition and subtraction of rectangles and squares to prove equalities and the construction of rectilineal areas satisfying given conditions. 1–3 and 8–10 are also applications of these methods; but why Euclid should choose to prove exactly those propositions does not seem to be fully explicable.

After setting out and comparing the five sides, Euclid continues, "I say next that, apart from the said five figures, there cannot be constructed any other figure which is contained by equilateral and equiangular figures equal to one another." This assertion is incorrect. A counterexample is provided by the hexahedron (fig. 7.33) formed from two regular pyramids with base ABC and vertices D and D' on opposite sides of ABC.[31] Clearly Euclid's tacit conception of a regular solid is inadequate; he needs to add as a condition of his assertion that each solid angle of the figure is contained by the same number of plane angles. With this condition added, his argument works. Its basis is the correct but unjustified assumption that two regular solids of different kinds will have solid angles contained by at least three regular polygons which differ either in number or in kind. But the angles of the five regular solids are contained by 3, 4, or 5 equilateral triangles (pyramid, octahedron, icosahedron), 3 squares (cube), or 3 pentagons (dodecahedron). However, by XI,21 the plane angles enclosing a solid angle are together less than 4 right angles. Hence one cannot have a solid angle contained by more than 5 equilateral triangles, 3 squares, or 3 regular pentagons. On the other hand, one cannot have a solid angle contained by even 3 regular polygons with more than 5 sides, because each angle of such a polygon is at least $\frac{4}{3}$ of a right angle, a fact which Euclid takes for granted.

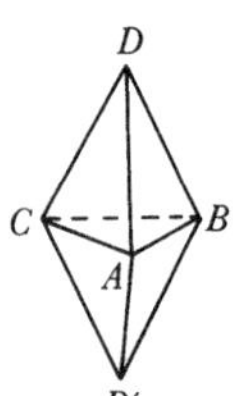

Figure 7.33

Figure 7.34 indicates the deductive structure of book XIII. In the diagram a broken line indicates that a deductive connection is relevant only to the determination of an edge value. I should also mention that 9 is used in 18 only in the proof that the edge of a regular icosahedron inscribed in a sphere is greater than the edge of a regular dodecahedron inscribed in the same sphere.

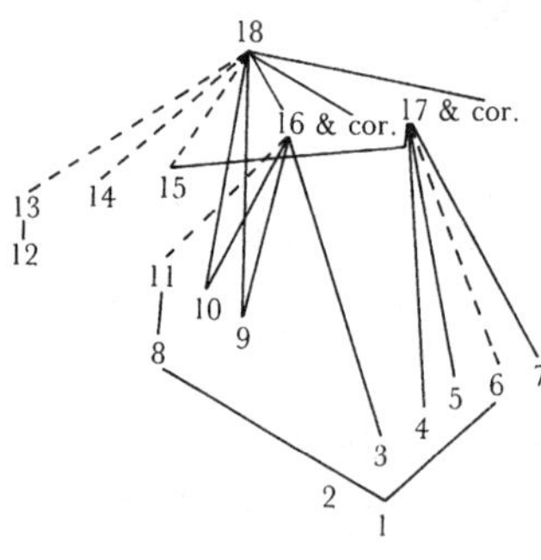

Figure 7.34

According to Proclus,[32] the construction of the five regular solids is Euclid's goal (*telos*) in the whole *Elements*. Proclus remark is clearly due to his desire to associate Euclid with Plato, who used the regular solids in his *Timaeus*. However although from the point of view of deductive structure the remark is a gross exaggeration, one can see how book XIII might have led Proclus to make it. For in book XIII Euclid makes direct use of material from every other book except the arithmetic books and book XII; and XIII is ultimately dependent on the arithmetic books because book X is. In this sense the treatment of the regular solids does constitute a synthesis of much of the *Elements*, a culmination of the Euclidean style in mathematics. However, the significance of the *Elements* lies less in its final destination than in the regions traveled through to reach it. To a greater extent than perhaps any other major work in the history of mathematics, the *Elements* are a mathematical world.

Notes for Chapter 7

Bibliographical Notes

On pp. 88–119 of Sachs's book on the five platonic solids there is an extensive discussion of book XIII, in which the principal question is the contribution of Theaetetus. Waterhouse's account of the relation between the recognition of the regular solids and their mathematical construction is interesting. The deductive structure of XIII is investigated by Neuenschwander in "Die stereometrischen Bücher . . . ."

One of the most frequently discussed topics in the history of Greek mathematics is the relationship between book X, the arithmetic books, and a brief mathematical passage in Plato's dialogue the *Theaetetus* (147C–148B). Knorr gives his own account of the subject and treats most of the previous literature. Other interesting discussions are to be found in Zeuthen, "Sur la constitution . . . ," and van der Waerden, "Die Arithmetik . . . ." The meaning of the passage itself is admirably expounded by Burnyeat.

Most commentaries on book X include an algebraic analysis of its content. Heath's commentary is very detailed in this respect. Dijksterhuis (vol. II, pp. 191–197) provides a briefer algebraic summary while also pointing out some of the ways in which such a representation might mislead. Junge's introduction to Pappus' commentary on book X is an example of the most extreme form of algebraic interpretation. The commentary itself is not very enlightening.

1. See, for example, on book XIII, Neuenschwander, "Die stereometrischen Bücher . . . ," pp. 103–109; and, on book X, Heath, vol. III, pp. 2–4. Knorr's discussion of the possible role of Theaetetus in the two books on pp. 273–285 of *The Evolution* . . . is very useful.

2. This suggestion was first made by Tannery in *La géométrie grecque*, p. 101; it is adopted by van der Waerden, p. 173.

3. Heiberg, "Paralipomena . . . ," p. 59. The correctness of this ascription is of no real importance for my purposes. For arguments in favor of ascribing 1–5 to Eudoxus, see Dijksterhuis, vol. I, pp. 70–71, vol. II, p. 269. Eudoxean authorship would not seem to me reconcilable with Euclid's failure to adopt the simpler proofs.

4. The treatment of the dodecahedron in XIII,17 is only an apparent exception to this characterization. Its construction is a continuation of the construction of the cube which is carried out in the way described.

5. Pappus (vol. I, pp. 132–162) gives analyses and syntheses for these inscriptions along these lines but involving construction procedures less similar to Euclid's. Pappus' reasoning is summarized by Heath in his commentary on XIII,13–17.

6. Euclid establishes (ii) as a general result for equilateral triangles BDC inscribed in circles with radius BG in XIII,12.

7. Euclid's method of construction and comprehension is perhaps easier in this case. He finds G on AA', constructs the right triangle ABA' with BG perpendicular to AA' and then a cube with edge equal to AB and diagonal A_1A_1'. He argues that the triangle determined by A_1A_1' and any vertex of the cube is right, so that the semicircle with A_1A_1' as diameter passes through each vertex and the cube is comprehended in a sphere. He then proves that $A_1A_1' \simeq AA'$.

8. Sachs (p. 103) suggests that an appropriately drawn figure would make clear that $RE'G'G$ is a square so that $h_s \simeq E'G' \simeq GG'$. She mentions the figure in Heath, vol. III, p. 487. Dijksterhuis (vol. II, p. 263) and Neuenschwander ("Die stereometrischen Bücher . . . ," p. 106) suggest that the congruence of the triangles AGB, BRE' is intuitively evident.

9. Dijksterhuis, vol. II, pp. 53, 254–256, argues strongly that Euclid does not recognize this connection.

10. Because Euclid proceeds by construction and comprehension he determines not G but the radius GB of the circle $BCDEF$ as one satisfying the condition $5 \cdot \mathbf{T}(GB) \simeq \mathbf{T}(AA')$. Since $\mathbf{T}(GB) \simeq \mathbf{T}(GG') \simeq 4 \cdot \mathbf{T}(GO)$ and $\mathbf{T}(AA') \simeq 4 \cdot \mathbf{T}(AO)$, it is clear that his procedure is equivalent to the one I have described. To determine such a GB Euclid takes (fig. 7.35) a point S on AA' such that $A'S \simeq 4 \cdot AS$, draws the semicircle ATA' on AA', erects TS perpendicular to AA', and draws the triangle ATA' in which, by VI,8, $(AA', AT) = (AT, AS)$, or $(\mathbf{T}(AA'), \mathbf{T}(AT)) = (AA', AS) = (5, 1)$; and AT gives the length of GB.

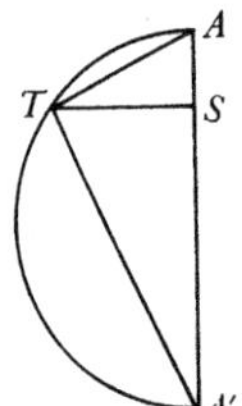

Figure 7.35

11. Becker, "Prinzip . . . ," p. 371ff.

12. In the proof I have given, it is assumed that $a' \succ b'$. C7–C9 enabl one to infer that $\neg(a' \simeq b')$; because if $a' \simeq b'$, then $\mathbf{O}(a', b') \simeq \mathbf{T}(\frac{1}{2}a$ and $a \simeq b$, contradicting the presupposition of C9 that $a \succ b$. T handle the case in which $b' \succ a'$ one need only interchange a' and b in the argument I have given. Geometrically it is only necessary to d one case, as Euclid does. I should perhaps also remark that many o the transformations of equalities requiring several elementary com binatorial steps are quite direct inferences from the geometric repre sentation of the objects involved.

13. For a possibility see Knorr, p. 282. This use of *elassōn* is not recorde in either Liddell-Scott-Jones or in Mugler.

14. Euclid repeats much of the argumentation for each proposition. should be apparent that his procedure of first defining the six

apotomes and six sides and then establishing a correspondence between them looks much more impressive than my analysis in which the definitions are derived from the correspondence.

15. Euclid's arguments again involve considerable repetition. Also, he proves first that c is an r-apotome and then that it is an nth r-apotome for the appropriate n.

16. The first additive side is, of course, the binomial. The second and third are called the first and second bimedial (*ek duo mesōn*), because each is the sum of two medial straight lines. Euclid refers to the fourth additive side with the unexplained term 'major' (*meidzon*), and calls the fifth and sixth the side of a rational plus a medial area and the side of the sum of two medial areas respectively. In the definiens of the last two, Euclid uses $\mathbf{O}(y, z)$ instead of $2 \cdot \mathbf{O}(y, z)$, which he uses in the definitions of the fifth and sixth sides, even though his nomenclature for the fifth and sixth additive sides is obviously derived from the fact that $\mathbf{T}(y + z) \simeq \mathbf{T}(y) + \mathbf{T}(z) + 2 \cdot \mathbf{O}(y, z)$.

17. I am not going to discuss the material which follows 115 in P and other manuscripts; it is included in Heiberg's edition of book X as appendices 24–28. This material contains in succession alternative proofs of 115, 106, and 107, two proofs that the side of a square is incommensurable in length with its diagonal, and a construction of incommensurable plane figures and solid figures.

18. To apply X,17, 18 in this analysis (but not in the corresponding synthesis) one needs an argument that $u \succ v$. Euclid provides one with a *lēmma* before X,60 which I discuss on p. 286.

19. Dijksterhuis (vol. II, p. 183) points out that Euclid does not give the most general solution to 30 (29), which requires only dissimilar numbers the difference between which is dissimilar (similar) to the greater. Euclid's reasoning is probably influenced by geometric considerations, since he constructs u and v as hypotenuse and side of a right triangle so that w is the remaining side.

20. For 35 what follows directly is $\text{COM}(u_1, u_2) \vee \text{TCOM}^*(u_1, u_2)$; but because of (f3) and the condition $\neg\text{COM}(u, v)$, it is impossible that $\text{COM}(u_1, u_2)$.

21. In Heiberg's text Euclid formulates condition (ii) of these propositions with $\text{COM}(u, w)$ in place of $\neg\text{COM}(u, w)$ and therefore uses X,29 instead of 30 in their proofs. At the end of these proofs, however, he remarks that he could also prove them in my formulation, the only formulation he subsequently uses. In the manuscript P the statement of 31, but not of 32, actually corresponds to my formulation; but Heiberg for some reason prefers the other manuscripts in this case.

22. Euclid is presumably relying on this *lēmma* in X,42, 43, 45, and 46 as well, because he assumes without mention that there is some difference between $\mathbf{T}(y_1) + \mathbf{T}(z_1)$ and $\mathbf{T}(y_2) + \mathbf{T}(z_2)$.

23. See Knorr, p. 269 and especially footnote 40.

24. Proclus, *In Platonis Rem Publicam*, vol. II, pp. 27–28. The passage and the algebraic interpretation are discussed by Heath, vol. I, pp. 398–401.

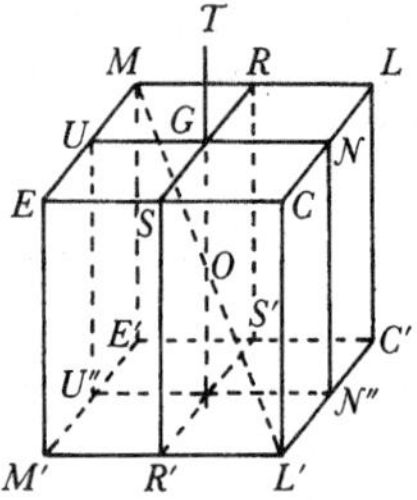

Figure 7.36

25. See Heath, vol. III, p. 246.

26. For Euclid, who construes the problem as a matter of comprehending the constructed dodecahedron in the sphere in which the cube has been comprehended, the task is to show that the straight line connecting the midpoints T of BF and G of UN passes through the center O of the sphere. He does this by assuming that TG extended (fig. 7.36) coincides with the common section of the planes through $NUU''N''$ and $SRS'R'$ and invoking a result proved just for this purpose, namely,

XI,39 If the sides of the opposite faces of a cube are bisected and planes carried through the points of section, the common section of the planes and the diagonal of the cube bisect one another.

Euclid's proof of this proposition is basically a plane argument based on the unjustified assumptions that there are in fact planes through the "points of section" and that the diagonal of the cube intersects the common section of these planes. See Heath, vol. III, pp. 362–363.

27. Heath (vol. III, p. 451) argues against the authenticity of this proposition, claiming that for Euclid it is an obvious consequence of XIII,1.

28. Hypsicles' proof is found on pp. 32–34 of vol. V of Euclid's *Opera Omnia*, Pappus' on p. 428 of vol. I of the *Collectio*.

29. These facts are mentioned by Heath, vol. I, p. 391. Frajese, pp. 175–176, conjectures about the algebraic use of II,8.

30. As indicated on p. 44, II,1 is used quite tacitly in book X; and II,3 is used nowhere except possibly in the problematic IX,15 discussed on p. 107. Euclid appears to cite II,2 in XIII,10.

31. There are in fact five convex counterexamples to Euclid's assertion, all with triangular faces. See Freudenthal and van der Waerden, "Over een bewering . . . ."

32. Proclus, 68.21–23.

Appendix 1 Symbols and Abbreviations

Variables: The letters i, j, k, l, m, and n are used to range over positive integers, in either the Greek or the modern sense. The letters o and, in chapter 2, p are used to range over units. Other lowercase letters are used to range over geometric objects of various kinds and over real numbers; the context suffices to determine the range exactly.

A
Basic equality and similarity assertions. 216

B
Basic assertions about the division of one figure into others. 216–217

C
Figures with equal heights and bases are equal. 217

c
Indicates that an assertion applies to right cylinders or cones. 229

C1
$c \simeq a - b.$ 268

C1′
$c \simeq a + b.$ 285

C2
$\mathbf{T}(x) \simeq \mathbf{O}(c, r).$ 268

C3
$x \simeq y - z.$ 268

C3′
$x \simeq y + z.$ 285

C4
$\mathbf{T}(y) + \mathbf{T}(z) \simeq \mathbf{O}(a, r).$ 268

C5
$2 \cdot \mathbf{O}(y, z) \simeq \mathbf{O}(b, r).$ 268

C6
$\mathbf{T}(y) \simeq \mathbf{O}(a', r)$ & $\mathbf{T}(z) \simeq \mathbf{O}(b', r).$ 269

C7
$a \simeq a' + b'.$ 269

C8
$\mathbf{O}(a', b') \simeq \mathbf{T}(\frac{1}{2}b).$ 269

C9
$\mathbf{T}(a) - \mathbf{T}(b) \simeq \mathbf{T}(e).$ 270

Ca
If C1–9 hold, then TCOM$(y, z) \leftrightarrow$ COM(a, e). 271

Cb
If C1–9 hold, then COM$(a, e) \leftrightarrow$ COM$(a, a') \leftrightarrow$ COM(a, b'). 271

Cc
If C1–9 hold and COM(a, e), then COM$(a, r) \leftrightarrow$ COM$(a', r) \leftrightarrow$ COM(b', r), and TCOM*$(a, r) \leftrightarrow$ TCOM*$(a', r) \leftrightarrow$ TCOM*(b', r). 271

Cd
If C1–9 hold and COM(a, e), then COM$(a, r) \leftrightarrow$ TCOM$(y, r) \leftrightarrow$ TCOM(y, z). 271–272

Ce
If C1–9 hold, then TCOM*$(a, b) \rightarrow \neg$ COM(y, z). 272

COM(x, y)
x and y are commensurable, have a common measure. 264

COMP(x, y, z, w)
For any multiples $m \cdot x$, $n \cdot y$, $m \cdot z$, $n \cdot w$, if $m \cdot x \succ n \cdot y$, then $m \cdot z \succ n \cdot w$; if $m \cdot x \simeq n \cdot y$, then $m \cdot z \simeq n \cdot w$; and if $m \cdot x \prec n \cdot y$, then $m \cdot z \prec n \cdot w$. 125

CPROP$(m_1, \ldots, m_{n+1})$
$(m_1, \ldots, m_{n+1}$ are continuously proportional, i.e., m_i is to m_{i+1} as m_{i+1} is to m_{i+2}. 84

k, l-CPROP$(m_1, \ldots, m_{n+1})$
m_i is to m_{i+1} as k is to l. 84

$k_1, \ldots, k_n, l_1, \ldots, l_n$-CPROP$(m_1, \ldots, m_{n+1})$
m_i is to m_{i+1} as k_i is to l_i. 84

D
Figures with equal heights are to one another as their bases. 217

E
Figures with equal bases are to one another as their heights. 217

F
Figures are equal if and only if their heights and bases reciprocate. 217

G
Similar figures are in the duplicate (triplicate) ratio of that of corresponding sides. 217

GCD $(m_1, \ldots, m_n)$
The greatest common divisor (measure) of $m_1, \ldots, m_n$. 69

H
If f_1, f_2, f_3, f_4 are similar rectilineal figures with corresponding sides s_1, s_2, s_3, s_4, then f_1 is to f_2 as f_3 is to f_4 if and only if s_1 is to s_2 as s_3 is to s_4. 217

J
Figures are to one another in the ratio compounded of the ratios of their bases and heights. 217

K
Similar figures inscribed in circles are to one another as the squares on the diameters of the circles. 217

L
Circles (spheres) are to one another as the squares (cubes) on their diameters. 217

LCM$(k_1, \ldots, k_n)$
The least common multiple of $k_1, \ldots, k_n$; in more Euclidean terminology, the least number measured by $k_1, \ldots, k_n$. 79

lcm$(k_1, \ldots, k_n)$
The least common multiple of $k_1, \ldots, k_n$; in more Euclidean terminology, the least number with k_1th, ..., k_nth parts. 79

LEAST $(k_1, \ldots, k_n)$
$k_1, \ldots, k_n$ are the least positive integers in their ratio, i.e., the least $m_1, \ldots, m_n$ satisfying the condition that m_i is to m_{i+1} as k_i is to k_{i+1}. 75

$l_1, \ldots, l_n$-LEAST $(k_1, \ldots, k_n)$
$k_1, \ldots, k_n$ are the least positive integers in the ratio of $l_1, \ldots, l_n$. 75

$\mathcal{M}(k, l, m)$
k measures l m times, i.e., $m \cdot k \simeq l$. 62

MED (x)
x is a medial straight line, i.e., the square on x is equal to a rectangle the sides of which are rational but commensurable only in square; or x is a medial area, i.e., the side of a square equal in area to x is a medial straight line. 269

m-MULTIPLE (k, l)
$m \cdot l \simeq k$. 62

O (x, y)
A rectangle with sides equal to the straight lines x and y. 42

P
A manuscript of the *Elements*. xiv

p
Indicates that an assertion applies to parallelepipeds. 220

PART (k, l)
k is less than l and divides it evenly. 61

m-PART (k, l)
k is an mth part of l, i.e., $m \cdot k \simeq l$. 62

PARTS (k, l)
k is parts of l, i.e., k is less than l and is not a part of l. 61

m-*n*-PARTS (k, l)
k is m nth parts of l, i.e., k is less than l and some positive integer divides k m times and l n times. 62

pr
Indicates that an assertion applies to triangular prisms. 226

PRIME $(k_1, \ldots, k_n)$
$k_1, \ldots, k_n$ are relatively prime, i.e., have no common divisors but 1. 75

PROPORTIONAL (k, l, m, n)
k is to l as m is to n. 64

py
Indicates that an assertion applies to pyramids. 228

r_1
An arbitrary rational straight line. 266–267

RAT (x)
x is a rational straight line, i.e., x is commensurable with a given straight line presupposed to be rational; or the square with x as side is commensurable with the square with the given straight line as side; or x is a rational area, i.e., the side of a square equal to x is a rational straight line. 266–267

*n*SIM (k, l)
k is a product $k_1 \cdot k_2 \cdot \ldots \cdot k_n$, and l is a product $l_1 \cdot l_2 \cdot \ldots \cdot l_n$, and k_i is to l_i as k_{i+1} is to l_{i+1}. 94

SIMPLANE (k, l)
k and l are similar plane numbers, i.e., k is a product $k_1 \cdot k_2$ and l is a product $l_1 \cdot l_2$, and k_1 is to l_1 as k_2 is to l_2. 91

SIMSOLID (k, l)
k and l are similar solid numbers, i.e., k is a product $k_1 \cdot k_2 \cdot k_3$ and l is a product $l_1 \cdot l_2 \cdot l_3$ and k_1 is to l_1 as k_2 is to l_2 and as k_3 is to l_3. 91

$\Sigma^n l_i$
The sum of $l_1, l_2, \ldots, l_n$; used for positive integers and also for other objects. 61

$\Sigma^n l_i \, (l_i \simeq l)$
The sum of n distinct objects each equal to l; used for positive integers and also for other objects. 61

T (x)
The square on a straight line equal to x. 45

TCOM (x, y)
x and y are commensurable in square, i.e., the squares on the straight lines x and y are commensurable. 264

TCOM* (x, y)
x and y are commensurable in square only, i.e., the squares on the straight lines x and y are commensurable, but x and y are not commensurable. 265

tpy
Indicates that an assertion applies to triangular pyramids. 226

$\forall$
For all. 53 n. 2 (to p. 1)

$\exists$
For at least one. 53 n. 2

$\neg$
It is not the case that. 53 n. 2

&
And. 53 n. 2

$\vee$
Or. 53 n. 2

$\rightarrow$
Only if. 53 n. 2

$\leftrightarrow$
If and only if. 53 n. 2

$=$
Is identical to. 53 n. 2

$\neq$
Is not identical to. 53 n. 2

$\simeq$
Equals. 33, 59

$\prec$
Is smaller than. (The symbol $<$ is used with a modern sense.) 33, 59

$\succ$
Is larger than. (The symbol $>$ is used with a modern sense.) 33

$+$
Plus. 33, 59

$-$
Minus. 33, 59

$m \cdot k \, (m \cdot x)$
The sum of m objects each equal to k (x). 59–61, 121

k/l
k measures l, i.e., divides it evenly. 61

$(k, l) = (m, n)$ or $(x, y) = (z, w)$
k is to l as m is to n or x is to y as z is to w. 66, 125

$(k_1, \ldots, k_n) = (l_1, \ldots, l_n)$
k_i is to k_{i+1} as l_i is to l_{i+1}. 75

Appendix 2 Euclidean Assumptions

Ia
$x \simeq x.$ 36

Ib
$x + y \simeq y + x.$ 37

Ic
$(x + y) + z \simeq x + (y + z).$ 37

Id
$x \simeq (x + y) - y.$ 37

Ie
$(x + y) - z \simeq (x - z) + y.$ 37

If
$x - (y + z) \simeq (x - y) - z.$ 37

Ig
$x \prec y \;\&\; y \prec z \rightarrow x \prec z.$ 37

Ih
$(x \prec y \vee x \simeq y \vee y \prec x) \;\&\; (x \prec y \rightarrow \neg (x \simeq y \vee y \prec x)).$
37

Ii
$x \simeq y \rightarrow (x \prec z \rightarrow y \prec z).$ 37

Ij
$x \prec x + \mathrm{y}.$ 37

Ik
$x - y \prec x.$ 37

Il
$x \prec y \rightarrow \exists z(z + x \simeq y).$ 37

CN4′
$x \simeq y \;\&\; z \prec w \rightarrow x + z \prec y + w.$ 37

CN5′
$x \prec y \;\&\; z \simeq w \rightarrow x - z \prec y - w.$ 37

CN5″
$x \simeq y \;\&\; z \prec w \rightarrow x - z \succ y - w.$ 37

Va
$\forall x \exists y(\mathrm{x} \simeq m \cdot y).$ 122, 138

Vb
(i) $x \simeq y \leftrightarrow m \cdot x \simeq m \cdot y;$ 123
(ii) $x \prec y \leftrightarrow m \cdot x \prec m \cdot y;$
(iii) $m \prec n \leftrightarrow m \cdot x \prec n \cdot x.$

Vc
$\exists w((x, y) = (z, w)).$ 127, 138–139, 173, 180, 232–234

Vd
$x \prec y \rightarrow \exists m(m \cdot x \succ y)$. 139–145, 233–234

VIIa
$\exists o \exists k \forall p(p \in k \leftrightarrow p = o)$ &
$\forall k \exists p \exists l(p \notin k \;\&\; p \in l \;\&\; \forall o(o \in k \rightarrow o \in l))$ &
$\forall o_1 \ldots \forall o_n \exists k \forall o(o \in k \leftrightarrow o = o_1 \vee \ldots \vee o = o_n)$. 61

VIIb
$k \simeq k' \;\&\; l \simeq l' \;\&\; m \simeq m' \rightarrow (\mathcal{M}(k, l, m) \rightarrow \mathcal{M}(k', l', m'))$.
63

VIIc
$k \simeq k' \;\&\; l \simeq l' \;\&\; \mathcal{M}(k, l, m) \;\&\; \mathcal{M}(k', l', m') \rightarrow m \simeq m'$.
63

VIId
$k \simeq k' \;\&\; m \simeq m' \;\&\; \mathcal{M}(k, l, m) \;\&\; \mathcal{M}(k', l', m') \rightarrow l \simeq l'$.
63

VIIe
$l \simeq l' \;\&\; m \simeq m' \;\&\; \mathcal{M}(k, l, m) \;\&\; \mathcal{M}(k', l', m') \rightarrow k \simeq k'$.
63

VIIf
(i) $k/l \;\&\; l/m \rightarrow k/m$; (ii) $k/l \rightarrow k \preceq l$;
(iii) $k/l \;\&\; k/m \rightarrow k/l + m$; (iv) $k/l \;\&\; k/m \rightarrow k/l - m$. 64, 77, 78, 81, 82

VIIg
$k \prec l \rightarrow (m \cdot k \prec m \cdot l \;\&\; k \cdot m \prec k \cdot l)$. 64

VIIh
$\Sigma^n k_i + \Sigma^n l_i \simeq \Sigma^n (k_i + l_i)$. 64, 70, 124

VIIh′
$k \cdot \Sigma^n l_i \simeq \Sigma^n (k \cdot l_i)$. 64, 123

VIIh″
$\Sigma_i^n (\Sigma_j^m k_{j_i}) \simeq \Sigma_j^m (\Sigma_i^n k_{j_i})$. 64, 123

VIIi
$\Sigma^n k_i - \Sigma^n l_i \simeq \Sigma^n (k_i - l_i)$. 64, 70–71, 123

VIIj
PROPORTIONAL (k, l, m, n) & PROPORTIONAL $(k, l, i, j) \rightarrow$
PROPORTIONAL (i, j, m, n). 65, 73

VIIk
(i) PROPORTIONAL $(k, l, m, n) \rightarrow$ PROPORTIONAL (m, n, k, l);
(ii) PROPORTIONAL $(k, l, m, n) \rightarrow (k \simeq m \leftrightarrow l \simeq n)$;
(iii) $k \simeq m \rightarrow$ (PROPORTIONAL $(k, l, m, n) \leftrightarrow l \simeq n)$);
(iv) $l \simeq n \rightarrow$ (PROPORTIONAL $(k, l, m, n) \leftrightarrow k \simeq m)$).
66, 73

XIa1
If two points lie on a plane, so does the straight line through them. 208–210

XIa2
Any three points not on a straight line determine a single plane. 208

XIa3
If two planes intersect, they intersect in a straight line. 208

XIa4
For any plane there is a point not on it. 208

Appendix 3 Additional Propositions

VA
$(x, y) = (z, w) \rightarrow (z, w) = (x, y)$. 128

VB
$\neg((x, y) = (z, w) \,\&\, (x, y) > (z, w))$. 130

VC
$\neg((x, y) > (z, w) \,\&\, (z, w) > (x, y))$. 130

VD
$(x, x) = (y, y)$. 130

VE
$(x, y) = (z, w) \,\&\, x \succ y \rightarrow z \succ w$. 130

VF
$(x, y) > (z, w) \,\&\, (z, w) = (u, v) \rightarrow (x, y) > (u, v)$. 131

VG
$(x_1, x_2) = (y_1, y_2) \,\&\, (x_1, x_3) = (y_1, y_3) \rightarrow$
$(x_1, x_2 + x_3) = (y_1, y_2 + y_3)$. 133

VH
(i) Equal ratios have equal duplicates and triplicates;
(ii) equal duplicates and triplicates are duplicates or triplicates of equal ratios. 135–136, 156, 173–174, 200

VJ
If (x, y) is compounded of (z, w) and (u, v), then
$x \simeq y \leftrightarrow (z, w) = (v, u)$. 136, 161

VK
(i) $y \simeq m \cdot x \,\&\, w \simeq m \cdot z \rightarrow (x, y) = (z, w)$;
(ii) $y \simeq n \cdot u \,\&\, x \simeq m \cdot u \,\&\, w \simeq n \cdot v \,\&\, z \simeq m \cdot v \rightarrow$
$(x, y) = (z, w)$;
(iii) $(x, y) = (z, w) \,\&\, y \simeq m \cdot x \rightarrow w \simeq m \cdot z$;
(iv) $(x, y) = (z, w) \,\&\, y \simeq n \cdot u \,\&\, x \simeq m \cdot u \,\&\, \text{LEAST}(m, n) \rightarrow$
$\exists v(w \simeq n \cdot v \,\&\, z \simeq m \cdot v)$. 137–138, 156, 160

VL
$$(x, y) > (z, w) \leftrightarrow \exists m \exists n(m \cdot x \succ n \cdot y \,\&\, m \cdot z \preceq n \cdot w)$$
$$\leftrightarrow \exists m \exists n(m \cdot x \succeq n \cdot y \,\&\, m \cdot z \prec n \cdot w)$$
$$\leftrightarrow \exists m \exists n(m \cdot x \succ n \cdot y \,\&\, m \cdot z \prec n \cdot w).$$
145, 233

VM
$(x, y) > (z, w) \leftrightarrow (w, z) > (y, x)$. 145

VIA
Of two triangles or parallelograms on unequal bases and in the same parallels the one on the greater base is the greater. 153, 156, 223, 224

VIB

Parallelograms are to one another in the ratio compounded of the ratio of their heights and the ratio of their bases. 154

VIC

Similar and equal rectilineal figures have equal corresponding sides. 173–174

VID

Of similar and unequal rectilineal figures the side of the greater is greater than its correspondent in the lesser figure. 173

VIE

Of similar rectilineal figures with unequal corresponding sides the one with the greater side is the greater. 174

VIIIA

(i) If, for $1 \leq i \leq n + 1$, $m_i = k_i \cdot \ldots \cdot k_n \cdot l_1 \cdot \ldots \cdot l_{i-1}$, then $k_1, \ldots, k_n, l_1, \ldots, l_n$-CPROP $(m_1, \ldots, m_{n+1})$;

(ii) if, for $1 \leq i \leq n + 1$, $m_i = k^{(n+1)-i} \cdot l^{i-1}$, then k, l-CPROP $(m_1, \ldots, m_{n+1})$;

(iii) CPROP $(k^n, k^{n-1} \cdot l, \ldots, k \cdot l^{n-1}, l^n)$;

(iv) CPROP $(1, l, l^2, \ldots, l^n)$;

(v) CPROP $(1, m_1, \ldots, m_n) \rightarrow m_i = m_1^i$. 84–90, 92, 98

VIIIA (iii)′

$\exists m_1 \ldots \exists m_{n-1}$ CPROP $(k^n, m_1, \ldots, m_{n-1}, l^n)$. 90

VIIIB

(CPROP $(m_1, \ldots, m_{n+1})$ & LEAST $(m_1, \ldots, m_{n+1})) \leftrightarrow$ $\exists k \exists l($LEAST (k, l) & $m_i = k^{(n+1)-i} \cdot l^{i-1})$. 85, 89

VIIIB′

CPROP $(m, m_1, \ldots, m_{n-1}, m')$ & LEAST $(m, m_1, \ldots, m_{n-1}, m') \rightarrow$ $\exists k \exists l(m \simeq k^n$ & $m' \simeq l^n)$. 90

VIIIC

CPROP $(m_1, \ldots, m_{n+1}) \leftrightarrow \exists j \exists k \exists l(m_i = j \cdot k^{(n+1)-i} \cdot l^{i-1}$ & LEAST $(k, l))$. 85, 88–89, 93

VIIID

SIMPLANE $(k^2, m) \rightarrow$ SQUARE (m). 93–94

VIIIE

SIMSOLID $(k^3, m) \rightarrow$ CUBE (m). 94

Appendix 4 The Contents of the *Elements*

I give here in an English translation, which varies in many minor ways from Heath's, all of the first principles and propositions of the *Elements* as they are given in the first hand in the body of the manuscript P. I also indicate the pages where a proposition or principle is referred to or discussed. Material which is added for clarity is put in parentheses; material excluded by Heiberg is put in brackets.

(Book) I

Definitions (*Horoi*)

(1) A point is that which has no part (*hou meros outhen*).

(2) A line is breadthless length.

(3) The extremities (*perata*) of a line are points.

(4) A straight line is one which lies evenly (*ex isou*) with the points on itself.

(5) A surface is that which has length and breadth only.

(6) The extremities of a surface are lines.

(7) A plane surface is one which lies evenly with the straight lines on itself.

(8) A plane angle is the inclination (*klisis*) to one another of two lines in a plane which meet one another and do not lie in a straight line.

(9) And when the lines containing the aforesaid angle are straight, the angle is called rectilineal.

(10) When a straight line set up on a straight line makes the adjacent (*ephexēs*) angles equal to one another, each of the equal angles is right, and the straight line standing on the other is called a perpendicular to that on which it stands. 38

(11) An obtuse angle is an angle greater than a right angle. 38

(12) An acute angle is an angle less than a right angle. 38

(13) A boundary (*horos*) is that which is an extremity (*peras*) of something.

(14) A figure is that which is contained by some boundary or some boundaries.

(15) A circle is a plane figure contained by one line [which is called its circumference (*periphereia*)] such that all the straight lines falling upon it [upon the circumference of the circle]

from one point of those lying inside the figure are equal to one another. 38, 179

(16) The point is called the center of the circle. 38

(17) A diameter of the circle is any straight line drawn through the center and terminated in both directions (*eph' hekatera ta merē*) by the circumference of the circle, and such a straight line also bisects the circle. 38

(18) A semicircle is the figure contained by the diameter and the circumference cut off by it.

[A segment (*tmēma*) of a circle is the figure, either greater or less than a semicircle, contained by a straight line and a circumference of a circle.]

(19) Rectilineal figures are those which are contained by straight lines; trilateral by three, quadrilateral by four, and multilateral those contained by more than four straight lines. 38

(20) Of trilateral figures, an equilateral triangle is that which has its three sides equal, an isosceles triangle that which has only two of its sides equal, a scalene triangle that which has its three sides unequal. 38

(21) Further, of trilateral figures, a right-angled triangle is that which has a right angle, an obtuse-angled that which has an obtuse angle, an acute-angled that which has three acute angles. 38

(22) Of quadrilateral figures, a square is that which is equilateral and right-angled, an oblong (*heteromēkes*) that which is right-angled but not equilateral, a rhombus that which is equilateral but not right-angled, a rhomboid that which has its opposite sides and angles equal to one another but which is neither equilateral nor right-angled. 38

(23) Parallel straight lines are those which, being in the same plane and being produced *ad infinitum* in both directions, do not meet each other in either direction. 38

Postulates (*Aitēmata*)

1. Let it be postulated (*aitēsthō*) to draw a straight line from any (*pas*) point to any (*pas*) point, 14, 24, 210

2. and to produce a limited straight line in a straight line, 24, 209

3. and to describe a circle with any center and distance, 24–25

4. and that all right angles are equal to one another, 22, 29–30

5. and that, if one straight line falling on two straight lines makes the interior angles in the same direction less than two right angles, the two straight lines, if produced *ad infinitum*, meet one another in that direction in which the angles less than two right angles are, 28, 30–31, 159

[6. and that two straight lines do not enclose a space.] 31–32, 210

Common Notions (*Koinai Ennoiai*)

(1) Things equal to the same thing are also equal to one another. 33–34

(2) And if equals are added to equals the wholes are equal. 33–34

(3) And if equals are subtracted from equals the remainders are equal. 33–34

[(4) And if equals are added to unequals the wholes are unequal.] 34

[(5) And if equals are subtracted from unequals the remainders are unequal.] 34

[(6) And doubles of the same thing are equal to one another.] 34–35

[(7) And halves of the same thing are equal to one another.] 34–35

(8) And things which coincide with one another (*ta epharmodzonta ep' allēla*) are equal to one another. 35

(9) And the whole is greater than the part. 35

(Propositions)

1. On a given straight line to construct an equilateral triangle. 11, 24, 26

2. To place at (*pros*) a given point a straight line equal to a given straight line. 24–26

3. Given two unequal straight lines, to cut off from the greater a straight line equal to the less. 24–26

4. If two triangles have the two sides equal to two sides respectively and have the angle contained by the equal straight lines equal to the angle, they will also have the base equal to the base, the triangle will be equal to the triangle, and the remaining angles will be equal to the remaining angles respectively, (namely) those which the equal sides subtend (*hupoteinein*). 21, 25–26, 31, 35, 216

5. The angles at the base of isosceles triangles are equal to one another, and if the equal straight lines are produced further,

the angles under the base will be equal to one another. 22, 25, 26

6. If two angles of a triangle are equal to one another, the sides which subtend the equal angles will also be equal to one another. 22, 26, 35

7. On the same straight line there cannot be constructed (*ou sustathēsontai*) two other straight lines equal to the same two straight lines (and) at (*pros*) a different point, in the same direction, (and) having the same extremities as the original straight lines. 22, 26

8. If two triangles have the two sides equal to two sides respectively and also have the base equal to the base, they will also have the angle contained by the equal straight lines equal to the angle. 20–22, 26, 35, 216

9. To bisect a given rectilineal angle. 20–21, 24, 26

10. To bisect a given limited straight line. 20–21, 24, 26

11. To draw a straight line at right angles to a given straight line from a given point on it. 20–21, 24, 26

12. To draw a straight line perpendicular to a given infinite straight line from a given point which is not on it. 20, 24, 26

13. When a straight line set up on a straight line makes angles, it will make either two right angles or angles equal to two right angles. 20, 26

14. If relative to (*pros*) some straight line and a point on it, two straight lines not lying in the same direction make the adjacent angles equal to two right angles, the straight lines will be in a straight line with one another. 20, 30

15. If two straight lines cut one another, they will make the vertical angles (*hai kata koruphēn gōniai*) equal to one another 20, 30

16. If one of the sides of any triangle is produced, the exterior angle is greater than each of the interior and opposite angles. 20

17. Two angles of any triangle taken in any way are less than two right angles. 20, 30

18. The greater side of any triangle subtends the greater angle. 20

19. The greater angle of any triangle is subtended by the greater side. 20

20· Two sides of any triangle taken in any way are greater than the remaining side. 20, 26, 36

21. If two straight lines are constructed inside (and) on one of the sides of a triangle from its extremities, the constructed straight lines will be less than the remaining two sides of the triangle but will contain a greater angle. 26

22. To construct a triangle out of three straight lines which are equal to three given straight lines; thus it is necessary that two taken in any way be greater than the remaining one [because also the two sides of any triangle taken in any way are greater than the remaining side]. 20, 26

23. To construct relative to a given straight line and a point on it a rectilineal angle equal to a given rectilineal angle. 19, 22–23, 24, 26

24. If two triangles have the two sides equal to two sides respectively but the angle contained by the equal straight lines greater than the angle, they will also have the base greater than the base. 19, 26

25. If two triangles have the two sides equal to two sides, respectively, but have the base greater than the base, they will also have the angle contained by the two equal straight lines greater than the angle. 19, 26

26. If two triangles have the two angles equal to two angles respectively and one side equal to one side, either the one adjoining (*pros*) the equal angles or the one subtending one of the equal angles, they will also have the remaining sides equal to the remaining sides respectively and the remaining angle to the remaining angle. 19, 21–22, 26

27. If a straight line falling on two straight lines makes the alternate (*enallax*) angles equal to one another, the straight lines will be parallel to one another. 19–20.

28. If a straight line falling on two straight lines makes the exterior angle equal to the interior and opposite angle in the same direction or the interior angles in the same direction equal to two right angles, the straight lines will be parallel to one another. 19, 20, 30

29. The straight line falling on parallel straight lines makes the alternate angles equal to one another and the exterior angle equal to the opposite and interior angle and the interior angles in the same direction equal to two right angles. 19, 20

30. Straight lines parallel to the same straight line are also parallel to one another. 19, 208

31. To draw a straight line parallel to a given straight line through a given point. 19, 24

32. If one of the sides of any triangle is produced, the exterior

angle is equal to the interior and opposite angle, and the three interior angles of the triangle are equal to two right angles. 19, 20, 178

33. Straight lines joining equal and parallel straight lines in the same direction are themselves also equal and parallel. 18, 19

34. The opposite sides and angles of parallelogrammic areas (*parallēlogramma chōria*) are equal to one another, and the diameter bisects them. 18, 19, 21, 154, 216

35. Parallelograms which are on the same base and in the same parallels are equal to one another. 18, 19, 217

36. Parallelograms which are on equal bases and in the same parallels are equal to one another. 18, 19, 217

37. Triangles which are on the same base and in the same parallels are equal to one another. 18, 19, 217

38. Triangles which are on equal bases and in the same parallels are equal to one another. 18, 19, 217

39. Equal triangles which are on the same base and in the same direction are also in the same parallels. 18, 19

40. Equal triangles which are on equal bases and in the same direction are also in the same parallels. 18, 19

41. If a parallelogram has the same base as a triangle and is in the same parallels, the parallelogram is double of the triangle. 18, 19

42. To construct in a given rectilineal angle a parallelogram equal to a given triangle. 17–19

43. The complements (*paraplērōmata*) of the parallelograms around the diameter of any parallelogram are equal. 18–19, 45, 47–48

44. To apply to (*parabalein para*) a given straight line in a given rectilineal angle a parallelogram equal to a given triangle. 17–19, 23, 24, 25, 293–294

45. To construct in a given rectilineal angle a parallelogram equal to a given rectilineal (figure or area). 16–19, 26, 45, 293–294

46. To describe a square on (*apo*) a given straight line. 26–27

47. In right-angled triangles the square on the side subtending the right angle is equal to the squares on the sides containing the right angle. 26–27, 45, 172–173, 178

48. If the square on one of the sides of a triangle is equal to the squares on the remaining two sides of the triangle, the

angle contained by the remaining two sides of the triangle is right. 26–27

(The) Second (Book) of Euclid's *Elements*

(Definitions)

(1) Any right-angled parallelogram is said to be contained by the straight lines containing the right angle.

(2) Let any one of the parallelograms around the diameter of any parallelogrammic area (together) with the two complements be called a gnomon. 166

(Propositions)

1. If there are two straight lines and one of them is cut into any number of segments, the rectangle (*orthogōnion*) contained by the two straight lines is equal to the rectangles contained by the uncut straight line and each of the segments. 41–47, 301–302

2. If a straight line is cut at random (*hōs etuchen*), the rectangle contained by the whole and both of the segments is equal to the square on the whole. 27, 45–46, 176 n. 24, 258, 301–302

3. If a straight line is cut at random, the rectangle contained by the whole and one of the segments is equal to the rectangle contained by the segments and the square on the aforesaid segment. 45–48, 108, 301–302

4. If a straight line is cut at random, the square on the whole is equal to the squares on the segments and twice the rectangle contained by the segments. 45, 47–49, 108, 253, 284, 285–286, 293, 301–302

5. If a straight line is cut into equal and unequal segments, the rectangle contained by the unequal segments of the whole with the square on the segment between the sections is equal to the square on the half. 45, 48–50, 178, 195, 269–270, 285, 293, 301–302

6. If a straight line is bisected and some straight line is added to it in a straight line, the rectangle contained by the whole with the added straight line and the added straight line is equal to the square on the straight line composed of the half and the added straight line. 45, 48–50, 108, 170, 178, 193–194, 253, 301–302

7. If a straight line is cut at random, the two squares together, that on the whole and that on one of the segments, are equal to twice the rectangle contained by the whole and the said segment and the square on the remaining segment. 45–49, 252, 266, 268, 277, 284, 286–287, 293, 301–302

8. If a straight line is cut at random, four times the rectangle

contained by the whole and one of the segments with the square on the remaining segment is equal to the square described on the whole and the aforesaid segment as on one straight line. 48, 300–302

9. If a straight line is cut into equal and unequal segments, the squares on the unequal segments of the whole are double of the square on the half and the square on the segment between the sections. 286–288, 293, 301–302

10. If a straight line is bisected and some straight line is added to it in a straight line, the two squares together, that on the whole with the added straight line and that on the added straight line, are double of the square on the half and the square described on the straight line composed of the half and the added straight line as on one straight line. 287–288, 301–302

11. To cut a given straight line so that the rectangle contained by the whole and one of the segments is equal to the square on the remaining segment. 169–170, 193–194, 301–302

12. In obtuse-angled triangles the square on the side subtending the obtuse angle is greater than the squares on the sides containing the obtuse angle by twice the rectangle contained by one of the sides around the obtuse angle, the one on which the perpendicular falls, and the straight line cut off outside (the triangle) by the perpendicular towards (*pros*) the obtuse angle. 244, 301–302

13. In acute-angled triangles the square on the side subtending the acute angle is less than the squares on the sides containing the acute angle by twice the rectangle contained by one of the sides containing the acute angle, the one on which the perpendicular falls, and the straight line cut off inside by the perpendicular towards the acute angle. 244, 301–302

14. To construct a square equal to a given rectilineal (figure or area). 16–17, 45, 161, 162, 293, 301–302

(Book) III

(Definitions)

(1) Equal circles are those the diameters of which are equal, or the radii (*hai ek tōn kentrōn*) of which are equal. 196–202

(2) A straight line is said to touch (*ephaptesthai*) a circle which, meeting (*haptesthai*) the circle and being produced, does not cut the circle [in either direction]. 183

(3) Circles are said to touch one another which, meeting one another, do not cut one another. 183

(4) Straight lines in a circle are said to be equally distant

from the center when the perpendiculars drawn from the center to them are equal. 181

(5) But (that straight line) is said to be at a greater distance to which the greater perpendicular falls. 181

(6) A segment of a circle is the figure contained by a straight line and an arc (*periphereia*) of a circle.

(7) An angle of a segment is that contained by a straight line and an arc of a circle. 178, 187

(8) An angle in a segment is the angle contained by the connected straight lines when some point is taken on the arc of the segment and straight lines are connected from it to the extremities of the straight line which is the base of the segment.

(9) But when the straight lines containing the angle cut off some arc, the angle is said to stand upon that arc.

(10) A sector of a circle is the figure contained by the straight lines containing the angle and the arc cut off by them when an angle is constructed at (*pros*) the center of the circle. 178

(11) Similar segments of circles are those which admit (*dechomai*) equal angles or in which the angles are equal to one another. 195–202

(Propositions)

1. To find the center of a given circle. 177–180

Corollary. From this it is manifest that if in a circle a straight line bisects a straight line and cuts it at right angles, the center of the circle is on the cutting straight line. 181

2. If two chance points are taken on the circumference of a circle, the straight line connecting the points will fall inside the circle. 177–179, 184

3. If in a circle some straight line through the center bisects some straight line not through the center, it also cuts it at right angles; and if it cuts it at right angles, it also bisects it. 177–178, 181

4. If in a circle two straight lines which are not through the center cut one another, they do not bisect one another. 177–178, 181

5. If two circles cut one another, they will not have the same center. 177–178, 183, 184

6. If two circles touch one another, they will not have the same center. 177–178, 183, 184

7. If some point which is not the center of the circle is taken on the diameter of a circle and some straight lines fall from

the point to the circle, the greatest will be the one on which the center is, the least the remainder (of the same straight line), and of the rest the nearer to the straight line through the center is always greater than the further away; and only two equal straight lines fall from the point to the circle, on each side (*eph' hekatera*) of the least straight line. 177–178, 182

8. If some point is taken outside a circle and some straight lines are drawn through to the circle, of which one is through the center but the remainder at random, of the straight lines which fall on the concave arc the greatest is the one through the center, [the least the one falling between the point and the diameter], and of the rest the nearer to the straight line through the center is always greater than the further away, but of those which fall on the convex arc the nearer to the least straight line is always greater than the further away; and only two equal straight lines will fall from the point to the circle, on each side of the least straight line. 177–178, 182

9. If some point is taken inside a circle and more than two equal straight lines fall from the point to the circle, the point taken is center of the circle. 177–178, 182–183

10. A circle does not cut a circle at more points than two. 177–178, 184

11. If two circles touch one another (internally) and their centers are taken, the straight line joining their centers, when produced, will fall on the contact (*sunaphē*) of the circles. 177–178, 184–185

12. If two circles touch one another externally, the straight line joining their centers will pass through their contact (*epaphē*). 177–178, 184

13. A circle does not touch a circle at more points than one, if it touches it internally or externally. 177–178, 184

14. In a circle equal straight lines are equally distant from the center, and straight lines equally distant from the center are equal. 177–178, 181, 182, 245, 299

15. In a circle the diameter is a greatest (straight line), and of the rest the nearer to the center is greater than the further away. 177–178, 181–182, 192

16. The straight line drawn at right angles to the diameter of a circle at its extremity will fall outside the circle, and another straight line will not fall into the region (*topos*) between the straight line and the circumference, and the angle of a semicircle is greater than every acute rectilineal angle, the remaining angle less. 33, 177, 178, 185–187, 192

(Corollary) From this it is manifest that the straight line drawn at right angles to the diameter of a circle at its extremity touches the circle. 177, 185–187, 203

17. To draw a straight line from a given point touching a given circle. 177, 185–187, 192

18. If some straight line touches a circle and some straight line is joined from the center to the contact (*aphē*), the joined straight line will be perpendicular to the tangent (*hē ephaptomenē*). 177, 178, 185–187, 192

19. If some straight line touches a circle and a straight line is drawn from the contact at right angles to the tangent, the center of the circle will be on the drawn straight line. 177, 185–187, 192

20. In a circle the angle at the center is double the angle at the circumference when the angles have the same arc as base. 177, 178, 187–188, 196

21. In a circle the angles in the same segment are equal to one another. 177, 187–188, 196

22. The opposite angles of quadrilaterals in circles are equal to two right angles. 177, 178, 187–188

23. Two similar and unequal segments of circles cannot be constructed on the same straight line and in the same direction. 177, 178, 195, 198

24. Similar segments of circles on equal straight lines are equal to one another. 35, 177, 195–196, 198

25. To describe the circle of which a given segment of a circle is a segment. 177, 178, 180, 183

26. In equal circles equal angles stand on equal arcs, if they are at the centers and if they are at the circumferences. 177, 191, 195–199, 203

27. In equal circles angles standing on equal arcs are equal to one another, if they stand at the centers and if they stand at the circumferences. 177, 191, 195–197, 200, 203

28. In equal circles equal straight lines cut off equal arcs, the greater equal to the greater, the less to the less. 177, 192, 195–197, 203, 259

29. In equal circles equal arcs are subtended by equal straight lines. 177, 178, 191, 195–197, 203

30. To bisect a given arc. 177, 178, 200, 203

31. In a circle the angle in a semicircle is right, that in a greater segment less than a right angle, that in a lesser segment greater than a right angle; and further, the angle of the greater

segment is greater than a right angle, the angle of the lesser segment less than a right angle. 33, 157, 177, 178, 187, 188

32. If some straight line touches a circle and some straight line cutting the circle is drawn from the contact into the circle, the angles which it makes with the tangent will be equal to the angles in the alternate segments of the circle. 177, 178, 188, 193

33. To describe on a given straight line a segment of a circle admitting an angle equal to a given rectilineal angle. 177, 178, 189

34. To cut off from a given circle a segment admitting an angle equal to a given rectilineal angle. 177, 178, 189

35. If in a circle two straight lines cut one another, the rectangle contained by the segments of one is equal to the rectangle contained by the segments of the other. 177, 178, 194–195

36. If some point is taken outside a circle and two straight lines fall from it to the circle and one of them cuts the circle, the other touches it, the rectangle contained by the whole straight line cutting the circle and the straight line cut off outside (the circle) between the point and the convex arc will be equal to the square on the tangent. 177, 189, 194

37. If some point is taken outside a circle and two straight lines fall from the point to the circle and one of them cuts the circle, the other falls to it, and if the rectangle contained by the whole straight line cutting the circle and the straight line cut off outside between the point and the convex arc is equal to the square on the straight line falling to the circle, the straight line falling to the circle will touch the circle. 177, 178, 189, 194

(Book) IV

Definitions

(1) A rectilineal figure is said to be inscribed in a rectilineal figure when each of the angles of the inscribed figure meets (*haptesthai*) each (i.e., some) side of that in which it is inscribed.

(2) Similarly a (rectilineal) figure is said to be circumscribed about a (rectilineal) figure when each side of the circumscribed figure meets each angle of that about which it is circumscribed.

(3) A rectilineal figure is said to be inscribed in a circle when each angle of the inscribed figure meets the circumference of the circle.

(4) A rectilineal figure is said to be circumscribed about a circle when each side of the circumscribed figure touches the circumference of the circle.

(5) Similarly a circle is said to be inscribed in a (rectilineal) figure when the circumference of the circle meets (touches) each side of that in which it is inscribed.

(6) A circle is said to be circumscribed about a (rectilineal) figure when the circumference of the circle meets each angle of that about which it is circumscribed.

(7) A straight line is said to be fitted into (*enarmodzesthai*) a circle when its extremities are on the circumference of the circle.

(Propositions)

1. To fit into a given circle a straight line equal to a given straight line which is not greater than the diameter of the circle. 189, 194, 202, 203

2. To inscribe in a given circle a triangle equiangular with a given triangle. 189, 190

3. To circumscribe about a given circle a triangle equiangular with a given triangle. 189, 190

4. To inscribe a circle in a given triangle. 189, 190

5. To circumscribe a circle about a given triangle. 189, 194

(Addition) And it is manifest that when the center of the circle falls inside the triangle (*BAC*), the angle *BAC*, being in a segment greater than a semicircle, is less than a right angle; but when the center falls on the straight line *BC*, the angle *BAC*, being in a semicircle, is right; and when the center of the circle falls outside the triangle, the angle *BAC*, being in a segment less than a semicircle, is greater than a right angle; [thus also, when the given angle (*BAC*) is less than a right angle, *DF*, *EF* (the perpendicular bisectors of *AB*, *AC*) will meet inside the triangle, but when it is right, they will meet on *BC*, and when it is greater than a right angle, they will meet outside *BC*].

6. To inscribe a square in a given circle. 189, 190

7. To circumscribe a square about a given circle. 189, 190

8. To inscribe a circle in a given square. 189, 190

9. To circumscribe a circle about a given square. 189, 190

10. To construct an isosceles triangle having each of the angles at the base double of the remaining one. 189, 192–194

11. To inscribe an equilateral and equiangular pentagon in a given circle. 189, 192–194

12. To circumscribe an equilateral and equiangular pentagon about a given circle. 189, 190–191

13. To inscribe a circle in a given pentagon which is equilateral and equiangular. 189, 190–191

14. To circumscribe a circle about a given pentagon which is equilateral and equiangular. 189, 190–191

15. To inscribe an equilateral and equiangular hexagon in a given circle. 189, 191

Corollary. From this it is manifest that the side of the hexagon is equal to the radius of the circle. Just as in the (propositions) concerning the pentagon, if we draw tangents to the circle through the (points of) division on the circle, there will be circumscribed about the circle an equilateral and equiangular hexagon in conformity with (*akolouthōs*) what was said in the case of the pentagon. And further we can inscribe a circle in a given hexagon and circumscribe one by means similar to those expressed in the case of the pentagon. 189, 203

16. To inscribe an equilateral and equiangular *pentekaidekagon* in a given circle. 189

(Addition) Just as in the (propositions) concerning the pentagon, if we draw tangents to the circle through the (points of) division on the circle, there will be circumscribed about the circle an equilateral and equiangular *pentekaidekagon*. Further, we can inscribe a circle in a given *pentekaidekagon* and circumscribe one also by proofs similar to those in the case of the pentagon. 190–191

(Book) V

(Definitions)

(1) A magnitude is part of a magnitude, the less of the greater, when it measures the greater.

(2) The greater is a multiple of the less when it is measured by the less.

(3) A ratio is a kind of relation with respect to size between two homogeneous magnitudes. (*Logos esti duo megethōn homogenōn hē kata pēlikotēta poia schesis.*) 126

(4) Magnitudes which, when multiplied, can exceed one another are said to have a ratio to one another. 141–145

(5) Magnitudes are said to be in the same ratio, first to second and third to fourth, when equal multiples of the first and the third at the same time exceed or at the same time are equal to or at the same time fall short of equal multiples of the second and the fourth when compared to one another, each to each whatever multiples are taken. 125

(7) When of the equal multiples the multiple of the first exceed the multiple of the second but the multiple of the third does no

exceed the multiple of the fourth, then the first is said to have a greater ratio to the second than the third to the fourth. 125

(6) Let magnitudes having the same ratio be called proportional.

(8) A proportion in three terms is least.

(9) When three magnitudes are proportional, the first is said to have to the third the duplicate (*diplasios*) ratio of that which it has to the second. 92

(10) When four magnitudes are (continuously) proportional, the first is said to have to the fourth the triplicate (*triplasios*) ratio of that which it has to the second, and similarly continually forever, whatever proportion holds (*kai aei hexēs homoiōs, hōs an hē analogia huparchēi*). 92

(11) Magnitudes are said to be corresponding, antecedents to antecedents, consequents to consequents. 94

(12) Alternate ratio is taking the antecedent to the antecedent and consequent to the consequent.

(13) Inverse ratio is taking the consequent as antecedent to the antecedent as consequent.

(14) Composition (*sunthesis*) of a ratio is taking the antecedent with the consequent as one thing to the consequent itself.

(15) Separation (*diairesis*) of a ratio is taking the excess by which the antecedent exceeds the consequent to the consequent itself.

(16) Conversion (*anastrophē*) of a ratio is taking the antecedent to the excess by which the antecedent exceeds the consequent.

(17) An *ex equali* (*di' isou*) ratio is when, there being magnitudes and others equal to them in multitude which taken in pairs are also in the same ratio, as in the first magnitudes the first is to the last, so in the second magnitudes the first is to the last; in other words, it is taking the extremes by eliminating the means.

(18) A perturbed (*tetaragmenos*) proportion is when, there being three magnitudes and others equal to them in multitude, it is the case that as in the first magnitudes antecedent is to consequent so in the second magnitudes antecedent is to consequent, but as in the first magnitudes consequent is to something else (i.e., the third magnitude) so in the second magnitudes something else is to the consequent.

(Propositions)

1. If any number of magnitudes are respectively equal multiples of any number of magnitudes equal in number, whatever

multiple one of the magnitudes is of one, the same multiple will all be of all. 123, 125, 131, 134

2. If a first is an equal multiple of a second and a third of a fourth, and a fifth is an equal multiple of the second and a sixth of the fourth, also first and fifth added together will be an equal multiple of the second, and third and sixth of the fourth. 124–125, 134

3. If a first is an equal multiple of a second and a third of a fourth, and equal multiples of the first and third are taken, also *ex equali* the things taken will be equal multiples respectively, the one of the second, the other of the fourth. 124–125, 134

4. If a first has to a second the same ratio as a third to a fourth, also equal multiples of the first and third will have the same ratio to equal multiples of the second and fourth when compared to one another, whatever multiples are taken. 124–125, 134

5. If a magnitude is an equal multiple of a magnitude as a (part) subtracted is of a (part) subtracted, also the remainder will be an equal multiple of the remainder, (namely) whatever multiple the whole is of the whole. 123, 125

6. If two magnitudes are equal multiples of two magnitudes and some things subtracted from them are equal multiples of the same things, also the remainders are either equal to these things or equal multiples of them. 124–125

7. Equals have the same ratio to the same thing, and the same thing has the same ratio to equals. 129, 131

Corollary. From this it is manifest that if some magnitudes are proportional, they will also be proportional inversely. 129

8. Of unequal magnitudes the greater has a greater ratio than the less to the same thing; and the same thing has to the less a greater ratio than to the greater. 130, 131, 139–142

9. Things having the same ratio to the same thing are equal to one another; and things to which the same thing has the same ratio are equal. 130, 131

10. Of things having a ratio to the same thing, that having a greater ratio is greater; but that to which the same thing has a greater ratio is less. 130, 131

11. Ratios which are the same with the same ratio are also the same with one another. 128, 129, 131

12. If any number of magnitudes are proportional, it will be the case that as one of the antecedents is to one of the consequents so are all the antecedents to all the consequents. 129, 131, 134

13. If a first has to a second the same ratio as a third to a fourth, but the third has to the fourth a greater ratio than a fifth to a sixth, also the first will have to the second a greater ratio than the fifth to the sixth. 130, 131

14. If a first has to a second the same ratio as a third to a fourth but the first is greater than the third, also the second will be greater than the fourth, and if equal, equal, and if less, less. 128–131, 134, 248 n. 21 (to p. 232)

15. Parts have the same ratio as the same multiples (of them) when compared to one another. 128, 129, 131, 134

16. If four magnitudes are proportional, they will also be alternately proportional. 128, 131, 134

17. If compounded magnitudes are proportional, they will also be proportional when divided. 132–134, 156

18. If divided magnitudes are proportional, they will also be proportional when compounded. 132–134, 139

19. If as whole is to whole so is (part) subtracted to (part) subtracted, the remainder will also be to the remainder as whole to whole. 132–134, 156

(Addition) From this it is manifest that if compounded magnitudes are proportional, they are also proportional when converted. 149 n. 12

20. If there are three magnitudes and others equal to them in multitude which taken in pairs are in the same ratio and *ex equali* the first is greater than the third, also the fourth will be greater than the sixth, and if equal, equal, and if less, less. 131, 134

21. If there are three magnitudes and others equal to them in multitude which taken in pairs are in the same ratio but the proportion of them is perturbed and *ex equali* the first is greater than the third, also the fourth will be greater than the sixth, and if equal, equal, and if less, less. 133–134

22. If there are any number of magnitudes and others equal to them in multitude which taken in pairs are in the same ratio, they will also be in the same ratio *ex equali*. 131–132, 135

23. If there are three magnitudes and others equal to them in multitude which taken in pairs are in the same ratio but the proportion of them is perturbed, they will also be in the same ratio *ex equali*. 133, 134, 156

24. If a first has to a second the same ratio as a third to a fourth and a fifth has to the second the same ratio as a sixth to a fourth, also the first and fifth added together will have the

same ratio to the second as the third and sixth to the fourth. 133, 134

25. If four magnitudes are proportional, the greatest and the least (together) are greater than the remaining two. 133, 134

(Book) VI

(Definitions)

(1) Similar rectilineal figures are such as have their angles equal one by one and the sides around the equal angles proportional. 157–158

[(2) Figures are reciprocal when in each of the figures there are antecedent and consequent ratios.]

(3) A straight line is said to be divided in extreme and mean ratio when as the whole is to the greater segment so is the greater to the less.

(4) A height of any figure is the perpendicular drawn from the vertex to the base. 153

(Propositions)

1. Triangles and parallelograms which are under the same height are to one another as the bases. 153–154, 217, 292

2. If some straight line is drawn parallel to one of the sides of a triangle, it will cut the sides of the triangle proportionally; and if the sides of the triangle are cut proportionally, the straight line connecting the sections will be parallel to the remaining side of the triangle. 158–160

3. If an angle of a triangle is bisected and the straight line cutting the angle also cuts the base, the sections of the base will have the same ratio as the remaining sides of the triangle; and if the sections of the base have the same ratio as the remaining sides of the triangle, the straight line connected from the vertex to the base will bisect the angle of the triangle. 159–160

4. The sides of equal-angled (*isogōnios*) triangles around the equal angles are proportional, and the sides subtending the equal angles correspond. 158–160

5. If two triangles have their sides proportional, the triangle will be equal-angled and will have the angles which the corresponding sides subtend equal. 158–160

6. If two triangles have one angle equal to one angle and the sides around the equal angles proportional, the triangles will be equal-angled and will have the angles which the corresponding sides subtend equal. 158–160

7. If two triangles have one angle equal to one angle, the side

about other angles proportional, and each of the remaining angles at the same time either less than or not less than a right angle, the triangles will be equal-angled and will have the angles around which the sides are proportional equal. 55 n. 35, 158–160

8. If a perpendicular is drawn in a right-angled triangle from the right angle to the base, the triangles at (*pros*) the perpendicular are similar to the whole and to one another. 159, 160

Corollary. From this it is manifest that if a perpendicular is drawn in a right-angled triangle from the right angle to the base, what is drawn is a mean proportional straight line between the segments of the base. [And further between the base and any one of the segments the side toward (*pros*) the segment is a mean proportional straight line.] 159, 160

9. To cut off a prescribed part from a given straight line. 134, 156, 160–161, 180

10. To cut a given uncut straight line similarly to a given cut one. 134, 160, 180

11. To find a third proportional to two given straight lines. 134, 160, 180

12. To find a fourth proportional to three given straight lines. 134, 160, 180, 293

13. To find a mean proportional between two given straight lines. 134, 135, 157, 160–162, 180, 293

14. The sides of equal and equal-angled parallelograms around the equal angles reciprocate; and those equal-angled parallelograms in which the sides around the equal angles reciprocate are equal. 160–161, 163, 169, 217, 292

15. The sides around the equal angles of equal triangles having one angle equal to one angle reciprocate; and those triangles having one angle equal to one angle in which the sides around the equal angles reciprocate are equal. 161, 163, 217

16. If four straight lines are proportional, the rectangle contained by the extremes is equal to the rectangle contained by the means; and if the rectangle contained by the extremes is equal to the rectangle contained by the means, the four straight lines will be proportional. 161, 163, 217, 292

17. If three straight lines are proportional, the rectangle contained by the extremes is equal to the square on the mean; and if the rectangle contained by the extremes is equal to the square on the mean, the three straight lines will be proportional. 161, 163, 217, 292

18. To describe on a given straight line a rectilineal figure

similar and similarly situated to a given rectilineal figure. 165, 166

19. Similar triangles are to one another in the duplicate ratio of the corresponding sides. 162–163, 166, 217, 292

Corollary. From this it is manifest that if three straight lines are proportional, as the first is to the third so is the figure on the first to that which is similar and similarly described on the second [inasmuch as it has been proved that as *CB* is to *BG* so is the triangle *ABC* to the triangle *ABG*, that is, to *DEF*]. 163

20. Similar polygons are divided into similar triangles equal in multitude to and in the same ratio (*homologos*) as the wholes, and polygon has to polygon the duplicate ratio of that which the corresponding side has to the corresponding side. 162–163, 166, 217, 292

(Addition) Similarly it can be proved in the case of (similar) quadrilaterals that they are in the duplicate ratio of the corresponding sides. And it was proved in the case of triangles, so that also universally similar rectilineal figures are to one another in the duplicate ratio of the corresponding sides. 163

21. Things similar to the same rectilineal (figure) are also similar to one another. 165, 166

22. If four straight lines are proportional, the rectilineal (figures) similar and similarly described on them will also be proportional; and if the rectilineal (figures) similar and similarly described on them are proportional, the straight lines themselves will be proportional. 156, 173–174, 217, 292

[(Lemma for the preceding) We will prove thus that if rectilineal (figures) are equal and similar, the corresponding sides are equal to one another.] 174

23. Equal-angled parallelograms have to one another the ratio compounded of (the ratios of) their sides. 88, 154–155, 162–163, 217

24. The parallelograms around the diameter of any parallelogram are similar to the whole and to one another. 166

25. To construct the same (figure) similar to a given rectilineal (figure) and equal to another given one. 165, 166

26. If from a parallelogram a parallelogram similar to the whole, similarly situated, and having a common angle with it is subtracted, it is around the same diameter as the whole. 166, 275

27. Of all the parallelograms applied to the same straight line and deficient by a parallelogrammic figure similar and similarly situated to one described on the half of the straight line, the

greatest is the one applied to the half and similar to the defect. 166, 167

28. To apply to a given straight line a parallelogram equal to a given rectilineal (figure) and deficient by a parallelogrammic figure similar to a given one; thus it is necessary that the (first) given rectilineal (figure) be not greater than the (parallelogram) described on the half and similar to the defect. 163–172, 269–270, 275, 293

29. To apply to a given straight line a parallelogram equal to a given rectilineal figure and exceeding by a parallelogrammic figure similar to a given one. 163–172, 193

30. To cut a given limited straight line in extreme and mean ratio. 156, 161, 168–169, 193

31. In right-angled triangles the (rectilineal) figure on the side subtending the right angle is equal to the similar and similarly described figures on the sides containing the right angle. 156, 172–173

32. If two triangles having two sides proportional to two sides are put together at one angle so that their corresponding sides are also parallel, the remaining sides of the triangles will be in a straight line. 172, 297

33. In equal circles angles have the same ratio as the arcs on which they stand, if they stand at the centers and if they stand at the circumferences. 155, 172, 203, 259

(Book) VII

(Definitions)

(1) A unit is that with respect to which each existing thing is called one (*kath' hēn hekaston tōn ontōn hen legetai*). 58

(2) A number is a multitude composed of units (*to ek monadōn sugkeimenon plēthos*). 58

(3) A number is part of a number, the less of the greater, when it measures the greater. 61

(4) But parts when it does not measure it. 61

(5) The greater (number) is a multiple of the less when it is measured by the less. 61

(6) An even number is one which is divisible into two (equal parts). 103

(7) An odd one is one which is not divisible into two (equal parts) or differs from an even number by a unit. 103

(8) An even-times even number is one which is measured by an even number according to an even number. 103

(9) An even-times odd number is one which is measured by an even number according to an odd number. 103

[(10) An odd-times even one is one which is measured by an odd number according to an even one.] 103

(11) An odd-times odd number is one which is measured by an odd number according to an odd number. 103

(12) A prime number is one which is measured by a unit only.

(13) Numbers prime to one another are ones measured by a unit alone as common measure. 75

(14) A composite number is one which is measured by some number.

(15) Numbers composite to one another are ones which are measured by some number as common measure.

(16) A number is said to multiply a number when the multiplied number is added to itself as many times as there are units in it (the multiplier), and some (number) comes-to-be (*genētai*). 59–60

(17) When two numbers having multiplied one another make some (number), what comes-to-be is called plane, and its sides are the numbers which have multiplied one another. 87

(18) When three numbers having multiplied one another make some (number), what comes-to-be is solid, and its sides are the numbers which have multiplied one another. 87

(19) A square number is one which is equal-times equal or contained by two equal numbers. 87

(20) A cube is one which is equal-times equal-times equal (*isakis isos isakis*) or contained by three equal numbers. 87

(21) Numbers are proportional when the first is an equal multiple of the second and the third of the fourth, or they are the same part or the same parts. 64–66

(22) Similar plane and solid numbers are ones which have their sides proportional. 91

(23) A perfect (or complete) number is one which is equal to its own parts. 101

(Propositions)

1. If two unequal numbers are set out and the lesser is always subtracted in turn from the greater (*anthuphairoumenou de aei tou elassonos apo tou meidzonos*), then, if the remainder never measures the number before it until a unit is left, the original numbers will be prime to one another. 78–80

2. Given two numbers not prime to one another, to find their greatest common measure. 69, 78–80

Corollary. From this it is manifest that if a number measures two numbers, it will also measure their greatest common measure. 78–79, 114 n. 19 (to p. 80)

3. Given three numbers not prime to one another, to find their greatest common measure. 69, 78–80

4. Any number is either part or parts of any number, the less of the greater. 61–62, 76, 79

5. If a number is part of a number and another is the same part of another, both together will be the same part of both together as the one is of the one. 69–72

6. If a number is parts of a number and another is the same parts of another, both together will be the same parts of both together as the one is of the one. 69–72

7. If a number is that part of a number which a number subtracted is of a number subtracted, the remainder will also be the same part of the remainder as the whole is of the whole. 69–72

8. If a number is those parts of a number which a number subtracted is of a number subtracted, the remainder will also be the same parts of the remainder as whole of whole. 69–72

9. If a number is a part of a number and another is the same part of another, then also alternately whatever part or parts the first is of the third, the second will also be the same part or the same parts of the fourth. 69–72, 113 n. 1 (to p. 58)

10. If a number is parts of a number and another is the same parts of another, then also, alternately, whatever parts or part the first is of the third, the second will also be the same parts or the same part of the fourth. 69–72

11. If as whole is to whole so is number subtracted to number subtracted, the remainder will also be to the remainder as whole to whole. 65, 67, 69–72, 98, 102

12. If any number of numbers are proportional, it will be the case that as one of the antecedents is to one of the consequents so are all the antecedents to all the consequents. 67, 69–72, 74, 98, 113 n. 2 (to p. 58)

13. If four numbers are proportional, they will also be alternately proportional. 67–72, 98

14. If there are any number of numbers and others equal to them in multitude which taken in pairs are in the same ratio,

they will also be in the same ratio *ex equali*. 65, 67, 69–72

15. If a unit measures some number and another number measures some other number equally many times, then also alternately the unit will measure the third number equally many times as the second the fourth. 73, 74, 98, 113 nn. 1, 2 (to p. 58)

16. If two numbers having multiplied one another make some (numbers), the (numbers) coming-to-be from them will be equal to one another. 62, 73, 74, 76, 98, 114 n. 10 (to p. 63)

17. If a number having multiplied two numbers makes some (numbers), the (numbers) coming-to-be from them will have the same ratio as those multiplied. 73, 74, 98

18. If two numbers having multiplied some number make some (numbers), the (numbers) coming-to-be from them have the same ratio as the multipliers. 73, 74, 98

19. If four numbers are proportional, the number which comes-to-be from first and fourth will be equal to the number which comes-to-be from second and third; and if the number which comes-to-be from first and fourth is equal to that from second and third, the four numbers will be proportional. 72, 74

20.[1] The least numbers of those having the same ratio with them measure those having the same ratio equally many times, the greater the greater and the less the less. 75–77, 98

21. Numbers prime to one another are the least of those having the same ratio with them. 75–77, 98

22. The least numbers of those having the same ratio with them are prime to one another. 75–77, 98

23. If two numbers are prime to one another, a number measuring one of them will be prime to the remaining one. 81, 98, 99

24. If two numbers are prime to some number, the (number) coming-to-be from them will be prime to the same. 81, 98, 99, 114 n. 10

25. If two numbers are prime to one another, the (number) coming-to-be from one of them (i.e., its square) will be prime to the remaining one. 81, 98, 99

26. If two numbers are prime to two numbers, both to each, the (numbers) which come-to-be from them will also be prime to one another. 81, 98, 99

27. If two numbers are prime to one another and each having multiplied itself makes some (number), the (numbers) coming-

to-be from them will be prime to one another; and if the original (numbers) having multiplied those which have come-to-be make some (numbers), these will be prime to one another [, and this always happens with the extremes]. 81, 98, 99

28. If two numbers are prime to one another, both together will also be prime to each of them; and if both together are prime to some one of them, the original numbers will also be prime to one another. 81, 82, 98, 99, 107

29. Any prime number is prime to any number which it does not measure. 81, 82, 98, 99

30. If two numbers having multiplied one another make some (number) and some prime number measures the (number) coming-to-be from them, it will also measure one of the original numbers. 81, 82, 98, 99,

31. Any composite number is measured by some prime number. 77, 80–81, 98, 99,

32. Any number is either prime or measured by some prime number. 80–81, 98, 99

33. Given any number of numbers, to find the least of those having the same ratio with them. 69, 74–80, 98

34. Given two numbers, to find the least number which they measure. 79–80, 98, 99

35. If two numbers measure some number, the least measured by them will also measure the same. 80, 98, 99

36. Given three numbers, to find the least number which they measure. 79–80, 99

37. If a number is measured by some number, the measured number will have a part called by the same name (*homōnumos*) as the measuring number. 74, 80

38. If a number has any part at all, it will be measured by a number called by the same name as the part. 74, 80

39. To find a number which is the least which will have given parts. 79–80

(Book) VIII

1. If any number of numbers are continuously proportional and their extremes are prime to one another, they are the least having the same ratio with them. 85, 90, 95

2. To find the least numbers which are continuously proportional in a given ratio, as many as may be prescribed. 81, 85, 90

Corollary. From this it is manifest that if three continuously proportional numbers are the least having the same ratio with them, their extremes are squares, and if four, cubes. 85, 90, 95

3. If any number of continuously proportional numbers are the least having the same ratio with them, their extremes are prime to one another. 81, 85, 90, 95

4. Given any number of ratios in least numbers, to find the least numbers which are continuously proportional in the given ratios. 86, 90, 111

5. Plane numbers have to one another the ratio compounded of (the ratios of) the sides. 87–88, 90, 92–93

6. If any number of numbers are continuously proportional and the first does not measure the second, neither will any other measure any other. 88, 89, 90

7. If any number of numbers are (continuously) proportional and the first measures the last, it will also measure the second. 89, 90

8. If numbers fall between two numbers in continuous proportion, as many numbers as fall in between them in continuous proportion will also fall in continuous proportion in between numbers having the same ratio (as the original two numbers). 89, 90, 95

9. If two numbers are prime to one another and numbers fall in between them in continuous proportion, as many numbers as fall in between them in continuous proportion will also fall in continuous proportion between either of them and a unit. 89, 90

10. If (the same number of) numbers fall in continuous proportion between each of two numbers and a unit, as many numbers as fall in continuous proportion between each of them and a unit will also fall between them (the original two numbers) in continuous proportion. 89

11. There is one mean proportional number between two square numbers, and the square has to the square the duplicate ratio of that which the side has to the side. 92–93, 97–98, 112

12. There are two mean proportional numbers between two cube numbers, and the cube has to the cube the triplicate ratio of that which the side has to the side. 92, 97–98

13. If any number of numbers are continuously proportional and each having multiplied itself makes some (number), the (numbers) coming-to-be from them will be proportional; and if the original (numbers) having multiplied those which have

come-to-be make some (numbers), these will also be proportional [, and this always happens with the extremes]. 98

14. If a square measures a square, the side will also measure the side; and if the side measures the side, the square will also measure the square. 98

15. If a cube number measures a cube number, the side will also measure the side; and if the side measures the side, the cube will measure the cube. 98

16. If a square number does not measure a square number, the side will not measure the side either; and if the side does not measure the side, the square will not measure the square either. 98

17. If a cube number does not measure a cube number, the side will not measure the side either; and if the side does not measure the side, the cube will not measure the cube either. 98

18. There is one mean proportional number between two similar plane numbers; and the plane has to the plane the duplicate ratio of that which the corresponding side has to the corresponding side. 94, 95

19. Two mean proportional numbers fall (between) two similar solid numbers; and the solid has to the similar solid the triplicate ratio of that which the corresponding side has to the corresponding side. 94, 95

20. If one mean proportional number falls (between) two numbers, they will be similar plane numbers. 93, 94, 95

21. If two mean proportional numbers fall (between) two numbers, they will be similar solid numbers. 94, 95

22. If three numbers are continuously proportional and the first is a square, the third will be a square. 93, 95

23. If four numbers are continuously proportional and the first is a cube, the fourth will also be a cube. 94, 95

24. If two numbers have to one another a ratio which a square number has to a square number and the first is square, the second will also be square. 95, 97, 111

25. If two numbers have to one another a ratio which a cube number has to a cube number and the first is a cube, the second will also be a cube. 95, 97

26. Similar plane numbers have to one another a ratio which a square number has to a square number. 95, 97, 112, 292

27. Similar solid numbers have to one another a ratio which a cube number has to a cube number. 95, 97

(Book) IX

1. If two similar plane numbers having multiplied one another make some (number), the (number) coming-to-be will be square. 90–95, 108, 111

2. If two numbers having multiplied one another make a square, they are similar plane numbers. 90–95, 108, 111

3. If a cube number having multiplied itself makes some (number), the (number) coming-to-be will be cube. 95

4. If a cube number having multiplied a cube number makes some (number), the (number) coming-to-be will be cube. 90–95

5. If a cube number having multiplied some number makes a cube, the number multiplied will also be a cube. 90–95

6. If a number having multiplied itself makes a cube, it will also be a cube. 96, 97

7. If a composite number having multiplied some number makes some (number), the (number) coming-to-be will be solid. 96–97

8. If any number of numbers are continuously proportional (beginning) from a unit, the third from the unit and (every) other one (*hoi hena dialeipontes*) will be square, the fourth and every third one cube, and the seventh and every sixth one will at the same time be cube and square. 96–97

9. If any number of successive numbers are in continuous proportion from a unit and the number after the unit is square, all the rest will also be square; and if the number after the unit is cube, all the rest will also be cube. 96–97

10. If any number of numbers are (continuously) proportional from a unit and the number after the unit is not square, no other one will be square except the third from the unit and every other one; and if the number after the unit is not cube, no other one will be cube except the fourth from the unit and every third one. 96–97

11. If any number of numbers are continuously proportional from a unit, the less measures the greater according to some number existing among the proportional numbers. 74, 98–100, 103, 105

(Corollary) And it is manifest that whatever position the measuring number has from the unit, the number according to which it is measured has the same position from the number measured forward. 100, 105

12. If any number of numbers are continuously proportional from a unit, the number next to the unit will be measured by

whatever prime numbers the last is measured by. 98–100, 103, 105

13. If any number of numbers are continuously proportional from a unit and the number after the unit is prime, the greatest will not be measured by any outside of those existing among the proportional numbers. 98, 100–101, 103

14. If a number is the least measured by prime numbers, it will be measured by no other prime number outside of those originally measuring it. 98–100, 103

15. If three continuously proportional numbers are the least having the same ratio with them, any two added together are prime to the remaining one. 86, 99, 107–111

16. If two numbers are prime to one another, it will not be the case that as the first is to the second so the second is to some other number. 98, 99, 103

17. If any number of numbers are continuously proportional and their extremes are prime to one another, it will not be the case that as the first is to the second so is the last to some other number. 98, 99, 103,

18. Given two numbers, to investigate if it is possible to find a third proportional to them. 98, 99, 103

19. Given three numbers, to investigate when it is possible to find a fourth proportional to them. 98, 99, 103

20. There are more prime numbers than any assigned multitude of prime numbers. 99, 103

21. If any number of even numbers are added, the whole is even. 103–104, 105

22. If any number of odd numbers are added and their multitude is odd, whole will be even. 104, 105

23. If any number of odd numbers are added and their multitude is odd, whole will also be odd. 104, 105

24. If an even is subtracted from an even number, the remainder will be even. 104, 105, 108, 111

25. If an odd is subtracted from an even number, the remainder will be odd. 104, 105

26. If an odd is subtracted from an odd number, the remainder will be even. 104, 105, 108, 111

27. If an even is subtracted from an odd number, the remainder will be odd. 104, 105

28. If an odd number having multiplied an even one makes

some (number), the (number) coming-to-be will be even. 104, 105

29. If an odd number having multiplied an odd number makes some (number), the (number) coming-to-be will be odd. 104, 105

30. If an odd number measures an even number, it will also measure its half. 104, 105

31. If an odd number is prime to some number, it will also be prime to its double. 105

32. Each of the numbers doubled from the dyad is even-times even only. 99, 105

33. If a number has its half odd, it is even-times odd only. 104–105

34. If a number is neither one of those doubled from the dyad nor has its half odd, it is even-times even and even-times odd. 105

35. If any number of numbers are continuously proportional and numbers equal to the first are subtracted from the second and the last, it will be the case that as the excess of the second (over the first) is to the first so is the excess of the last (over the first) to all its predecessors. 99, 102, 105

36. If any number of numbers are set out continuously in double proportion from a unit until the entire sum becomes prime and the sum multiplied into the last makes some (number), the (number) coming-to-be will be perfect. 65, 99, 101–106

(Book) X of Euclid's *Elements*

(Definitions)

(1) Magnitudes measured by the same measure are called commensurable, those for which no common measure can come-to-be incommensurable. 263–264

(2) Straight lines are commensurable in square (*dunamei*) when the squares on them are measured by the same area, incommensurable when no area can come-to-be a common measure for the squares on them. 264

(3) These things being laid down, it is proved that there exist infinitely many straight lines commensurable and also incommensurable with an assigned straight line, some in length only, some also in square. Let then the assigned straight line be called rational, and let the straight lines commensurable with it either in length and in square or in square only, be called rational, those incommensurable with it irrational. 266

(4) And let the square on the assigned straight line be called

rational and the things commensurable with it rational, but the things incommensurable with it irrational and the straight lines productive (*dunamenos*) of these things irrational; if they are squares, the sides themselves (are productive of them), but if they are some other rectilineal figure, the straight lines which describe squares equal to them. 266

(Propositions)

1. If two unequal magnitudes are set out and if there is subtracted from the greater more than its half and from the remainder more than its half and this goes on forever (*touto aei gignetai*), there will be left some magnitude which will be less than the lesser magnitude set out. 139–143, 234, 265

(Remark) It can be proved similarly even if the things subtracted are halves.

2. If, if two unequal magnitudes (are set out) and the lesser is always subtracted in turn from the greater, the remainder never measures the magnitude before it, then the magnitudes will be incommensurable. 265

3. Given two commensurable magnitudes to find their greatest common measure. 179–180, 265

Corollary. From this it is manifest that if a magnitude measures two magnitudes, it will also measure their greatest common measure.

4. Given three commensurable magnitudes, to find their greatest common measure. 265

Corollary. From this it is manifest that if a magnitude measures three magnitudes, it will also measure their greatest common measure.

(Remark) In the case of more magnitudes the greatest common measure can be taken similarly, and the corollary can be extended (*proschōrein*).

5. Commensurable magnitudes have to one another a ratio which a number has to a number. 111, 136, 265, 292

6. If two magnitudes have to one another a ratio which a number has to a number, the magnitudes will be commensurable. 111, 265, 292

Corollary. From this it is manifest that if there are two numbers, as D, E, and a straight line, as A, it is possible to make it that as the number D is to the number E so is the straight line (A) to a straight line (F). And if a mean proportional, as B, is taken between A and F, it will be the case that as A is to F so is the square on A to the square on B, that is, as the first is to the third so is the (figure) on the first to the similar and similarly described (figure) on the second. But as A is to F so is the number D to the number E. Therefore it has come-to-be that as the number D is to the number E so is the square on the straight line A to the square on the straight line B. 278, 282

7. Incommensurable magnitudes do not have to one another a ratio which a number has to a number. 111, 265, 292

8. If two magnitudes do not have to one another a ratio which a number has to a number, the magnitudes will be incommensurable. 111, 265, 292

9. The squares on straight lines commensurable in length have to one another a ratio which a square number has to a square number; and squares having to one another a ratio which a square number has to a square number will also have their sides commensurable in length. But the squares on straight lines incommensurable in length do not have to one another a ratio which a square number has to a square number; and squares not having to one another a ratio which a square number has to a square number will not have their sides commensurable in length either. 112–113, 265, 292

Corollary. And it will be manifest from what has been proved that straight lines commensurable in length are always also commensurable in square, but those in square not always also in length...[2]

(Lemma) It has been proved in the arithmetic (books) that similar plane numbers have to one another a ratio which a square number has to a square number and that if two numbers have to one another a ratio which a square number has to a square number, they are similar plane numbers. And it is clear from these things that non-similar-plane numbers, that is, those not having their sides proportional, do not have to one another a ratio which a square number has to a square number; for if they have they will be similar plane numbers, but they are assumed not to be. Therefore, non-similar-plane numbers do not have to one another a ratio which a square number has to a square number. 112

10. To find two straight lines which are incommensurable with an assigned straight line, one in length only, the other also in square. 112–113, 180, 265, 276, 278, 293

11.[3] If four magnitudes are proportional and the first is commensurable with the second, the third will also be commensurable with the fourth; and if the first is incommensurable with the second, the third will also be incommensurable with the fourth. 265, 292

12. Things commensurable with the same magnitude are also commensurable with one another. 265, 292

13. If two magnitudes are commensurable and one of them is incommensurable with some magnitude, the other one will also be incommensurable with it. 265, 292

Lēmma. Given two unequal straight lines, to find by what the greater is greater than the less in square.

14. If four straight lines are proportional and the first is greater than the second in square by a square on a straight line commensurable with it (the first straight line), the third will also be greater than the fourth in square by the square on a straight line commensurable with it (the third); and if the first is greater than the second in square by the square on a straight line incommensurable with it, the third will also be greater than the fourth in square by the square on a straight line incommensurable with it. 266, 292

15. If two commensurable magnitudes are added, the whole will also be commensurable with each of them; and if the whole is commensurable with one of them, the original magnitudes will be commensurable. 265, 292

16. If two incommensurable magnitudes are added, the whole will also be incommensurable with each of them; and if the whole is incommensurable with one of them, the original magnitudes will be incommensurable. 265, 292

Lēmma. If a parallelogram is applied to some straight line and is deficient by a square figure, the applied parallelogram is equal to the rectangle contained by the segments of the straight line produced by the application.

17. If there are two unequal straight lines and there is applied to the greater a (parallelogram) equal to the fourth part of the square on the less and deficient by a square figure and it divides the straight line into parts commensurable in length, the greater will be greater than the less in square by the square on a straight line commensurable with it; and if the greater is greater than the less in square by the square on a straight line commensurable with it and there is applied to the greater a (parallelogram) equal to a fourth of the square on the less and deficient by a square figure, it will divide the straight line into parts commensurable in length. 269, 270, 279, 293

18. If there are two unequal straight lines and there is applied to the greater a (parallelogram) equal to the fourth part of the square on the less and deficient by a square figure and it divides the straight line into incommensurable parts, the greater will be greater than the less in square by the square on a straight

line incommensurable with it; and if the greater is greater than the less in square by the square on a straight line incommensurable with it and there is applied to the greater a (parallelogram) equal to the fourth of the square on the less and deficient by a square figure, it will divide the straight line into incommensurable parts. 269–270, 279, 293

(Remark) Since it has been proved that straight lines commensurable in length are always commensurable in square also, but those commensurable in square not always in length, but they can be both commensurable in length and incommensurable in length, it is manifest that if some straight line is commensurable in length with a given (*ekkeimenos*) rational straight line, it is called commensurable with it not only in length but also in square, since straight lines commensurable in length are always commensurable in square also. But if some straight line is commensurable in square with a given rational straight line, if it is also commensurable in length, it is then also called rational and commensurable with it in length and in square; but again if some straight line which is commensurable in square with a given rational straight line is incommensurable in length with it, it is then called rational and commensurable in square only.[4]

19. The rectangle contained by rational straight lines commensurable in length in any of the aforesaid ways is rational. 268, 293

20. If a rational (area) is applied to a rational straight line, it makes a rational breadth commensurable in length with the straight line to which it is applied. 268, 293

[*Lēmma*. The straight line the square on which is equal to an irrational area (*hē dunamenē alogon chōrion*) is irrational.]

21. The rectangle contained by rational straight lines commensurable in square only is irrational, and the side of a square equal to it will be irrational; let it be called medial. 268, 293

Lēmma. If there are two straight lines, as the first is to the second, so is the square on the first to the rectangle contained by the two straight lines.

22. The square on a medial applied to a rational makes a rational breadth incommensurable in length with the straight line to which it is applied. 161, 272, 293

23. A straight line commensurable with a medial is medial. 272–273, 293

Corollary. From this it is manifest that what is commensurable with a medial area is medial. [For straight lines commensurable in square of which one is medial produce them.] Just as was said in the case of rationals, it follows in the case of medials that a straight line commensurable in length with a medial is called medial and commensurable with it not only in length but also in square, inasmuch as, universally, straight lines commensurable in length are also commensurable in square. But if some straight line is commensurable in square with a medial, if it is also commensurable in length, (the straight lines) are then called medial and commensurable in length and square, but if commensurable in square only, they are called medial and commensurable in square only. [There are again other straight lines which are incommensurable in length with a medial but commensurable in square only, and they are again called medials because they are commensurable in square with a medial and commensurable with one another, as medials, but commensurable with one another either in length and, of course, in square or in square only. And if they are commensurable in length, they are also called medials commensurable in length, with the understanding that they are also commensurable in square; and if they are commensurable in square only, they are then called medials commensurable in square only.][5] 272

24. The rectangle contained by medial straight lines commensurable in length in any of the aforesaid ways is medial. 277, 281

25. The rectangle contained by medial straight lines commensurable in square only is either rational or medial. 277, 281

26. A medial area does not exceed a medial area by a rational one. 277, 284, 293

27. To find medial straight lines commensurable in square only containing a rational (area). 276, 281–282, 293

28. To find medial straight lines commensurable in square only containing a medial (area). 276, 278, 281–282, 293

Lēmma. To find two square numbers such that their sum is also square. 108–109, 280, 292

(Remark) And it is manifest that in addition two square numbers, the square of *BD* and that of *CD*, have been found such that their difference, the product of *AB*, *BC*, is a square, when *AB*, *BC* are similar plane numbers; and when they are not similar plane numbers, two square numbers, the square of *BD* and that of *DC*, of which the difference, the product of *AB*, *BC*, is not a square have been found.

Lēmma. To find two square numbers such that their sum is not a square. 108–109, 280, 292

29. To find two rational straight lines commensurable in square only such that the greater is greater than the less in square by a square on a straight line commensurable with it. 111, 276, 279–280, 293

30. To find two rational straight lines commensurable in square only such that the greater is greater than the less in square by the square on a straight line incommensurable with it. 111, 276, 279–280, 282, 293

31. To find two medial straight lines commensurable in square only, containing a rational (area), and such that the greater is greater than the less in square by the square on a straight line [in]commensurable with it. 276, 280–281, 282, 293

(Remark) The case "by the square on a commensurable straight line" can also be proved similarly.

32. To find two medial straight lines commensurable in square only, containing a medial (area), and such that the greater is greater than the less in square by the square on a straight line commensurable with it. 276, 278, 280–282, 293

(Remark) The case "by the square on an incommensurable straight line" can also be proved similarly.

Lēmma. Let *ABC* be a right-angled triangle having the angle *A* right, and let the perpendicular *AD* be drawn; I say that the rectangle contained by *CBD* (*CB*, *BD*) is equal to the square on *BA*, that contained by *BCD* is equal to the square on *CA* and that contained by *BD*, *DC* equal to the square on *AD*, and further, that contained by *BC*, *AD* is equal to that contained by *BA*, *AC*. 279

33. To find two straight lines incommensurable in squar making the sum of the squares on them rational, the rectangl contained by them medial. 276, 278–283, 293

34. To find two straight lines incommensurable in squar making the sum of the squares on them medial, the rectang contained by them rational. 276, 278–283, 293

35. To find two straight lines incommensurable in squar making the sum of the squares on them medial and the rectang contained by them medial and, further, incommensurable wit the sum of the squares on them. 276, 278–283, 293

36. If two rational straight lines commensurable in square on are added, the whole is irrational, and it is called a binomi (*ek duo honomatōn*). 276–277, 284, 293

37. If two medial straight lines commensurable in square on and containing a rational (area) are added, the whole

irrational; let it be called a first bimedial (*ek duo mesōn*). 276–277, 284, 293

38. If two medial straight lines commensurable in square only and containing a medial (area) are added, the whole is irrational; let it be called a second bimedial. 276–277, 284, 293

39. If two straight lines incommensurable in square and making the sum of the squares on them rational, the rectangle contained by them medial, are added, the whole straight line is irrational; let it be called a major (*meidzōn*). 276–277, 284, 293

40. If two straight lines incommensurable in square and making the sum of the squares on them medial, the rectangle contained by them rational, are added, the whole straight line is irrational; let it be called productive of a rational and a medial (area). 276–277, 284, 293

41. If two straight lines incommensurable in square and making the sum of the squares on them medial and the rectangle contained by them medial and, further, incommensurable with the sum of the squares on them are added, the whole straight line is irrational; let it be called productive of two medial (areas). 276–277, 284, 293

(Lemma) We will show that the described irrational straight lines are divided into the straight lines of which they are composed to make the proposed species, first setting out a little *lēmma* of the following kind: let a straight line *AB* be set out, and let the whole be cut into unequal parts at each of *C*, *D*, and let *AC* be supposed greater than *DB*; I say that the squares on *AC*, *CB* are greater than those on *AD*, *DB*. 285–286

42. The binomial is divided into terms (*honomata*) at one point only. 276–277, 284, 293

43. The first bimedial is divided at one point only. 276–277, 284, 293

44. The second bimedial is divided at one point only. 276–277, 284–286, 293

45. The major is divided at the same point only. 276–277, 284, 293

46. A straight line productive of a rational and a medial (area) is divided at one point only. 276–277, 284, 293

47. A straight line productive of two medial (areas) is divided at one point only. 276–277, 284–286, 293

Second Definitions

(1) Given a rational straight line and a binomial divided into

its terms of which the greater term is greater than the less in square by the square on a straight line commensurable in length with it, if the greater term is commensurable in length with the given rational line, let it (the binomial) be called a first binomial;

(2) if the lesser term is commensurable in length with the given rational line, let it be called a second binomial;

(3) if neither of the terms is commensurable in length with the given rational line, let it be called a third binomial;

(4) in addition, if the greater term is greater in square by the square on a straight line incommensurable in length with it, if the greater term is commensurable in length with the given rational straight line, let it be called a fourth binomial;

(5) if the less is commensurable with it, a fifth;

(6) if neither, a sixth.

(Propositions Continued)

48. To find a first binomial. 111, 276, 277, 284, 293

49. To find a second binomial. 111, 276, 277, 284, 293

50. To find a third binomial. 111, 276, 277, 284, 293

51. To find a fourth binomial. 111, 276, 277, 284, 293

52. To find a fifth binomial. 111, 276, 277, 284, 293

53. To find a sixth binomial. 111, 276, 277, 284, 293

Lēmma. Let *AB* (*ADBF*), *BC* (*BECG*) be two squares, and le[t] them be placed so that *DB* is in a straight line with *BE*; there-fore *FB* is also in a straight line with *BG*; and let the parallelo-gram *AC* be completed. I say that *AC* is a square and that th[e] (figure) *DG* is a mean proportional between *AB*, *BC* and furthe[r] that *DC* is a mean proportional between *AC*, *CB*. 275, 285 293

54. If an area is contained by a rational straight line and [a] first binomial, the irrational straight line producing the are[a] is the one called binomial. 276, 277, 284, 285, 288, 293

55. If an area is contained by a rational straight line and [a] second binomial, the irrational straight line producing the are[a] is the one called a first bimedial. 276, 277, 284, 285, 288, 29[3]

56. If an area is contained by a rational straight line and third binomial, the irrational straight line producing the are[a] is the one called a second bimedial. 276, 277, 284, 285, 28[8] 293

57. If an area is contained by a rational straight line and

fourth binomial, the irrational straight line producing the area is the one called major. 276, 277, 284, 285, 288, 293

58. If an area is contained by a rational straight line and a fifth binomial, the irrational straight line producing the area is the one called productive of a rational and a medial (area). 276, 277, 284, 285, 288, 293

59. If an area is contained by a rational straight line and a sixth binomial, the irrational straight line producing the area is the one called productive of two medial (areas). 276, 277, 284, 285, 288, 293

(Lemma) If a straight line is cut into unequal parts, the squares on the unequal parts are greater than twice the rectangle contained by the unequal parts. 286, 293

60. The square on a binomial straight line applied to a rational one makes as breadth a first binomial. 276, 277, 284, 288, 293

61. The square on a first bimedial straight line applied to a rational one makes as breadth a second binomial. 276, 277, 284, 288, 293

62. The square on a second bimedial straight line applied to a rational one makes as breadth a third binomial. 276, 277, 284, 288, 293

63. The square on a major straight line applied to a rational one makes as breadth a fourth binomial. 276, 277, 284, 288, 293

64. The square on a straight line productive of a rational and a medial (area) applied to a rational one makes as breadth a fifth binomial. 276, 277, 284, 288, 293

65. The square on a straight line productive of two medial (areas) applied to a rational one makes as breadth a sixth binomial. 276, 277, 284, 288, 293

66. A straight line commensurable in length with a binomial is itself also a binomial and the same in order. 276, 277, 284, 293

67. A straight line commensurable with a bimedial is itself also a bimedial and the same in order. 276, 277, 284, 293

68. A straight line commensurable with a major is itself also a major. 276, 277, 284, 293

69. A straight line commensurable with one productive of a rational and a medial (area) is productive of a rational and a medial (area). 276, 277, 284, 293

70. A straight line commensurable with one productive of two

medial (areas) is productive of two medial (areas). 276, 277, 284, 293

71. When a rational and a medial (area) are added, four irrational (straight lines) come-to-be (as the straight lines productive of the wholes); either a binomial or a first bimedial or a major or one productive of a rational and a medial (area). 276, 277, 284, 294

72. When two medial (areas) incommensurable with one another are added, the remaining two irrational (straight lines) come-to-be, either a second bimedial or one productive of two medial (areas). 276, 277, 284, 294

(Remark) The binomial and the irrational (straight lines) after it are not the same with the medial or with one another. . . . And the said breadths (the six binomials) differ from the first (a rational straight line) and from one another . . . so that the irrationals themselves differ from one another. 277, 294

73.[6] If a rational straight line which is commensurable in square only with the whole is subtracted from a rational straight line, the remainder is irrational; let it be called an apotome. 266, 267, 276, 293

74. If a medial straight line which is commensurable in square only with the whole and contains with the whole a rational (area) is subtracted from a medial straight line, the remainder is irrational; let it be called a first apotome of a medial. 273, 276, 277, 293

75. If a medial straight line which is commensurable in square only with the whole and contains with the whole a medial (area) is subtracted from a medial straight line, the remainder is irrational; let it be called a second apotome of a medial 273, 276, 278, 293

76. If a straight line which is incommensurable in square with the whole and with the whole makes the square on them together rational, the rectangle contained by them medial, is subtracted from a straight line, the remainder is irrational; let it be called a minor (*elassōn*). 273, 276, 277, 293

77. If a straight line which is incommensurable in square with the whole and with the whole makes the sum of the squares on them medial, twice the rectangle contained by them rational is subtracted from a straight line, the remainder is irrational; let it be called a (straight line) making with a rational (straight line) a medial whole. 273, 276, 277, 293

78. If a straight line which is incommensurable in square with the whole and with the whole makes the sum of the squares on them medial, twice the rectangle contained by them medial

and, further, the squares on them incommensurable with twice the rectangle contained by them, is subtracted from a straight line, the remainder is irrational; let it be called a (straight line) making with a medial (straight line) a medial whole. 273, 276, 278, 293

79. (Only) one rational straight line which is commensurable in square only with the whole can be annexed (*prosarmodzein*) to an apotome. 276, 277, 284, 293

80. Only one medial straight line which is commensurable in square only with the whole and contains with the whole a rational (area) can be annexed to a first apotome of a medial. 276, 277, 284, 293

81. Only one medial straight line commensurable in square only with the whole and containing with the whole a medial (area) can be annexed to a second apotome of a medial. 276, 284, 293

82. Only one straight line which is incommensurable in square with the whole and with the whole makes the sum of the squares on them rational, twice the rectangle contained by them medial, can be annexed to a minor. 276, 277, 284, 293

83. Only one straight line which is incommensurable in square with the whole and with the whole makes the sum of the squares on them medial, twice the rectangle contained by them rational, can be annexed to a straight line making with a rational straight line a medial whole. 276, 277, 284, 293

84. Only one straight line which is incommensurable in square with the whole and with the whole makes the sum of the squares on them medial, twice the rectangle contained by them medial and further incommensurable with the sum of the squares on them, can be annexed to a straight line making with a medial straight line a medial whole. 276, 284, 293

Third Definitions

(1) Given a rational straight line and an apotome, if the whole is greater than the annexed straight line in square by the square on a straight line commensurable in length with it and the whole is commensurable in length with the given rational line, let it be called a first apotome;

(2) if the annexed straight line is commensurable with the given rational line and the whole is greater than the annexed straight line in square by the square on a straight line commensurable with it, let it be called a second apotome;

(3) if neither is commensurable with the given rational line and the whole is greater than the annexed straight line in

square by the square on a straight line commensurable with it, let it be called a third apotome;

(4) in addition, if the whole is greater than the annexed straight line in square by the square on a straight line incommensurable with it, if the whole is commensurable in length with the given rational line, let it be called a fourth apotome;

(5) if the annexed straight line is commensurable with it, a fifth;

(6) if neither, a sixth. 267–271

(Propositions Continued)

85. To find a first apotome. 111, 276, 282, 283, 293

86. To find a second apotome. 111, 276, 282, 283, 293

87. To find a third apotome. 111, 276, 283, 293

88. To find a fourth apotome. 111, 276, 282, 283, 293

89. To find a fifth apotome. 111, 276, 282, 283, 293

90. To find a sixth apotome. 111, 276, 283, 293

91. If an area is contained by a rational straight line and a first apotome, the straight line producing the area is an apotome. 274, 276, 277, 288, 293, 294

92. If an area is contained by a rational straight line and a second apotome, the straight line producing the area is a first apotome of a medial. 274–276, 288, 293, 294

93. If an area is contained by a rational straight line and a third apotome, the straight line producing the area is a second apotome of a medial. 274–276, 288, 293, 294

94. If an area is contained by a rational straight line and a fourth apotome, the straight line producing the area is a minor. 274–276, 288, 293, 294

95. If an area is contained by a rational straight line and a fifth apotome, the straight line producing the area is the one making with a rational straight line a medial whole. 274–276, 288, 293, 294

96. If an area is contained by a rational straight line and a sixth apotome, the straight line producing the area is one making with a medial straight line a medial whole. 274–276, 288, 293, 294

97. The square on an apotome applied to a rational straight line makes as breadth a first apotome. 275–276, 288, 293, 294

98. The square on a first apotome of a medial applied to

rational straight line makes as breadth a second apotome. 275–276, 288, 293, 294

99. The square on a second apotome of a medial applied to a rational straight line makes as breadth a third apotome. 275–276, 288, 293, 294

100. The square on a minor applied to a rational straight line makes as breadth a fourth apotome. 275–276, 288, 293, 294

101. The square on a straight line making with a rational straight line a medial whole applied to a rational straight line makes as breadth a fifth apotome. 275–276, 288, 293, 294

102. The square on a straight line making with a medial straight line a medial whole applied to a rational straight line makes as breadth a sixth apotome. 275–276, 288, 293, 294

103. A straight line commensurable in length with an apotome is an apotome and the same in order. 276, 283, 293

104. A straight line commensurable with an apotome of a medial is an apotome of a medial and the same in order. 276, 283, 293

105. A straight line commensurable with a minor is a minor. 276, 283, 293

106. A straight line commensurable with one making with a rational straight line a medial whole is one making with a rational straight line a medial whole. 276, 283, 293

107. A straight line commensurable with one making with a medial straight line a medial whole is itself also one making with a medial straight line a medial whole. 276, 283, 293

108. When a medial area is subtracted from a rational one, the straight line productive of the area remaining comes-to-be one of two irrational straight lines, either an apotome or a minor. 276, 283, 294

109. When a rational area is subtracted from a medial one, two other irrational straight lines come-to-be, either a first apotome of a medial or one making with a rational straight line a medial whole. 276, 283, 294

110. When a medial area incommensurable with the whole is subtracted from a medial area, the remaining two irrationals come-to-be, either a second apotome of a medial or one making with a medial straight line a medial whole. 276, 283, 294

111. An apotome is not the same with a binomial. 277, 288, 294

(Remark) The apotome and the irrational (straight lines)

after it are not the same with the medial or with one another. . . . Therefore, the straight lines following the apotome are different and those following the binomial are different, so that there are, in order, 13 irrational straight lines in all . . .

112. The square on a rational straight line applied to a binomial makes as breadth an apotome of which the terms are commensurable with the terms of the binomial and further in the same ratio; and further the apotome which comes-to-be has the same order as the binomial. 289, 294

113. The square on a rational straight line applied to an apotome makes as breadth a binomial of which the terms are commensurable with those of the apotome and in the same ratio; further the binomial which comes-to-be has the same order as the apotome. 289, 294

114. If an area is contained by an apotome and a binomial of which the terms are commensurable with the terms of the apotome and in the same ratio, the straight line producing the area is rational. 289, 294

(Porism) And it has also come-to-be manifest to us through this that it is possible for a rational area to be contained by irrational straight lines.

115. There come-to-be from a medial straight line infinitely many irrational straight lines and the same with none of their predecessors. 289, 294

[A straight line commensurable with a minor is a minor.] 305 n. 17

[A straight line commensurable with one making with a rational straight line a medial whole is one making with a rational straight line a medial whole.] 305 n. 17

[Let it be assigned to us to prove that in the case of square figures the diameter is incommensurable in length with the side.] 305 n. 17

(Book) XI of Euclid's *Elements*

(Definitions)

(1) A solid is that which has length, breadth, and depth. 211

(2) An extremity of a solid is a surface. 211

(3) A straight line is orthogonal (*orthos*) to a plane when it makes right angles in relation to (*pros*) all the straight lines meeting it and in the plane. 211, 213

(4) A plane is orthogonal to a plane when the straight lines in one of the planes drawn at right angles to the common section of the planes are at right angles to the remaining plane. 21

(5) The inclination of a straight line to a plane is, when a perpendicular is drawn from the higher extremity of the straight line to the plane and a straight line is joined from the point which comes-to-be and the extremity of the straight line in the plane, the angle contained by the straight line last drawn and the straight line standing up (*hē periechomenē gōnia hupo tēs achtheisēs kai tes ephestōsēs*).

(6) The inclination of a plane to a plane is the acute angle contained by the straight lines drawn to the same point at right angles to the common section in each of the planes.

(7) A plane is said to be inclined to a plane similarly as another to another when the said angles of inclination are equal to one another.

(8) Parallel planes are those which do not meet (*ta asumptōta*).

(9) Similar solid figures are those contained by similar plane (figures) equal in multitude. 218–220, 237

(10) Equal and similar solid figures are those contained by similar plane (figures) equal in multitude and magnitude. 218–220, 225–226, 237

(11) A solid angle is the inclination of more than two lines which meet one another and are not in the same plane to all the lines. Otherwise, a solid angle is the (angle) contained by more than two plane angles which are not in one plane and are constructed to one point. 212

(12) A pyramid is a solid figure contained by planes constructed from one plane to one point.

(13) A prism is a solid figure contained by plane (figures) of which two, the opposite ones, are equal and similar and parallel, the rest parallelograms.

(14) A sphere is the figure comprehended when, the diameter of the semicircle remaining fixed, the semicircle is carried around and restored again to the same (position) from which it began to be moved (*hēmikukliou menousēs tēs diametrou perienechthen to hēmikuklion eis to auto palin apokatastathēi, hothen ērxato pheresthai*). 29, 211, 254

(15) The axis of the sphere is the straight line which remains fixed and about which the semicircle is turned.

(16) The center of the sphere is the same as that of the semicircle.

(17) A diameter of the sphere is some straight line drawn through the center and limited in both directions by the surface of the sphere.

(18) A cone is the figure comprehended when, one of the sides around the right angle of a right-angled triangle remaining fixed, the triangle is carried around and restored again to the same (position) from which it began to be moved. And if the straight line which remains fixed is equal to the remaining (side) around the right angle which is carried around, the cone will be right-angled; if less, obtuse-angled; if greater, acute-angled.

(19) The axis of the cone is the straight line which remains fixed and about which the triangle is turned.

(20) The base is the circle described by the straight line which is carried around.

(21) A cylinder is the figure comprehended when, one of the sides around the right angle of a right-angled parallelogram remaining fixed, the parallelogram is carried around and restored again to the same (position) from which it began to be moved.

(22) The axis of the cylinder is the straight line which remains fixed and about which the parallelogram is turned.

(23) The bases are the circles described by the two opposite sides which are carried around.

(24) Similar cones and cylinders are those of which the axes and the diameters of the bases are proportional. 229

(25) A cube is a solid figure contained by six equal squares.

(26) An octahedron is a solid figure contained by eight equal and equilateral triangles.

(27) An icosahedron is a solid figure contained by twenty equal and equilateral triangles.

(28) A dodecahedron is a solid figure contained by twelve equal and equilateral and equiangular pentagons.

(Propositions)

1. Some part of a straight line is not in the plane of reference and some part in a higher plane. 209–212

2. If two straight lines cut one another, they are in one plane, and every triangle is in one plane. 209–212, 213

3. If two planes cut one another, their common section is a straight line. 31, 209–212

4. If a straight line is set up at right angles to two straight lines which cut one another, at their common section, it will also be at right angles to the plane through them. 211–212, 214

5. If a straight line is set up at right angles to three straight

lines which meet one another, at their common section, the three straight lines are in the same plane. 212–214

6. If two straight lines are at right angles to the same plane, the straight lines will be parallel. 212–214, 297

7. If two straight lines are parallel and chance points are taken on each of them, the straight line connecting the points is in the same plane as the parallels. 31, 208–212

8. If two straight lines are parallel and one of them is at right angles to some plane, the remaining one will also be at right angles to the same plane. 212–213

9. Straight lines parallel to the same straight line and not in the same plane with it are also parallel to one another. 212–214

10. If two straight lines meeting one another are parallel to two straight lines meeting one another and not in the same plane, they will contain equal angles. 213–214

11. To draw a straight line perpendicular to a given plane from a given higher point. 212–213

12. To set up a straight line at right angles to a given plane from a given point on (*pros*) it. 212–213

13. Two straight lines cannot be set up in the same direction at right angles to the same plane from the same point. 212 (and 247 n. 5), 213 (and 247 n. 6)

14. The planes to which the same straight line is orthogonal will be parallel. 212

15. If two straight lines meeting one another are parallel to two straight lines meeting one another and not in the same plane, the planes through them will be parallel. 212

16. If two parallel planes are cut by some plane, their common sections are parallel. 212, 219

17. If two straight lines are cut by parallel planes, they will be cut in the same ratios. 212, 238

18. If a straight line is at right angles to some plane, all the planes through it will also be at right angles to the same plane. 212

19. If two planes cutting one another are at right angles to some plane, their common section will also be at right angles to the same plane. 212

20. If a solid angle is contained by three plane angles, any two taken together in any way are greater than the remaining one. 212, 216

21. Every solid angle is contained by plane angles (together) less than four right angles. 212, 216, 302

22. If there are three plane angles of which two taken together in any way are greater than the remaining one and they are contained by equal straight lines, it is possible to construct a triangle from the (straight lines) joining the equal straight lines. 212, 215

23. To construct a solid angle out of three plane angles of which two taken together in any way are greater than the remaining one; thus it is necessary that the three angles be less than four right angles. 212, 214–216

(Lemma for the preceding) We will show as follows the way in which it is possible to take (*OR* so that) the square on *OR* is equal to that by which the square on *AB* is greater than the square on *LO*.

24. If a solid is contained by parallel plane (figures), its opposite plane (figures) are equal and parallelogrammic. 207, 219 225

25. If a parallelepipedal solid is cut by a plane which is parallel to the opposite planes, it will be the case that as the base is to the base so is the solid to the solid. 219, 220, 223, 224, 225 248, n. 17

26. To construct relative to a given straight line and a point on it a solid angle equal to a given solid angle. 212, 221, 225

27. To describe on a given straight line a parallelepipedal solid similar and similarly situated to a given parallelepipedal solid. 219, 221, 225

28. If a parallelepipedal solid is cut by a plane through the diagonals of opposite planes, the solid will be bisected by the plane. 218–219, 225–226

29. Parallelepipedal solids which are on the same base and under the same height, the (points) standing up (i.e., the vertices not on the bases) of which are on the same straight lines, are equal to one another. 218, 219, 220–222, 225–226

30. Parallelepipedal solids which are on the same base and under the same height, the (points) standing up of which are not on the same straight lines, are equal to one another. 220–222, 225–226

31. Parallelepipedal solids which are on equal bases and under the same heights are equal to one another. 219, 220, 222–226, 238

32. Parallelepipedal solids which are under the same heights are to one another as their bases. 220, 224–226

33. Similar parallelepipedal solids are to one another in the triplicate ratio of the corresponding sides. 219, 220, 224, 225, 226

Corollary. From this it is manifest that if four straight lines are (continuously) proportional it will be the case that as the first is to the fourth so is the parallelepipedal solid on the first to the similar and similarly described one on the second, inasmuch as the first has to the fourth the triplicate ratio of what it has to the second.

34. The bases of equal parallelepipedal solids reciprocate with the heights; and those parallelepipedal solids of which the bases reciprocate with the heights are equal. 219, 220, 225–226

35. If there are two equal plane angles and at their vertices there are set up higher straight lines containing equal angles respectively with the original straight lines and chance points are taken on the higher straight lines and from them perpendiculars are drawn to the planes in which the original angles are and if from the points which come-to-be in the planes straight lines are connected to the (vertices of) the original angles, they will contain equal angles with the higher straight lines. 220, 225

Corollary. From this it is manifest that if there are two equal plane angles and there are set up on (the vertices of) them higher equal straight lines containing equal angles respectively with the original straight lines, the perpendiculars drawn from (the extremities) of them to the planes in which the original angles are are equal to one another. 225

36. If three straight lines are proportional, the parallelepipedal solid (formed) from the three is equal to the parallelepipedal solid on the mean, which is equilateral and equiangular with the aforesaid solid. 220, 225

37. If four straight lines are proportional, the parallelepipedal solids similar and similarly described on them will also be proportional; and if the parallelepipedal solids similar and similarly described on them are proportional, the straight lines themselves will also be proportional. 173, 220, 224–225

[38. If a plane is perpendicular to a plane and a perpendicular is drawn from some point in one of the planes to the other plane, the perpendicular drawn will fall on the common section of the planes.] 221, 250 n. 34

39. If the sides of opposite planes of a cube are bisected and planes are carried through the sections, the common section of the planes and the diagonal of the cube bisect one another. 220, 221, 306 n. 26

40. If there are two prisms of equal height and one has as base a parallelogram, the other a triangle and the parallelogram is double the triangle, the prisms will be equal. 221, 249 n. 28

(Book) XII of Euclid's *Elements*

1. Similar polygons (inscribed) in circles are to one another as the squares on the diameters. 200, 217, 236

2. Circles are to one another as the squares on their diameters. 200–202, 217, 230–233, 236

(Lemma for the preceding) I say that when the area *S* is greater than the circle *EFGH*, it is the case that as the area *S* is to the circle *ABCD* so is the circle *EFGH* to some area less than the circle *ABCD*. 248 n. 21

3. Every pyramid having as base a triangle is divided into two pyramids equal and similar to one another and to the whole and having as bases triangles and into two equal prisms; and the two prisms are greater than half of the whole pyramid. 221, 227, 236–238

4. If there are two pyramids under the same height and having as bases triangles and each of them is divided into two pyramids equal to one another and similar to the whole and into two equal prisms, it will be the case that as the base of one pyramid is to the base of the other pyramid so are all the prisms in the one pyramid to all the equally many prisms in the other pyramid. 221, 227, 239

Lēmma (for the preceding). It is to be shown thus that it is the case that as the triangle *LOC* is to the triangle *RVF* so is the prism of which the triangle *LOC* is the base, *PMN* its opposite, to the prism of which *RVF* is the base, *STU* its opposite. 238 (and 249 nn. 29, 30)

5. Pyramids which are under the same height and have as bases triangles are to one another as their bases. 221, 227, 230–233, 236–240

6. Pyramids which are under the same height and have as bases polygons are to one another as the bases. 221, 227–228, 241

7. Every prism having as base a triangle is divided into three pyramids equal to one another and having as bases triangles. 221, 226–228

Corollary. From this it is manifest that every pyramid is a third part of the prism having the same base as it and an equal height, inasmuch as if the base has some other rectilineal shape, its opposite will be of the same kind, and (the prism) is divided into prisms having as bases and opposite (faces) triangles, and as the whole base is to each . . . (lacuna) 228, 241

8. Similar pyramids also having as bases triangles are in the triplicate ratio of the corresponding sides. 221, 226–228, 242, 250 n. 32

9. The bases of equal pyramids also having as bases triangles reciprocate with their heights; and those pyramids having as bases triangles of which the bases reciprocate with the heights are equal. 221, 226–229

10. Every cone is a third part of a cylinder having the same base as it and an equal height. 202, 221, 229–233, 240–242

11. Cones and cylinders which are under the same height are to one another as their bases. 202, 221, 229–233, 240–242

12. Similar cones and cylinders are to one another in the triplicate ratio of the diameters in their bases. 202, 221, 230–233, 240–242

13. If a cylinder is cut by a plane which is parallel to its opposite planes, it will be the case that the cylinder is to the cylinder as the axis to the axis. 221, 229, 230

14. Cones and cylinders which are on equal bases are to one another as their heights. 221, 229, 230

15. The bases of equal cones and cylinders reciprocate with their heights; and those cones and cylinders the bases of which reciprocate with their heights are equal. 221, 230

16. Two circles being around the same center, to inscribe in the greater circle an equilateral polygon with an even number of sides not meeting the lesser circle. 221, 245

17. Two spheres being about the same center, to inscribe in the greater sphere a polyhedral solid not meeting the lesser sphere at its surface. 221, 243

(Remark) And if in another sphere there is inscribed a polyhedral solid similar to the polyhedral solid inscribed in the sphere *BCDE*, the polyhedral solid in the sphere *BCDE* has to the polyhedral solid inscribed in the other sphere the ratio triplicate of that which the diameter of the sphere *BCDE* has to the diameter of the other sphere. 221, 243

18. Spheres are to one another in the triplicate ratio of their own diameters. 221, 230–233, 242–246

(Book) XIII of Euclid's *Elements*

1. If a straight line is cut in extreme and mean ratio, the greater segment added to half of the whole is five times greater in square than the square on the half. 251–254, 262, 297, 302

2. If a straight line is five times a segment of it in square, then when the double of the said segment is cut in extreme and mean ratio, the greater segment is the remaining segment of the original straight line. 251–254, 302

(Lemma for the preceding) That the double of AC (the lesser segment when AB is cut in extreme and mean ratio) is greater than BC is to be proved thus.

3. If a straight line is cut in extreme and mean ratio, the lesser segment added to half of the greater segment is five times greater in square than the square on half of the greater segment. 48, 251–254, 260, 302

4. If a straight line is cut in extreme and mean ratio, the square on the whole and on the lesser segment, the two squares together, are triple of the square on the greater segment. 251–254, 296, 300, 302

5. If a straight line is cut in extreme and mean ratio and a straight line equal to the greater segment added to it, the whole straight line has been cut in extreme and mean ratio and the greater segment is the original straight line. 251–254, 296, 302

6. If a straight line is cut in extreme and mean ratio, each of the segments is an irrational straight line, the one called apotome.[7] 251, 294, 297–298, 302

7. If three angles of an equilateral pentagon, either successive or not successive, are equal, the pentagon will be equiangular. 251, 297, 302

8. If straight lines subtend two successive angles of an equilateral and equiangular pentagon, they cut one another in extreme and mean ratio and their greater segments are equal to the side of the pentagon. 251, 262, 295, 302

9. If the side of a (regular) hexagon and that of a (regular) decagon which are inscribed in the same circle are added, the whole straight line has been cut in extreme and mean ratio and its greater segment is the side of the hexagon. 251, 258–260, 261, 302

10. If an equilateral pentagon is inscribed in a circle, the side of the pentagon is equal in square to that of the (regular) hexagon and that of the (regular) decagon inscribed in the same circle. 251, 258–261, 302

11. If an equilateral pentagon is inscribed in a circle having a rational diameter, the side of the pentagon is an irrational straight line, the one called minor. 251, 262, 271, 274, 294, 302

12. If an equilateral triangle is inscribed in a circle, the side of the triangle is triple in square the radius of the circle. 251, 255 (and 304 n. 6), 302

13. To construct a (regular) pyramid and to comprehend it in a given sphere and to prove that the diameter of the sphere is one and a half times in square the side of the pyramid. 251, 254–255, 302

(Lemma for the preceding) It is to be proved that (if *ABD* is a right triangle and *DC* is a perpendicular from the right angle at *D* to *AB*) it is the case that as *AB* is to *BC* so is the square on *AD* to the square on *DC*.

14. To construct an octahedron and to comprehend it in a sphere, as with the preceding (figures),[8] and to prove that the diameter of the sphere is double in square the side of the octahedron. 251, 254–256, 302

15. To construct a cube and comprehend it in a sphere, as with the pyramid, and to prove that the diameter of the sphere is triple in square the side of the cube. 251, 254–255, 256, 302

16. To construct an icosahedron and comprehend it in a sphere, as with the aforesaid figures, and to prove that the side of the icosahedron is an irrational straight line, the one called minor. 251, 254–255, 257–263, 266, 271, 274, 302

Corollary. From this it is manifest that the diameter of the sphere is five times in square the radius of the circle from which the icosahedron has been described, and that the diameter of the sphere is composed of the side of the (regular) hexagon and two sides of the (regular) decagon inscribed in the same circle. 299, 302

17. To construct a dodecahedron and comprehend it in a sphere, as with the aforesaid figures, and to prove that the side of the dodecahedron is an irrational straight line, the one called apotome. 251, 254–255 (and 304 n. 4) 295–298, 302

(Corollary) From this it is manifest that when the side of the cube is cut in extreme and mean ratio, the greater segment is the side of the dodecahedron. 298, 302

18. To set out the sides of the five figures and compare them to one another. 180–181, 251, 298–301, 302

I say next that it is not possible to construct beside the said five figures another figure contained by equilateral and equiangular (plane figures) equal to one another. 302

(Lemma for the preceding) But that an angle of an equilateral and equiangular pentagon is a right angle and a fifth is to be proved thus.

Notes for Appendix 4

1. P shows traces of reworking at this point and through the rest of book VII. Proposition 20 is numbered 21 in a later hand, and in the margin there appears in this hand the assertion and proof of

If three numbers are proportional, the (number contained) by the extremes is equal to the (square) on the mean; and if the (number contained) by the extremes is equal to the (square) on the mean, the three numbers are proportional,

and, with the number 22,

If there are three numbers and others equal to them in multitude which, taken in pairs, are in the same ratio but the proportion of them is perturbed, they will also be in the same ratio *ex equali*.

The subsequent propositions are then given, in the later hand, numbers two higher than those assigned them by Heiberg.

2. The corollary continues with a lengthy argument, bracketed by Heiberg, for this assertion.

3. The numeral 11 is a correction in a later hand for 10. There are similar changes through proposition 16, but thereafter the numbering is in the first hand and as indicated.

4. There follows less intelligible matter of the same kind in which Euclid is referred to in the third person.

5. There follows a fallacious argument that medials are commensurable because commensurable with the same thing, namely, a medial.

6. P and other manuscripts contain the heading 'second ordering of other rationals, those involving subtraction' (*deutera taxis heterōn logōn* (*alogōn?*) *tōn kata aphairesin*).

7. There follows in P an alternative proof of XIII, 5, brief definitions of analysis and synthesis, and analyses and syntheses of propositions 1–5 "without a diagram."

8. The formulation of this proposition and the following three in P suggests that 15 originally preceded 14.

Bibliography

*Books marked with an asterisk (*) are available from Dover Publications. For information and prices, go to **doverpublications.com.***

Al-Narizi. See 'Euclid' and '*Codex Leidensis*'

Apollonius Pergaeus, *Opera Graeca* (ed. J. Heiberg), 2 vols., Leipzig, 1891, 1893

Archimedes, *Archimède* (ed. and trans. C. Mugler), 4 vols., Paris, 1970–1972 (volume 4 is cited as 'Eutocius')

Aristarchus, *On the Sizes and Distances of the Sun and Moon* (ed. and trans. T. L. Heath), in T. L. Heath, *Aristarchus*, Oxford, 1913

Aristotle, *Opera* (ed. I. Bekker), 5 vols., Berlin, 1831–1870

* Baron, M. E., *The Origins of the Infinitesimal Calculus*, Oxford, 1969

Becker, O., "Die Lehre vom Geraden und Ungeraden im neunten Buch der euklidischen *Elemente*," *Quellen und Studien zur Geschichte der Mathematik, Astronomie, und Physik*, Abteilung B, 3 (1936), 533–553

——— "Eudoxos-Studien I. Eine voreudoxische Proportionenlehre und ihre Spuren bei Aristoteles und Euklid," *Quellen und Studien zur Geschichte der Mathematik, Astronomie, und Physik*, Abteilung B, 2 (1933), 311–333

——— "Eudoxos-Studien II. Warum haben die Griechen die Existenz der vierten Proportionale angenommen?" *Quellen und Studien zur Geschichte der Mathematik, Astronomie, und Physik*, Abteilung B, 2 (1933), 369–387

——— "Eudoxos-Studien III. Spuren eines Stetigkeitsaxioms in der Art des Dedekindschen zur Zeit des Eudoxos," *Quellen und Studien zur Geschichte der Mathematik, Astronomie, und Physik*, Abteilung B, 3 (1936), 236–244

——— "Eudoxos-Studien IV. Das Prinzip des ausgeschlossenen Dritten in der griechischen Mathematik," *Quellen und Studien zur Geschichte der Mathematik, Astronomie, und Physik*, Abteilung B, 3 (1936), 370–388

Beckman, F., "Neue Gesichtspunkte zum 5. Buch Euklids," *Archive for History of Exact Sciences*, 4 (1967), 1–144

Beth, E. W., *The Foundations of Mathematics*, Amsterdam, 1959

Boltyanskii, V. G., *Equivalent and Equidecomposable Figures* (trans. E. K. Herrn and C. E. Watts), Boston, 1963

* Bonola, R., *Non-Euclidean Geometry* (trans. H. S. Carslaw), Chicago, 1912

Bourbaki, N. (pseudonym), *Éléments d'histoire des mathématiques*, 2nd ed., Paris, 1974

Boyer, C. B., *The Concepts of the Calculus*, New York, 1949

Burnyeat, M. F., "The philosophical sense of Theaetetus' mathematics," *Isis*, 69 (1978), 489–513

Chasles, M., *Les trois Livres des Porismes d'Euclide*, Paris, 1860

Codex Leidensis 399, I. Euclidis *Elementa* ex interpretatione Al-Hadschschadschii cum commentariis Al-Narizii (ed. and trans. R. O. Besthorn and J. L. Heiberg), 6 vols., Copenhagen, 1893–1930 (cited as '*Codex Leidensis*')

Curry, H., *Outlines of a Formalist Philosophy of Mathematics*, Amsterdam, 1951

* Dedekind, R., "Continuity and irrational numbers," in *Essays on the Theory of Numbers* (trans. W. Beman), Chicago and London, 1901

Dehn, M., "Beziehungen zwischen der Philosophie und der Grundlagen der Mathematik im Altertum," *Quellen und Studien zur Geschichte der Mathematik, Astronomie, und Physik*, Abteilung B, 4 (1937), 1–28

Dijksterhuis, E. J., *Archimedes*, Copenhagen, 1956

——— *De Elementen van Euclides*, 2 vols., Groningen, 1929–1930 (cited as 'Dijksterhuis')

* Eisenhart, L. P., *Co-ordinate Geometry*, Boston, 1939

Eutocius. See 'Archimedes'

Euclid, *Die Elemente* (trans. C. Thaer), Darmstadt, 1969

——— *Gli Elementi* (trans. A. Frajese and L. Maccioni), Turin, 1970

——— *Opera Omnia* (eds. J. L. Heiberg and H. Menge), 9 vols., Leipzig, 1883–1916 (volumes I-IV cited as 'Heiberg', volume V as 'Scholia', and volume IX as 'Al-Narizi, Commentary')

* Eves, H. W., and Newsom, C. V., *An Introduction to the Foundations and Fundamental Concepts of Mathematics*, revised ed., New York, 1965

Frajese, A., "Sur la signification des postulats euclidiens," *Archives internationales d'histoire des sciences*, 4 (1951), 383–392

——— and Maccioni, L. See 'Euclid'

Frege, G., *The Foundations of Arithmetic* (trans. J. L. Austin), Oxford, 1950

Freudenthal, H., "What is algebra and what has been its history?" *Archive for History of Exact Sciences* 16 (1977), 189–200

——— "Zur Geschichte der Grundlagen der Geometrie," *Nieuw Archief voor Wiskunde*, 4 (1957), 105–142

——— "Zur Geschichte der vollständige Induktion," *Archives internationales d'histoire des sciences*, 6 (1953), 17–37

——— and van der Waerden, B. L., "Over een bewering van Euclides," *Simon Stevin*, 25 (1947), 115–121

Fritz, K. von, "Die APXAI in der griechischen Mathematik," *Archiv für Begriffsgeschichte*, 1 (1955), 13–103

——— "Gleichheit, Kongruenz, und Ähnlichkeit in der antiken Mathematik bis auf Euklid," *Archiv für Begriffsgeschichte*, 4 (1959), 7–81

Gandz, S., "Origin and development of quadratic equations in Babylonian, Greek, and early Arabic algebra," *Osiris*, 3 (1938), 405–557

* Gillings, R. J., *Mathematics in the Time of the Pharaohs*, Cambridge, Mass., 1972

Gödel, K., "Russell's mathematical logic," in P. Benacerraf and H. Putnam (eds.), *Philosophy of Mathematics*, Englewood Cliffs, N. J., 1964, pp. 258–273

——— "What is Cantor's continuum hypothesis?" in P. Benacerraf and H. Putnam (eds.), *Philosophy of Mathematics*, Englewood Cliffs, N. J., 1964, pp. 211–232

Goetsch, H., "Die Algebra der Babylonier," *Archive for History of Exact Sciences*, 5 (1968–1969), 79–153

Hasse, H., and Scholz, H., "Die Grundlagenkrisis der griechischen Mathematik," *Kantstudien*, 33 (1928), 4–34

* Heath, T. L., *A History of Greek Mathematics*, 2 vols., Oxford, 1921

——— "Greek geometry with special reference to infinitesimals," *The Mathematical Gazette*, 11 (1922–1923), 248–259

——— *Mathematics in Aristotle*, Oxford, 1949

* ——— *The Thirteen Books of Euclid's Elements*, 2nd ed., 3 vols., Cambridge, England, 1926 (cited as 'Heath')

Heyting, A., *Intuitionism: An Introduction*, 2nd ed., Amsterdam, 1966

Heiberg, J. L., "Paralipomena zu Euklid," *Hermes*, 38 (1903), 46–74, 161–201, 321–356. See also 'Euclid'

Heller, S., "Ueber Euklids Definitionen ähnlicher und kongruenter Polyeder," *Janus*, 51 (1964), 277–290

Hilbert, D., *Grundlagen der Geometrie*, 10th ed. with revisions and additions by P. Bernays, Stuttgart, 1968

——— "Ueber das Unendliche," *Mathematische Annalen*, 95 (1926), 161–90

——— *The Foundations of Geometry* (trans. E. J. Townsend), Chicago, 1902

——— and Bernays, P., *Grundlagen der Mathematik*, 2nd ed., 2 vols., Berlin, 1968–1970

Itard, J., *Les Livres arithmétiques d'Euclide*, Paris, 1961

Kleene, S. C., *Introduction to Metamathematics*, Amsterdam, 1952

* Klein, F., *Elementary Mathematics from an Advanced Standpoint: Geometry* (trans. E. R. Hedrick and C. A. Noble), New York, 1939

Kneale, W. and M., *The Development of Logic*, Oxford, 1962

Knorr, W., *The Evolution of the Euclidean Elements*, Dordrecht, 1975 (cited as 'Knorr')

——— "Problems in the interpretation of Greek number theory," *Studies in the History and Philosophy of Science*, 7 (1976), 353–368

Landau, E. G. H., *Foundations of Analysis* (trans. F. Steinhardt), 2nd ed., New York, 1960

Liddell, H. G., Scott, R., and Jones, H. S., *A Greek-English Lexicon*, Oxford, 1940

Malmendier, N., "Eine Axiomatik zum 7. Buch der *Elemente* von Euklid," *Mathematisch-Physikalische Semesterberichte*, 22 (1975), 240–254

Mueller, I., "Greek mathematics and Greek logic," in J. Corcoran (ed.), *Ancient Logic and Its Modern Interpretations*, Dordrecht, 1974, pp. 35–70

——— "Homogeneity in Eudoxus's theory of proportion," *Archive for History of Exact Sciences*, 7 (1970), 1–6

Mugler, C., *Dictionnaire historique de la terminologie géométrique des Grecs*, Paris, 1958

Neuenschwander, E., "Die ersten vier Bücher der *Elemente* Euklids," *Archive for History of Exact Sciences*, 9 (1973), 325–380

——— "Die stereometrischen Bücher der *Elemente* Euklids," *Archive for History of Exact Sciences*, 14 (1974–1975), 91–125

Neugebauer, O., "The survival of Babylonian methods in the exact sciences of antiquity and the middle ages," *Proceedings of the American Philosophical Society*, 107 (1963), 528–535

——— "Ueber babylonische Mathematik und ihre Stellung zur ägyptischen und griechischen," *Atti de XIX Congresso Internazionali degli Orientalisti*, Rome, 1936, 64–69

——— "Ueber griechische Mathematik und ihr Verhältnis zur vorgriechischen," *Comptes rendus du congrès international des mathématiciens*, vol. I, Oslo, 1937, 157–170

——— "Ueber vorgriechische Mathematik," *Abhandlungen aus dem mathematischen Seminar der Hamburgischen Universität*, 7 (1930), 107–124

——— *Vorgriechische Mathematik*, 2nd ed., Heidelberg and New York, 1969

——— "Zur geometrischen Algebra," *Quellen und Studien zur Geschichte der Mathematik, Astronomie, und Physik*, Abteilung B, 3 (1936), 245–259

——— "Zur vorgriechischen Mathematik," *Erkenntnis*, 2 (1931), 122–134

Niebel, E., *Untersuchungen über die Bedeutung der geometrischen Konstruktion in der Antike* (*Kantstudien Ergänzungsheft* 76), Cologne, 1959

Pappus, *Collectio* (ed. F. Hultsch), 3 vols., Leipzig, 1875–1878

——— *Commentary on Book X of Euclid's Elements* (ed. and trans. G. Junge and W. Thomson), Cambridge, Mass., 1930

Plato, *Opera* (ed. J. Burnet), 5 vols., Oxford, 1900–1907

Poincaré, H., Review of D. Hilbert, *Grundlagen der Geometrie*, *Bulletin des sciences mathématiques*, 2nd series, 26 (1902), 249–272

Proclus, *In Primum Euclidis Elementorum Commentarii* (ed. G. Friedlein), Leipzig, 1873 (cited as 'Proclus')

——— *In Platonis Rem Publicam Commentarii* (ed. W. Kroll), 2 vols., Leipzig, 1899–1901

* Robbin, J. W., *Mathematical Logic: A First Course*, New York, 1969

* Russell, B., *Introduction to Mathematical Philosophy*, London, 1919

Sachs, E., *Die fünf platonischen Körper*, Berlin, 1917

Schmidt, O. Review of J. Itard, *Les Livres . . .*, *Centaurus*, 10 (1964–1965), 198–201

Scholia. See 'Euclid'

Seidenberg, A., "Did Euclid's *Elements* book I develop geometry axiomatically?" *Archive for History of Exact Sciences*, 14 (1974–1975), 263–295

——— "On the area of a semicircle," *Archive for History of Exact* Sciences, 9 (1972–1973), 171–211

——— "Peg and cord in ancient Greek geometry," *Scripta Mathematica*, 24 (1959), 107–122

Simplicius, *Commentary on Aristotle's Physics* (ed. H. Diels), 2 vols., Berlin, 1882, 1895

Steele, A. D., "Ueber die Rolle von Zirkel und Lineal in der griechischen Mathematik," *Quellen und Studien zur Geschichte der Mathematik, Astronomie, und Physik*, Abteilung B, 3 (1936), 288–369

Szabò, A., *Anfänge der griechischen Mathematik*, Budapest, 1969

Taisbak, C. M., *Division and Logos*, Odense, 1971 (cited as 'Taisbak')

——— "Perfect numbers, a mathematical pun?" *Centaurus*, 20 (1976), 269–275

Tannery, P., *La Géométrie grecque*, Paris, 1887

——— "Sur l'authenticité des axiomes d'Euclide," in *Mémoires scientifiques* (ed. J. L. Heiberg and H. G. Zeuthen), vol. II, Paris, 1912, pp. 48–63

Tarski, A., "The concept of truth in formalized languages," in *Logic, Semantics, and Metamathematics*, Oxford, 1956, pp. 153–278

Thaer, C., "Die Euklid-Ueberlieferung durch At-Tûsî," *Quellen und Studien zur Geschichte der Mathematik, Astronomie, und Physik*, Abteilung B, 3 (1934), 116–121. See also 'Euclid'

Theodosius, *Spherica* (ed. J. L. Heiberg), *Abhandlungen von der Akademie der Wissenschaften zu Göttingen, Philologisch-historische Klasse*, new series, 19, no. 3 (1927)

Toth, I., "Das Parallelenproblem in Corpus Aristotelicum," *Archive for History of Exact Sciences*, 3 (1967), 249–422

Unguru, S., "On the need to rewrite the history of Greek mathematics," *Archive for History of Exact Sciences*, 15 (1975–1976), 67–114

van der Waerden, B. L., "Defense of a 'shocking' point of view," *Archive for History of Exact Sciences*, 15 (1975–1976), 199–210

——— "Die Arithmetik der Pythagoreer," *Mathematische Annalen*, 120 (1947–1949), 127–153, 676–700

——— "Die Postulate und Konstruktionen in der frühgriechischen Geometrie," *Archive for History of Exact Sciences*, 18 (1978), pp. 343–357

——— *Science Awakening* (trans. A. Dresden), Groningen, 1954 (cited as 'van der Waerden')

——— "Zenon und die Grundlagenkrise der griechischen Mathematik," *Mathematische Annalen*, 117 (1940–1941), 141–161

Vogel, K., *Beitrage zur griechischen Logistik* (*Sitzungsberichte der Bayerischen Akademie der Wissenschaften, Mathematisch-naturwissenschaftliche Abteilung*), Munich, 1936

Waterhouse, W. C., "The discovery of the regular solids," *Archive for History of Exact Sciences*, 9 (1972–1973), 212–221

Weil, A., "Who betrayed Euclid?" *Archive for History of Exact Sciences*, 14 (1978), 91–93

Zeuthen, H. G., "Die geometrische Konstruction als 'Existenzbeweis' in der antiken Geometrie," *Mathematische Annalen*, 47 (1896), 222–228

——— *Die Lehre von den Kegelschnitten im Altertum*, Copenhagen, 1896

——— *Histoire des mathématiques dans l'antiquité et le moyen âge* (trans. J. Mascart), Paris, 1902

——— "Sur la constitution des livres arithmétiques des *Éléments* d'Euclide et leur rapport à la question de l'irrationalité," *Oversigt over det Kongelige Danske Videnskabernes Selskabs Forhandlinger*, 1910, 395–435

——— "Sur les connaissances géométriques des grecs avant la reforme platonicienne," *Oversigt over det Kongelige Danske Videnskabernes Selskabs Forhandlinger*, 1913, 431–473

——— *Sur l'Origine de l'Algèbre* (*Det Kongelige Videnskabernes Selskab, Mathematisk-Fysiske Meddelelser*, vol. II, no. 4), Copenhagen, 1919

——— "Sur l'origine historique de la connaissance des quantités irrationelles," *Oversigt over det Kongelige Danske Videnskabernes Selskabs Forhandlinger*, 1915, 333–362

Index